Abbreviations for Units

A	ampere	H	henry	nm	nanometer (10^{-9} m)
Å	angstrom (10^{-10} m)	h	hour	pt	pint
atm	atmosphere	Hz	hertz	qt	quart
Btu	British thermal unit	in	inch	rev	revolution
Bq	becquerel	J	joule	R	roentgen
C	coulomb	K	kelvin	Sv	seivert
°C	degree Celsius	kg	kilogram	s	second
cal	calorie	km	kilometer	T	tesla
Ci	curie	keV	kilo-electron volt	u	unified mass unit
cm	centimeter	lb	pound	V	volt
dyn	dyne	L	liter	W	watt
eV	electron volt	m	meter	Wb	weber
°F	degree Fahrenheit	MeV	mega-electron volt	y	year
fm	femtometer, fermi (10^{-15} m)	Mm	megameter (10^6 m)	yd	yard
ft	foot	mi	mile	μm	micrometer (10^{-6} m)
Gm	gigameter (10^9 m)	min	minute	μs	microsecond
G	gauss	mm	millimeter	μC	microcoulomb
Gy	gray	ms	millisecond	Ω	ohm
g	gram	N	newton		

Some Conversion Factors

Length

1 m = 39.37 in = 3.281 ft = 1.094 yd

1 m = 10^{15} fm = 10^{10} Å = 10^9 nm

1 km = 0.6214 mi

1 mi = 5280 ft = 1.609 km

1 lightyear = 1 $c \cdot$y = 9.461×10^{15} m

1 in = 2.540 cm

Volume

1 L = 10^3 cm^3 = 10^{-3} m^3 = 1.057 qt

Time

1 h = 3600 s = 3.6 ks

1 y = 365.24 d = 3.156×10^7 s

Speed

1 km/h = 0.278 m/s = 0.6214 mi/h

1 ft/s = 0.3048 m/s = 0.6818 mi/h

Angle–angular speed

1 rev = 2π rad = 360°

1 rad = 57.30°

1 rev/min = 0.1047 rad/s

Force–pressure

1 N = 10^5 dyn = 0.2248 lb

1 lb = 4.448 N

1 atm = 101.3 kPa = 1.013 bar = 76.00 cmHg = 14.70 lb/in^2

Mass

1 u = $[(10^{-3}\ \text{mol}^{-1})/N_A]$ kg = 1.661×10^{-27} kg

1 tonne = 10^3 kg = 1 Mg

1 slug = 14.59 kg

1 kg weighs about 2.205 lb

Energy–power

1 J = 10^7 erg = 0.7376 ft·lb = 9.869×10^{-3} L·atm

1 kW·h = 3.6 MJ

1 cal = 4.184 J = 4.129×10^{-2} L·atm

1 L·atm = 101.325 J = 24.22 cal

1 eV = 1.602×10^{-19} J

1 Btu = 778 ft·lb = 252 cal = 1054 J

1 horsepower = 550 ft·lb/s = 746 W

Thermal conductivity

1 W/(m·K) = 6.938 Btu·in/(h·ft^2·°F)

Magnetic field

1 T = 10^4 G

Viscosity

1 Pa·s = 10 poise

Prefixes for Powers of 10*

Multiple	Prefix	Abbreviation
10^{24}	yotta	Y
10^{21}	zetta	Z
10^{18}	exa	E
10^{15}	peta	P
10^{12}	tera	T
10^{9}	giga	G
10^{6}	mega	M
10^{3}	kilo	k
10^{2}	hecto	h
10^{1}	deka	da
10^{-1}	deci	d
10^{-2}	centi	c
10^{-3}	milli	m
10^{-6}	micro	μ
10^{-9}	nano	n
10^{-12}	pico	p
10^{-15}	femto	f
10^{-18}	atto	a
10^{-21}	zepto	z
10^{-24}	yocto	y

*Commonly used prefixes are in bold. All prefixes are pronounced with the accent on the first syllable.

Terrestrial and Astronomical Data*

Acceleration of gravity at Earth's surface	g	$9.81\ \text{m/s}^2 = 32.2\ \text{ft/s}^2$
Radius of Earth R_E	R_E	6371 km = 3959 mi
Mass of Earth	M_E	5.97×10^{24} kg
Mass of the Sun		1.99×10^{30} kg
Mass of the moon		7.35×10^{22} kg
Escape speed at Earth's surface		11.2 km/s = 6.95 mi/s
Standard temperature and pressure (STP)		0°C = 273.15 K 1 atm = 101.3 kPa
Earth–moon distance†		3.84×10^{8} m $= 2.39 \times 10^{5}$ mi
Earth–Sun distance (mean)†		1.50×10^{11} m $= 9.30 \times 10^{7}$ mi
Speed of sound in dry air (at STP)		331 m/s
Speed of sound in dry air (20°C, 1 atm)		343 m/s
Density of dry air (STP)		$1.29\ \text{kg/m}^3$
Density of dry air (20°C, 1 atm)		$1.20\ \text{kg/m}^3$
Density of water (4°C, 1 atm)		$1000\ \text{kg/m}^3$
Heat of fusion of water (0°C, 1 atm)	L_f	333.5 kJ/kg
Heat of vaporization of water (100°C, 1 atm)	L_v	2.257 MJ/kg

*Additional data on the solar system can be found in Appendix B and at http://nssdc.gsfc.nasa.gov/planetary/planetfact.html.
†Center to center.

The Greek Alphabet

Alpha	A	α	Nu	N	ν
Beta	B	β	Xi	Ξ	ξ
Gamma	Γ	γ	Omicron	O	o
Delta	Δ	δ	Pi	Π	π
Epsilon	E	ϵ, ε	Rho	P	ρ
Zeta	Z	ζ	Sigma	Σ	σ
Eta	H	η	Tau	T	τ
Theta	Θ	θ	Upsilon	Y	υ
Iota	I	ι	Phi	Φ	ϕ
Kappa	K	κ	Chi	X	χ
Lambda	Λ	λ	Psi	Ψ	ψ
Mu	M	μ	Omega	Ω	ω

Mathematical Symbols

$=$	is equal to
$\equiv$	is defined by
$\neq$	is not equal to
$\approx$	is approximately equal to
$\sim$	is of the order of
$\propto$	is proportional to
$>$	is greater than
$\geq$	is greater than or equal to
$>>$	is much greater than
$<$	is less than
$\leq$	is less than or equal to
$<<$	is much less than
Δx	change in x
dx	differential change in x
$\lvert x \rvert$	absolute value of x
$\lvert \vec{v} \rvert$	magnitude of $\vec{v}$
$n!$	$n(n-1)(n-2)\ldots1$
Σ	sum
lim	limit
$\Delta t \to 0$	Δt approaches zero
$\frac{dx}{dt}$	derivative of x with respect to t
$\frac{\partial x}{\partial t}$	partial derivative of x with respect to t
$\int_{x_1}^{x_2} f(x)dx$	definite integral $= F(x)\Big\rvert_{x_1}^{x_2} = F(x_2) - F(x_1)$

SIXTH EDITION

PHYSICS FOR SCIENTISTS AND ENGINEERS

VOLUME 3

MODERN PHYSICS: QUANTUM MECHANICS, RELATIVITY, AND THE STRUCTURE OF MATTER

Paul A. Tipler
Gene Mosca

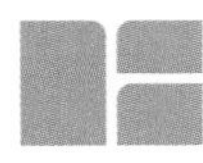

W. H. Freeman and Company • New York

PT: For Claudia

GM: For Vivian

Publisher: Susan Finnemore Brennan
Executive Editor: Clancy Marshall
Marketing Manager: Anthony Palmiotto
Senior Developmental Editor: Kharissia Pettus
Media Editor: Jeanette Picerno
Editorial Assistants: Janie Chan, Kathryn Treadway
Photo Editor: Ted Szczepanski
Photo Researcher: Dena Digilio Betz
Cover Designer: Blake Logan
Text Designer: Marsha Cohen/Parallelogram Graphics
Senior Project Editor: Georgia Lee Hadler
Copy Editor: Connie Parks
Illustrations: Network Graphics
Illustration Coordinator: Bill Page
Production Coordinator: Susan Wein
Composition: Preparé Inc.
Printing and Binding: RR Donnelly

Library of Congress Control Number: 2007010418

ISBN-10: 1-4292-0134-7 (Volume 3, Chapters 34–41)
ISBN-13: 978-1-4292-0134-6
ISBN-10: 1-4292-0132-0 (Volume 1, Chapters 1–20, R)
ISBN-10: 1-4292-0133-9 (Volume 2, Chapters 21–33)
ISBN-10: 1-4292-0124-X (Standard, Chapters 1–33, R)
ISBN-10: 0-7167-8964-7 (Extended, Chapters 1–41, R)

Printed in the United States of America
First printing

W. H. Freeman and Company
41 Madison Avenue
New York, NY 10010
Houndmills, Basingstoke
RG21 6XS, England
www.whfreeman.com

Contents in Brief

Thinkstock/Alamy

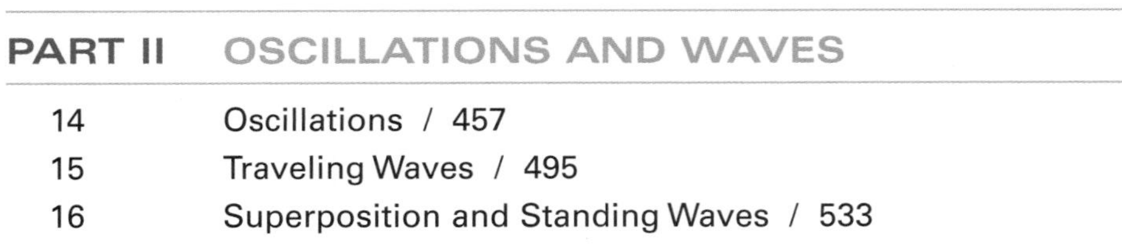

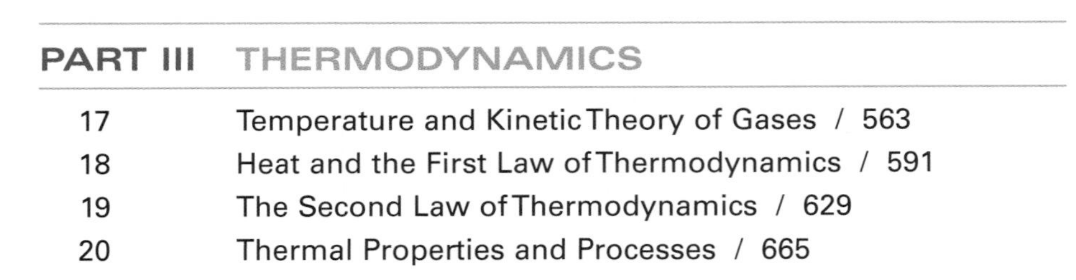

PART IV ELECTRICITY AND MAGNETISM

PART V LIGHT

PART VI MODERN PHYSICS: QUANTUM MECHANICS, RELATIVITY, AND THE STRUCTURE OF MATTER

APPENDICES

Contents

Volume 3

* optional material

PART VI MODERN PHYSICS: QUANTUM MECHANICS, RELATIVITY, AND THE STRUCTURE OF MATTER

NASA/CXC/Cfa/S. Wolk et al.

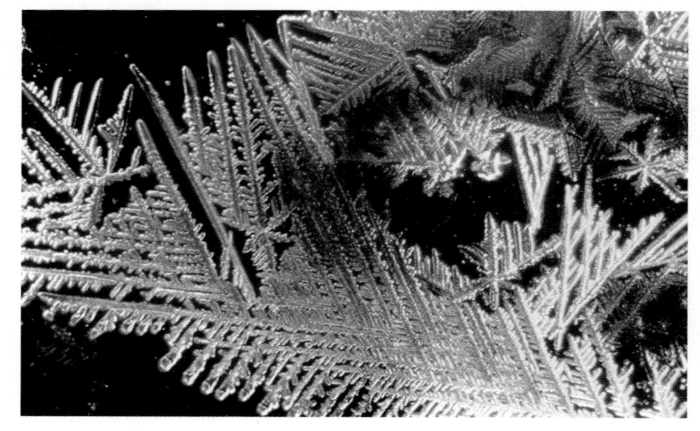

Photo Researchers

NASA

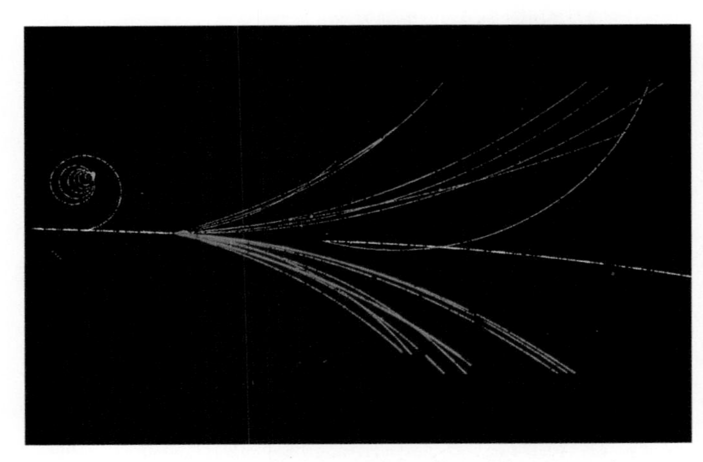

Lawrence Livermore Laboratory/Science Photo Library/Photo Researchers

Preface

The sixth edition of *Physics for Scientists and Engineers* offers a completely integrated text and media solution that will help students learn most effectively and will enable professors to customize their classrooms so that they teach most efficiently.

The text includes a new strategic problem-solving approach, an integrated Math Tutorial, and new tools to improve conceptual understanding. New Physics Spotlights feature cutting-edge topics that help students relate what they are learning to real-world technologies.

The new online learning management system enables professors to easily customize their classes based on their students' needs and interests by using the new interactive Physics Portal, which includes a complete e-book, student and instructor resources, and a robust online homework system. Interactive Exercises in the Physics Portal give students the opportunity to learn from instant feedback, and give instructors the option to track and grade each step of the process. Because no two physics students or two physics classes are alike, tools to help make each physics experience successful are provided.

KEY FEATURES

PROBLEM-SOLVING STRATEGY

The sixth edition features a new problem-solving strategy in which Examples follow a consistent **Picture**, **Solve**, and **Check** format. This format walks students through the steps involved in analyzing the problem, solving the problem, and then checking their answers. Examples often include helpful **Taking It Further** sections which present alternative ways of solving problems, interesting facts, or additional information regarding the concepts presented. Where appropriate, Examples are followed by **Practice Problems** so students can assess their mastery of the concepts.

In this edition, the problem-solving steps are again juxtaposed with the necessary equations so that it's easier for students to see a problem unfold.

After each problem statement, students are asked to **Picture** the problem. Here, the problem is analyzed both conceptually and visually.

In the **Solve** sections, each step of the solution is presented with a written statement in the left-hand column and the corresponding mathematical equations in the right-hand column.

Check reminds students to make sure their results are accurate and reasonable.

Taking It Further suggests a different way to approach an Example or gives additional information relevant to the Example.

A **Practice Problem** often follows the solution of an Example, allowing students to check their understanding. Answers are included at the end of the chapter to provide immediate feedback.

Example 3-4 Rounding a Curve

A car is traveling east at 60 km/h. It rounds a curve, and 5.0 s later it is traveling north at 60 km/h. Find the average acceleration of the car.

PICTURE We can calculate the average acceleration from its definition, $\vec{a}_{av} = \Delta\vec{v}/\Delta t$. To do this, we first calculate $\Delta\vec{v}$, which is the vector that when added to $\vec{v}_i$, results in $\vec{v}_f$.

SOLVE

1. The average acceleration is the change in velocity divided by the elapsed time. To find $\vec{a}_{av}$, we first find the change in velocity: $\vec{a}_{av} = \dfrac{\Delta\vec{v}}{\Delta t}$
2. To find $\Delta\vec{v}$, we first specify $\vec{v}_i$ and $\vec{v}_f$. Draw $\vec{v}_i$ and $\vec{v}_f$ (Figure 3-7*a*), and draw the vector addition diagram (Figure 3-7*b*) corresponding to $\vec{v}_f = \vec{v}_i + \Delta\vec{v}$:
3. The change in velocity is related to the initial and final velocities: $\vec{v}_f = \vec{v}_i + \Delta\vec{v}$
4. Substitute these results to find the average acceleration: $\vec{a}_{av} = \dfrac{\vec{v}_f - \vec{v}_i}{\Delta t} = \dfrac{60\text{ km/h }\hat{j} - 60\text{ km/h }\hat{i}}{5.0\text{ s}}$
5. Convert 60 km/h to meters per second: $60\text{ km/h} \times \dfrac{1\text{ h}}{3600\text{ s}} \times \dfrac{1000\text{ m}}{1\text{ km}} = 16.7\text{ m/s}$
6. Express the acceleration in meters per second squared: $\vec{a}_{av} = \dfrac{\vec{v}_f - \vec{v}_i}{\Delta t} = \dfrac{16.7\text{ m/s }\hat{j} - 16.7\text{ m/s }\hat{i}}{5.0\text{ s}} = \boxed{-3.4\text{ m/s}^2\,\hat{i} + 3.4\text{ m/s}^2\,\hat{j}}$

CHECK The eastward component of the velocity decreases from 60 km/h to zero, so we expect a negative acceleration component in the *x* direction. The northward component of the velocity increases from zero to 60 km/h, so we expect a positive acceleration component in the *y* direction. Our step 6 result meets both of these expectations.

TAKING IT FURTHER Note that the car is accelerating even though its speed remains constant.

PRACTICE PROBLEM 3-1 Find the magnitude and direction of the average acceleration vector.

FIGURE 3-7

A boxed **Problem-Solving Strategy** is included in almost every chapter to reinforce the **Picture**, **Solve**, and **Check** format for successfully solving problems.

> ***PROBLEM-SOLVING STRATEGY***
>
> ***Relative Velocity***
>
> **PICTURE** The first step in solving a relative-velocity problem is to identify and label the relevant reference frames. Here, we will call them reference frame A and reference frame B.
>
> **SOLVE**
>
> 1. Using $\vec{v}_{pB} = \vec{v}_{pA} + \vec{v}_{AB}$ (Equation 3-9), relate the velocity of the moving object (particle *p*) relative to frame A to the velocity of the particle relative to frame B.
> 2. Sketch a vector addition diagram for the equation $\vec{v}_{pB} = \vec{v}_{pA} + \vec{v}_{AB}$. Use the head-to-tail method of vector addition. Include coordinate axes on the sketch.
> 3. Solve for the desired quantity. Use trigonometry where appropriate.
>
> **CHECK** Make sure that you solve for the velocity or position of the moving object relative to the proper reference frame.

INTEGRATED MATH TUTORIAL

This edition has improved mathematical support for students who are taking calculus concurrently with introductory physics or for students who need a math review.

The comprehensive **Math Tutorial**

- reviews basic results of algebra, geometry, trigonometry, and calculus,
- links mathematical concepts to physics concepts in the text,
- provides Examples and Practice Problems so students may check their understanding of mathematical concepts.

Example M-13 **Radioactive Decay of Cobalt-60**

The half-life of cobalt-60 (^{60}Co) is 5.27 y. At $t = 0$ you have a sample of ^{60}Co that has a mass equal to 1.20 mg. At what time t (in years) will 0.400 mg of the sample of ^{60}Co have decayed?

PICTURE When we derived the half-life in exponential decay, we set $N/N_0 = 1/2$. In this example, we are to find the time at which two-thirds of a sample remains, and so the ratio N/N_0 will be 0.667.

SOLVE

1. Express the ratio N/N_0 as an exponential function: $\frac{N}{N_0} = 0.667 = e^{-\lambda t}$
2. Take the reciprocal of both sides: $\frac{N_0}{N} = 1.50 = e^{\lambda t}$
3. Solve for t: $t = \frac{\ln 1.50}{\lambda} = \frac{0.405}{\lambda}$
4. The decay constant is related to the half-life by $\lambda = (\ln 2)/t_{1/2}$ (Equation M-70). Substitute $(\ln 2)/t_{1/2}$ for λ and evaluate the time: $t = \frac{\ln 1.5}{\ln 2} t_{1/2} = \frac{\ln 1.5}{\ln 2} \times 5.27\text{ y} = 3.08\text{ y}$

CHECK It takes 5.27 y for the mass of a sample of ^{60}Co to decrease to 50 percent of its initial mass. Thus, we expect it to take less than 5.27 y for the sample to lose 33.3 percent of its mass. Our step-4 result of 3.08 y is less than 5.27 y, as expected.

PRACTICE PROBLEMS

27. The discharge time constant τ of a capacitor in an RC circuit is the time in which the capacitor discharges to e^{-1} (or 0.368) times its charge at $t = 0$. If $\tau = 1$ s for a capacitor, at what time t (in seconds) will it have discharged to 50.0% of its initial charge?
28. If the coyote population in your state is increasing at a rate of 8.0% a decade and continues increasing at the same rate indefinitely, in how many years will it reach 1.5 times its current level?

M-12 INTEGRAL CALCULUS

Integration can be considered the inverse of differentiation. If a function $f(t)$ is *integrated*, a function $F(t)$ is found for which $f(t)$ is the derivative of $F(t)$ with respect to t.

THE INTEGRAL AS AN AREA UNDER A CURVE; DIMENSIONAL ANALYSIS

The process of finding the area under a curve on the graph illustrates integration. Figure M-27 shows a function $f(t)$. The area of the shaded element is approximately $f_i \Delta t_i$, where f_i is evaluated anywhere in the interval Δt_i. This approximation is highly accurate if Δt_i is very small. The total area under some stretch of the curve is found by summing all the area elements it covers and taking the limit as each Δt_i approaches zero. This limit is called the **integral** of f over t and is written

$$\int f\,dt = \text{area}_i = \lim_{\Delta t_i \to 0} \sum_i f_i \Delta t_i \qquad \text{M-74}$$

The *physical dimensions* of an integral of a function $f(t)$ are found by multiplying the dimensions of the *integrand* (the function being integrated) and the dimensions of the integration variable t. For example, if the integrand is a velocity function

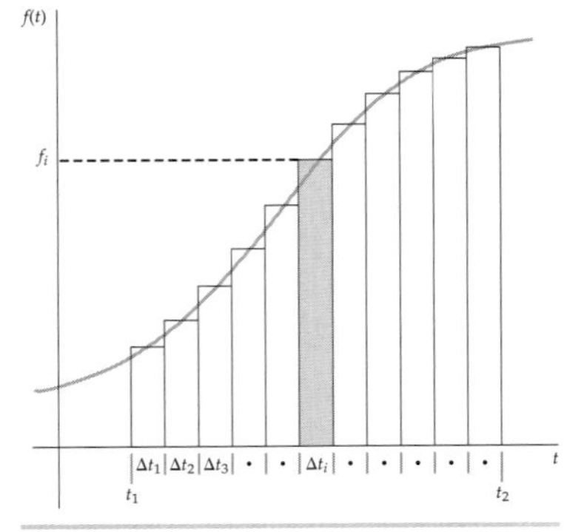

FIGURE M-27 A general function $f(t)$. The area of the shaded element is approximately $f_i \Delta t_i$, where f_i is evaluated anywhere in the interval.

In addition, margin notes allow students to easily see the links between physics concepts in the text and math concepts.

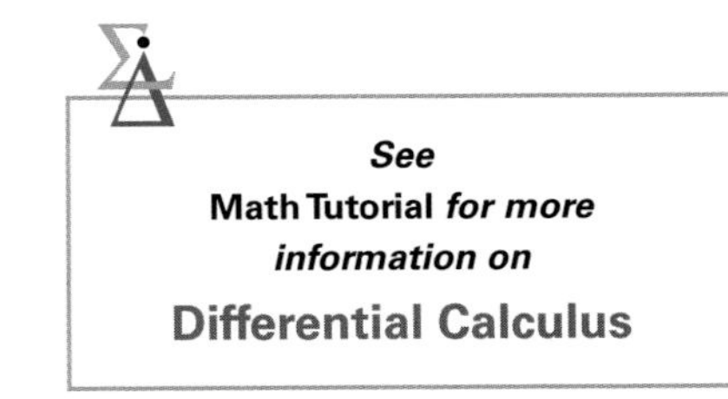

NEW! PEDAGOGY TO ENSURE CONCEPTUAL UNDERSTANDING

Student-friendly tools have been added to allow for better conceptual understanding of physics.

- New **Conceptual Examples** are introduced, where appropriate, to help students fully understand essential physics concepts. These Examples use the **Picture**, **Solve**, and **Check** strategy so that students not only gain fundamental conceptual understanding but must evaluate their answers.

Example 8-12 **Collisions with Putty** *Conceptual*

Mary has two small balls of equal mass, a ball of plumber's putty and a one of Silly Putty. She throws the ball of plumber's putty at a block suspended by strings shown in Figure 8-20. The ball strikes the block with a "thonk" and falls to the floor. The block subsequently swings to a maximum height *h*. If she had thrown the ball of Silly Putty (instead of the plumber's putty), would the block subsequently have risen to a height greater than *h*? Silly Putty, unlike plumber's putty, is elastic and would bounce back from the block.

PICTURE During impact the change in momentum of the ball–block system is zero. The greater the magnitude of the change in momentum of the ball, the greater, the magnitude of the change in momentum of the block. Does magnitude of the change in momentum of the ball increase more if the ball bounces back than if it does not?

SOLVE

The ball of plumber's putty loses a large fraction of its forward momentum. The ball of Silly Putty would lose all of its forward momentum and then gain momentum in the opposite direction. It would undergo a larger change in momentum than did the ball of plumber's putty.

The block would swing to a greater height after being struck with the ball of Silly Putty than it did after being struck with the ball of plumbers putty.

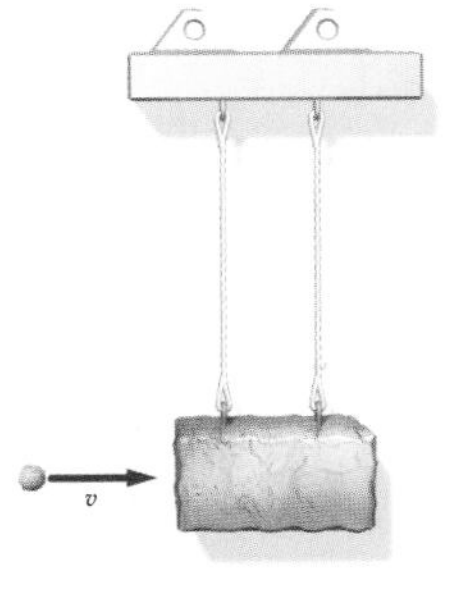

FIGURE 8-20

CHECK The block exerts a backward impulse on the ball of plumber's putty to slow the ball to a stop. The same backward impulse on the ball of Silly Putty would also bring it to a stop, and an additional backward impulse on the ball would give it momentum in the backward direction. Thus, the block exerts the larger backward impulse on the Silly-Putty ball. In accord with Newton's third law, the impulse of a ball on the block is equal and opposite to the impulse of the block on the ball. Thus, the Silly-Putty ball exerts the larger forward impulse on the block, giving the block a larger forward change in momentum.

- New **Concept Checks** enable students to check their conceptual understanding of physics concepts while they read chapters. Answers are located at the end of chapters to provide immediate feedback. Concept Checks are placed near relevant topics so students can immediately reread any material that they do not fully understand.

- New **Pitfall Statements**, identified by exclamation points, help students avoid common misconceptions. These statements are placed near the topics that commonly cause confusion, so that students can immediately address any difficulties.

CONCEPT CHECK 3-1

Figure 3-9 is a motion diagram of the bungee jumper before, during, and after time t_6, when she momentarily come to rest at the lowest point in her descent. During the part of her ascent shown, she is moving upward with increasing speed. Use this diagram to determine the direction of the jumper's acceleration (*a*) at time t_6 and (*b*) at time t_9.

where U_0, the arbitrary constant of integration, is the value of the potential energy at $y = 0$. Because only a change in the potential energy is defined, the actual value of U is not important. For example, if the gravitational potential energy of the Earth–skier system is chosen to be zero when the skier is at the bottom of the hill, its value when the skier is at a height h above that level is mgh. Or we could choose the potential energy to be zero when the skier is at point P halfway down the ski slope, in which case its value at any other point would be mgy, where y is the height of the skier above point P. On the lower half of the slope, the potential energy would then be negative.

! We are free to choose U to be zero at any convenient reference point.

PHYSICS SPOTLIGHTS

Physics Spotlights at the end of appropriate chapters discuss current applications of physics and connect applications to concepts described in chapters. These topics range from wind farms to molecular thermometers to pulse detonation engines.

Physics Spotlight

Blowing Warmed Air

Wind farms dot the Danish coast, the plains of the upper Midwest, and hills from California to Vermont. Harnessing the kinetic energy of the wind is nothing new. Windmills have been used to pump water, ventilate mines,[*] and grind grain for centuries.

Today, the most visible wind turbines run electrical generators. These turbines transform kinetic energy into electromagnetic energy. Modern turbines range widely in size, cost, and output. Some are very small, simple machines that cost under $500/turbine, and put out less than 100 watts of power.[†] Others are complex behemoths that cost over $2 million and put out as much as 2.5 MW/turbine.[‡] All of these turbines take advantage of a widely available energy source—the wind.

The theory behind the windmill's conversion of kinetic energy to electromagnetic energy is straightforward. The moving air molecules push on the turbine blades, driving their rotational motion. The rotating blades then turn a series of gears. The gears, in turn, step up the rotation rate, and drive the rotation of a generator rotor. The generator sends the electromagnetic energy out along power lines.

A wind farm converting the kinetic energy of the air to electrical energy. (*Image Slate.*)

But the conversion of the wind's kinetic energy to electromagnetic energy is not 100 percent efficient. The most important thing to remember is that it *cannot* be 100 percent efficient. If turbines converted 100 percent of the kinetic energy of the air into electrical energy, the air would leave the turbines with zero kinetic energy. That is, the turbines would stop the air. If the air were completely stopped by the turbine, it would flow around the turbine, rather than through the turbine.

So the theoretical efficiency of a wind turbine is a trade-off between capturing the kinetic energy of the moving air, and preventing most of the wind from flowing around the turbine. Propeller-style turbines are the most common, and their theoretical efficiency at transforming the kinetic energy of the air into electromagnetic energy varies from 30 percent to 59 percent.[§] (The predicted efficiencies vary because of assumptions made about the way the air behaves as it flows through and around the propellers of the turbine.)

So even the most efficient turbine cannot convert 100 percent of the theoretically available energy. What happens? Upstream from the turbine, the air moves along straight streamlines. After the turbine, the air rotates and is turbulent. The rotational component of the air's movement beyond the turbine takes energy. Some dissipation of energy occurs because of the viscosity of air. When some of the air slows, there is friction between it and the faster moving air flowing by it. The turbine blades heat up, and the air itself heats up.[°] The gears within the turbine also convert kinetic energy into thermal energy through friction. All this thermal energy needs to be accounted for. The blades of the turbine vibrate individually—the energy associated with those vibrations cannot be used. Finally, the turbine uses some of the electricity it generates to run pumps for gear lubrication, and to run the yaw motor that moves the turbine blades into the most favorable position to catch the wind.

In the end, most wind turbines operate at between 10 and 20 percent efficiency.[#] They are still attractive power sources, because of the free fuel. One turbine owner explains, "The bottom line is we did it for our business to help control our future."[**]

* Agricola, Gorgeus, *De Re Metallic*. (Herbert and Lou Henry Hoover, Transl.) Reprint Mineola, NY: Dover, 1950, 200–203.
† Conally, Abe, and Conally, Josie, "Wind Powered Generator," *Make*, Feb. 2006, Vol. 5, 90–101.
‡ "Why Four Generators May Be Better than One," *Modern Power Systems*, Dec. 2005, 30.
§ Gorban, A. N., Gorlov, A. M., and Silantyev, V. M., "Limits of the Turbine Efficiency for Free Fluid Flow." *Journal of Energy Resources Technology*, Dec. 2001, Vol. 123, 311–317.
° Roy, S. B., S. W. Pacala, and R. L. Walko. "Can Large Wind Farms Affect Local Meteorology?" *Journal of Geophysical Research (Atmospheres)*, Oct. 16, 2004, 109, D19101.
Gorban, A. N., Gorlov, A. M., and Silantyev, V. M., "Limits of the Turbine Efficiency for Free Fluid Flow." *Journal of Energy Resources Technology*, December 2001, Vol. 123, 311–317.
** Wilde, Matthew, "Colwell Farmers Take Advantage of Grant to Produce Wind Energy." *Waterloo-Cedar Falls Courier*, May 1, 2006, B1+.

PHYSICS PORTAL

www.whfreeman.com/physicsportal

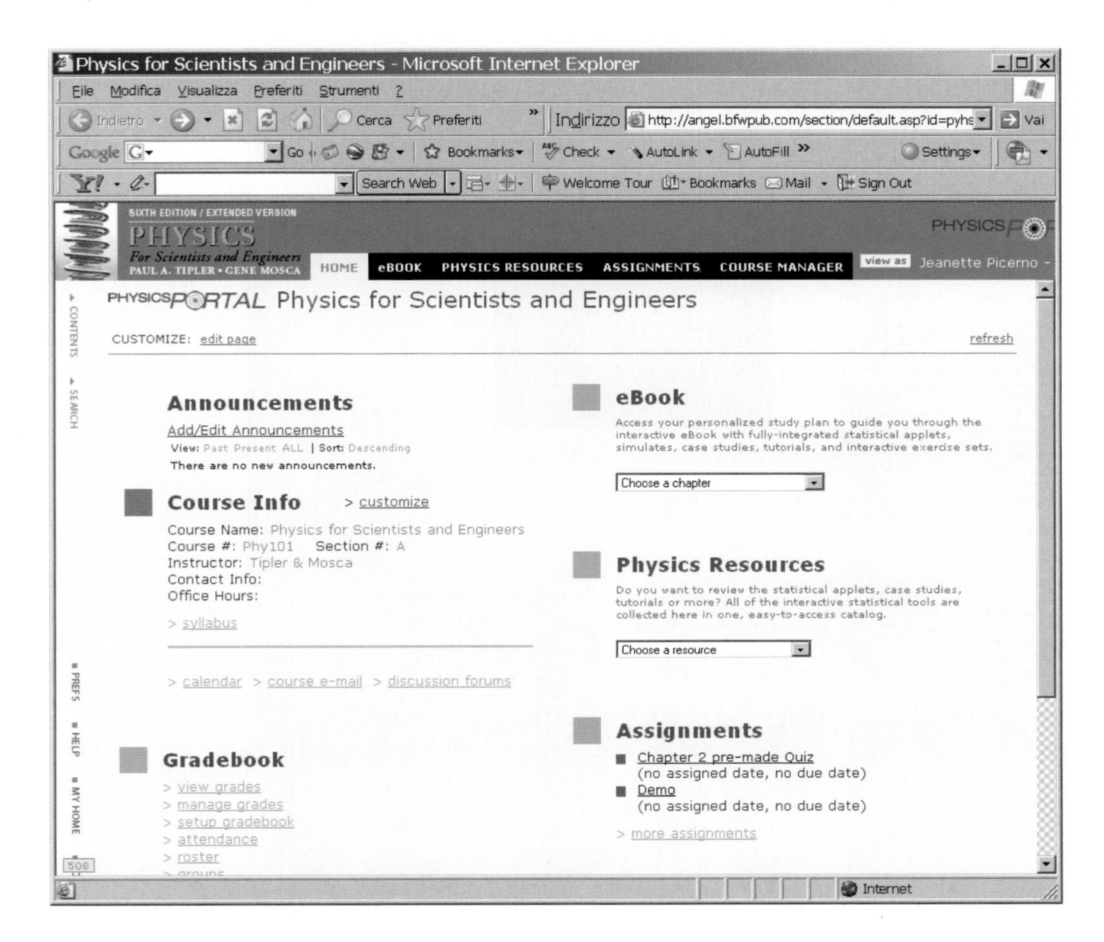

Physics Portal is a complete learning management system that includes a complete e-book, student and instructor resources, and an online homework system. Physics Portal is designed to enrich any course and enhance students' study.

All Resources in One Place

Physics Portal creates a powerful learning environment. Its three central components—the **Interactive e-Book**, the **Physics Resources** library, and the **Assignment Center** —are conceptually tied to the text and to one another, and are easily accessed by students with a single log-in.

Flexibility for Teachers and Students

From its home page to its text content, Physics Portal is fully customizable. Instructors can customize the home page, set course announcements, annotate the e-book, and edit or create new exercises and tutorials.

Interactive e-Book

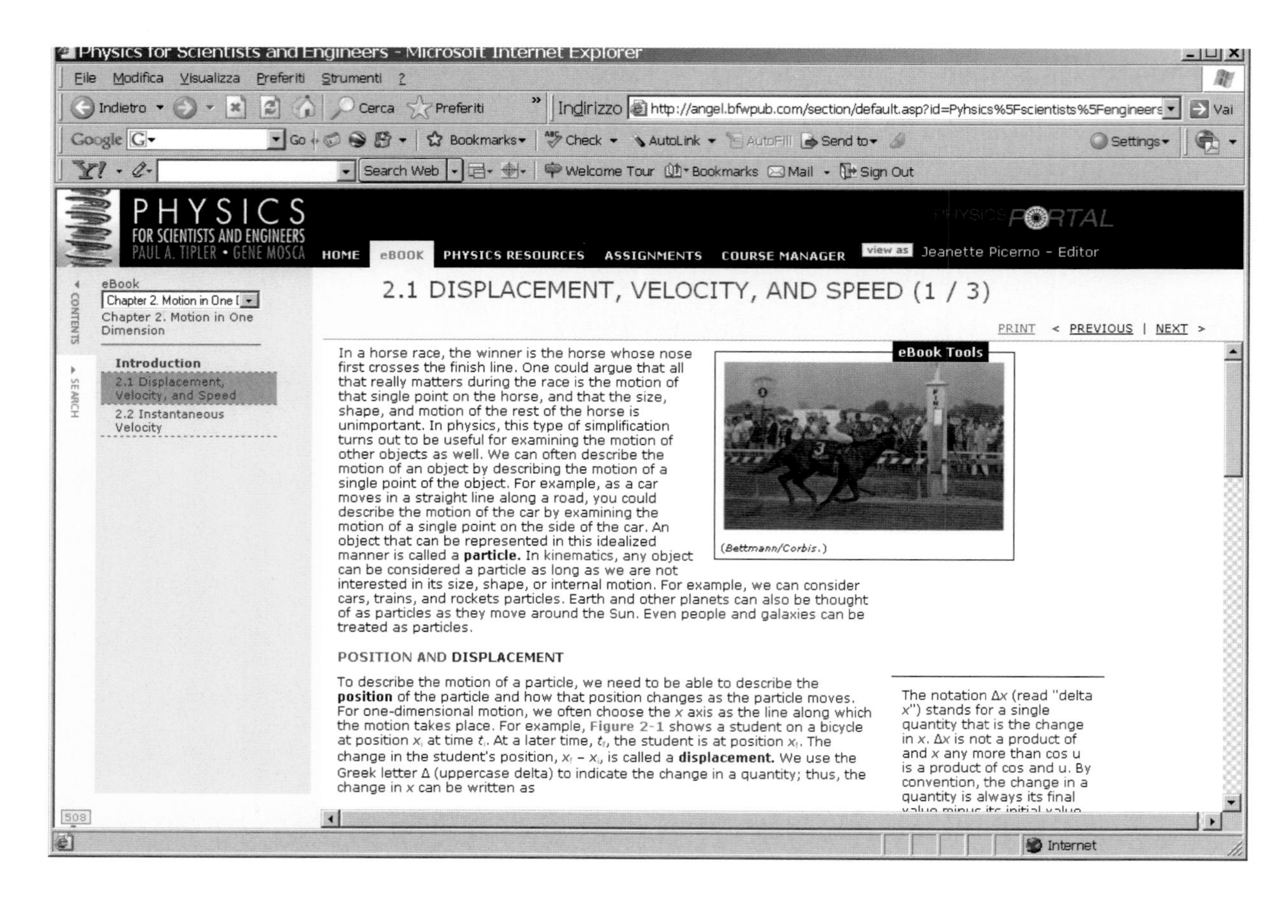

The complete text is integrated with the following:

- Conceptual animations
- Interactive exercises
- Video illustrations of key concepts

Study resources include

- **Notetaking and highlighting** Student notes can be collectively viewed and printed for a personalized study guide.
- **Bookmarking** for easy navigation and quick return to important locations.
- **Key terms with links** to definitions, Wikipedia, and automated Google Search
- **Full text search** for easy location of *every* resource for each topic

Instructors can customize their students' texts through annotations and supplementary links, providing students with a guide to reading and using the text.

Physics Resources

For the student, the wide range of resources focuses on interactivity and conceptual examples, engaging the student and addressing different learning styles.

- **Flashcards** Key terms from the text can be studied and used as self-quizzes.
- **Concept Tester—Picture It** Students input values for variables and see resulting graphs based on values.
- **Concept Tester—Solve It** Provides additional questions within interactive animations to help students visualize concepts.
- **Applied Physics Videos** Show physics concepts in real-life scenarios.
- **On-line quizzing** Provides immediate feedback to students and quiz results can be collected for the instructor in a gradebook.

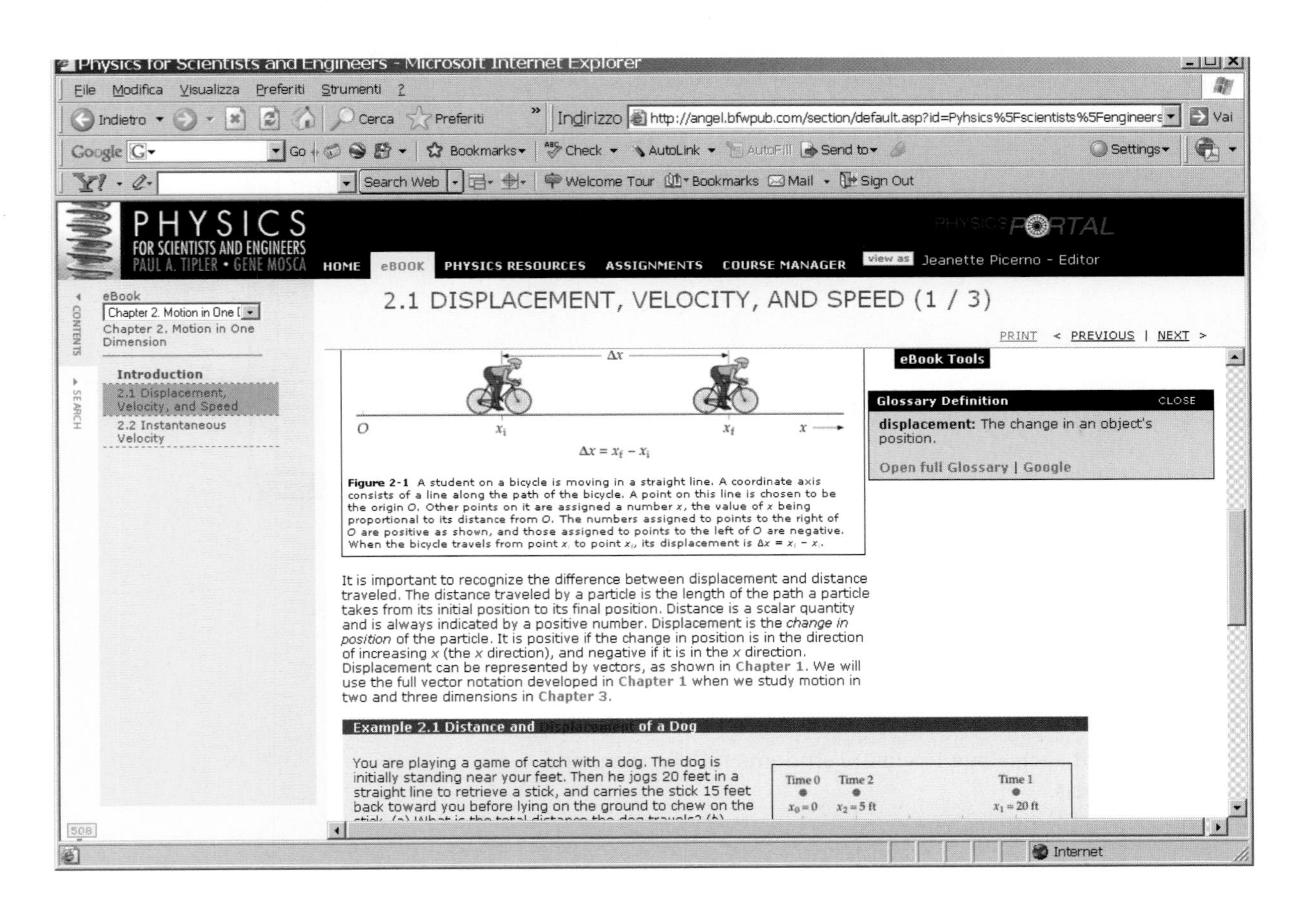

Assignment Center

Homework and Branched-Tutorials for Student Practice and Success

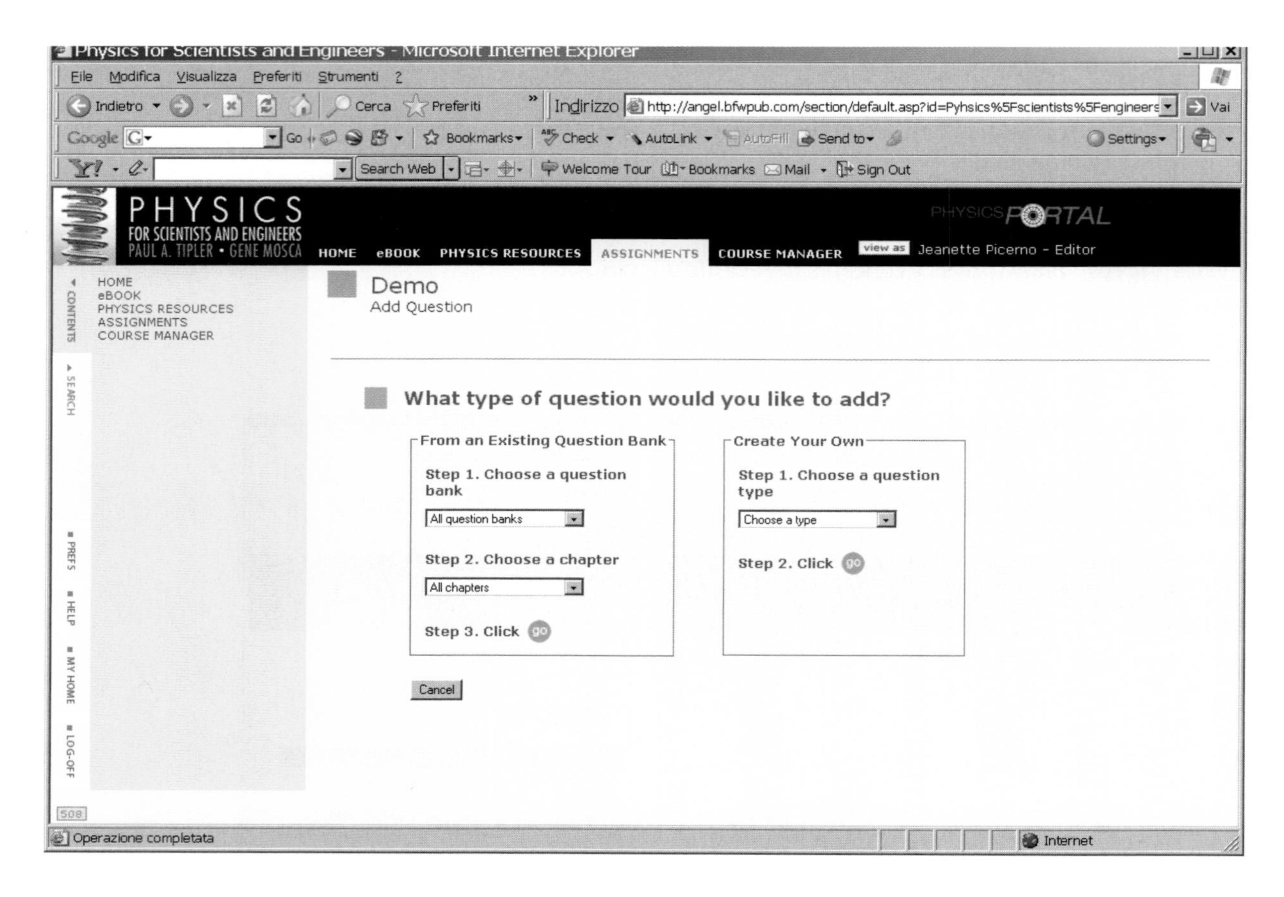

The **Assignment Center** manages and automatically grades homework, quizzes, and guided practice.

- All aspects of **Physics Portal** can be assigned, including e-book sections, simulations, tutorials, and homework problems.
- **Interactive Exercises** break down complex problems into individual steps.
- **Tutorials** offer guidance at each stage to ensure students fully understand the problem-solving process.
- **Video Analysis Exercises** enable students to investigate real-world motion.

Student progress is tracked in a single, easy-to-use gradebook.

- Details tracked include completion, time spent, and type of assistance.
- Instructors can choose grade criteria.
- The gradebook is easily exported into alternative course management systems.

Homework services End-of-chapter problems are available in WebAssign and on Physics Portal. A list of all the sixth edition problems included in WebAssign is posted on the instructor's section of the Tipler Web site at **www.whfreeman.com/tiplerphysics6e.**

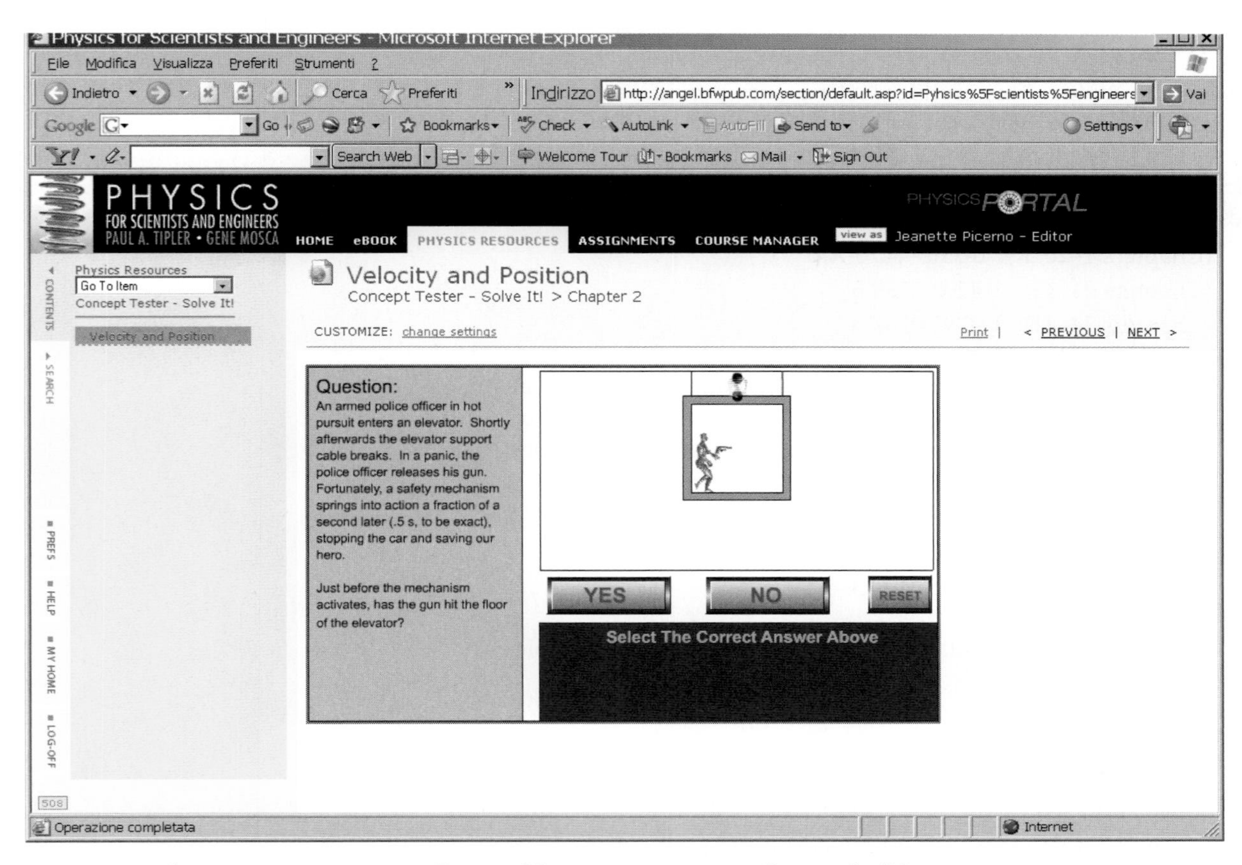

MEDIA AND PRINT SUPPLEMENTS

FOR THE STUDENT

Student Solutions Manual The new manual, prepared by David Mills, professor emeritus at the College of the Redwoods in California, provides solutions for selected odd-numbered end-of-chapter problems in the textbook and uses the same side-by-side format and level of detail as the Examples in the text.

- **Volume 1** (Chapters 1–20, R) 1-4292-0302-1
- **Volume 2** (Chapters 21–33) 1-4292-0303-X
- **Volume 3** (Chapters 34–41) 1-4292-0301-3

Study Guide The Study Guide provides students with key physical quantities and equations, misconceptions to avoid, questions and practice problems to gain further understanding of physics concepts, and quizzes to test student knowledge of chapters.

- **Volume 1** (Chapters 1–20, R) 0-7167-8467-X
- **Volume 2** (Chapters 21–33) 1-4292-0410-9
- **Volume 3** (Chapters 34–41) 1-4292-0411-7

Student Web Site

- **On-line quizzing** Multiple-choice quizzes are available for each chapter. Students will receive immediate feedback, and the quiz results are collected for the instructor in a grade book.
- **Concept Tester Questions**
- **Flashcards**

FOR THE INSTRUCTOR

Instructor's Resource CD-ROM This multifaceted resource provides instructors with the tools to make their own Web sites and presentations. The CD contains illustrations from the text in .jpg format, Powerpoint Lecture Slides for each chapter of the book, i-clicker questions, a problem conversion guide, and a complete test bank that includes more than 4000 multiple-choice questions.

- **Volume 1** (Chapters 1–20, R) 0-7167-8470-X
- **Volume 2** (Chapters 21–33) 1-4292-0268-8
- **Volume 3** (Chapters 34–41) 1-4292-0267-X

Answer Booklet with Solution CD Resource This book contains answers to all end-of-chapter problems and CD-ROMs of fully worked solutions for all end-of-chapter problems. Solutions are available in editable Word files on the Instructor's CD-ROM and are also available in print.

- **Volume 1** (Chapters 1–20, R) 0-7167-8479-3
- **Volume 2** (Chapters 21–33) 1-4292-0457-5
- **Volume 3** (Chapters 34–41) 1-4292-0461-3

Transparencies 7167-8469-6 More than 100 full color acetates of figures and tables from the text are included, with type enlarged for projection.

FLEXIBILITY FOR PHYSICS COURSES

We recognize that not all physics courses are alike, so we provide instructors with the opportunity to create the most effective resource for their students.

Custom-Ready Content and Design

Physics for Scientists and Engineers was written and designed to allow maximum customization. Instructors are invited to create specific volumes (such as a five-volume set), reduce the text's depth by selecting only certain chapters, and add additional material. To make using the textbook easier, W. H. Freeman encourages instructors to inquire about our custom options.

Versions Accomodate Common Course Arrangements

To simplify the review and use of the text, *Physics for Scientists and Engineers* is available in these versions:

Volume 1 *Mechanics/Oscillations and Waves/Thermodynamics* (Chapters 1–20, R) 1-4292-0132-0
Volume 2 *Electricity and Magnetism/Light* (Chapters 21–33) 1-4292-0133-9
Volume 3 *Elementary Modern Physics* (Chapters 34–41) 1-4292-0134-7
Standard Version (Chapters 1-33, R) 1-4292-0124-X
Extended Version (Chapters 1-41, R) 0-7167-8964-7

Acknowledgments

We are grateful to the many instructors, students, colleagues, and friends who have contributed to this edition and to earlier editions.

Anthony J. Buffa, professor emeritus at California Polytechnic State University in California, wrote many new end-of-chapter problems and edited the end-of-chapter problems sections. Laura Runkle wrote the Physics Spotlights. Richard Mickey revised the Math Review of the fifth edition, which is now the Math Tutorial of the sixth edition. David Mills, professor emeritus at the College of the Redwoods in California, extensively revised the Solutions Manual. We received invaluable help in creating text and checking the accuracy of text and problems from the following professors:

Thomas Foster
Southern Illinois University

Karamjeet Arya
San Jose State University

Mirley Bala
Texas A&M University—Corpus Christi

Michael Crivello
San Diego Mesa College

Carlos Delgado
Community College of Southern Nevada

David Faust
Mt. Hood Community College

Robin Jordan
Florida Atlantic University

Jerome Licini
Lehigh University

Dan Lucas
University of Wisconsin

Laura McCullough
University of Wisconsin, Stout

Jeannette Myers
Francis Marion University

Marian Peters
Appalachian State University

Todd K. Pedlar
Luther College

Paul Quinn
Kutztown University

Peter Sheldon
Randolph-Macon Woman's College

Michael G. Strauss
University of Oklahoma

Brad Trees
Ohio Wesleyan University

George Zober
Yough Senior High School

Patricia Zober
Ringgold High School

Many instructors and students have provided extensive and helpful reviews of one or more chapters of this edition. They have each made a fundamental contribution to the quality of this revision, and deserve our gratitude. We would like to thank the following reviewers:

Ahmad H. Abdelhadi
James Madison University

Edward Adelson
Ohio State University

Royal Albridge
Vanderbilt University

J. Robert Anderson
University of Maryland, College Park

Toby S. Anderson
Tennessee State University

Wickram Ariyasinghe
Baylor University

Yildirim Aktas
University of North Carolina, Charlotte

Eric Ayars
California State University

James Battat
Harvard University

Eugene W. Beier
University of Pennsylvania

Peter Beyersdorf
San Jose State University

Richard Bone
Florida International University

Juliet W. Brosing
Pacific University

Ronald Brown
California Polytechnic State University

Richard L. Cardenas
St. Mary's University

Troy Carter
University of California, Los Angeles

Alice D. Churukian
Concordia College

N. John DiNardo
Drexel University

Jianjun Dong
Auburn University

Fivos R Drymiotis
Clemson University

Mark A. Edwards
Hofstra University

James Evans
Broken Arrow Senior High

Nicola Fameli
University of British Columbia

N. G. Fazleev
University of Texas at Arlington

Thomas Furtak
Colorado School of Mines

Richard Gelderman
Western Kentucky University

Yuri Gershtein
Florida State University

Paolo Gondolo
University of Utah

Benjamin Grinstein
University of California, San Diego

Parameswar Hari
University of Tulsa

Joseph Harrison
University of Alabama—Birmingham

Patrick C. Hecking
Thiel College

Kristi R. G. Hendrickson
University of Puget Sound

Linnea Hess
Olympic College

Mark Hollabaugh
Normandale Community College

Daniel Holland
Illinois State University

Richard D. Holland II
Southern Illinois University

Eric Hudson
Massachusetts Institute of Technology

David C. Ingram
Ohio University

Colin Inglefield
Weber State University

Nathan Israeloff
Northeastern University

Donald J. Jacobs
California State University, Northridge

Erik L. Jensen
Chemeketa Community College

Colin P Jessop
University of Notre Dame

Ed Kearns
Boston University

Alice K. Kolakowska
Mississippi State University

Douglas Kurtze
Saint Joseph's University

Eric T. Lane
University of Tennessee at Chattanooga

Christie L. Larochelle
Franklin & Marshall College

Mary Lu Larsen
Towson University

Clifford L. Laurence
Colorado Technical University

Bruce W. Liby
Manhattan College

Ramon E. Lopez
Florida Institute of Technology

Ntungwa Maasha
Coastal Georgia Community Collegee and University Center

Jane H MacGibbon
University of North Florida

A. James Mallmann
Milwaukee School of Engineering

Rahul Mehta
University of Central Arkansas

R. A. McCorkle
University of Rhode Island

Linda McDonald
North Park University

Kenneth McLaughlin
Loras College

Eric R. Murray
Georgia Institute of Technology

Jeffrey S. Olafsen
University of Kansas

Richard P. Olenick
University of Dallas

Halina Opyrchal
New Jersey Institute of Technology

Russell L. Palma
Minnesota State University—Mankato

Todd K. Pedlar
Luther College

Daniel Phillips
Ohio University

Edward Pollack
University of Connecticut

Michael Politano
Marquette University

Robert L. Pompi
SUNY Binghamton

Damon A. Resnick
Montana State University

Richard Robinett
Pennsylvania State University

John Rollino
Rutgers University

Daniel V. Schroeder
Weber State University

Douglas Sherman
San Jose State University

Christopher Sirola
Marquette University

Larry K. Smith
Snow College

George Smoot
University of California at Berkeley

Zbigniew M. Stadnik
University of Ottawa

Kenny Stephens
Hardin-Simmons University

Daniel Stump
Michigan State University

Jorge Talamantes
California State University, Bakersfield

Charles G. Torre
Utah State University

Brad Trees
Ohio Wesleyan University

John K. Vassiliou
Villanova University

Theodore D. Violett
Western State College

Hai-Sheng Wu
Minnesota State University—Mankato

Anthony C. Zable
Portland Community College

Ulrich Zurcher
Cleveland State University

We also remain indebted to the reviewers of past editions. We would therefore like to thank the following reviewers, who provided immeasurable support as we developed the fourth and fifth editions:

Edward Adelson
The Ohio State University

Michael Arnett
Kirkwood Community College

Todd Averett
The College of William and Mary

Yildirim M. Aktas
University of North Carolina at Charlotte

Karamjeet Arya
San Jose State University

Alison Baski
Virginia Commonwealth University

William Bassichis
Texas A&M University

Joel C. Berlinghieri
The Citadel

Gary Stephen Blanpied
University of South Carolina

Frank Blatt
Michigan State University

Ronald Brown
California Polytechnic State University

Anthony J. Buffa
California Polytechnic State University

John E. Byrne
Gonzaga University

Wayne Carr
Stevens Institute of Technology

George Cassidy
University of Utah

Lay Nam Chang
Virginia Polytechnic Institute

I. V. Chivets
Trinity College, University of Dublin

Harry T. Chu
University of Akron

Alan Cresswell
Shippensburg University

Robert Coakley
University of Southern Maine

Robert Coleman
Emory University

Brent A. Corbin
UCLA

Andrew Cornelius
University of Nevada at Las Vegas

Mark W. Coffey
Colorado School of Mines

Peter P. Crooker
University of Hawaii

Jeff Culbert
London, Ontario

Paul Debevec
University of Illinois

Ricardo S. Decca
Indiana University-Purdue University

Robert W. Detenbeck
University of Vermont

N. John DiNardo
Drexel University

Bruce Doak
Arizona State University

Michael Dubson
University of Colorado at Boulder

John Elliott
University of Manchester, England

William Ellis
University of Technology — Sydney

Colonel Rolf Enger
U.S. Air Force Academy

John W. Farley
University of Nevada at Las Vegas

David Faust
Mount Hood Community College

Mirela S. Fetea
University of Richmond

David Flammer
Colorado School of Mines

Philip Fraundorf
University of Missouri, Saint Louis

Tom Furtak
Colorado School of Mines

James Garland
Retired

James Garner
University of North Florida

Ian Gatland
Georgia Institute of Technology

Ron Gautreau
New Jersey Institute of Technology

David Gavenda
University of Texas at Austin

Patrick C. Gibbons
Washington University

David Gordon Wilson
Massachusetts Institute of Technology

Christopher Gould
University of Southern California

Newton Greenberg
SUNY Binghamton

John B. Gruber
San Jose State University

Huidong Guo
Columbia University

Phuoc Ha
Creighton University

Richard Haracz
Drexel University

Clint Harper
Moorpark College

Michael Harris
University of Washington

Randy Harris
University of California at Davis

Tina Harriott
Mount Saint Vincent, Canada

Dieter Hartmann
Clemson University

Theresa Peggy Hartsell
Clark College

Kristi R.G. Hendrickson
University of Puget Sound

Michael Hildreth
University of Notre Dame

Robert Hollebeek
University of Pennsylvania

David Ingram
Ohio University

Shawn Jackson
The University of Tulsa

Madya Jalil
University of Malaya

Monwhea Jeng
University of California — Santa Barbara

James W. Johnson
Tallahassee Community College

Edwin R. Jones
University of South Carolina

Ilon Joseph
Columbia University

David Kaplan
University of California — Santa Barbara

William C. Kerr
Wake Forest University

John Kidder
Dartmouth College

Roger King
City College of San Francisco

James J. Kolata
University of Notre Dame

Boris Korsunsky
Northfield Mt. Hermon School

Thomas O. Krause
Towson University

Eric Lane
University of Tennessee, Chattanooga

Andrew Lang (graduate student)
University of Missouri

David Lange
University of California — Santa Barbara

Donald C. Larson
Drexel University

Paul L. Lee
California State University, Northridge

Peter M. Levy
New York University

Jerome Licini
Lehigh University

Isaac Leichter
Jerusalem College of Technology

William Lichten
Yale University

Robert Lieberman
Cornell University

Fred Lipschultz
University of Connecticut

Graeme Luke
Columbia University

Dan MacIsaac
Northern Arizona University

Edward McCliment
University of Iowa

Robert R. Marchini
The University of Memphis

Peter E. C. Markowitz
Florida International University

Daniel Marlow
Princeton University

Fernando Medina
Florida Atlantic University

Howard McAllister
University of Hawaii

John A. McClelland
University of Richmond

Laura McCullough
University of Wisconsin at Stout

M. Howard Miles
Washington State University

Matthew Moelter
University of Puget Sound

Eugene Mosca
U.S. Naval Academy

Carl Mungan
U.S. Naval Academy

Taha Mzoughi
Mississippi State University

Charles Niederriter
Gustavus Adolphus College

John W. Norbury
University of Wisconsin at Milwaukee

Aileen O'Donughue
St. Lawrence University

Jack Ord
University of Waterloo

Jeffry S. Olafsen
University of Kansas

Melvyn Jay Oremland
Pace University

Richard Packard
University of California

Antonio Pagnamenta
University of Illinois at Chicago

George W. Parker
North Carolina State University

John Parsons
Columbia University

Dinko Pocanic
University of Virginia

Edward Pollack
University of Connecticut

Robert Pompi
The State University of New York at Binghamton

Bernard G. Pope
Michigan State University

John M. Pratte
Clayton College and State University

Brooke Pridmore
Clayton State College

Yong-Zhong Qian
University of Minnesota

David Roberts
Brandeis University

Lyle D. Roelofs
Haverford College

R. J. Rollefson
Wesleyan University

Larry Rowan
University of North Carolina at Chapel Hill

Ajit S. Rupaal
Western Washington University

Todd G. Ruskell
Colorado School of Mines

Lewis H. Ryder
University of Kent, Canterbury

Andrew Scherbakov
Georgia Institute of Technology

Bruce A. Schumm
University of California, Santa Cruz

Cindy Schwarz
Vassar College

Mesgun Sebhatu
Winthrop University

Bernd Schuttler
University of Georgia

Murray Scureman
Amdahl Corporation

Marllin L. Simon
Auburn University

Scott Sinawi
Columbia University

Dave Smith
University of the Virgin Islands

Wesley H. Smith
University of Wisconsin

Kevork Spartalian
University of Vermont

Zbigniew M. Stadnik
University of Ottawa

G. R. Stewart
University of Florida

Michael G. Strauss
University of Oklahoma

Kaare Stegavik
University of Trondheim, Norway

Jay D. Strieb
Villanova University

Dan Styer
Oberlin College

Chun Fu Su
Mississippi State University

Jeffrey Sundquist
Palm Beach Community College — South

Cyrus Taylor
Case Western Reserve University

Martin Tiersten
City College of New York

Chin-Che Tin
Auburn University

Oscar Vilches
University of Washington

D. J. Wagner
Grove City College
Columbia University

George Watson
University of Delaware

Fred Watts
College of Charleston

David Winter

John A. Underwood
Austin Community College

John Weinstein
University of Mississippi

Stephen Weppner
Eckerd College

Suzanne E. Willis
Northern Illinois University

Frank L. H. Wolfe
University of Rochester

Frank Wolfs
University of Rochester

Roy C. Wood
New Mexico State University

Ron Zammit
California Polytechnic State University

Yuriy Zhestkov
Columbia University

Dean Zollman
Kansas State University

Fulin Zuo
University of Miami

Of course, our work is never done. We hope to receive comments and suggestions from our readers so that we can improve the text and correct any errors. If you believe you have found an error, or have any other comments, suggestions, or questions, send us a note at **asktipler@whfreeman.com**. We will incorporate corrections into the text during subsequent reprinting.

Finally, we would like to thank our friends at W. H. Freeman and Company for their help and encouragement. Susan Brennan, Clancy Marshall, Kharissia Pettus, Georgia Lee Hadler, Susan Wein, Trumbull Rogers, Connie Parks, John Smith, Dena Digilio Betz, Ted Szczepanski, and Liz Geller were extremely generous with their creativity and hard work at every stage of the process.

We are also grateful for the contributions and help of our colleagues Larry Tankersley, John Ertel, Steve Montgomery, and Don Treacy.

About the Authors

Paul Tipler was born in the small farming town of Antigo, Wisconsin, in 1933. He graduated from high school in Oshkosh, Wisconsin, where his father was superintendent of the public schools. He received his B.S. from Purdue University in 1955 and his Ph.D. at the University of Illinois in 1962, where he studied the structure of nuclei. He taught for one year at Wesleyan University in Connecticut while writing his thesis, then moved to Oakland University in Michigan, where he was one of the original members of the physics department, playing a major role in developing the physics curriculum. During the next 20 years, he taught nearly all the physics courses and wrote the first and second editions of his widely used textbooks *Modern Physics* (1969, 1978) and *Physics* (1976, 1982). In 1982, he moved to Berkeley, California, where he now resides, and where he wrote *College Physics* (1987) and the third edition of *Physics* (1991). In addition to physics, his interests include music, hiking, and camping, and he is an accomplished jazz pianist and poker player.

Gene Mosca was born in New York City and grew up on Shelter Island, New York. He studied at Villanova University, the University of Michigan, and the University of Vermont, where he received his Ph.D. in physics. Gene recently retired from his teaching position at the U.S. Naval Academy, where as coordinator of the core physics course he instituted numerous enhancements to both the laboratory and classroom. Proclaimed by Paul Tipler "the best reviewer I ever had," Mosca became his coauthor beginning with the fifth edition of this book.

PART VI MODERN PHYSICS: Quantum Mechanics, Relativity, and the Structure of Matter

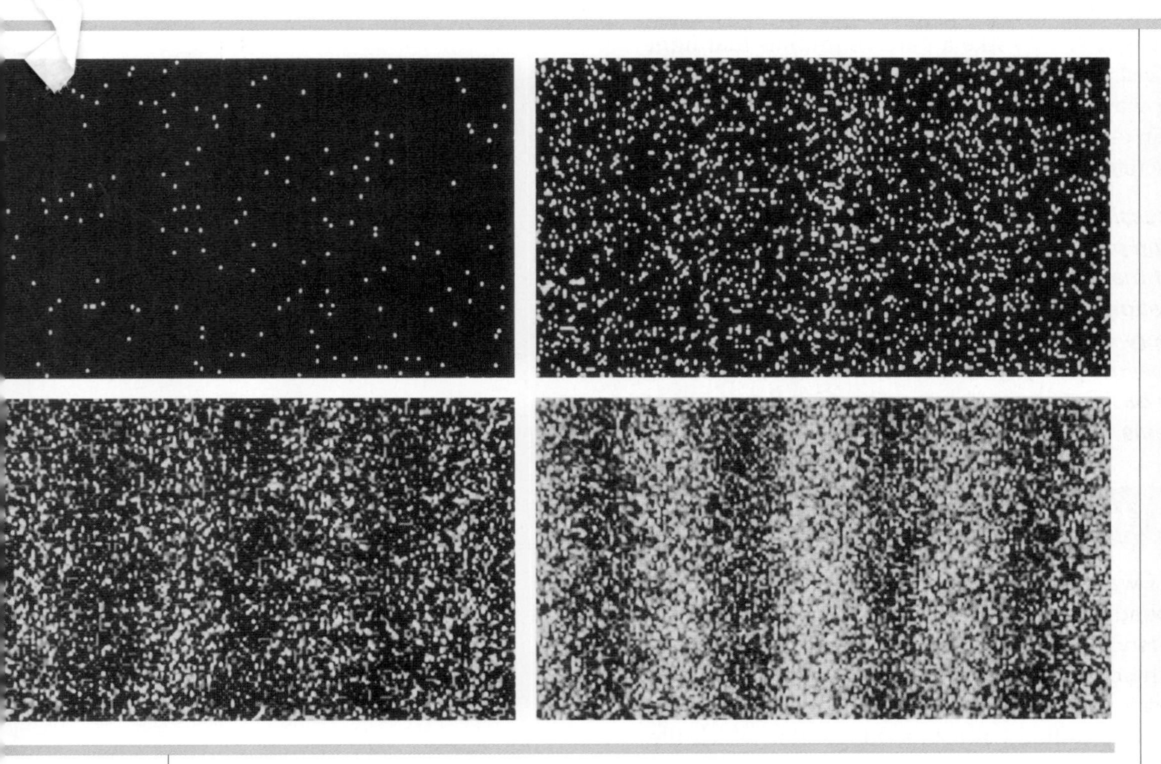

ELECTRON INTERFERENCE PATTERN PRODUCED BY ELECTRONS INCIDENT ON A BARRIER CONTAINING TWO SLITS: (*A*) 10 ELECTRONS, (*B*) 100 ELECTRONS, (*C*) 3000 ELECTRONS, AND (*D*) 70,000 ELECTRONS. THE MAXIMA AND MINIMA DEMONSTRATE THE WAVE NATURE OF THE ELECTRON AS IT TRAVERSES THE SLITS. INDIVIDUAL DOTS ON THE SCREEN INDICATE THE PARTICLE NATURE OF THE ELECTRON AS IT EXCHANGES ENERGY WITH THE DETECTOR. THE PATTERN IS THE SAME WHETHER ELECTRONS OR PHOTONS (PARTICLES OF LIGHT) ARE USED. *(Courtesy of Akira Tononmura, Advanced Research Laboratory, Hitachi, Ltd.)*

Wave–Particle Duality and Quantum Physics

? How do you calculate the wavelength of an electron? (See Example 34-4.)

At the beginning of the twentieth century, it was thought that sound, light, and other electromagnetic radiation (such as radio waves) were waves, and electrons, protons, atoms, and similar units were particles. The first 30 years of that century revealed startling developments in theoretical and experimental physics, such as the finding that light actually exchanges energy in discrete lumps or quanta, just like particles, and the finding that an electron exhibits diffraction and interference as it propagates through space, just like a wave. The fact that light exchanges energy like a particle implies that light energy is not continuous but is *quantized*. Similarly, the wave nature of the electron, along with the fact that the standing wave condition requires a discrete set of frequencies, implies that the energy of an electron in a confined region of space is not continuous, but is quantized to a discrete set of values.

In this chapter, we begin by discussing some basic properties of light and electrons, examining their wave and particle characteristics. We then consider some of the detailed properties of matter waves, showing, in particular, how standing waves imply the quantization of energy. Finally, we discuss some of the important features of the theory of quantum physics, which was developed in the 1920s and which has been extremely successful in describing nature. Quantum physics is now the basis of our understanding of both atomic and subatomic systems and systems that have very low temperatures.

34-1 WAVES AND PARTICLES

We have seen that the propagation of waves through space is quite different from the propagation of particles. Waves bend around corners (diffraction) and interfere with one another, producing interference patterns. If a wave encounters a small aperture, the wave spreads out on the other side as if the aperture were a point source. The propagation of particles is quite unlike the propagation of waves. Particles travel in straight lines until they collide with something, after which the particles again travel in straight lines. If two particle beams meet in space, they never produce an interference pattern.

Particles and waves also exchange energy differently. Particles exchange energy in collisions that occur at specific points in space and in time. The energy of waves, on the other hand, is spread out in space and deposited continuously as the wave fronts interact with matter.

Sometimes the propagation of a wave cannot be distinguished from the propagation of a beam of particles. If the wavelength λ is very small compared to distances from the edges of objects, diffraction effects are negligible and the wave travels in straight lines. Also, interference maxima and minima are so close together in space as to be unobservable. The result is that the wave interacts with a detector, like a beam of numerous small particles each exchanging a small amount of energy; the exchange cannot distinguish particles from waves. For example, you do not observe the individual air molecules bouncing off your face if the wind blows on it. Instead, the interaction of billions of particles appears to be continuous, like that of a wave.

34-2 LIGHT: FROM NEWTON TO MAXWELL

The question of whether light consists of a beam of particles or waves in motion is one of the most interesting in the history of science (see Chapter 31). Isaac Newton used a particle theory of light to explain the laws of reflection and refraction; however, for refraction, Newton needed to assume that light travels faster in water or glass than in air, an assumption later shown to be incorrect. The chief early proponents of the wave theory were Robert Hooke and Christian Huygens, who explained refraction by assuming that light travels more slowly in glass or water than it does in air.* Newton favored the theory that light consists of particles and does not consist of waves because, in his time, light was believed to travel through a medium only in straight lines—diffraction had not yet been observed.

Because of Newton's great reputation and authority, his particle theory of light was accepted for more than a century. Then, in 1801, Thomas Young demonstrated the wave nature of light in a famous experiment in which two coherent light sources are produced by illuminating a pair of narrow, parallel slits with a single source (Figure 34-1*a*).† In Chapter 33, we saw that when light encounters a small opening, the opening acts as a point source of waves (Figure 33-7). In Young's experiment, each slit acts as a line source, which is equivalent to a point source in two dimensions.‡ The interference pattern is observed on a screen placed behind the slits. Interference maxima occur at angles so that the path difference is an integral number of wavelengths. Similarly, interference minima occur if the path difference is one-half the wavelength or any odd number of half wavelengths. Figure 34-1*b* shows the intensity pattern as seen on the screen. Remember that if two coherent waves of equal intensity I_0 meet in space, the result can be a wave of intensity $4I_0$ (constructive interference), an intensity of zero (destructive interference), or a wave of intensity between zero and $4I_0$, depending on the phase difference between the waves at the observation point. Young's experiment and many other experiments demonstrate that light propagates like a wave.

(a)

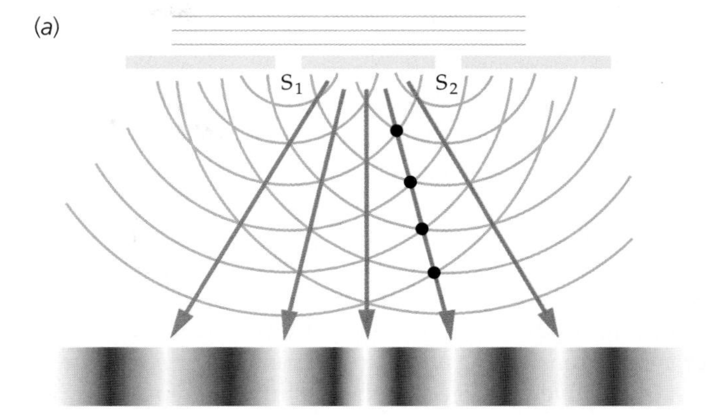

(b)

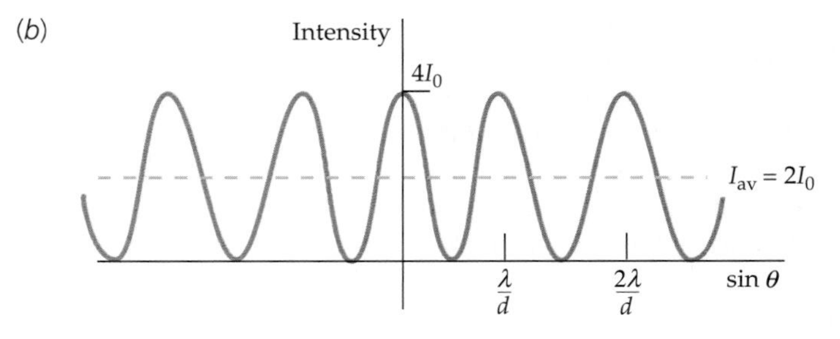

FIGURE 34-1 (*a*) Two slits act as coherent sources of light for the observation of interference in Young's experiment. Cylindrical waves from the slits overlap and produce an interference pattern on a screen far away. (*b*) The intensity pattern produced in Figure 34-1*a*. The intensity is maximum at points where the path difference is an even number of half wavelengths, and the intensity is zero where the path difference is an odd number of half wavelengths.

In the early nineteenth century, the French physicist Augustin Fresnel (1788–1827) performed extensive experiments on interference and diffraction and put the wave theory on a rigorous mathematical basis. Fresnel showed that the observed straight-line propagation of light is a result of the very short wavelengths of visible light.

The classical wave theory of light culminated in 1860 when James Clerk Maxwell published his mathematical theory of electromagnetism. This theory yielded a wave equation that predicted the existence of electromagnetic waves that propagate with a speed that can be calculated from the laws of electricity and magnetism.# The fact that the result of this calculation was $c \approx 3 \times 10^8$ m/s, the same as the speed of light, suggested to Maxwell that light is an electromagnetic wave. The eye is sensitive to electromagnetic waves that have wavelengths in the range from approximately 400 nm (1 nm $= 10^{-9}$ m) to approximately 700 nm. This range is called *visible light*. Other electromagnetic waves (for example, microwave waves, radio waves, and X rays) differ from visible light waves only in wavelength and in frequency.

34-3 THE PARTICLE NATURE OF LIGHT: PHOTONS

The diffraction of light and the existence of an interference pattern in the two-slit experiment give clear evidence that light has wave properties. However, early in the twentieth century, it was found that light energy comes in discrete amounts.

* See Section 5 of Chapter 31.
† See Section 3 of Chapter 33.
‡ See Section 4 of Chapter 33.
\# See Section 3 of Chapter 30

THE PHOTOELECTRIC EFFECT

The quantum nature of light and the quantization of energy were suggested by Albert Einstein in 1905 in his explanation of the photoelectric effect. Einstein's work marked the beginning of quantum theory, and for his work, Einstein received the Nobel Prize in Physics. Figure 34-2 shows a schematic diagram of the basic apparatus for studying the photoelectric effect. Light of a single frequency enters an evacuated chamber and falls on a clean metal surface C (C for cathode), causing electrons to be emitted. Some of these electrons strike the second metal plate A (A for anode), constituting an electric current between the plates. Plate A is negatively charged, so the electrons are repelled by it, with only the most energetic electrons reaching plate A. The maximum kinetic energy of the emitted electrons is measured by slowly increasing the voltage until the current becomes zero. Experiments give the surprising result that the maximum kinetic energy of the emitted electrons is *independent of the intensity* of the incident light. Classically, we would expect that increasing the rate at which light energy falls on the metal surface would increase the energy absorbed by individual electrons and, therefore, would increase the maximum kinetic energy of the electrons emitted. Experiments show that this classical result does not happen. The maximum kinetic energy of the emitted electrons is the same for a given wavelength of incident light, no matter how intense the light is. Einstein demonstrated that this experimental result can be explained if light energy is quantized in small bundles called **photons.** The energy E of each photon is given by

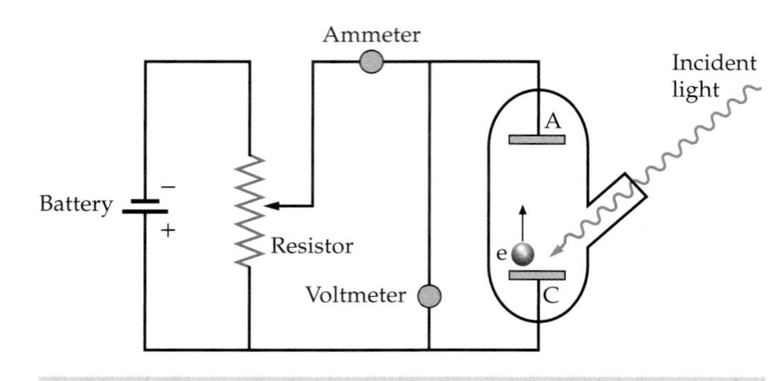

FIGURE 34-2 A schematic drawing of the apparatus for studying the photoelectric effect. Light of a single frequency enters an evacuated chamber and strikes the cathode C, which then ejects electrons (electron in figure is not drawn to scale). The current in the ammeter measures the number of these electrons that reach the anode A per unit time. The anode is made electrically negative with respect to the cathode to repel the electrons. Only those electrons that have enough initial kinetic energy to overcome the repulsion can reach the anode. The voltage between the two plates is slowly increased until the current becomes zero, which happens when even the most energetic electrons do not make it to plate A.

$$E = hf = \frac{hc}{\lambda} \qquad 34\text{-}1$$

EINSTEIN EQUATION FOR PHOTON ENERGY

where f is the frequency, and h is a constant now known as **Planck's constant.*** The measured value of this constant is

$$h = 6.626 \times 10^{-34}\,\text{J}\cdot\text{s} = 4.136 \times 10^{-15}\,\text{eV}\cdot\text{s} \qquad 34\text{-}2$$

PLANCK'S CONSTANT

Equation 34-1 is sometimes called the **Einstein equation.**

A light beam consists of a beam of particles—photons—each having energy hf. The intensity (power per unit area) of a monochromatic light beam is the number of photons per unit area per unit of time, multiplied by the energy per photon. The interaction of the light beam with the metal surface consists of collisions between photons and electrons. During each of these collisions, the photon gives all its energy to an electron and the photon no longer exists. The electron is emitted from the surface after it receives the energy from a single photon. If the intensity of light is increased, more photons fall on the surface per unit time, and more electrons are emitted per unit time. However, each photon still has the same energy hf, so the energy absorbed by each electron is unchanged.

* In 1900, the German physicist Max Planck introduced this constant to explain discrepancies between the theoretical curves and experimental data on the spectrum of blackbody radiation. Planck also assumed that the radiation was emitted and absorbed by a blackbody in quanta of energy hf, but he considered his assumption to be just a computational device rather than a fundamental property of electromagnetic radiation. (Blackbody radiation is discussed in Chapter 20.)

If ϕ is the minimum energy necessary to remove an electron from a metal surface, the maximum kinetic energy of the electrons emitted is given by

$$K_{max} = \left(\tfrac{1}{2}mv^2\right)_{max} = hf - \phi \qquad 34\text{-}3$$

EINSTEIN'S PHOTOELECTRIC EQUATION

where f is the frequency of the photons. The quantity ϕ, called the **work function,** is a characteristic of the particular metal. (Some electrons will have kinetic energies less than $hf - \phi$, because of the loss of energy from traveling through the metal.)

According to Einstein's photoelectric equation, a plot of K_{max} versus frequency f should be a straight line that has the slope h. This was a bold prediction, because, at the time, no evidence existed that Planck's constant had any application outside of blackbody radiation. In addition, there was no experimental data on K_{max} versus frequency f, because no one before had even suspected that the frequency of the light was related to K_{max}. This prediction was difficult to verify experimentally, but careful experiments by R. A. Millikan approximately 10 years later showed that Einstein's equation was correct. Figure 34-3 shows a plot of Millikan's data.

Photons that have frequencies less than a **threshold frequency** f_t, and therefore have wavelengths greater than a **threshold wavelength** $\lambda_t = c/f_t$, do not have enough energy to eject an electron from a particular metal. The threshold frequency and the corresponding threshold wavelength can be related to the work function ϕ by setting the maximum kinetic energy of the electrons equal to zero in Equation 34-3. Then

$$\phi = hf_t = \frac{hc}{\lambda_t} \qquad 34\text{-}4$$

Work functions for metals are typically a few electron volts. Because wavelengths are usually given in nanometers and energies in electron volts, it is useful to have the value of hc in electron volt–nanometers:

$$hc = (4.1357 \times 10^{-15}\ \text{eV}\cdot\text{s})(2.9979 \times 10^{8}\ \text{m/s}) = 1.240 \times 10^{-6}\ \text{eV}\cdot\text{m}$$

or

$$hc = 1240\ \text{eV}\cdot\text{nm} \qquad 34\text{-}5$$

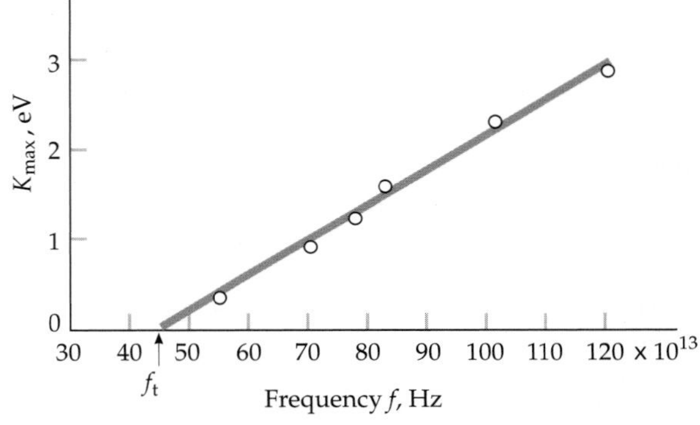

FIGURE 34-3 Millikan's data for the maximum kinetic energy K_{max} versus frequency f for the photoelectric effect. The data fall on a straight line that has a slope h, as predicted by Einstein approximately a decade before the experiment was performed.

Example 34-1 Photon Energies for Visible Light

Calculate the photon energies for light that has a wavelength equal to 400 nm (violet) and light that has a wavelength equal to 700 nm (red). (The wavelengths of 400 nm and 700 nm are the approximate wavelengths for the two extremes of the visible light spectrum.)

PICTURE Photon energies are related to photon frequencies and wavelengths by $E = hf = hc/\lambda$ (Equation 34-1).

SOLVE

1. The energy is related to the wavelength by Equation 34-1: $$E = hf = \frac{hc}{\lambda}$$
2. For $\lambda = 400$ nm, the energy is $$E = \frac{hc}{\lambda} = \frac{1240\ \text{eV}\cdot\text{nm}}{400\ \text{nm}} = \boxed{3.10\ \text{eV}}$$
3. For $\lambda = 700$ nm, the energy is $$E = \frac{hc}{\lambda} = \frac{1240\ \text{eV}\cdot\text{nm}}{700\ \text{nm}} = \boxed{1.77\ \text{eV}}$$

CHECK The shorter the wavelength of light, the greater the energy and 3.10 eV for 400 nm is greater than 1.77 eV for 700 nm.

TAKING IT FURTHER We can see from these calculations that visible light has photons that have energies which range from approximately 1.8 eV to 3.1 eV. X rays, which have much shorter wavelengths, have photons that have energies of the order of keV. Gamma rays emitted by nuclei have even shorter wavelengths and photons that have energies of the order of MeV.

PRACTICE PROBLEM 34-1 Find the energy of a photon corresponding to electromagnetic radiation in the FM radio band of wavelength 3.00 m.

PRACTICE PROBLEM 34-2 Find the wavelength of a photon whose energy is (*a*) 0.100 eV, (*b*) 1.00 keV, and (*c*) 1.00 MeV.

Example 34-2 The Number of Photons per Second in Sunlight *Try It Yourself*

The intensity of sunlight at Earth's surface is approximately 1400 W/m^2. Assuming the average photon energy is 2.00 eV (corresponding to a wavelength of approximately 600 nm), calculate the number of photons that strike an area of 1.00 cm^2 each second.

PICTURE The intensity (power per unit area) is given, as is the area. From these given quantities, we can calculate the power, which is the energy per unit time.

SOLVE

Cover the column to the right and try these on your own before looking at the answers.

Steps	Answers
1. The energy ΔE is related to the number N of photons and the energy per photon $hf = 2.00$ eV:	$\Delta E = Nhf$
2. The intensity I (power per unit area) and the area A are given, so we can find the power:	$I = \frac{P}{A}$
3. Knowing the power (energy per unit time) and the time, we can find the energy:	$\Delta E = P\Delta t$
4. Combine the results from steps 1–3 and solve for N (take care to get the units to cancel):	$N = \boxed{4.38 \times 10^{17}}$

CHECK This is an enormous number of photons. However, in everyday situations we do not notice that the energy of sunlight arrives in discrete amounts. Thus, an enormous number is expected.

PRACTICE PROBLEM 34-3 Calculate the photon density (in photons per cubic centimeter) of the sunlight in Example 34-2. The number arriving on an area of 1.00 cm^2 in one second is the number in a column whose cross section is 1.00 cm^2 and whose height is the distance light travels in one second.

COMPTON SCATTERING

The first use of the photon concept was to explain the results of photoelectric-effect experiments. In the photoelectric effect, all the energy of the photon is transferred to an electron. However, in Compton scattering only some of the energy of the photon is transferred to an electron. The photon concept was also used by Arthur H. Compton to explain the results of his measurements of the scattering of X rays by free electrons in 1923. According to classical theory, if an electromagnetic wave of frequency f_i is incident on material containing free charges, the charges will oscillate with this frequency and reradiate electromagnetic waves of the same frequency.

Compton considered these reradiated waves as scattered photons, and he pointed out that if the scattering process were a collision between a photon and an electron (Figure 34-4), the electron would recoil and thus absorb energy. The scattered photon would then have less energy, and therefore a lower frequency and longer wavelength, than the incident photon.

According to classical electromagnetic theory (see Section 30-3), the energy and momentum of an electromagnetic wave are related by

$$E = pc \qquad 34\text{-}6$$

The momentum of a photon is thus related to its wavelength λ by $p = E/c = hf/c = h/\lambda$.

$$p = \frac{h}{\lambda} \qquad 34\text{-}7$$

MOMENTUM OF A PHOTON

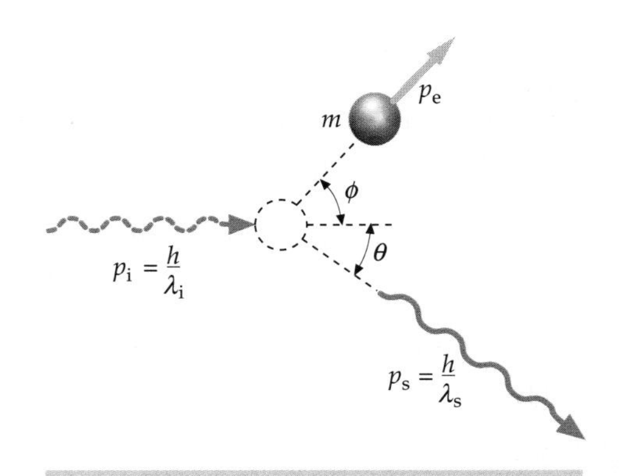

FIGURE 34-4 The scattering of light by an electron is considered as a collision of a photon of momentum h/λ_i and a stationary electron. The scattered photon has less energy and therefore has a longer wavelength than does the incident photon.

Compton applied the laws of conservation of momentum and energy to the collision of a photon and an electron to calculate the momentum p_s and thus the wavelength $\lambda_s = h/p_s$ of the scattered photon (see Figure 34-4). Applying conservation of momentum to the collision gives

$$\vec{p}_i = \vec{p}_s + \vec{p}_e \qquad 34\text{-}8$$

where $\vec{p}_i$ is the momentum of the incident photon and $\vec{p}_e$ is the momentum of the electron after the collision. The initial momentum of the electron is zero. Rearranging Equation 34-8, we have $\vec{p}_e = \vec{p}_i - \vec{p}_s$. Taking the dot product of each side with itself gives

$$p_e^2 = p_i^2 + p_s^2 - 2p_i p_s \cos\theta \qquad 34\text{-}9$$

where θ is the angle the direction of motion of the scattered photon makes with the direction of motion of the incident photon. Because the kinetic energy of the electron after the collision can be a significant fraction of the rest energy of an electron, the relativistic expression relating the total energy E of the electron to its momentum is used (see Chapter R). This expression (Equation R-17) is

$$E = \sqrt{p_e^2c^2 + (m_ec^2)^2}$$

where m_e is the mass of the electron. Applying conservation of energy to the collision gives

$$p_ic + m_ec^2 = p_sc + \sqrt{p_e^2c^2 + (m_ec^2)^2} \qquad 34\text{-}10$$

where pc (Equation 34-6) has been used to express the energies of the photons. Eliminating p_e^2 from Equations 34-9 and 34-10 gives

$$\frac{1}{p_s} - \frac{1}{p_i} = \frac{1}{m_ec}(1 - \cos\theta)$$

and substituting for p_i and p_s, using Equation 34-7, gives

$$\lambda_s - \lambda_i = \frac{h}{m_ec}(1 - \cos\theta) \qquad 34\text{-}11$$

COMPTON EQUATION

The increase in wavelength is independent of the wavelength λ_i of the incident photon. The quantity $h/(m_ec)$ has dimensions of length and is called the *Compton wavelength* λ_C. Its value is

$$\lambda_C = \frac{h}{m_ec} = \frac{hc}{m_ec^2} = \frac{1240\ \text{eV}\cdot\text{nm}}{5.110 \times 10^5\ \text{eV}} = 2.426 \times 10^{-12}\ \text{m} = 2.426\ \text{pm} \qquad 34\text{-}12$$

Because $\lambda_s - \lambda_i$ is small, it is difficult to observe unless λ_i is so small that the fractional change $(\lambda_s - \lambda_i)/\lambda_i$ is appreciable.

Compton used X rays that have wavelengths equal to 71.1 pm (1 pm $= 10^{-12}$ m $= 10^{-3}$ nm). The energy of a photon of this wavelength is $E = hc/\lambda = (1240\text{ eV}\cdot\text{nm})/(0.0711\text{ nm}) = 17.4$ keV. The electrons in the experiment can be considered essentially free because the energy of the X rays is much greater than the binding energies of the valence electrons in atoms (which are of the order of a few eV). Compton's measurements of $\lambda_s - \lambda_i$ as a function of scattering angle θ agreed with Equation 34-11, thereby confirming the correctness of the photon concept (the particle nature of light).

Example 34-3 Finding the Increase in Wavelength

An X-ray photon of wavelength 6.00 pm makes a head-on collision with an electron, so that the scattered photon goes in a direction opposite to that of the incident photon. The electron is initially at rest. (*a*) How much longer is the wavelength of the scattered photon than the wavelength of the incident photon? (*b*) What is the kinetic energy of the recoiling electron?

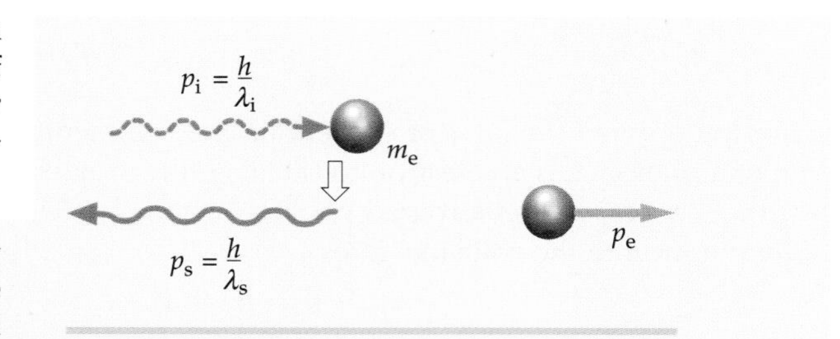

FIGURE 34-5

PICTURE We can calculate the increase in wavelength, and thus the new wavelength, from the Compton equation (Equation 34-11). We then use the new wavelength to find the energy of the scattered photon and then to find the kinetic energy of the recoiling electron from conservation of energy (Figure 34-5).

SOLVE

(*a*) Use Equation 34-11 to calculate the increase in wavelength:

$$\Delta\lambda = \lambda_s - \lambda_i = \frac{h}{m_e c}(1 - \cos\theta)$$
$$= (2.43\text{ pm})(1 - \cos 180°) = \boxed{4.86\text{ pm}}$$

(*b*) 1. The kinetic energy of the recoiling electron equals the energy of the incident photon E_i minus the energy of the scattered photon E_s:

$$K_e = E_i - E_s = hf_i - hf_s = \frac{hc}{\lambda_i} - \frac{hc}{\lambda_s}$$

2. Calculate λ_s from the given wavelength of the incident photon and the change found in Part (*a*):

$$\lambda_s = \lambda_i + \Delta\lambda = 6.00\text{ pm} + 4.86\text{ pm}$$
$$= 10.86\text{ pm}$$

3. Substitute the values of λ_i and λ_s into the Part (*b*), step-1 result to find the energy of the recoiling electron:

$$K_e = \frac{hc}{\lambda_i} - \frac{hc}{\lambda_s}$$
$$= \frac{1240\text{ eV}\cdot\text{nm}}{6.00\text{ pm}} - \frac{1240\text{ eV}\cdot\text{nm}}{10.86\text{ pm}}$$
$$= \frac{1.240\text{ keV}\cdot\text{nm}}{6.00\times 10^{-3}\text{ nm}} - \frac{1.240\text{ keV}\cdot\text{nm}}{10.86\times 10^{-3}\text{ nm}}$$
$$= 207\text{ keV} - 114\text{ keV} = \boxed{93\text{ keV}}$$

TAKING IT FURTHER The kinetic energy of the scattered electron is 93 keV and the rest energy of an electron is 511 keV, so the kinetic energy is 18 percent of the rest energy. Thus, the nonrelativistic formula for the kinetic energy $(\frac{1}{2}m_e v^2)$ is not valid.

PRACTICE PROBLEM 34-4 What is the speed of the scattered electron given by the nonrelativistic formula for the kinetic energy $(\frac{1}{2}m_e v^2)$?

34-4 ENERGY QUANTIZATION IN ATOMS

Ordinary white light has a continuous spectrum; that is, it contains *all* the wavelengths in the visible spectrum. But if atoms in a gas at low pressure are excited by an electric discharge, they emit light of specific wavelengths that are characteristic of the element or the compound. Because the energy of a photon is related to its wavelength by $E = hf = hc/\lambda$, a discrete set of wavelengths implies a discrete set of energies. Conservation of energy then implies that if an atom absorbs a photon, its internal energy increases by a discrete amount, an amount equal to the energy of the photon. (It also implies that if an atom emits a photon, its internal energy decreases by a discrete amount that is equal to the energy of the photon.) In 1913, this led Niels Bohr to postulate that the internal energy of an atom can have only a discrete set of values. That is, the internal energy of an atom is quantized. If an excited atom radiates light of frequency f, the atom makes a transition from one allowed level to another level that has less energy by $|\Delta E| = hf$. Bohr was able to construct a semiclassical model of the hydrogen atom that had a discrete set of energy levels consistent with the observed spectrum of emitted light.* However, the *reason* for the quantization of energy levels in atoms and other systems remained a mystery until the wave nature of electrons was discovered a decade later.

34-5 ELECTRONS AND MATTER WAVES

In 1897, J. J. Thomson showed that the rays of a cathode-ray tube (Figure 34-6) can be deflected by electric and magnetic fields and therefore must consist of electrically charged particles. By measuring the deflections of these particles, Thomson showed that all the particles have the same charge-to-mass ratio q/m. He also showed that particles with this charge-to-mass ratio can be obtained using any material for the cathode, which means that these particles, now called electrons, are a fundamental constituent of all matter.

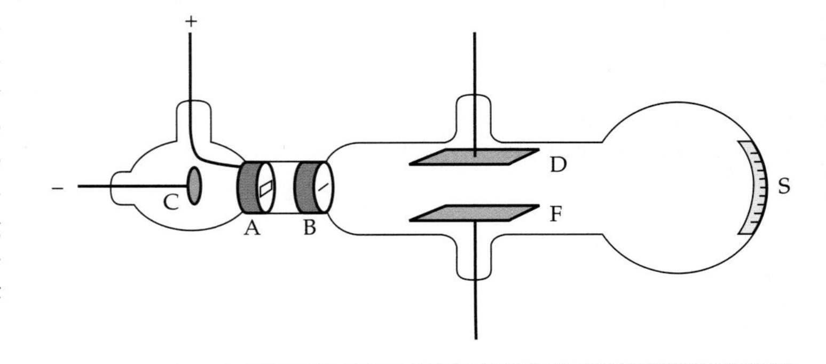

FIGURE 34-6 Schematic diagram of the cathode-ray tube Thomson used to measure q/m for the particles that comprise cathode rays (electrons). Electrons from the cathode C pass through the slits at A and B and strike a phosphorescent screen S. The beam can be deflected by an electric field between plates D and F or by a magnetic field (not shown).

THE DE BROGLIE HYPOTHESIS

Because light seems to have both wave and particle properties, it is natural to ask whether matter (for example, electrons and protons) might also have both wave and particle characteristics. In 1924, a French physics student, Louis de Broglie, suggested this idea in his doctoral dissertation. de Broglie's work was highly speculative, because no evidence existed at that time of any wave aspects of matter.

For the wavelength of electron waves, de Broglie chose

$$\lambda = \frac{h}{p} \qquad \text{34-13}$$

DE BROGLIE RELATION FOR THE WAVELENGTH OF ELECTRON WAVES

where p is the momentum of the electron. Note that this is the same as Equation 34-7 for a photon. For the frequency of electron waves, de Broglie chose the Einstein equation relating the frequency and energy of a photon.

$$f = \frac{E}{h} \qquad \text{34-14}$$

DE BROGLIE RELATION FOR THE FREQUENCY OF ELECTRON WAVES

* The Bohr model is reviewed in Chapter 36.

These equations are thought to apply to all matter. However, for macroscopic objects, the wavelengths calculated from Equation 34-13 are so small that it is impossible to observe the usual wave properties of interference or diffraction. Even a dust particle that has a mass as small as 1 μg is much too massive for any wave characteristics to be noticed, as we see in the following example.

Example 34-4 The de Broglie Wavelength — *Try It Yourself*

Find the de Broglie wavelength of a 1.00×10^{-6} g particle moving with a speed of 1.00×10^{-6} m/s.

PICTURE The wavelength λ and the momentum p of a particle are related by $\lambda = h/p$.

SOLVE

Cover the column to the right and try this on your own before looking at the answers.

Steps	Answers
Write the definition of the de Broglie wavelength and substitute the given data.	$\lambda = \frac{h}{p} = \frac{h}{mv} = \frac{6.63 \times 10^{-34}\,\text{J}\cdot\text{s}}{(1.00 \times 10^{-9}\,\text{kg})(1.00 \times 10^{-6}\,\text{m/s})} = \boxed{6.63 \times 10^{-19}\,\text{m}}$

CHECK As expected, this wavelength, which is four or five orders of magnitude smaller than the diameter of an atomic nucleus, is too small to be observed.

Because the wavelength found in Example 34-4 is so small, much smaller than any possible apertures or obstacles, diffraction or interference of such waves cannot be observed. In fact, the propagation of waves of very small wavelengths is indistinguishable from the propagation of particles. The momentum of the particle in Example 34-4 is only 10^{-15} kg·m/s. A macroscopic particle that has a greater momentum would have an even smaller de Broglie wavelength. We therefore do not observe the wave properties of such macroscopic objects as baseballs and billiard balls.

PRACTICE PROBLEM 34-5

Find the de Broglie wavelength of a baseball of mass 0.17 kg moving at 100 km/h.

The situation is different for low-energy electrons and other subatomic particles. Consider a particle with kinetic energy K. Its momentum is found from

$$K = \frac{p^2}{2m}$$

or

$$p = \sqrt{2mK}$$

Its wavelength is then

$$\lambda = \frac{h}{p} = \frac{h}{\sqrt{2mK}}$$

If we multiply the numerator and the denominator by c, we obtain

$$\lambda = \frac{hc}{\sqrt{2mc^2K}} = \frac{1240\,\text{eV}\cdot\text{nm}}{\sqrt{2mc^2K}} \qquad \text{34-15}$$

WAVELENGTH ASSOCIATED WITH A PARTICLE OF MASS m

where we have used $hc = 1240$ eV·nm. For electrons, $mc^2 = 0.5110$ MeV. Then,

$$\lambda = \frac{1240\ \text{eV}\cdot\text{nm}}{\sqrt{2mc^2K}} = \frac{1240\ \text{eV}\cdot\text{nm}}{\sqrt{2(0.5110 \times 10^6\ \text{eV})K}}$$

or

$$\lambda = \frac{1.226}{\sqrt{K}}\ \text{nm} \qquad (K \text{ in electron volts}) \qquad 34\text{-}16$$

ELECTRON WAVELENGTH

Equation 34-15 and Equation 34-16 do not hold for relativistic particles whose kinetic energies are a significant fraction of their rest energies mc^2. (Rest energies are discussed in Chapter 7 and in Chapter R.)

PRACTICE PROBLEM 34-6

Find the wavelength of an electron whose kinetic energy is 10.0 eV.

ELECTRON INTERFERENCE AND DIFFRACTION

The observation of diffraction and interference of electron waves would provide the crucial test of the existence of wave properties of electrons. This observation was first seen serendipitously in 1927 by C. J. Davisson and L. H. Germer as they were studying electron scattering from a nickel target at the Bell Telephone Laboratories. After heating the target to remove an oxide coating that had accumulated during an accidental break in the vacuum system, they found that the scattered electron intensity as a function of the scattering angle showed maxima and minima. Their target had crystallized, and they had observed electron diffraction by accident. Davisson and Germer then prepared a target consisting of a single crystal of nickel and investigated this phenomenon extensively. Figure 34-7*a* illustrates their experiment. Electrons from an electron gun are directed at a crystal and detected at some angle ϕ that can be varied. Figure 34-7*b* shows a typical pattern observed. There is a strong scattering maximum at an angle of 50°. The angle for maximum scattering of waves from a crystal depends on the wavelength of the waves and the spacing of the atoms in the crystal. Using the known spacing of atoms in their crystal, Davisson and Germer calculated the wavelength that could

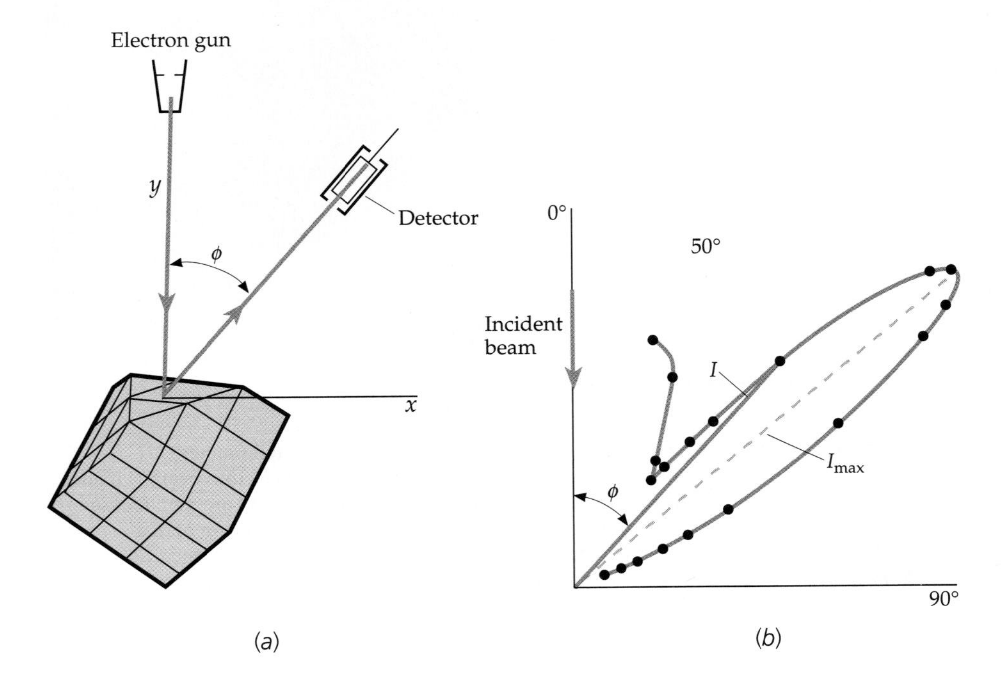

FIGURE 34-7 The Davisson–Germer experiment. (*a*) Electrons are scattered from a nickel crystal into a detector. (*b*) A polar plot of the intensity I of scattered electrons versus scattering angle. The maximum intensity I_{max} is at the angle predicted by the diffraction of waves of wavelength λ given by the de Broglie formula.

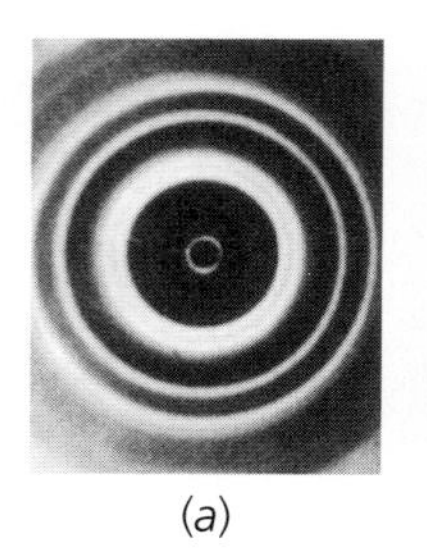
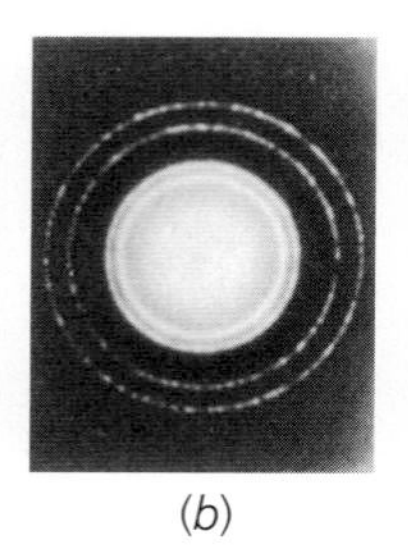
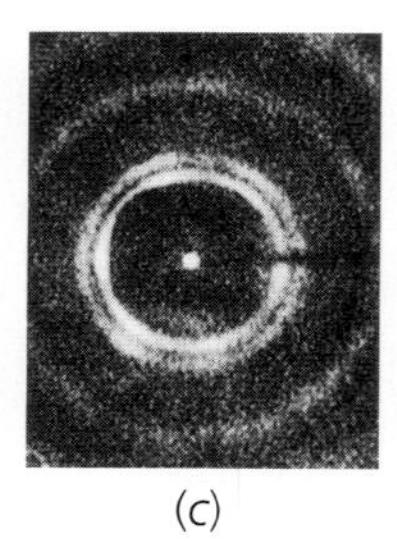
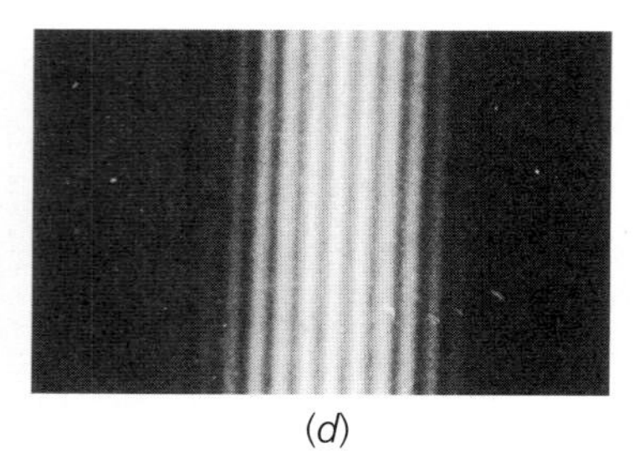

(a) (b) (c) (d)

FIGURE 34-8 (*a*) The diffraction pattern produced by X rays of wavelength 0.071 nm on an aluminum foil target. (*b*) The diffraction pattern produced by 600-eV electrons (λ = 0.050 nm) on an aluminum foil target. (*c*) The diffraction of 0.0568 eV neutrons (λ = 0.12 nm) incident on a copper foil. (*d*) A two-slit electron diffraction–interference pattern. *((a) and (b) PSSC Physics, 2nd ed., 1965. D.C. Heath & Co., and Education Development Center, Inc., Newton, MA, (c) C.G. Shull, (d) Claus Jönsson.)*

produce such a maximum and found that it agreed with the de Broglie equation (Equation 34-16) for the electron energy they were using. By varying the energy of the incident electrons, they could vary the electron wavelengths and produce maxima and minima at different locations in the diffraction patterns. In all cases, the measured wavelengths agreed with de Broglie's hypothesis.

Another demonstration of the wave nature of electrons was provided in the same year by G. P. Thomson (son of J. J. Thomson) who observed electron diffraction in the transmission of electrons through thin metal foils. A metal foil consists of tiny, randomly oriented crystals. The diffraction pattern resulting from such a foil is a set of concentric circles. Figure 34-8*a* and Figure 34-8*b* show the diffraction pattern observed using X rays and electrons on an aluminum foil target. Figure 34-8*c* shows the diffraction patterns of neutrons on a copper foil target. Note the similarity of the patterns. The diffraction of hydrogen and helium atoms was observed in 1930. In all cases, the measured wavelengths agree with the de Broglie predictions. Figure 34-8*d* shows a diffraction pattern produced by electrons incident on two narrow slits. This experiment is equivalent to Young's famous double-slit experiment with light. The pattern is identical to the pattern observed with photons of the same wavelength. (Compare with Figure 34-1.)

Shortly after the wave properties of the electron were demonstrated, it was suggested that electrons rather than light might be used to *see* small objects. As discussed in Chapter 33, reflected waves or transmitted waves can resolve details of objects only if the details are larger than the wavelength of the reflected wave. Beams of electrons, which can be focused by electric and magnetic fields, can have very small wavelengths—much shorter than visible light. Today, the electron microscope (Figure 34-9) is an important research tool used to visualize specimens at scales far smaller than those possible with a light microscope.

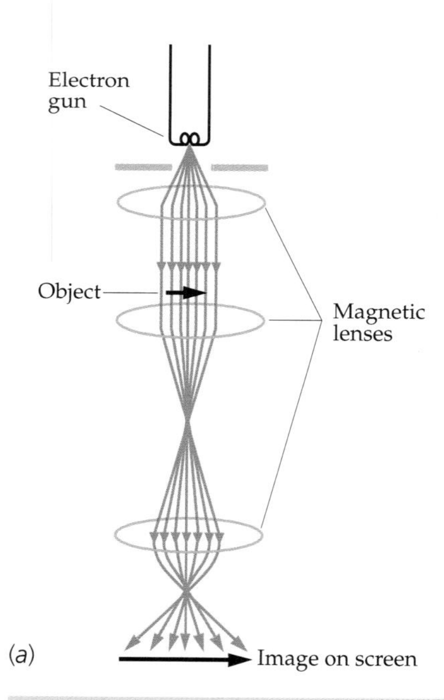

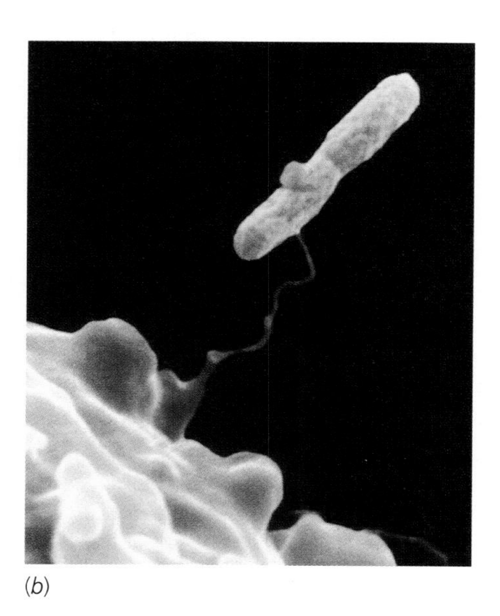

FIGURE 34-9 (*a*) An electron microscope. Electrons from a heated filament (the electron gun) are accelerated by a large potential difference. The electron beam is made parallel by a magnetic focusing lens. The electrons strike a thin target and are then focused by a second magnetic lens. The third magnetic lens projects the electron beam onto a fluorescent screen to produce the image. (*b*) An electron micrograph of an amoeba (*Hartmannella vermiformis*) that uses an extended pseudopod to entrap a bacterium (*Legionella pneumophila*). *((b) CDC/Dr. Barry S. Fields.)*

STANDING WAVES AND ENERGY QUANTIZATION

Given that electrons have wavelike properties, it should be possible to produce standing electron waves. If energy is associated with the frequency of a standing wave, as in $E = hf$ (Equation 34-14), then standing waves imply quantized energies.

The idea that the discrete energy states in atoms could be explained by standing waves led to the development of a detailed mathematical theory known as quantum theory, quantum mechanics, or wave mechanics by Erwin Schrödinger and others in 1926. In this theory, the electron is described by a wave function ψ that obeys a wave equation called the **Schrödinger equation.** The form of the Schrödinger equation of a particular system depends on the forces acting on the particle, which are described by the potential energy functions associated with those forces. In Chapter 35, we discuss this equation, which is somewhat similar to the classical wave equations for sound or for light. Schrödinger solved the standing wave problem for the hydrogen atom, the simple harmonic oscillator, and other systems of interest. He found that the allowed frequencies, combined with $E = hf$, resulted in the set of energy levels found experimentally for the hydrogen atom, thereby demonstrating that quantum theory provides a general method of finding the quantized energy levels for a given system. Quantum theory is the basis for our modern understanding of the world—from the inner workings of the atomic nucleus to the radiation spectra of distant galaxies.

! Do not think energy is always quantized. It is not unless the system is bound. The energy of a system consisting of a proton and an electron is quantized only if the electron is bound to the proton—as it is in the hydrogen atom. If the electron is not bound to the proton, then the energy of the system is not quantized.

34-6 THE INTERPRETATION OF THE WAVE FUNCTION

The wave function for waves on a string is the string displacement y. The wave function for sound waves can be either the displacement of the air molecules s, or the pressure P. The wave function for electromagnetic waves is the electric field $\vec{E}$ and the magnetic field $\vec{B}$. What is the wave function for electron waves? The symbol we use for this wave function is ψ (the Greek letter psi). When Schrödinger published his wave equation, neither he nor anyone else knew just how to interpret the wave function ψ. We can get a hint about how to interpret ψ by considering the quantization of light waves. For classical waves, such as sound or light, the energy per unit volume in the wave is proportional to the square of the wave function. Because the energy of a light wave is quantized, the energy per unit volume is proportional to the number of photons per unit volume. We might therefore expect the square of the photon's wave function to be proportional to the number of photons per unit volume in a light wave. But suppose we have a very low-energy source of light that emits just one photon at a time. In any unit volume, there is either one photon or none. The square of the wave function must then describe the *probability* of finding a photon in some unit volume.

The Schrödinger equation describes a single particle. The square of the wave function for a particle must then describe the *probability density*, which is the probability per unit volume, of finding the particle at a location. The probability of finding the particle in some volume element must also be proportional to the size of the volume element dV. Thus, in one dimension, the probability of finding a particle in a region of length dx at the position x is $\psi^2(x)\,dx$. If we call this probability $P(x)\,dx$, where $P(x)$ is the **probability density,** we have

$$P(x) = \psi^2(x) \tag{34-17}$$

PROBABILITY DENSITY

Generally the wave function depends on time as well as position, and is written $\psi(x,t)$. However, for standing waves, the probability density is independent of time. Because we will be concerned mostly with standing waves in this chapter, we omit the time dependence of the wave function and write it $\psi(x)$ or just ψ.

The probability of finding the particle either in the region between x_1 and $x_1 + dx$ or in the region between x_2 and $x_2 + dx$ is the sum of the separate probabilities $P(x_1)\,dx + P(x_2)\,dx$. If we have a particle at all, the probability of finding the particle somewhere must be 1. Then the sum of the probabilities over all the possible values of x must equal 1. That is,

$$\int_{-\infty}^{\infty} \psi^2\, dx = 1 \qquad \text{34-18}$$

NORMALIZATION CONDITION

Equation 34-18 is called the **normalization condition.** If ψ is to satisfy the normalization condition, it must approach zero as $|x|$ approaches infinity. This condition places restrictions on the possible solutions of the Schrödinger equation. There are mathematical solutions to the Schrödinger equation that do not approach zero as $|x|$ approaches infinity. However, these solutions are not acceptable as wave functions.

Example 34-5 Probability Calculation for a Classical Particle

It is known that a classical point particle moves back and forth with constant speed between two walls at $x = 0$ and $x = 8.0$ cm (Figure 34-10). No additional information about the location of the particle is known. (*a*) What is the probability density $P(x)$? (*b*) What is the probability of finding the particle at the point where x equals exactly 2 cm? (*c*) What is the probability of finding the particle between $x = 3.0$ cm and $x = 3.4$ cm?

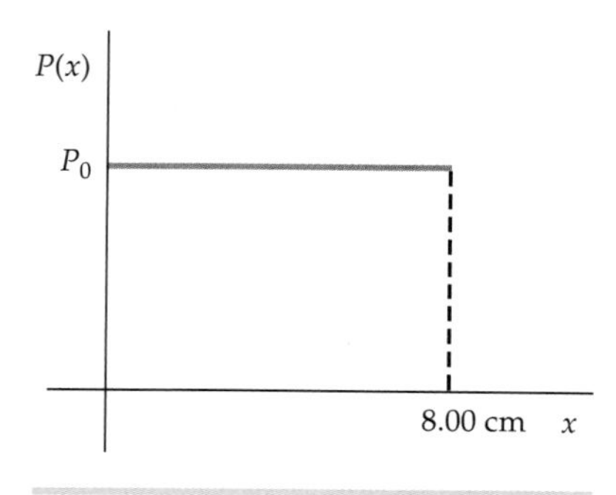

FIGURE 34-10 The probability function $P(x)$.

PICTURE We do not know the initial position of the particle. Because the particle moves with constant speed, it is equally likely to be anywhere in the region $0 < x < 8.0$ cm. The probability density $P(x)$ is therefore independent of x, for $0 < x < 8.0$ cm, and zero outside of this range. We can find $P(x)$, for $0 < x < 8.0$ cm, by normalization, that is, by requiring that the probability that the particle is somewhere between $x = 0$ and $x = 8.0$ cm is 1.

SOLVE

(*a*) 1. The probability density $P(x)$ is uniform between the walls and zero elsewhere:

$$P(x) = \begin{cases} 0 & x < 0 \\ P_0 & 0 < x < 8.0\text{ cm} \\ 0 & x > 8.0\text{ cm} \end{cases}$$

2. Apply the normalization condition:

$$\int_{-\infty}^{+\infty} P(x)\,dx = \int_{-\infty}^{0} P(x)\,dx + \int_{0}^{8.0\text{ cm}} P(x)\,dx + \int_{8.0\text{ cm}}^{\infty} P(x)\,dx$$

$$= 0 + \int_{0}^{8.0\text{ cm}} P_0\,dx + 0 = P_0\,(8.0\text{ cm}) = 1$$

3. Solve for P_0:

$$P_0 = \boxed{\frac{1}{8.0\text{ cm}}}$$

(*b*) On the interval $0 < x < 8.0$ cm, the probability of finding the particle in some range Δx is proportional to $P_0\Delta x = \Delta x/(8\text{ cm})$. The probability of finding the particle at the point $x = 2$ cm is zero because Δx is zero (no range exists). Alternatively, because an infinite number of points exists between $x = 0$ and $x = 8$ cm, and the particle is equally likely to be at any point, the chance that the particle will be at any one particular point must be zero.

The probability of finding the particle at the point where x equals exactly 2 cm is 0.

(c) Because the probability density is uniform, the probability of a particle being in some range Δx in the region $0 < x < 8.0$ cm is $P_0\Delta x$. The probability of the particle being in the region $3.0 \text{ cm} < x < 3.4$ cm is thus:

$$P_0\Delta x = \left(\frac{1}{8.0 \text{ cm}}\right)0.4 \text{ cm} = \boxed{0.05}$$

CHECK The length of the interval $3.0 \text{ cm} < x < 3.4$ cm is 0.4 cm, which is 5 percent of $L = 8.0$ cm. Because the particle is moving at constant speed v, we expect it to be in the interval $3.0 \text{ cm} < x < 3.4$ cm during 5 percent of the time, provided the total time is much much longer than the time L/v (the time required for the particle to travel 8.0 cm). Our Part (c) result meets this expectation.

34-7 WAVE–PARTICLE DUALITY

We have seen that light, which we ordinarily think of as a wave, exhibits particle properties when it interacts with matter, as in the photoelectric effect or in Compton scattering. Electrons, which we usually think of as particles, exhibit the wave properties of interference and diffraction when they pass near the edges of obstacles. All carriers of momentum and energy (for example, electrons, atoms, or photons) exhibit both wave and particle characteristics. It might be tempting to say that an electron, for example, is both a wave and a particle, but what does this mean? In classical physics, the concepts of waves and particles are mutually exclusive. A **classical particle** behaves like a piece of shot; it can be localized and scattered, it exchanges energy suddenly at a point in space, and it obeys the laws of conservation of energy and momentum in collisions. It does *not* exhibit interference or diffraction. A **classical wave,** on the other hand, behaves like a sound or light wave; it exhibits diffraction and interference, and its energy is spread out continuously in space and time. A classical wave and a classical particle are mutually exclusive. Nothing can be both a classical particle and a classical wave at the same time.

After Thomas Young observed the two-slit interference pattern by using light in 1801, light was thought to be a classical wave. On the other hand, the electrons discovered by J. J. Thomson were thought to be classical particles. We now know that these classical concepts of waves and particles do not adequately describe the complete behavior of any phenomenon.

Everything propagates like a wave and exchanges energy like a particle.

Often the concepts of the classical particle and the classical wave give the same results. If the wavelength is very small, diffraction effects are negligible, so the waves travel in straight lines like classical particles. Also, interference is not seen for waves of very short wavelength, because the interference fringes are too closely spaced to be observed. It then makes no difference which concept we use. If diffraction is negligible, we can think of light as a wave propagating along rays, as in geometrical optics, or as a beam of photon particles. Similarly, we can think of an electron as a wave propagating in straight lines along rays or, more commonly, as a particle.

We can also use either the wave or particle concept to describe exchanges of energy if we have a large number of particles and we are interested only in the average values of energy and momentum exchanges.

THE TWO-SLIT EXPERIMENT REVISITED

The wave–particle duality of nature is illustrated by the analysis of the experiment in which a single electron is incident on a barrier that has two slits. The analysis is virtually the same whether we use an electron or a photon (light). To describe the propagation of an electron, we must use wave theory. Let us assume the source is a point source, such as a needle point, so we have spherical waves spreading out from the source. After passing through the two slits, the wavefronts spread out—as if each slit were a source of wavefronts. The wave function ψ at a point on a screen or film far from the slits depends on the difference in path lengths from the source to the point, one path through one slit, and the other path through the other slit. At points on the screen for which the difference in path lengths is either zero or an integral number of wavelengths, the amplitude of the wave function is a maximum. Because the probability of detecting the electron is proportional to ψ^2, the electron is very likely to arrive near these points. At points for which the path difference is an odd number of half wavelengths, the wave function ψ is zero, so the electron is very unlikely to arrive near these points. The chapter opening photos show the interference pattern produced by 10 electrons, 100 electrons, 3000 electrons, and 70,000 electrons. Note that, although the electron propagates through the slits like a wave, the electron interacts with the screen at a single point—like a particle.

THE UNCERTAINTY PRINCIPLE

An important principle consistent with the wave–particle duality of nature is the **uncertainty principle.** It states that, in principle, it is impossible to simultaneously measure both the position and the momentum of a particle with unlimited precision. A common way to measure the position of an object is to look at the object by using light. If we do this, we scatter light from the object and determine the position by the direction of the scattered light. If we use light of wavelength λ, we can measure the position x only to an uncertainty Δx of the order of λ because of diffraction effects.

$$\Delta x \sim \lambda$$

To reduce the uncertainty in position, we therefore use light of very short wavelength, perhaps even X rays. In principle, there is no limit to the accuracy of such a position measurement, because there is no limit on how small the wavelength λ can be.

We can determine the momentum p_x of the object if we know the mass and can determine its velocity. The momentum of the object can be found by measuring the object's position at two nearby times and computing its velocity. If we use light of wavelength λ, the photons carry momentum h/λ. If these photons are scattered by the object we are looking at, the scattering changes the momentum of the object in an uncontrollable way. Each photon carries momentum h/λ, so the uncertainty in the momentum Δp_x of the object is of the order of h/λ:

$$\Delta p_x \sim \frac{h}{\lambda}$$

If the wavelength of the radiation is small, the momentum of each photon will be large and the momentum measurement will have a large uncertainty. Reducing the intensity of light cannot eliminate this uncertainty; such a reduction merely reduces the number of photons in the beam. To see the object, we must scatter at least one photon. Therefore, the uncertainty in the momentum measurement of the object will be large if λ is small, and the uncertainty in the position measurement of the object will be large if λ is large.

Of course, we could always look at the objects by scattering electrons instead of photons, but the same difficulty remains. If we use low-momentum electrons to reduce the uncertainty in the momentum measurement, we have a large uncertainty in the position measurement because of diffraction of the electrons. The relation between the wavelength and momentum $\lambda = h/p_x$ is the same for electrons as it is for photons.

The product of the intrinsic uncertainties in position and momentum is

$$\Delta x\,\Delta p_x \sim \lambda \times \frac{h}{\lambda} = h$$

This relation between the uncertainties in position and momentum is called the uncertainty principle. If we define precisely what we mean by uncertainties in measurement, we can give a precise statement of the uncertainty principle. If Δx and Δp are defined to be the standard deviations in the measurements of position and momentum, it can be shown that their product must be greater than or equal to $\hbar/2$.

$$\Delta x\,\Delta p_x \geq \tfrac{1}{2}\hbar \qquad \text{34-19}$$

where $\hbar = h/2\pi$.*

Equation 34-19 provides a statement of the uncertainty principle first given by Werner Heisenberg in 1927. In practice, the experimental uncertainties are usually much greater than the intrinsic lower limit that results from wave–particle duality.

34-8 A PARTICLE IN A BOX

We can illustrate many of the important features of quantum physics without solving the Schrödinger equation by considering a simple problem of a particle of mass m confined to a one-dimensional box of length L, like the particle in Example 34-5. This can be considered a crude description of an electron that is constrained to be within an atom or a proton that is constrained to be within a nucleus. If a classical particle bounces back and forth between the walls of the box, the particle's energy and momentum can have any values. However, according to quantum theory, the particle is described by a wave function ψ, whose square describes the probability of finding the particle in some region. Because we are assuming that the particle is indeed inside the box, the wave function must be zero everywhere outside the box. If the box is between $x = 0$ and $x = L$, we have

$$\psi = 0 \qquad \text{for } x \leq 0 \text{ and for } x \geq L$$

In particular, if we assume the wave function to be continuous everywhere, it must be zero at the end points of the box $x = 0$ and $x = L$. This is the same condition as the condition for standing waves on a string fixed at $x = 0$ and $x = L$, and the results are the same. The allowed wavelengths for a particle in the box are those where the length L equals an integral number of half wavelengths (Figure 34-11).

$$L = n\frac{\lambda_n}{2} \qquad n = 1, 2, 3, \ldots \qquad \text{34-20}$$

STANDING-WAVE CONDITION FOR A PARTICLE IN A BOX OF LENGTH L

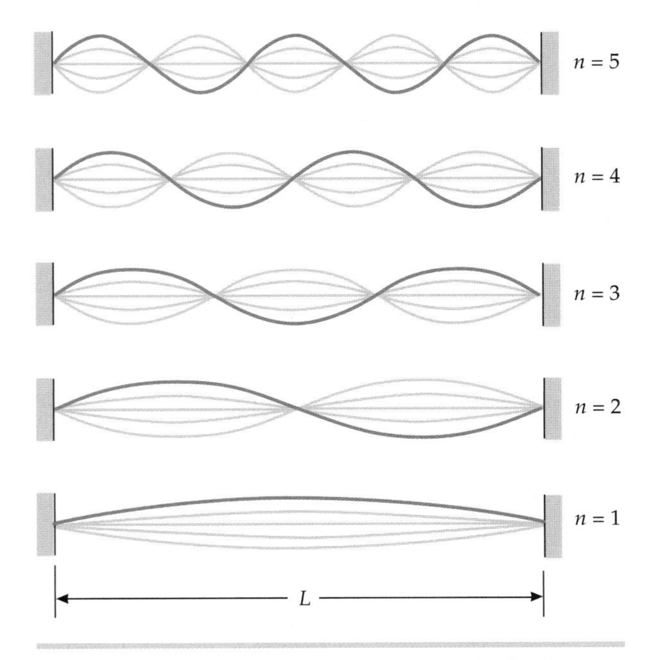

FIGURE 34-11 Standing waves on a string fixed at both ends. The standing-wave condition is the same as for standing electron waves in a box.

The total energy E of the particle is its kinetic energy

$$E = \frac{1}{2}mv^2 = \frac{p^2}{2m}$$

Substituting the de Broglie relation $p_n = h/\lambda_n$,

$$E_n = \frac{p_n^2}{2m} = \frac{(h/\lambda_n)^2}{2m} = \frac{h^2}{2m\lambda_n^2}$$

* The combination $h/2\pi$ occurs so often it is given a special symbol, somewhat analogous to giving the special symbol ω for $2\pi f$, which occurs often in oscillations.

Then the standing-wave condition $\lambda_n = 2L/n$ gives the allowed energies.

$$E_n = n^2\frac{h^2}{8mL^2} = n^2E_1 \qquad 34\text{-}21$$

ALLOWED ENERGIES FOR A PARTICLE IN A BOX

where

$$E_1 = \frac{h^2}{8mL^2} \qquad 34\text{-}22$$

GROUND-STATE ENERGY FOR A PARTICLE IN A BOX

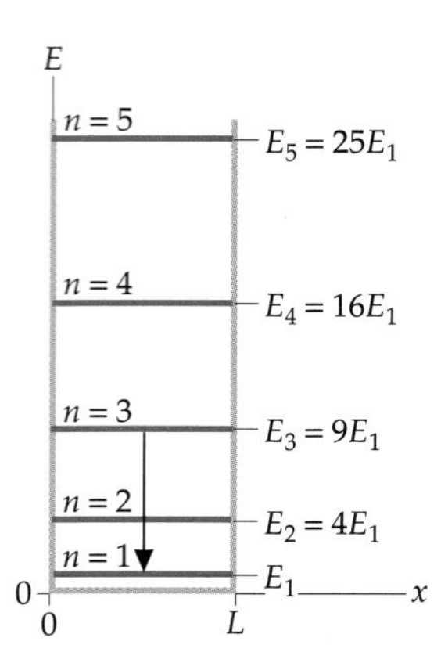

FIGURE 34-12 Energy-level diagram for a particle in a box. Classically, a particle can have any energy value. Quantum mechanically, only those energy values given by Equation 34-21 are allowed. A transition between the state $n = 3$ and the ground state $n = 1$ is indicated by the vertical arrow.

is the energy of the lowest state, which is the ground state.

The condition $\psi = 0$ at $x = 0$ and $x = L$ is called a **boundary condition.** Boundary conditions in quantum theory lead to energy quantization. Figure 34-12 shows the energy-level diagram for a particle in a box. Note that the lowest energy is not zero. This result is a general feature of quantum theory. If a particle is confined to some region of space, the particle has a minimum kinetic energy, called the **zero-point energy** that is greater than zero. The smaller the region of space the particle is confined to, the greater its zero-point energy. In Equation 34-22, this is indicated by the fact that E_1 varies as $1/L^2$.

If an electron is confined (bound to an atom) in some energy state E_i, the electron can make a transition to another energy state E_f by the emission of a photon if E_f is less than E_i. (If E_f is greater than E_i, the system absorbs a photon.) The transition from state 3 to the ground state is indicated in Figure 34-12 by the vertical arrow. The frequency of the emitted photon is found from the conservation of energy*

$$hf = E_i - E_f \qquad 34\text{-}23$$

The wavelength of the photon is then

$$\lambda = \frac{c}{f} = \frac{hc}{E_i - E_f} \qquad 34\text{-}24$$

STANDING-WAVE FUNCTIONS

The amplitude of a vibrating string that is fixed at $x = 0$ and $x = L$ is given by Equation 16-15:

$$A_n(x) = A_n \sin k_n x \qquad n = 1, 2, 3, \ldots$$

where A_n is a constant, $k_n = 2\pi/\lambda_n$ is the wave number, and $\lambda_n = 2L/n$. The wave functions for a particle in a box (which can be obtained by solving the Schrödinger equation, as we will see in Chapter 35) are the same:

$$\psi_n(x) = A_n \sin k_n x \qquad n = 1, 2, 3, \ldots$$

where $k_n = 2\pi/\lambda_n$. Using $\lambda_n = 2L/n$, we have

$$k_n = \frac{2\pi}{\lambda_n} = \frac{2\pi}{2L/n} = \frac{n\pi}{L}$$

The wave functions can thus be written

$$\psi_n(x) = A_n \sin\left(n\pi\frac{x}{L}\right)$$

* This equation was first proposed by Niels Bohr in his semiclassical model of the hydrogen atom in 1913, about 10 years before de Broglie's suggestion that electrons have wave properties. The Bohr model is presented in Chapter 36.

The constant A_n is determined by the normalization condition (Equation 34-18):

$$\int_{-\infty}^{\infty} \psi^2\,dx = \int_0^L A_n^2 \sin^2\left(n\pi\frac{x}{L}\right) dx = 1$$

Note that we need integrate only from $x = 0$ to $x = L$ because $\psi(x)$ is zero everywhere else. The result of evaluating the integral and solving for A_n is

$$A_n = \sqrt{\frac{2}{L}}$$

which is independent of n. The normalized standing-wave functions for a particle in a box are thus

$$\psi_n(x) = \sqrt{\frac{2}{L}} \sin\left(n\pi\frac{x}{L}\right) \qquad n = 1, 2, 3, \ldots \qquad \text{34-25}$$

STANDING-WAVE FUNCTIONS FOR A PARTICLE IN A BOX

The standing-wave functions for $n = 1$, $n = 2$, and $n = 3$ are shown in Figure 34-13.

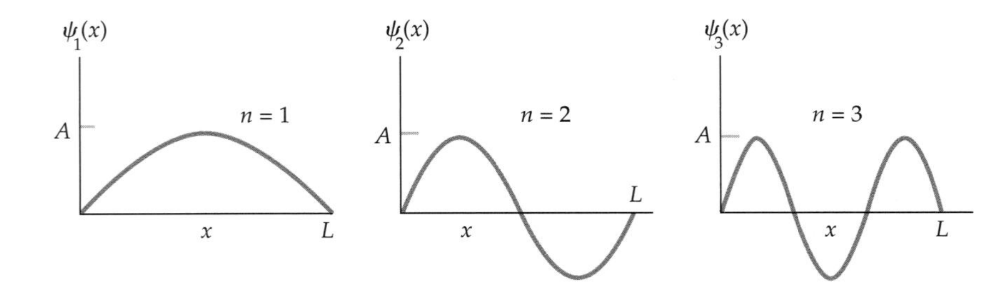

FIGURE 34-13 Standing-wave functions for $n = 1$, $n = 2$, and $n = 3$.

The number n is called a **quantum number.** It characterizes the wave function for a particular state and for the energy of that state. In our one-dimensional problem, a quantum number arises from the boundary condition on the wave function that it must be zero at $x = 0$ and $x = L$. In three-dimensional problems, three quantum numbers arise, one associated with a boundary condition in each dimension.

Figure 34-14 shows plots of ψ^2 for the ground state $n = 1$, the first excited state $n = 2$, the second excited state $n = 3$, and the state $n = 10$. In the ground state, the particle is most likely to be found near the center of the box, as indicated by

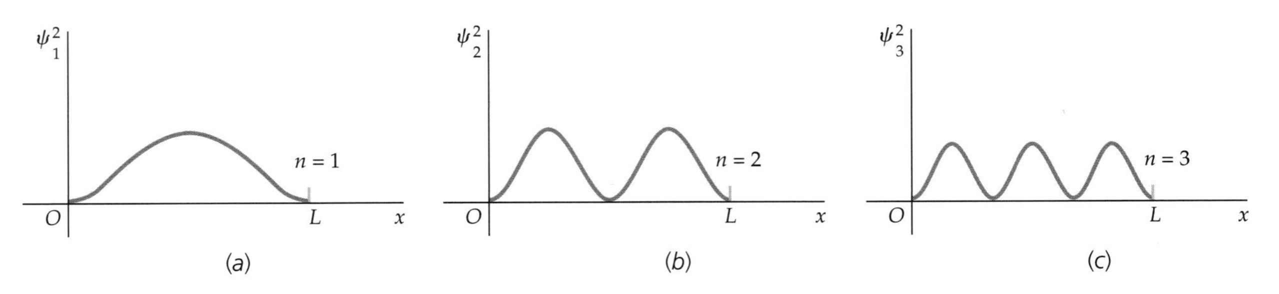

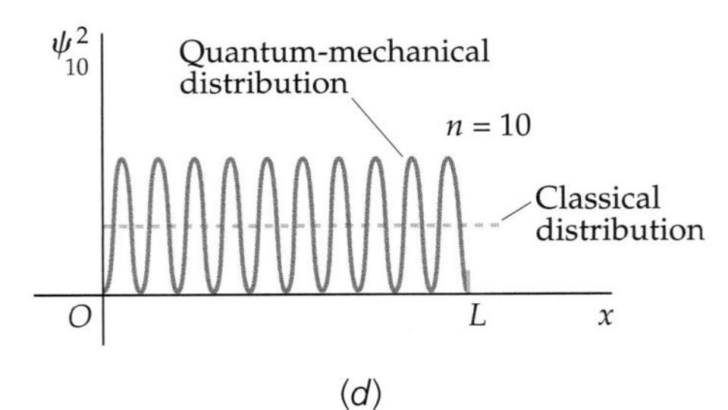

FIGURE 34-14 ψ^2 versus x for a particle in a box of length L for (*a*) the ground state, $n = 1$; (*b*) the first excited state, $n = 2$; (*c*) the second excited state, $n = 3$; and (*d*) the state $n = 10$. For $n = 10$, the maxima and minima of ψ^2 are so close together that individual maxima may be hard to distinguish. The dashed line indicates the average value of ψ^2. It gives the classical prediction that the particle is equally likely to be found near any point in the box.

the maximum value of ψ^2 at $x = L/2$. In the first excited state, the particle is least likely to be found near the center of the box because ψ^2 is small near $x = L/2$. For very large values of n, the maxima and minima of ψ^2 are very close together, as illustrated for $n = 10$. The average value of ψ^2 is indicated in this figure by the dashed line. For very large values of n, the maxima and minima are so closely spaced that ψ^2 cannot be distinguished from its average value. The fact that $(\psi^2)_{av}$ is constant across the whole box means that the particle is equally likely to be found anywhere in the box—the same as in the classical result. This is an example of **Bohr's correspondence principle:**

> In the limit of very large quantum numbers, the classical calculation and the quantum calculation must yield the same results.
>
> BOHR'S CORRESPONDENCE PRINCIPLE

The region of very large quantum numbers is also the region of very large energies. For large energies, the percentage change in energy between adjacent quantum states is very small, so energy quantization is not important (see Problem 71).

We are so accustomed to thinking of the electron as a classical particle that we tend to think of an electron in a box as a particle bouncing back and forth between the walls. But the probability distributions shown in Figure 34-14 are stationary; that is, they do not depend on time. A better picture for an electron in a bound state is a cloud of charge that has the charge density proportional to ψ^2. Figure 34-14 can then be thought of as plots of the charge density versus x for the various states. In the ground state, $n = 1$, the electron cloud is centered in the middle of the box and is spread out over most of the box, as indicated in Figure 34-14*a*. In the first excited state, $n = 2$, the charge density of the electron cloud has two maxima, as indicated in Figure 34-14*b*. For very large values of n, there are many closely spaced maxima and minima in the charge density resulting in an average charge density that is approximately uniform throughout the box. This electron-cloud picture of an electron is very useful in understanding the structure of atoms and molecules. However, it should be noted that whenever an electron is observed to interact with matter or radiation, it is always observed as a whole unit charge.

Example 34-6 Photon Emission by a Particle in a Box

An electron is in a one-dimensional box of length 0.100 nm. (*a*) Find the ground-state energy. (*b*) Find the energies of the four lowest-energy states that have energies above the ground-state energy, and then sketch an energy-level diagram. (*c*) Find the wavelength of the photon emitted for each transition from the state $n = 3$ to a lower-energy state.

PICTURE For Part (*a*) the ground state is the $n = 1$ state, and $E_1 = h^2/8mL^2$ (Equation 34-22). For Part (*b*), the energies are given by $E_n = n^2E_1$ (Equation 34-21), where, $n = 2, 3, 4,$ and 5. For Part (*c*), the photon wavelengths are given by $\lambda = hc/(E_i - E_f)$ (Equation 34-24).

SOLVE

(*a*) Use $hc = 1240\ \text{eV}\cdot\text{nm}$ and $mc^2 = 0.5110\ \text{MeV}$ to calculate E_1:

$$E_1 = \frac{h^2}{8mL^2} = \frac{(hc)^2}{8(mc^2)L^2}$$

$$= \frac{(1240\ \text{eV}\cdot\text{nm})^2}{8(5.110 \times 10^5\ \text{eV})(0.100\ \text{nm})^2} = \boxed{37.6\ \text{eV}}$$

(*b*) 1. Calculate $E_n = n^2E_1$ for $n = 2, 3, 4,$ and 5:

$$E_2 = (2)^2(37.6\ \text{eV}) = \boxed{150\ \text{eV}}$$

$$E_3 = (3)^2(37.6\ \text{eV}) = \boxed{338\ \text{eV}}$$

$$E_4 = (4)^2(37.6\ \text{eV}) = \boxed{602\ \text{eV}}$$

$$E_5 = (5)^2(37.6\ \text{eV}) = \boxed{940\ \text{eV}}$$

2. Sketch an energy-level diagram using the values for the five energy states (Figure 34-15).

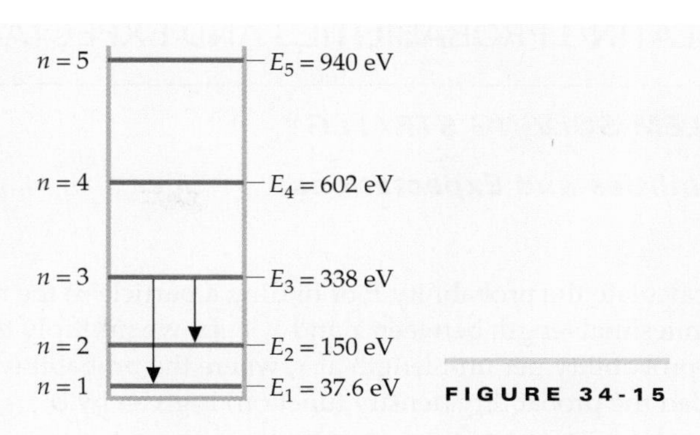

FIGURE 34-15

(*c*) 1. Use the energies found in Part (*b*) to calculate the wavelength for a transition from state 3 to state 2:

$$\lambda = \frac{hc}{E_3 - E_2} = \frac{1240 \text{ eV} \cdot \text{nm}}{338 \text{ eV} - 150 \text{ eV}} = \boxed{6.60 \text{ nm}}$$

2. Then use the energies in Part (*a*) and Part (*b*) to calculate the wavelength for a transition from state 3 to state 1:

$$\lambda = \frac{hc}{E_3 - E_1} = \frac{1240 \text{ eV} \cdot \text{nm}}{338 \text{ eV} - 37.6 \text{ eV}} = \boxed{4.13 \text{ nm}}$$

CHECK The wavelength of the photon emitted during the transition from the $n = 3$ to the $n = 1$ state is shorter than the wavelength of the photon emitted during the transition from the $n = 3$ to the $n = 2$ state. This result is expected—the greater the energy of the photon the shorter its wavelength.

TAKING IT FURTHER The energy-level diagram is shown in Figure 34-15. The transitions from $n = 3$ to $n = 2$ and from $n = 3$ to $n = 1$ are indicated by the vertical arrows. The ground-state energy of 37.6 eV is on the same order of magnitude as the kinetic energy of the electron in the ground state of the hydrogen atom, which is 13.6 eV. In the hydrogen atom, the electron has potential energy of −27.2 eV in the ground state, giving it a total ground-state energy (potential energy plus kinetic energy) of −13.6 eV.

PRACTICE PROBLEM 34-7 Calculate the wavelength of the photon emitted if the electron in the box makes a transition from $n = 4$ to $n = 3$.

34-9 EXPECTATION VALUES

The solution of a classical mechanics problem is typically specified by giving the position of a particle as a function of time. But the wave nature of matter prevents us from doing this for microscopic systems. The most that we can know is the relative probability of measuring a certain value of the position x. If we measure the position for a large number of identical systems, we get a range of values corresponding to the probability distribution. The average value of x obtained from such measurements is called the **expectation value** and is written $\langle x \rangle$. The expectation value of x is the same as the average value of x that we would expect to obtain from a measurement of the positions of a large number of particles that have the same wave function $\psi(x)$.

Because $\psi^2(x)\,dx$ is the probability of finding a particle in the region dx, the expectation value of x is

See **Math Tutorial** *for more information on* **Integrals**

$$\langle x \rangle = \int_{-\infty}^{+\infty} x\psi^2(x)\,dx \qquad 34\text{-}26$$

EXPECTATION VALUE OF x DEFINED

The expectation value of any function $F(x)$ is given by

$$\langle F(x) \rangle = \int_{-\infty}^{+\infty} F(x)\psi^2(x)\,dx \qquad 34\text{-}27$$

EXPECTATION VALUE OF $F(x)$ DEFINED

CALCULATING PROBABILITIES AND EXPECTATION VALUES

PROBLEM-SOLVING STRATEGY

Probabilities and Expectations

SOLVE

1. To calculate the probability P of finding a particle in the region of infinitesimal length between x and $x + dx$, we multiply the length dx by the probability per unit length at x, where the probability per unit length (called the probability density function) is given by ψ^2.
2. To calculate the probability P of finding a particle in the region $x_1 < x < x_2$, we, in principle, divide the region into an infinite number of regions of infinitesimal length dx, calculate the probability P of finding the particle in each infinitesimal length, and then sum the probabilities. That is, we evaluate the integral $\int_{x_1}^{x_2} \psi^2\, dx$.
3. To calculate the expected value of a function $F(x)$, we evaluate the integral $\int_{-\infty}^{+\infty} F(x)\psi^2(x)\, dx$. The result of this calculation is called the expected value of $F(x)$.

The problem of a particle in a box allows us to illustrate the calculation of the probability of finding the particle in various regions of the box and the expectation values for various energy states. We give two examples, using the wave functions given by Equation 34-25.

Example 34-7 The Probability of the Particle Being Found in a Specified Region of a Box

A particle in a one-dimensional box of length L is in the ground state. Find the probability of finding the particle (*a*) in the region that has a length $\Delta x = 0.01L$ and is centered at $x = \frac{1}{2}L$ and (*b*) in the region $0 < x < \frac{1}{4}L$.

PICTURE The probability P of finding the particle in some infinitesimal range dx is $\psi^2\, dx$. For a particle in the nth state, the wave function is given by $\psi_n = \sqrt{2/L}\,\sin(n\pi x/L)$ (Equation34-25). For a particle in the ground state, $n = 1$; and ψ_1^2 is illustrated in Figure 34-14. The probability of finding x in some region is just the area under this curve for the region. For Part (*a*), the region is $\Delta x = 0.01L$, centered at $x = L/2$, and the area under the ψ_1^2 versus x curve is shown in Figure 34-16*a*. This area is $\sim\psi_1^2\Delta x$. For Part (*b*), the region is $0 < x < L/4$, and the area under the curve is shown in Figure 34-16*b*. To calculate this area, we must integrate ψ_1^2 from $x = 0$ to $x = L/4$.

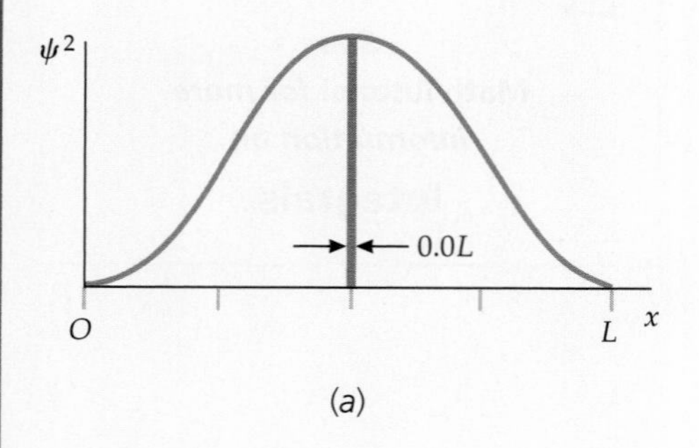

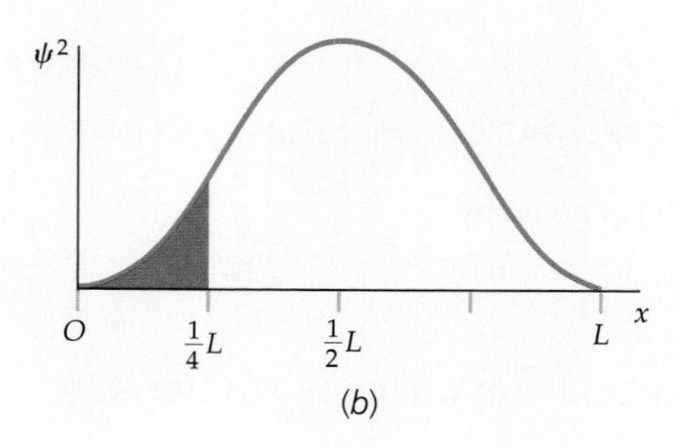

FIGURE 34-16

SOLVE

(*a*) 1. The probability of finding the particle is the area under the curve shown in Figure 34-16*a*. To calculate this area, we need to calculate the height of curve at $x = \frac{1}{2}L$:

$$\psi(x) = \psi_1(x) = \sqrt{\frac{2}{L}}\sin\left(\pi\frac{x}{L}\right)$$

so

$$\psi^2(\tfrac{1}{2}L) = \frac{2}{L}\sin^2\frac{\pi}{2} = \frac{2}{L}$$

2. The area is the height multiplied by the width, and the width is $\Delta x = 0.01L$:

$$P = \psi^2(\tfrac{1}{2}L)\Delta x = \frac{2}{L}\times 0.01L = \boxed{0.02}$$

(*b*) 1. The probability of finding the particle is the area under the curve shown in Figure 34-16*b*. To calculate this area, we need to integrate from $x = 0$ to $x = L/4$:

$$P = \int_0^{L/4}\psi^2(x)\,dx = \int_0^{L/4}\frac{2}{L}\sin^2\frac{\pi x}{L}\,dx$$

2. The integral can be evaluated a number of ways. If a table of integrals is used, a change in the integration variable in required. Changing the integration variable to $\theta = \pi x/L$ gives:

$$P = \frac{2}{\pi}\int_0^{\pi/4}\sin^2\theta\,d\theta$$

3. The integral can be found in tables:

$$\int_0^{\pi/4}\sin^2\theta\,d\theta = \left(\frac{\theta}{2} - \frac{\sin 2\theta}{4}\right)\Bigg|_0^{\pi/4} = \left(\frac{\pi}{8} - \frac{1}{4}\right)$$

4. Use the result from Part (*b*), step 3 to calculate the probability:

$$P = \frac{2}{\pi}\left(\frac{\pi}{8} - \frac{1}{4}\right) = \boxed{0.091}$$

CHECK If ψ_1^2 were uniformly distributed on the interval $0 < x < L$, the step-4 result would be 0.25. However, instead of ψ_1^2 being uniformly distributed, it is relatively small on the interval $0 < x < \frac{1}{4}L$, so a step-4 result that is less than 0.25 is expected.

TAKING IT FURTHER An integral was not necessary for Part (*a*) because the area of interest could be well approximated by a rectangle of height ψ^2 and width Δx. The chance of finding the particle in the region $\Delta x = 0.01L$ at $x = \frac{1}{2}L$ is approximately 2 percent. The chance of finding the particle in the region $0 < x < \frac{1}{4}L$ is about 9.1 percent.

CONCEPT CHECK 34-1

A fair six-sided die has the number 1 printed on four faces and the number 6 printed on the other two faces. What is the probability that a 1 comes up when the die is thrown? *Hint: The probability that a specific value comes up for one throw is the fraction of the throws that that value comes up after a large number of throws.*

Example 34-8 Calculating Expectation Values

Find (*a*) $\langle x\rangle$ and (*b*) $\langle x^2\rangle$ for a particle in its ground state in a box of length L.

PICTURE We use $\langle F(x)\rangle = \int F(x)\psi^2(x)\,dx$, with $\psi_n(x) = \sqrt{\frac{2}{L}}\sin\frac{n\pi x}{L}$.

SOLVE

(*a*) 1. Write $\langle x\rangle$ using the ground-state wave function given by Equation 34-25, with $n = 1$:

$$\langle x\rangle = \int_{-\infty}^{+\infty}x\psi^2(x)\,dx = \frac{2}{L}\int_0^L x\sin^2\left(\frac{\pi x}{L}\right)dx$$

2. To evaluate this integral by using a table of integrals, first change the integration variable to $\theta = \pi x/L$:

$$\langle x\rangle = \frac{2}{L}\left(\frac{L}{\pi}\right)^2\int_0^{\pi}\theta\sin^2\theta\,d\theta$$

$$= \frac{2L}{\pi^2}\int_0^{\pi}\theta\sin^2\theta\,d\theta$$

3. The table of integrals gives:

$$\int_0^{\pi}\theta\sin^2\theta\,d\theta = \left[\frac{\theta^2}{4} - \frac{\theta\sin 2\theta}{4} - \frac{\cos 2\theta}{8}\right]_0^{\pi} = \frac{\pi^2}{4}$$

4. Substitute this value into the expression in step 2:

$$\langle x\rangle = \frac{2L}{\pi^2}\int_0^{\pi}\theta\sin^2\theta\,d\theta = \frac{2L}{\pi^2}\frac{\pi^2}{4} = \boxed{\frac{L}{2}}$$

(*b*) 1. Repeat step 1 and step 2 of Part (*a*) for $\langle x^2 \rangle$:

$$\langle x^2 \rangle = \int_{-\infty}^{+\infty} x^2 \psi^2(x)\,dx = \int_0^L x^2 \frac{2}{L}\sin^2(\pi x/L)\,dx$$
$$= \frac{2}{L}\left(\frac{L}{\pi}\right)^3 \int_0^{\pi} \theta^2 \sin^2\theta\,d\theta = \frac{2L^2}{\pi^3}\int_0^{\pi} \theta^2 \sin^2\theta\,d\theta$$

2. Evaluating the integral using a table of integrals gives:

$$\int_0^{\pi} \theta^2 \sin^2\theta\,d\theta = \left[\frac{\theta^3}{6} - \left(\frac{\theta^2}{4} - \frac{1}{8}\right)\sin 2\theta - \frac{\theta\cos 2\theta}{4}\right]\Bigg|_0^{\pi}$$
$$= \frac{\pi^3}{6} - \frac{\pi}{4}$$

3. Substitute this value into the expression in step 1 of Part (*b*):

$$\langle x^2 \rangle = \frac{2L^2}{\pi^3}\left(\frac{\pi^3}{6} - \frac{\pi}{4}\right) = \left(\frac{1}{3} - \frac{1}{2\pi^2}\right)L^2 = \boxed{0.283L^2}$$

CHECK The expectation value of x is $L/2$, as we would expect, because the probability distribution is symmetric about the midpoint of the box.

TAKING IT FURTHER Note that $\langle x^2 \rangle$ is greater than $\langle x \rangle^2$.

CONCEPT CHECK 34-2

A fair six-sided die has the number 1 printed on four faces and the number 6 printed on the other two faces. Let N be the number that comes up when the die is thrown. What is the expectation value of N? What is the expectation value of N^2? *Hint: The expectation value of a quantity is the average value of that quantity after a large number of throws.*

34-10 ENERGY QUANTIZATION IN OTHER SYSTEMS

The quantized energies of a system are generally determined by solving the Schrödinger equation for that system. The form of the Schrödinger equation depends on the potential energy of the particle. The potential energy for a one-dimensional box from $x = 0$ to $x = L$ is shown in Figure 34-17. This potential energy function is called an **infinite square-well potential,** and it is described mathematically by

$$U(x) = \begin{cases} \infty & x < 0 \\ 0 & 0 < x < L \\ \infty & x > L \end{cases} \qquad \text{34-28}$$

The particle moves freely inside the box, so the potential energy is uniform. For convenience, we choose the value of this potential energy to be zero. Outside the box the potential energy is infinite, so the particle cannot exist outside the box no matter what its energy. We did not need to solve the Schrödinger equation for this potential because the wave functions and quantized frequencies are the same as for a string fixed at both ends, which we studied in Chapter 16. Although this problem seems artificial, actually it is useful for some physical problems, such as a neutron that is constrained to a nucleus that has a large number of protons and neutrons.

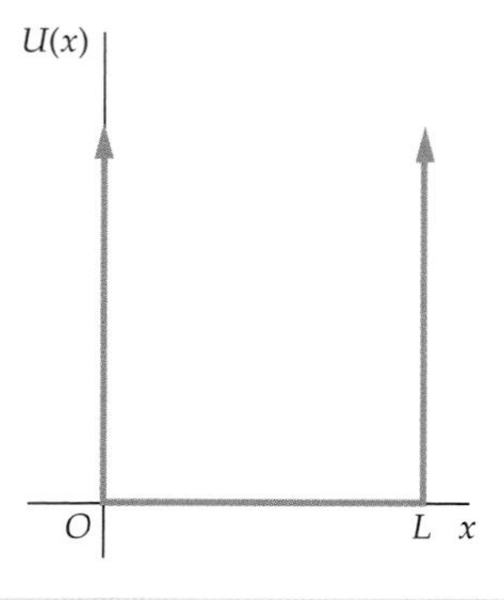

FIGURE 34-17 The infinite square-well potential energy. For $x < 0$ and $x > L$, the potential energy $U(x)$ is infinite. The particle is confined to the region in the well ($0 < x < L$).

THE HARMONIC OSCILLATOR

More realistic than the particle in a box is the harmonic oscillator, which applies to an object of mass m on a spring that has a force constant k or to any system undergoing small oscillations about a stable equilibrium. Figure 34-18 shows the potential energy function

$$U(x) = \tfrac{1}{2}kx^2 = \tfrac{1}{2}m\omega_0^2 x^2$$

where $\omega_0 = \sqrt{k/m}$ is the natural frequency of the oscillator. Classically, the object oscillates between $x = +A$ and $x = -A$. Its total energy is $E = \frac{1}{2}m\omega_0^2 A^2$, which can have any nonnegative value, including zero.

In quantum theory, the particle is represented by the wave function $\psi(x)$, which is determined by solving the Schrödinger equation for this potential. Normalizable wave functions $\psi_n(x)$ occur only for discrete values of the energy E_n given by

$$E_n = (n + \tfrac{1}{2})hf_0 \qquad n = 0, 1, 2, 3, \ldots \qquad \text{34-29}$$

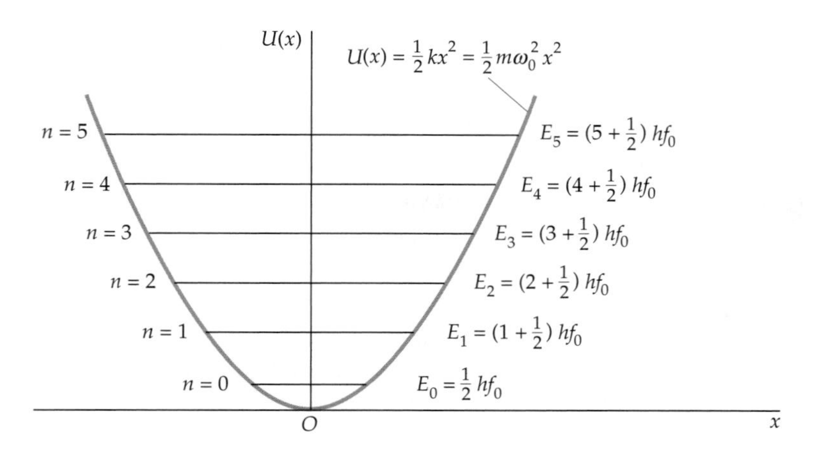

FIGURE 34-18 Harmonic oscillator potential energy function. The allowed energy levels are indicated by the equally spaced horizontal lines. Also, $\omega_0 = 2\pi f_0$.

where $f_0 = \omega_0/2\pi$ is the classical frequency of the oscillator. Note that the energy levels of a harmonic oscillator are evenly spaced with separation hf_0, as indicated in Figure 34-18. Compare this with the uneven spacing of the energy levels for the particle in a box, as shown in Figure 34-12. If a harmonic oscillator makes a transition from energy level n to the next lowest energy level $n - 1$, the frequency f of the photon emitted is given by $hf = E_i - E_f$ (Equation 34-23). Applying this equation gives

$$hf = E_n - E_{n-1} = (n + \tfrac{1}{2})hf_0 - (n - 1 + \tfrac{1}{2})hf_0 = hf_0$$

The frequency f of the emitted photon is therefore equal to the classical frequency f_0 of the oscillator.

THE HYDROGEN ATOM

In the hydrogen atom, an electron is bound to a proton by the electrostatic force of attraction (discussed in Chapter 21). This force varies inversely as the square of the separation distance (exactly like the gravitational attraction of Earth and the Sun). The potential energy of the electron–proton system therefore varies inversely with separation distance (Equation 23-9). As in the case of gravitational potential energy, the potential energy of the electron–proton system is chosen to be zero if the electron is an infinite distance from the proton. Then for all finite distances, the potential energy is negative. Like the case of an object orbiting Earth, the electron–proton system is a bound system if its total energy is negative. Like the energies of a particle in a box and of a harmonic oscillator, the energies are described by a quantum number n. As we will see in Chapter 36, the allowed energies of the hydrogen atom are given by

$$E_n = -\frac{13.6\text{ eV}}{n^2} \qquad n = 1, 2, 3, \ldots \qquad 34\text{-}30$$

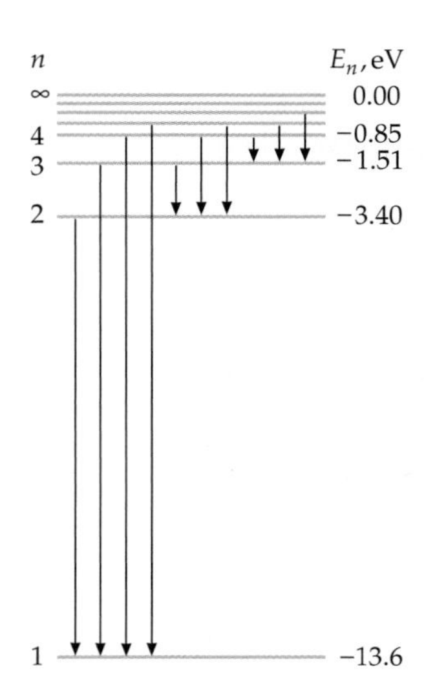

FIGURE 34-19 Energy-level diagram for the hydrogen atom. The energy of the ground state is −13.6 eV. As n approaches infinity, the energy approaches 0, which is the highest energy state for which an electron is bound to the nucleus.

The lowest energy corresponds to $n = 1$. The ground-state energy is thus −13.6 eV. The energy of the first excited state is $-(13.6\text{ eV})/2^2 = -3.40$ eV. Figure 34-19 shows the energy-level diagram for the hydrogen atom. The vertical arrows indicate transitions from a higher state to a lower state that accompany the emission of electromagnetic radiation. Only those transitions ending at the first excited state ($n = 2$) involve energy differences in the range of visible light of 1.77 eV to 3.10 eV, as calculated in Example 34-1.

Other atoms are more complicated than the hydrogen atom, but their energy levels are similar in many ways to those of hydrogen. Their ground-state energies are of the order of −1 eV to −10 eV, and many transitions involve energies corresponding to photons in the visible range.

! Do not think an electron orbits a proton in a classical orbit like Earth's orbit around the Sun. It doesn't.

Summary

1. All phenomena propagate like waves and interact like particles.
2. The quantum of light is called a photon and has energy $E = hf$, where h is Planck's constant.
3. The relation between wavelength and momentum of electrons, photons, and other particles is given by the de Broglie relation $\lambda = h/p$.
4. Energy quantization in bound systems arises from standing-wave conditions, which are equivalent to boundary conditions on the wave function.
5. The uncertainty principle is a fundamental law of nature that places theoretical restrictions on the precision of a simultaneous measurement of the position and momentum of a particle. It follows from the general properties of waves.

TOPIC	RELEVANT EQUATIONS AND REMARKS	
1. Constants and Values		
Planck's constant	$h = 6.626 \times 10^{-34}\ \text{J}\cdot\text{s} = 4.136 \times 10^{-15}\ \text{eV}\cdot\text{s}$	34-2
hc	$hc = 1240\ \text{eV}\cdot\text{nm}$	34-5
2. The Particle Nature of Light: Photons	Energy is quantized.	
Photon energy and momentum	$E = hf$ and $E = pc$	34-1 and 34-6
3. Frequency–Wavelength (Energy–Momentum) Relations		
Photons and material particles (de Broglie relations)	$E = hf$ and $p = \dfrac{h}{\lambda}$	34-14 and 34-13
Nonrelativistic particles	$K = \dfrac{p^2}{2m}$ so $\lambda = \dfrac{hc}{\sqrt{2mc^2K}}$	34-15
Photoelectric effect	$K_{\text{max}} = \left(\frac{1}{2}mv^2\right)_{\text{max}} = hf - \phi$ where ϕ is the work function of the cathode.	34-3
Compton scattering	$\lambda_s - \lambda_i = \dfrac{h}{m_e c}(1 - \cos\theta) = \lambda_C(1 - \cos\theta) = 2.426\ \text{pm}\,(1 - \cos\theta)$	34-11
4. Quantum Mechanics	The state of a particle, such as an electron, is described by its wave function ψ, which is the solution of the Schrödinger wave equation.	
Probability density	The probability of finding the particle in some region of space dx is given by $P(x) = \psi^2(x)\,dx$	34-17
Normalization condition	$\int_{-\infty}^{\infty} \psi^2\,dx = 1$	34-18
Quantum number	The wave function for a particular energy state is characterized by a quantum number n. In three dimensions there are three quantum numbers—one associated with a boundary condition in each dimension.	
Expectation value	The expectation value of x is the same as the average value of x that we would expect to obtain from a measurement of the positions of a large number of particles with the same wave function $\psi(x)$. $\langle x \rangle = \int_{-\infty}^{+\infty} x\psi^2(x)\,dx$	34-26
	$\langle F(x) \rangle = \int_{-\infty}^{+\infty} F(x)\psi^2(x)\,dx$	34-27

TOPIC	RELEVANT EQUATIONS AND REMARKS
5. **Wave–Particle Duality**	Photons, electrons, neutrons, and all other carriers of momentum and energy exhibit both wave and particle properties. Everything propagates like a classical wave, exhibiting diffraction and interference, but exchanges energy in discrete lumps like a classical particle. Because the wavelength of macroscopic objects is so small, diffraction and interference are not observed. Also, if a macroscopic amount of energy is exchanged, so many quanta are involved that the particle nature of the energy is not evident.
6. **Uncertainty Principle**	The wave–particle duality of nature leads to the uncertainty principle, which states that the product of the uncertainty in a measurement of position and the uncertainty in a measurement of momentum must be greater than or equal to $\frac{1}{2}\hbar$, where $\hbar$ is Planck's constant divided by 2π. $\Delta x\,\Delta p_x \geq \frac{1}{2}\hbar$ (34-19)

Answers to Concept Checks

34-1 2/3

34-2 $\langle N\rangle = 8/3$ $\langle N^2\rangle = 38/3$

Answers to Practice Problems

34-1 4.13×10^{-7} eV

34-2 (*a*) 12.4 μm, (*b*) 1.24 nm, (*c*) 1.24 pm

34-3 $1.46 \times 10^7\ \text{cm}^{-3}$

34-4 $0.6c$

34-5 1.4×10^{-34} m

34-6 0.388 nm. From this result, we see that a 10-eV electron has a de Broglie wavelength of about 0.4 nm. This quantity is of the same order of magnitude as the size of the atom and the spacing of atoms in a crystal.

34-7 4.70 nm

Problems

In a few problems, you are given more data than you actually need; in a few other problems, you are required to supply data from your general knowledge, outside sources, or informed estimate.

Interpret as significant all digits in numerical values that have trailing zeros and no decimal points.

- • Single-concept, single-step, relatively easy
- •• Intermediate-level, may require synthesis of concepts
- ••• Challenging
- SSM Solution is in the *Student Solutions Manual*
- Consecutive problems that are shaded are paired problems.

CONCEPTUAL PROBLEMS

1 • The quantized character of electromagnetic radiation is observed by (*a*) the Young double-slit experiment, (*b*) diffraction of light by a small aperture, (*c*) the photoelectric effect, (*d*) the J. J. Thomson cathode-ray experiment. SSM

2 •• Two monochromatic light sources, A and B, emit the same number of photons per second. The wavelength of A is $\lambda_A = 400$ nm and the wavelength of B is $\lambda_B = 600$ nm. The power radiated by source B (*a*) is equal to the power of source A, (*b*) is less than the power of source A, (*c*) is greater than the power of source A, (*d*) cannot be compared to the power from source A using the available data.

3 • The work function of a surface is ϕ. The threshold wavelength for emission of photoelectrons from the surface is equal to (*a*) hc/ϕ, (*b*) ϕ/hf, (*c*) hf/ϕ, (*d*) none of the above. SSM

4 •• When light of wavelength λ_1 is incident on a certain photoelectric cathode, no electrons are emitted, no matter how intense the incident light is. Yet, when light of wavelength $\lambda_2 < \lambda_1$ is incident, electrons are emitted, even when the incident light has low intensity. Explain this observation.

5 • True or false: (*a*) The wavelength of an electron's matter wave varies inversely with the momentum of the electron. (*b*) Electrons can undergo diffraction. (*c*) Neutrons can undergo diffraction.

6 • If the wavelength of an electron is equal to the wavelength of a proton, then (*a*) the speed of the proton is greater than the speed of the electron, (*b*) the speeds of the proton and the electron are equal, (*c*) the speed of the proton is less than the speed of the electron, (*d*) the energy of the proton is greater than the energy of the electron, (*e*) both (*a*) and (*d*) are correct.

7 • A proton and an electron have equal kinetic energies. It follows that the wavelength of the proton is (*a*) greater than the wavelength of the electron, (*b*) equal to the wavelength of the electron, (*c*) less than the wavelength of the electron.

8 • The parameter x represents the position of a particle. Can the expectation value of x ever have a value such that the probability density function $P(x)$ is zero? Give a specific example.

9 •• It was once believed that if two identical experiments are done on identical systems under the same conditions, the results must be identical. Explain how this statement can be modified so that it is consistent with quantum physics.

10 •• A six-sided die has the numeral 1 painted on three sides and the numeral 2 painted on the other three sides. (*a*) What is the probability of a 1 coming up when the die is thrown? (*b*) What is the expectation value of the numeral that comes up when the die is thrown? (*c*) What is the expectation value of the cube of the numeral that comes up when the die is thrown?

ESTIMATION AND APPROXIMATION

11 •• During an advanced physics lab, students measure the Compton wavelength, λ_C. The students obtain the following wavelength shifts $\lambda_s - \lambda_i$ as a function of scattering angle θ.

θ	45°	75°	90°	135°	180°
$\lambda_s - \lambda_i$	0.647 pm	1.67 pm	2.45 pm	3.98 pm	4.95 pm

Use their data to estimate the value for the Compton wavelength. Compare this number with the accepted value. SSM

12 •• **SPREADSHEET** Students in a physics lab are trying to determine the value of Planck's constant h, using a photoelectric apparatus similar to the one shown in Figure 34-2. The students are using a helium–neon laser that has a tunable wavelength as the light source. The data that the students obtain for the maximum electron kinetic energies are

λ	544 nm	594 nm	604 mn	612 nm	633 mn
K_{max}	0.360 eV	0.199 eV	0.156 eV	0.117 eV	0.062 eV

(*a*) Using a spreadsheet program or graphing calculator, plot K_{max} versus light frequency. (*b*) Use the graph to estimate the value of Planck's constant. (*Note:* You may wish to use a feature of your spreadsheet program or graphing calculator to obtain the best straight-line fit to the data.) (*c*) Compare your result with the accepted value for Planck's constant.

13 •• **SPREADSHEET** The cathode that was used by the students in the experiment described in Problem 12 is constructed from one of the following metals:

Metal	Tungsten	Silver	Potassium	Cesium
Work function	4.58 eV	2.4 eV	2.1 eV	1.9 eV

Determine which metal composes the cathode by using the same data given in Problem 12. (*a*) Using a spreadsheet program or graphing calculator, plot K_{max} versus frequency. (*b*) Use the graph to estimate the value of the work function based on the students' data. (*Note:* You may wish to use a feature of your spreadsheet program or graphing calculator to obtain the best straight-line fit to the data.) (*c*) Which metal was most likely used for the cathode in their experiment?

THE PARTICLE NATURE OF LIGHT: PHOTONS

14 • Find the photon energy in electron volts for light of wavelength (*a*) 450 nm, (*b*) 550 nm, and (*c*) 650 nm.

15 • Find the photon energy in electron volts for an electromagnetic wave of frequency (*a*) 100 MHz in the FM radio band and (*b*) 900 kHz in the AM radio band.

16 • What are the frequencies of photons that have the following energies: (*a*) 1.00 eV, (*b*) 1.00 keV, and (*c*) 1.00 MeV?

17 • Find the photon energy in electron volts if the wavelength is (*a*) 0.100 nm (about 1 atomic diameter) and (*b*) 1.00 fm (1 fm $= 10^{-15}$ m, about 1 nuclear diameter).

18 •• The wavelength of red light emitted by a 3.00-mW helium–neon laser is 633 nm. If the diameter of the laser beam is 1.00 mm, what is the density of photons in the beam? Assume that the intensity is uniformly distributed across the beam.

19 • **ENGINEERING APPLICATION** Lasers used in a telecommunications network typically produce light that has a wavelength near 1.55 μm. How many photons per second are being transmitted if such a laser has an output power of 2.50 mW? SSM

THE PHOTOELECTRIC EFFECT

20 • The work function for tungsten is 4.58 eV. (*a*) Find the threshold frequency and wavelength for the photoelectric effect to occur when monochromatic electromagnetic radiation is incident on the surface of a sample of tungsten. Find the maximum kinetic energy of the electrons if the wavelength of the incident light is (*b*) 200 nm and (*c*) 250 nm.

21 • When monochromatic ultraviolet light that has a wavelength equal to 300 nm is incident on a sample of potassium, the emitted electrons have maximum kinetic energy of 2.03 eV. (*a*) What is the energy of an incident photon? (*b*) What is the work function for potassium? (*c*) What would be the maximum kinetic energy of the electrons if the incident electromagnetic radiation had a wavelength of 430 nm? (*d*) What is the maximum wavelength of incident electromagnetic radiation that will result in the photoelectric emission of electrons by a sample of potassium?

22 • The maximum wavelength of electromagnetic radiation that will result in the photoelectric emission of electrons from a sample of silver is 262 nm. (*a*) Find the work function for silver. (*b*) Find the maximum kinetic energy of the electrons if the incident radiation has a wavelength of 175 nm.

23 • The work function for cesium is 1.90 eV. (*a*) Find the minimum frequency and maximum wavelength of electromagnetic radiation that will result in the photoelectric emission of electrons from a sample of cesium. Find the maximum kinetic energy of the electrons if the wavelength of the incident radiation is (*b*) 250 nm and (*c*) 350 nm.

24 •• When a surface is illuminated with electromagnetic radiation of wavelength 780 nm, the maximum kinetic energy of the emitted electrons is 0.37 eV. What is the maximum kinetic energy if the surface is illuminated using radiation of wavelength 410 nm?

COMPTON SCATTERING

25 • Find the shift in wavelength of photons scattered by free stationary electrons at $\theta = 60°$. (Assume that the electrons are initially moving with negligible speed and are virtually free of (unattached to) any atoms or molecules.)

26 • When photons are scattered by electrons in a carbon sample, the shift in wavelength is 0.33 pm. Find the scattering angle. (Assume that the electrons are initially moving with negligible speed and are virtually free of (unattached to) any atoms or molecules.)

27 • The photons in a monochromatic beam are scattered by electrons. The wavelength of the photons that are scattered at an angle of 135° with the direction of the incident photon beam is 2.3 percent less than the wavelength of the incident photons. What is the wavelength of the incident photons?

28 • Compton used photons of wavelength 0.0711 nm. (*a*) What is the energy of one of those photons? (*b*) What is the wavelength of the photons scattered in the direction opposite to the direction of the incident photons? (*c*) What is the energy of the photon scattered in that direction?

29 • For the photons used by Compton (see Problem 28), find the momentum of the incident photon and the momentum of the photon scattered in the direction opposite to the direction of the incident photons. Use the conservation of momentum to find the momentum of the recoil electron in this case.

30 •• A beam of photons that have a wavelength equal to 6.00 pm is scattered by electrons initially at rest. A photon in the beam is scattered in a direction perpendicular to the direction of the incident beam. *(a)* What is the change in wavelength of the photon? *(b)* What is the kinetic energy of the electron?

ELECTRONS AND MATTER WAVES

31 • An electron is moving at 2.5×10^5 m/s. Find the electron's wavelength.

32 • An electron has a wavelength of 200 nm. Find *(a)* the magnitude of its momentum and *(b)* its kinetic energy.

33 •• An electron, a proton, and an alpha particle each have a kinetic energy of 150 keV. Find *(a)* the magnitudes of their momenta and *(b)* their de Broglie wavelengths.

34 • A neutron in a reactor has a kinetic energy of approximately 0.020 eV. Calculate the wavelength of the neutron.

35 • Find the wavelength of a proton that has a kinetic energy of 2.00 MeV.

36 • What is the kinetic energy of a proton whose wavelength is *(a)* 1.00 nm and *(b)* 1.00 fm?

37 • The kinetic energy of the electrons in the electron beam in a run of Davisson and Germer's experiment was 54 eV. Calculate the wavelength of the electrons in the beam.

38 • The distance between Li^+ and Cl^- ions in a LiCl crystal is 0.257 nm. Find the energy of electrons that have a wavelength equal to that spacing.

39 • An electron microscope uses electrons that have energies equal to 70 keV. Find the wavelength of the electrons. SSM

40 • What is the wavelength of a neutron that has a speed of 1.00×10^6 m/s?

A PARTICLE IN A BOX

41 •• *(a)* Find the energy of the ground state ($n = 1$) and the first two excited states of a neutron in a one-dimensional box of length $L = 1.00 \times 10^{-15}$ m $= 1.00$ fm (about the diameter of an atomic nucleus). Make an energy-level diagram for the system. Calculate the wavelength of electromagnetic radiation emitted when the neutron makes a transition from *(b)* $n = 2$ to $n = 1$, *(c)* $n = 3$ to $n = 2$, and *(d)* $n = 3$ to $n = 1$.

42 •• *(a)* Find the energy of the ground state ($n = 1$) and the first two excited states of a neutron in a one-dimensional box of length 0.200 nm (about the diameter of a H_2 molecule). Calculate the wavelength of electromagnetic radiation emitted when the neutron makes a transition from *(b)* $n = 2$ to $n = 1$, *(c)* $n = 3$ to $n = 2$, and *(d)* $n = 3$ to $n = 1$.

CALCULATING PROBABILITIES AND EXPECTATION VALUES

43 •• A particle is in the ground state of a one-dimensional box that has length L. (The box has one end at the origin and the other end on the positive x axis.) Determine the probability of finding the particle in the interval of length $\Delta x = 0.002L$ and centered at *(a)* $\frac{1}{4}x = L$, *(b)* $x = \frac{1}{2}L$, and *(c)* $x = \frac{3}{4}L$. (Because Δx is very small you need not do any integration.)

44 •• A particle is in the second excited state ($n = 3$) of a one-dimensional box that has length L. (The box has one end at the origin and the other end on the positive x axis.) Determine the probability of finding the particle in the interval of length $\Delta x = 0.002L$ and centered at *(a)* $x = \frac{1}{3}L$, *(b)* $x = \frac{1}{2}L$, and *(c)* $x = \frac{2}{3}L$. (Because Δx is very small you need not do any integration.)

45 •• A particle is in the first excited ($n = 2$) state of a one-dimensional box that has length L. (The box has one end at the origin and the other end on the positive x axis.) Find *(a)* $\langle x \rangle$ and *(b)* $\langle x^2 \rangle$.

46 •• A particle in a one-dimensional box that has length L is in the first excited state ($n = 2$). (The box has one end at the origin and the other end on the positive x axis.) *(a)* Sketch $\psi^2(x)$ versus x for this state. *(b)* What is the expectation value $\langle x \rangle$ for this state? *(c)* What is the probability of finding the particle in some small region dx centered at $x = L/2$? *(d)* Are your answers for Part *(b)* and Part *(c)* contradictory? If not, explain why your answers are not contradictory.

47 •• A particle of mass m has a wave function given by $\psi(x) = Ae^{-|x|/a}$, where A and a are positive constants. *(a)* Find the normalization constant A. *(b)* Calculate the probability of finding the particle in the region $-a \leq x \leq a$.

48 •• A one-dimensional box is on the x axis in the region of $0 \leq x \leq L$. A particle in this box is in its ground state. Calculate the probability that the particle will be found in the region *(a)* $0 < x < \frac{1}{2}L$, *(b)* $0 < x < \frac{1}{3}L$, and *(c)* $0 < x < \frac{3}{4}L$.

49 •• A one-dimensional box is on the x axis in the region of $0 \leq x \leq L$. A particle in this box is in its first excited state. Calculate the probability that the particle will be found in the region *(a)* $0 < x < \frac{1}{2}L$, *(b)* $0 < x < \frac{1}{3}L$, and *(c)* $0 < x < \frac{3}{4}L$.

50 •• The classical probability distribution function for a particle in a one-dimensional box on the x axis in the region of $0 < x < L$ is given by $P(x) = 1/L$. Use this expression to show that $\langle x \rangle = \frac{1}{2}L$ and $\langle x^2 \rangle = \frac{1}{3}L^2$ for a classical particle in the box.

51 •• A one-dimensional box is on the x axis in the region of $0 \leq x \leq L$. *(a)* The wave functions for a particle in the box are given by

$$\psi_n(x) = \sqrt{\frac{2}{L}} \sin \frac{n\pi x}{L} \qquad n = 1, 2, 3, \ldots$$

For a particle in the nth state, show that $\langle x \rangle = \frac{1}{2}L$ and $\langle x^2 \rangle = L^2/3 - L^2/(2n^2\pi^2)$. *(b)* Compare these expressions for $\langle x \rangle$ and $\langle x^2 \rangle$, for $n \gg 1$, with the expressions for $\langle x \rangle$ and $\langle x^2 \rangle$ for the classical distribution of Problem 50.

52 •• **SPREADSHEET** *(a)* Use a spreadsheet program or graphing calculator to plot $\langle x^2 \rangle$ as a function of the quantum number n for the particle in the box described in Problem 48 and for values of n from 1 to 100. Assume $L = 1.00$ m for your graph. Refer to Problem 51. *(b)* Comment on the significance of any asymptotic limits that your graph shows.

53 •• The wave functions for a particle of mass m in a one-dimensional box of length L *centered at the origin* (so that the ends are at $x = \pm\frac{1}{2}L$) are given by

$$\psi(x) = \sqrt{\frac{2}{L}} \cos \frac{n\pi x}{L} \qquad n = 1, 3, 5, 7, \ldots$$

and

$$\psi(x) = \sqrt{\frac{2}{L}} \sin \frac{n\pi x}{L} \qquad n = 2, 4, 6, 8, \ldots$$

Calculate $\langle x \rangle$ and $\langle x^2 \rangle$ for the ground state ($n = 1$).

54 •• Calculate $\langle x \rangle$ and $\langle x^2 \rangle$ for the first excited state ($n = 2$) of the box described in Problem 53.

GENERAL PROBLEMS

55 • Photons in a uniform 4.00-cm-diameter light beam have wavelengths equal to 400 nm and the beam has an intensity of 100 W/m^2. (*a*) What is the energy of each photon in the beam? (*b*) How much energy strikes an area of 1.00 cm^2 perpendicular to the beam in 1.00 s? (*c*) How many photons strike this area in 1.00 s? SSM

56 • A 1-μg particle is moving with a speed of approximately 1 mm/s in a one-dimensional box that has a length equal to 1 cm. Calculate the approximate value of the quantum number n of the state occupied by the particle.

57 • (*a*) For the particle and box of Problem 56, find Δx and Δp_x, assuming that these uncertainties are given by $\Delta x/L = 0.01$ percent and $\Delta p_x/p_x = 0.01$ percent. (*b*) What is $(\Delta x\,\Delta p_x)/\hbar$?

58 • In 1987, a laser at Los Alamos National Laboratory produced a flash that lasted 1×10^{-12} s and had a power of 5×10^{15} W. Estimate the number of emitted photons, assuming they all had wavelengths equal to 400 nm.

59 • **ENGINEERING APPLICATION** You cannot "see" anything smaller than the wavelength of the wave used to make the observation. What is the minimum energy of an electron needed in an electron microscope to "see" an atom that has a diameter of about 0.1 nm?

60 • A common flea that has a mass of 0.008 g can jump vertically as high as 20 cm. Estimate the wavelength for the flea immediately after takeoff.

61 •• **BIOLOGICAL APPLICATION** A 100-W source radiates light of wavelength 600 nm uniformly in all directions. An eye that has been adapted to the dark has a 7-mm-diameter pupil and can detect the light if at least 20 photons per second enter the pupil. How far from the source can the light be detected under these rather extreme conditions? SSM

62 •• **BIOLOGICAL APPLICATION** The diameter of the pupil of an eye under room-light conditions is approximately 5 mm. Find the intensity of light that has a wavelength equal to 600 nm so that 1 photon per second passes through the pupil.

63 •• A 100-W incandescent lightbulb radiates 2.6 W of visible light uniformly in all directions. (*a*) Find the intensity of the light from the bulb at a distance of 1.5 m. (*b*) If the average wavelength of the visible light is 650 nm, and counting only those photons in the visible spectrum, find the number of photons per second that strike a surface that has an area equal to 1.0 cm^2, is oriented so that the line to the bulb is perpendicular to the surface, and is a distance of 1.5 m from the bulb.

64 •• When light of wavelength λ_1 is incident on the cathode of a photoelectric tube, the maximum kinetic energy of the emitted electrons is 1.8 eV. If the wavelength is reduced to $\frac{1}{2}\lambda_1$, the maximum kinetic energy of the emitted electrons is 5.5 eV. Find the work function ϕ of the cathode material.

65 •• An incident photon of energy E_i undergoes Compton scattering at an angle of θ. Show that the energy E_s of the scattered photon is given by

$$E_s = \frac{E_i}{1 + (E_i/m_e c^2)(1 - \cos\theta)}$$

66 •• A particle is confined to a one-dimensional box. While the particle makes a transition from the state n to the state $n - 1$, radiation of 114.8 nm is emitted. While the particle makes the transition from the state $n - 1$ to the state $n - 2$, radiation of wavelength 147 nm is emitted. The ground-state energy of the particle is 1.2 eV. Determine n.

67 •• The Pauli exclusion principle states that no more than one electron may occupy a particular quantum state at a time. Electrons intrinsically occupy two spin states. Therefore, if we wish to model an atom as a collection of electrons trapped in a one-dimensional box, no more than two electrons in the box can have the same value of the quantum number n. Calculate the energy that the most energetic electron(s) would have for the uranium atom that has an atomic number 92. Assume the box has a length of 0.050 nm and the electrons are in the lowest possible energy states. How does this energy compare to the rest energy of the electron? SSM

68 •• A beam of electrons that each have the same kinetic energy illuminates a pair of slits separated by a distance 54 nm. The beam forms bright and dark fringes on a screen located a distance 1.5 m beyond the two slits. The arrangement is otherwise identical to that used in the optical two-slit interference experiment described in Chapter 33 and in Figure 33-7 and the fringes have the appearance shown in Figure 34-8*d*. The bright fringes are found to be separated by a distance of 0.68 mm. What is the kinetic energy of the electrons in the beam?

69 •• When a surface is illuminated by light of wavelength λ, the maximum kinetic energy of the emitted electrons is 1.20 eV. If the wavelength $\lambda' = 0.800\lambda$ is used, the maximum kinetic energy increases to 1.76 eV. For wavelength $\lambda' = 0.600\lambda$, the maximum kinetic energy of the emitted electrons is 2.676 eV. Determine the work function of the surface and the wavelength λ.

70 •• A simple pendulum has a length equal to 1.0 m and has a bob that has a mass equal to 0.30 kg. The energy of this oscillator is quantized, and the allowed values of the energy are given by $E_n = \left(n + \frac{1}{2}\right)hf_0$, where n is an integer and f_0 is the frequency of the pendulum. (*a*) Find n if the angular amplitude is 1.0°. (*b*) Find n such that E_{n+1} exceeds E_n by 0.010 percent.

71 •• (*a*) Show that for large n, the fractional difference in energy between state n and state $n + 1$ for a particle in a one-dimensional box is given approximately by

$$(E_{n+1} - E_n)/E_n \approx 2/n$$

(*b*) What is the approximate percentage energy difference between the states $n_1 = 1000$ and $n_2 = 1001$? (*c*) Comment on how this result is related to Bohr's correspondence principle. SSM

72 •• A mode-locked, titanium–sapphire laser has a wavelength of 850 nm and produces 100 million pulses of light each second. Each pulse has a duration of 125 femtoseconds (1 fs = 10^{-15} s) and consists of 5×10^9 photons. What is the average power produced by the laser?

73 •• This problem estimates the time lag in the photoelectric effect that is expected classically but not observed. Let the intensity of the incident radiation falling on an atom be 0.010 W/m^2. (*a*) If the area presented by an atom is 0.010 nm^2, find the energy per second falling on an atom. (*b*) If the work function is 2.0 eV, how long would it take for this much energy to fall on the atom if the radiation energy was distributed uniformly rather than in compact packets (photons)?

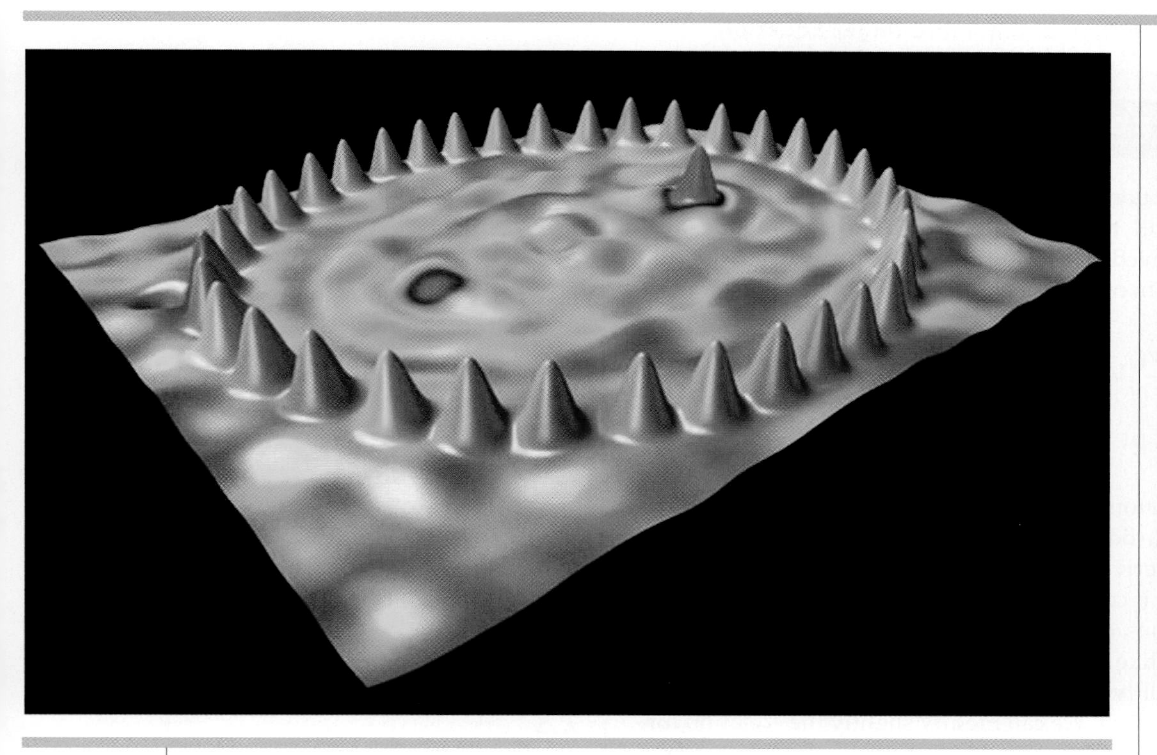

A QUANTUM MIRAGE. THE SCANNING TUNNELING MICROSCOPE (STM) ALLOWS ONE TO PUSH INDIVIDUAL ATOMS AROUND ON A SURFACE AND TO IMAGE THEM. ESPECIALLY INTRIGUING ARE IMAGES OF QUANTUM CORRALS, WHICH ARE CIRCULAR OR ELLIPTICAL ARRANGEMENTS ON A SURFACE INSIDE OF WHICH THE WAVES CORRESPONDING TO ELECTRONS NEAR THE SUBSTRATE SURFACE CAN BE REVEALED. THIS IMAGE COMES FROM IBM, WHERE PHYSICISTS PLACED THIRTY-SIX COBALT ATOMS IN AN ELLIPTICAL "STONEHENGE" PATTERN ON A COPPER SURFACE. AN EXTRA MAGNETIC COBALT ATOM WAS PLACED AT ONE OF THE TWO FOCI OF THE ELLIPSE, CAUSING VISIBLE INTERACTIONS WITH THE SURFACE ELECTRON WAVES. BUT THE WAVES ALSO SEEM TO BE INTERACTING WITH A PHANTOM COBALT ATOM AT THE OTHER FOCUS, AN ATOM THAT IS NOT REALLY THERE. *(Courtesy of IBM and the IBM Almadin Laboratories.)*

Applications of the Schrödinger Equation

In Chapter 34, we found that electrons and other particles have wave properties and are described by wave functions in the form $\Psi(x, t)$. We also mentioned that the wave function is a solution of the Schrödinger equation, and we discussed some solutions qualitatively without reference to the equation itself. In particular, we showed how the standing-wave conditions lead to quantization of energy for a particle confined to a one-dimensional box.

In this chapter we continue our discussion of the material introduced in Chapter 34. We discuss the Schrödinger equation and apply the equation to the particle-in-the-box problem and to several other situations in which a particle is confined to a region of space to illustrate how boundary conditions lead to energy quantization. We then show how the Schrödinger equation leads to barrier penetration and discuss the extension of the Schrödinger equation to more than one dimension and to more than one particle.

? Could the phantom cobalt atom described above be caused by reflections of waves from the corral of cobalt atoms? (See Section 35-4).

35-1 THE SCHRÖDINGER EQUATION

Like the classical wave equation (Equation 15-10*b*), the Schrödinger equation is a partial differential equation in space and time. Like Newton's laws of motion, the Schrödinger equation cannot be derived. Its validity, like that of Newton's laws, lies in its agreement with experiment. In one dimension, the Schrödinger equation is*

$$-\frac{\hbar^2}{2m}\frac{\partial^2\Psi(x,t)}{\partial x^2} + U\Psi(x,t) = i\hbar\frac{\partial\Psi(x,t)}{\partial t} \qquad 35\text{-}1$$

TIME-DEPENDENT SCHRÖDINGER EQUATION

where U is the potential energy function and $\Psi(x, t)$ is a wave function. Equation 35-1 is called the **time-dependent Schrödinger equation.** Unlike the classical wave equation, it relates the *second* space derivative of the wave function to the *first* time derivative of the wave function, and it contains the imaginary number $i = \sqrt{-1}$. The wave functions that are solutions of this equation are not necessarily real. $\Psi(x, t)$ is not a measurable function like the classical wave functions for sound or electromagnetic waves. The probability of finding a particle in some region of space dx certainly has a real value, though. We can modify slightly the equation for probability density given in Chapter 34 (Equation 34-17) to determine the probability of finding a particle in some region dx

See Math Tutorial *for more information on* Complex Numbers

$$P(x,t)\,dx = |\Psi(x,t)|^2\,dx = \Psi^*\Psi\,dx \qquad 35\text{-}2$$

where Ψ^*, the complex conjugate of Ψ, is identical to Ψ, except that $-i$ is substituted for i wherever i appears in the expression for Ψ.†

In classical mechanics, the standing-wave solutions to the wave equation (Equation 16-16) are of great interest and value. Not surprisingly, standing-wave solutions to the Schrödinger wave equation are also of great interest and value. The wave function for the standing-wave motion of a uniform taut string is $A\sin(kx)\cos(\omega t + \delta)$, which is representative of all standing waves. A standing wave function can always be expressed as a function of position multiplied by a function of time, where the function of time is one that varies sinusoidally with time. Standing-wave solutions to the one-dimensional Schrödinger wave equation are thus expressed

$$\Psi(x,t) = \psi(x)e^{-i\omega t} \qquad 35\text{-}3$$

where $e^{-i\omega t} = \cos(\omega t) - i\sin(\omega t)$. The right side of Equation 35-1 is then

$$i\hbar\frac{\partial\Psi(x,t)}{\partial t} = i\hbar(-i\omega)\psi(x)e^{-i\omega t} = \hbar\omega\psi(x)e^{-i\omega t} = E\psi(x)e^{-i\omega t}$$

where $E = \hbar\omega$ is the energy of the particle.

The Schrödinger wave equation has standing-wave solutions only if the potential energy function U depends on position x alone. Substituting $\psi(x)e^{-i\omega t}$ into Equation 35-1 and canceling the common factor $e^{-i\omega t}$, we obtain an equation for $\psi(x)$, called the **time-independent Schrödinger equation:**

$$-\frac{\hbar^2}{2m}\frac{d^2\psi(x)}{dx^2} + U(x)\psi(x) = E\psi(x) \qquad 35\text{-}4$$

TIME-INDEPENDENT SCHRÖDINGER EQUATION

---

* Although we simply state the Schrödinger equation, Schrödinger himself had a vast knowledge of classical wave theory that led him to this equation.

† Every complex number can be written in the form $z = a + bi$, where a and b are real numbers and $i = \sqrt{-1}$. The complex conjugate of z is $z^* = a - bi$, so $z^*z = (a + bi)(a - bi) = a^2 + b^2 = |z|^2$.

where we have written U as $U(x)$ to emphasize that while U may depend on position, U does not depend on time. The function $U(x)$ represents the interaction between the environment and the particle being observed. Different environments require different expressions for the potential energy function U in the Schrödinger equation.

The calculation of the allowed energy levels in a system involves only the time-independent Schrödinger equation, whereas finding the probabilities of transition between these levels requires the solution of the time-dependent equation. In this book, we will be concerned only with the time-independent Schrödinger equation (Equation 35-4).

The solution of Equation 35-4 depends on the form of the potential energy function $U(x)$. When $U(x)$ is such that the particle is confined to some region of space, only certain discrete energies E_n give solutions ψ_n that can satisfy the normalization condition (Equation 34-18):

$$\int_{-\infty}^{\infty} |\psi_n|^2 dx = 1$$

The complete time-dependent wave functions are then given, from Equation 35-3, by

$$\Psi_n(x, t) = \psi_n(x)e^{-i\omega_n t} = \psi_n(x)e^{-i(E_n/\hbar)t} \qquad 35\text{-}5$$

A PARTICLE IN AN INFINITE SQUARE-WELL POTENTIAL

We will illustrate the use of the time-independent Schrödinger equation by solving it for the problem of a particle in a box. The potential energy for a one-dimensional box from $x = 0$ to $x = L$ is shown in Figure 35-1. It is called an **infinite square-well potential** and is described mathematically by

$$U(x) = \begin{cases} \infty & x < 0 \\ 0 & 0 < x < L \\ \infty & x > L \end{cases} \qquad 35\text{-}6$$

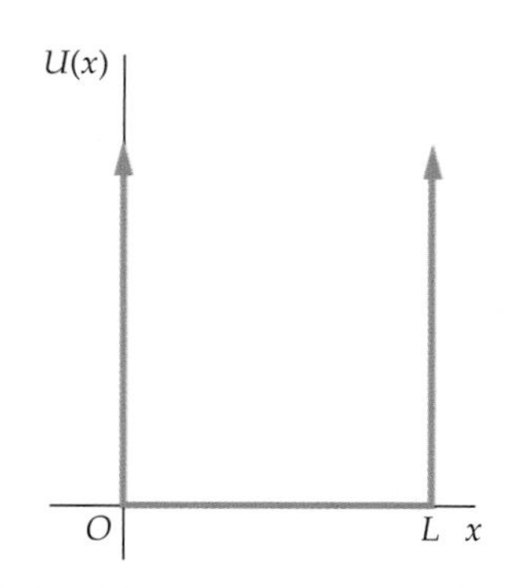

FIGURE 35-1 The infinite square-well potential energy function. For both $x < 0$ and $x > L$, the potential energy $U(x)$ is infinite. The particle is confined to the region in the well $0 < x < L$.

Inside the box, the potential energy is zero, whereas outside the box it is infinite. Because we require the particle to be in the box, we have $\psi(x) = 0$ everywhere outside the box. We then need to solve the Schrödinger equation inside the box for wave functions $\psi(x)$ that must be zero at $x = 0$ and at $x = L$.

Inside the box $U(x) = 0$, so the Schrödinger equation is written

$$-\frac{\hbar^2}{2m}\frac{d^2\psi(x)}{dx^2} = E\psi(x)$$

or

$$\frac{d^2\psi(x)}{dx^2} + k^2\psi(x) = 0 \qquad 35\text{-}7$$

where

$$k^2 = \frac{2mE}{\hbar^2} \qquad 35\text{-}8$$

The general solution of Equation 35-7 can be written as

$$\psi(x) = A\sin kx + B\cos kx \qquad 35\text{-}9$$

where A and B are constants. At $x = 0$, we have

$$\psi(0) = A\sin(0) + B\cos(0) = 0 + B$$

The boundary condition $\psi(x) = 0$ at $x = 0$ thus gives $B = 0$, and Equation 35-9 becomes

$$\psi(x) = A \sin kx \qquad 35\text{-}10$$

The wave function is thus a sine wave where the wavelength λ is related to the wave number k in the usual way, $\lambda = 2\pi/k$. The boundary condition $\psi(x) = 0$ at $x = L$ restricts the possible values of k and therefore the values of the wavelength λ, and (from Equation 35-8) the energy $E = \frac{1}{2}\hbar^2k^2/m$. We have

$$\psi(L) = A \sin kL = 0 \qquad 35\text{-}11$$

This condition is satisfied if kL is π or any integer multiplied by π, that is, if k is restricted to the values k_n given by

$$k_n = n\frac{\pi}{L} \qquad n = 1, 2, 3, \ldots \qquad 35\text{-}12$$

The condition (Equation 35-11) is also satisfied for $n = 0$. The function $\psi(x) = A \sin 0 = 0$ for all values of x, in the interval $0 < x < L$, is also a solution to the wave equation. However, if the wave function has a value of zero everywhere inside the box, then the box is empty. Furthermore, the wave function cannot be normalized and cannot be a wave function for a particle. Substituting $n\pi/L$ for k_n into Equation 35-8 and solving for E gives us the allowed energy values:

$$E_n = \frac{\hbar^2k_n^2}{2m} = \frac{\hbar^2}{2m}\left(n\frac{\pi}{L}\right)^2 = n^2\left(\frac{h^2}{8mL^2}\right) = n^2E_1 \qquad 35\text{-}13$$

where

$$E_1 = \frac{h^2}{8mL^2} \qquad 35\text{-}14$$

Equation 35-14 is the same as Equation 34-22, which we obtained by fitting an integral number of half-wavelengths into the box.

For each value of n, there is wave function $\psi_n(x)$ given by

$$\psi_n(x) = \begin{cases} 0 & x < 0 \\ A_n \sin\dfrac{n\pi x}{L} & 0 < x < L \\ 0 & x > L \end{cases} \qquad 35\text{-}15$$

which is the same as Equation 34-25, where the constant $A_n = \sqrt{2/L}$ is determined by normalization.*

CONCEPT CHECK 35-1

The function below is not an acceptable wave function. What makes the function unacceptable?

$$\psi_n(x) = A_n \sin\frac{n\pi x}{L} \qquad -\infty < x < \infty$$

35-2 A PARTICLE IN A FINITE SQUARE WELL

The quantization of energy that we found for a particle in an infinite square well is a result that follows from the general solution of the Schrödinger equation for any particle confined to some region of space. We will illustrate this by considering the qualitative behavior of the wave function for a slightly more general potential energy function, the finite square well, which is shown in Figure 35-2. This potential energy function is described mathematically by

$$U(x) = \begin{cases} U_0 & x < 0 \\ 0 & 0 < x < L \\ U_0 & x > L \end{cases} \qquad 35\text{-}16$$

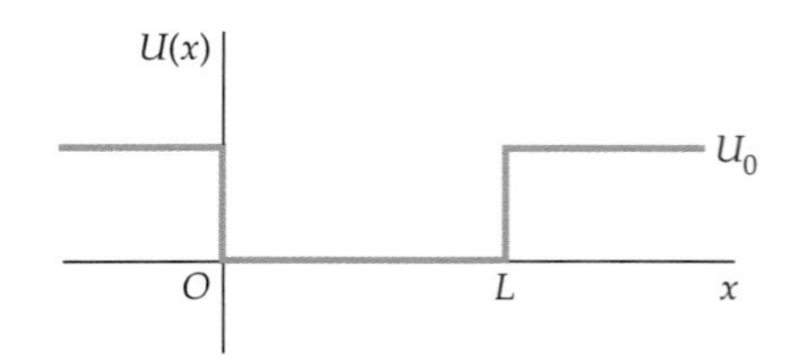

FIGURE 35-2 The finite square-well potential energy function.

* See Equation 34-18.

This potential energy function is discontinuous at $x = 0$ and $x = L$, but it is finite everywhere. The solutions of the Schrödinger equation for this type of potential energy function depend on whether the total energy E is greater or less than U_0. We will not discuss the case of $E > U_0$, except to remark that in that case the particle is not confined and any value of the energy is allowed. That is, there is no energy quantization when $E > U_0$. Here, we assume that $0 \leq E < U_0$.

Inside the well, $U(x) = 0$, and the time-independent Schrödinger equation is the same as for the infinite well (Equation 35-7):

$$-\frac{\hbar^2}{2m}\frac{d^2\psi(x)}{dx^2} = E\psi(x) \qquad 0 \leq x \leq L$$

or

$$\frac{d^2\psi(x)}{dx^2} + k^2\psi(x) = 0$$

where $k^2 = 2mE/\hbar^2$. The general solution is of the form

$$\psi(x) = A \sin kx + B \cos kx$$

In this case, $\psi(x)$ is not required to be zero at $x = 0$ (the particle is not required to be inside the box), so B is not zero. Outside the well, the time-independent Schrödinger equation is

$$-\frac{\hbar^2}{2m}\frac{d^2\psi(x)}{dx^2} + U_0\psi(x) = E\psi(x) \qquad x < 0 \text{ and } x > L$$

or

$$\frac{d^2\psi(x)}{dx^2} - \alpha^2\psi(x) = 0 \qquad \text{35-17}$$

where

$$\alpha^2 = \frac{2m}{\hbar^2}(U_0 - E) \qquad U_0 > E \qquad \text{35-18}$$

The wave functions and allowed energies for the particle can be found by solving Equation 35-17 for $\psi(x)$ outside the well and then requiring that both $\psi(x)$ and $d\psi(x)/dx$ be continuous at the boundaries $x = 0$ and $x = L$. The solution of Equation 35-17 is not difficult [in the region $x > L$, it is of the form $\psi(x) = Ce^{-\alpha x}$ and in the region $x < L$ it is of the form $\psi(x) = Ce^{+\alpha x}$], but applying the boundary conditions involves much tedious algebra and is not important for our purpose. The important feature of Equation 35-17 is that $d^2\psi/dx^2$ has the same sign as ψ. Thus, if ψ is positive, $d^2\psi/dx^2$ is also positive and the wave function curves away from the axis as x approaches either $+\infty$ or $-\infty$, as shown in Figure 35-3*a*. Similarly, if ψ is negative, $d^2\psi/dx^2$ is negative and ψ again curves away from the axis as x approaches either $+\infty$ or $-\infty$, as shown in Figure 35-3*b*. This behavior is very different from the behavior inside the well, where ψ and $d^2\psi/dx^2$ have opposite signs so that ψ always curves toward the axis like a sine or cosine function. Because of this behavior outside the well, for most values of the energy E in Equation 35-17, $\psi(x)$ becomes infinite as x approaches $\pm\infty$; that is, most wave functions $\psi(x)$ are not well behaved outside the well. Though they satisfy the Schrödinger equation, such functions are not proper wave functions because they cannot be normalized. The solutions of the Schrödinger equation are well behaved (that is, they approach 0 as $|x|$ becomes very large) only for certain values of the energy. These energy values are the allowed energies for the finite square well.

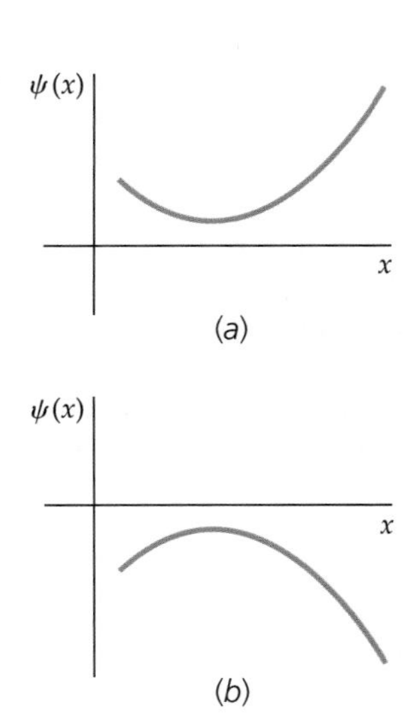

FIGURE 35-3 (*a*) A function ψ that has both a positive value and a positive concavity throughout the region shown. (Concavity is the sign of $d\psi/dx$.) (*b*) A function ψ that has both a negative value and a negative concavity throughout the region shown.

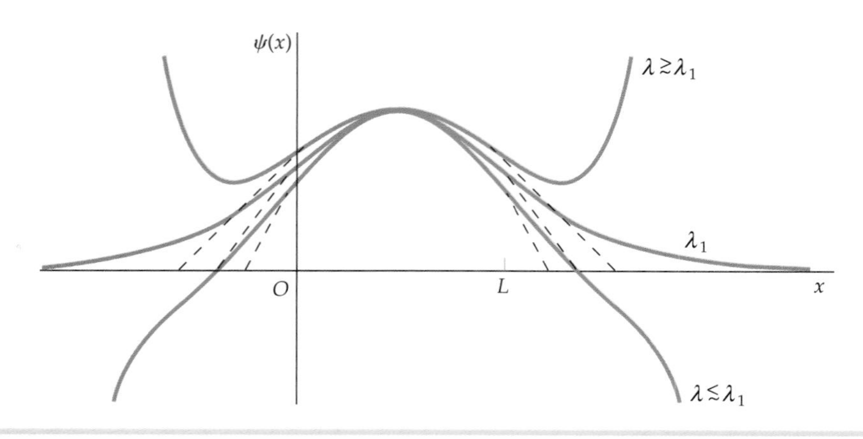

FIGURE 35-4 Functions satisfying the Schrödinger equation that have a wavelength λ that is almost equal to the wavelength λ_1, which is the wavelength that corresponds to the ground-state energy $E_1 = \hbar^2/2m\lambda_1^2$ in the finite well. If λ is slightly greater than λ_1, the function approaches plus infinity as $|x|$ approaches infinity, like the function in Figure 35-3*a*. At the critical wavelength λ_1, the function and its slope approach zero together as $|x|$ approaches infinity. If λ is slightly less than λ_1, the function crosses the x axis while the slope is still negative. The slope then becomes more negative because its rate of change $d^2\psi/dx^2$ is now negative. This function approaches negative infinity as $|x|$ approaches infinity.

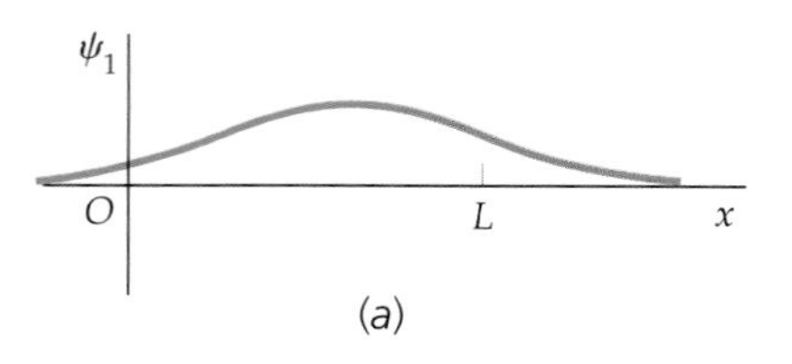

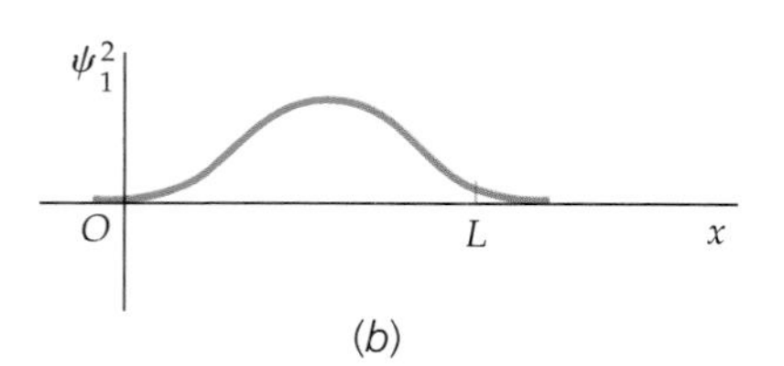

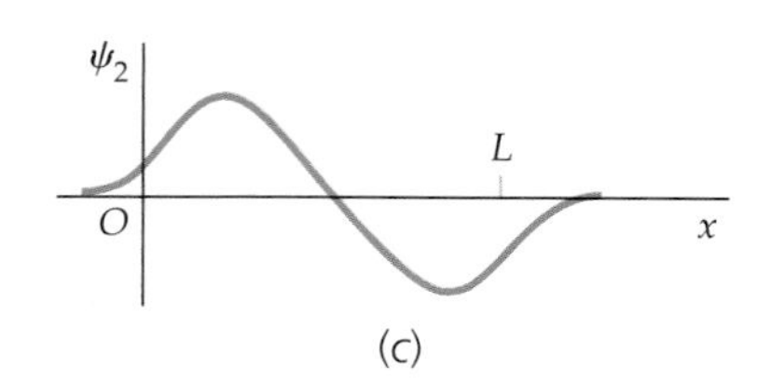

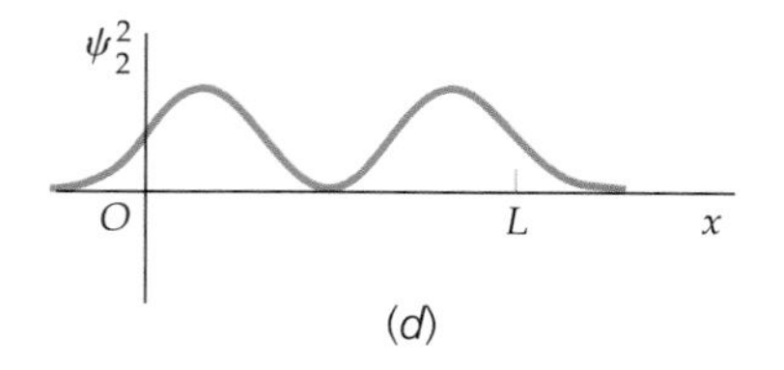

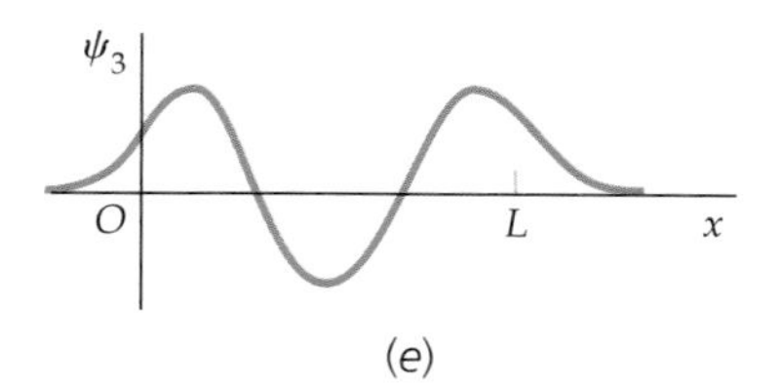

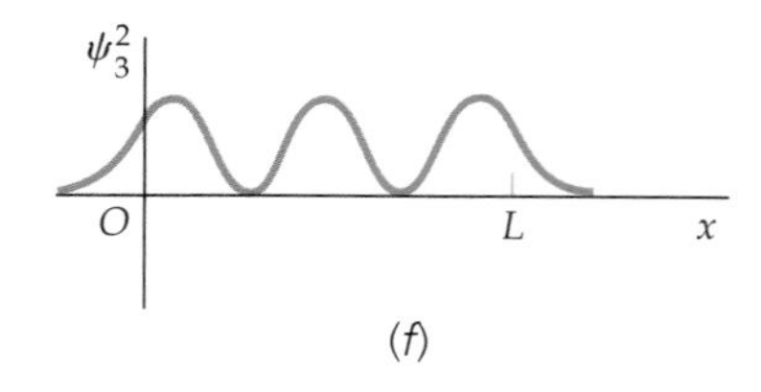

FIGURE 35-5 Graphs of the wave functions $\psi_n(x)$ and probability distributions $\psi^2(x)$ for $n = 1$, $n = 2$, and $n = 3$ for the finite square well. Compare these graphs with those of Figure 34-14 for the infinite square well, where the wave functions are zero at $x = 0$ and $x = L$. The wavelengths here are slightly longer than the corresponding wavelengths for the infinite well, so the allowed energies are somewhat smaller.

Figure 35-4 shows a well-behaved wave function, a wave function that has a wavelength λ_1 inside the well that corresponds to the ground-state energy. The behavior of the wave functions corresponding to nearby wavelengths and energies is also shown. Figure 35-5 shows the wave functions and probability distributions for the ground state and first two excited states. From this figure, we can see that the wavelengths inside the well are slightly longer than the corresponding wavelengths for the infinite well (Figure 34-14), so the corresponding energies are slightly less than those for the infinite well. Another feature of the finite-well problem is that there are only a finite number of allowed energies. For very small values of U_0, there is only one allowed energy.

Note that the wave function penetrates beyond the edges of the well at $x = L$ and $x = 0$, indicating that there is some small probability of finding the particle in the region in which its total energy E is less than its potential energy U_0. This region is called the *classically forbidden region* because the kinetic energy, $E - U_0$, would be negative when $U_0 > E$. Because negative kinetic energy has no meaning in classical physics, it is interesting to speculate on the result of an attempt to observe the particle in the classically forbidden region. It can be shown from the uncertainty principle that if an attempt is made to localize the particle in the classically forbidden region, such a measurement introduces an uncertainty in the momentum of the particle corresponding to a minimum kinetic energy that is greater than $U_0 - E$. This is just great enough to prevent us from measuring a negative kinetic energy. The penetration of the wave function into a classically forbidden region does have important consequences in barrier penetration, which will be discussed in Section 35-4.

Much of our discussion of the finite-well problem applies to any problem in which $E > U(x)$ in some region and $E < U(x)$ outside that region, as we see in the next section.

35-3 THE HARMONIC OSCILLATOR

The potential energy for a particle that has mass m and is attached to a spring that has force constant k is

$$U(x) = \tfrac{1}{2}kx^2 = \tfrac{1}{2}m\omega_0^2x^2 \qquad \text{35-19}$$

where $\omega_0 = \sqrt{k/m}$ is the natural frequency of the oscillator. Classically, the object oscillates between $x = +A$ and $x = -A$. The object's total energy is $E = \frac{1}{2}m\omega_0^2A^2$, which can have any positive value or zero.

This potential energy function, shown in Figure 35-6, applies to virtually any system undergoing small oscillations about a position of stable equilibrium. For example, it could apply to the oscillations of the atoms of a diatomic molecule (such as H_2 or HCl) in which the atoms are oscillating about their equilibrium positions. Between the classical turning points $(-A < x < A)$, the total energy is greater than the potential energy, and the Schrödinger equation can be written

$$\frac{d^2\psi(x)}{dx^2} = -k^2\psi(x) \qquad -A < x < A \qquad 35\text{-}20$$

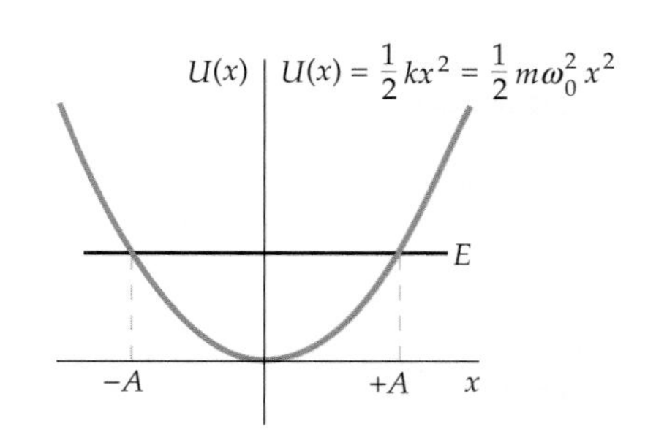

FIGURE 35-6 Harmonic oscillator potential.

where $k^2 = (2m/\hbar^2)[E - U(x)]$ now depends on x. The solutions of this equation are no longer simple sine or cosine functions because the wave number $k = 2\pi/\lambda$ now varies with x; but because $d^2\psi/dx^2$ and ψ have opposite signs throughout the region $-A < x < A$, ψ will always curve toward the axis and the solutions will oscillate.

Outside the classical turning points $(|x| > A)$, the potential energy is greater than the total energy and the Schrödinger equation is similar to Equation 35-17:

$$\frac{d^2\psi(x)}{dx^2} - \alpha^2\psi(x) = 0 \qquad |x| > A \qquad 35\text{-}21$$

except that here $\alpha^2 = (2m/\hbar^2)[U(x) - E] > 0$, where $U(x) > E$, depends on x. For $|x| > A$, $d^2\psi/dx^2$ and ψ have the same sign, so ψ will curve away from the axis and there will be only certain values of E for which solutions exist that approach zero as x approaches infinity.

For the harmonic oscillator potential energy function, the Schrödinger equation is

$$-\frac{\hbar^2}{2m}\frac{d^2\psi(x)}{dx^2} + \tfrac{1}{2}m\omega_0^2x^2\psi(x) = E\psi(x) \qquad 35\text{-}22$$

WAVE FUNCTIONS AND ENERGY LEVELS

Rather than pursue a general solution to the Schrödinger equation for this system, we simply present the solution for the ground state and the first excited state.

The ground-state wave function $\psi_0(x)$ is found to be a Gaussian function centered at the origin:

$$\psi_0(x) = A_0e^{-ax^2} \qquad 35\text{-}23$$

where A_0 and a are positive constants. This wave function and the wave function for the first excited state are shown in Figure 35-7.

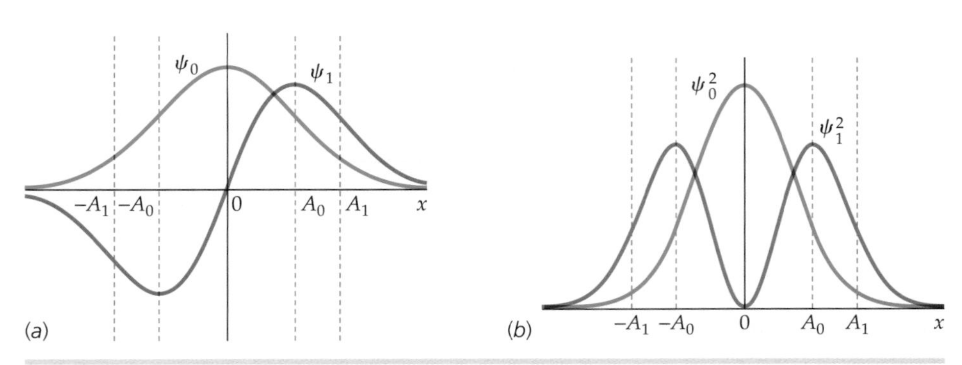

FIGURE 35-7 (*a*) The ground-state ψ_0 and first excited state ψ_1 wave functions for the harmonic oscillator potential. Classically, the motion of a harmonic oscillator with the ground-state energy E_0 would be restricted to the region $-A_0 \leq x \leq +A_0$ and the motion of a harmonic oscillator with the first excited state energy E_1 would be restricted to the region $-A_1 \leq x \leq +A_1$. (*b*) The ground-state ψ_0^2 and first excited state ψ_1^2 probability density functions for the harmonic oscillator potential.

Example 35-1 Verifying the Ground-State Wave Function

Verify that $\psi_0(x) = A_0 e^{-ax^2}$, where A_0 and a are positive constants, is a solution of the Schrödinger equation for the harmonic oscillator.

PICTURE We calculate the second derivative of ψ_0 with respect to x and substitute into Equation 35-22. Because this expression is the ground-state wave function, we write E_0 for the energy E.

SOLVE

1. Compute $d\psi_0/dx$:

$$\frac{d\psi_0(x)}{dx} = \frac{d}{dx}(A_0 e^{-ax^2}) = -2axA_0 e^{-ax^2}$$

2. Compute $d^2\psi_0/dx^2$:

$$\frac{d^2\psi_0(x)}{dx^2} = -2aA_0 e^{-ax^2} + 4a^2x^2A_0 e^{-ax^2}$$

$$= (4a^2x^2 - 2a)A_0 e^{-ax^2}$$

3. Substitute into the Schrödinger equation (Equation 35-22):

$$-\frac{\hbar^2}{2m}\frac{d^2\psi(x)}{dx^2} + \frac{1}{2}m\omega_0^2x^2\psi(x) = E\psi(x)$$

$$-\frac{\hbar^2}{2m}(4a^2x^2 - 2a)A_0 e^{-ax^2} + \frac{1}{2}m\omega_0^2x^2A_0 e^{-ax^2} = E_0A_0 e^{-ax^2}$$

4. Cancel the common factor $A_0 e^{-ax^2}$ and show the result in standard polynomial form:

$$-\frac{\hbar^2}{2m}(4a^2x^2 - 2a) + \frac{1}{2}m\omega_0^2x^2 = E_0$$

so

$$\left(\frac{1}{2}m\omega_0^2 - \frac{2\hbar^2a^2}{m}\right)x^2 + \left(\frac{\hbar^2a}{m} - E_0\right) = 0$$

5. The equation in step 4 must hold for all x. Set $x = 0$ and solve for E_0:

$$0 + \left(\frac{\hbar^2a}{m} - E_0\right) = 0$$

so

$$E_0 = \frac{\hbar^2a}{m}$$

6. Substitute this result for E_0 into the equation in step 4 and simplify:

$$\left(-\frac{2\hbar^2a^2}{m} + \frac{1}{2}m\omega_0^2\right)x^2 + 0 = 0$$

7. It follows that the coefficient of x^2 must equal zero:

$$-\frac{2\hbar^2a^2}{m} + \frac{1}{2}m\omega_0^2 = 0$$

8. Solve for a:

$$a = \frac{m\omega_0}{2\hbar}$$

9. Substitute this result into the equation for E_0 in step 5:

$$E_0 = \frac{\hbar^2a}{m} = \frac{1}{2}\hbar\omega_0$$

We have shown that the given function, $\psi_0(x) = A_0e^{-ax^2}$, satisfies the Schrödinger equation for any value of A_0, as long as the energy is given by $E_0 = \frac{1}{2}\hbar\omega_0$.

CHECK Planck's constant has units of joules multiplied by seconds, and angular frequency has units of reciprocal seconds, so the step-9 expression $\frac{1}{2}\hbar\omega_0$ has the dimensions of energy, as expected.

TAKING IT FURTHER The step-4 equation is a polynomial that is equal to zero. A theorem that would have simplified the solution is "If a polynomial is equal to zero over a continuous range of values of x, then each of the polynomial coefficients is equal to zero. For example, if $Ax^3 + Bx^2 + Cx + D = 0$ on the interval $1 < x < 2$, then $A = B = C = D = 0$."

We see from this example that the ground-state energy is given by

$$E_0 = \frac{\hbar^2 a}{m} = \frac{1}{2}\hbar\omega_0 \qquad 35\text{-}24$$

The first excited state has a node in the center of the potential well, just as with the particle in a box.* The wave function $\psi_1(x)$ is

$$\psi_1(x) = A_1 x e^{-ax^2} \qquad 35\text{-}25$$

where $a = \frac{1}{2}m\omega_0/\hbar$, as in Example 35-1. This function is also shown in Figure 35-7. Substituting $\psi_1(x)$ into the Schrödinger equation, as was done for $\psi_0(x)$ in Example 35-1, yields the energy of the first excited state,

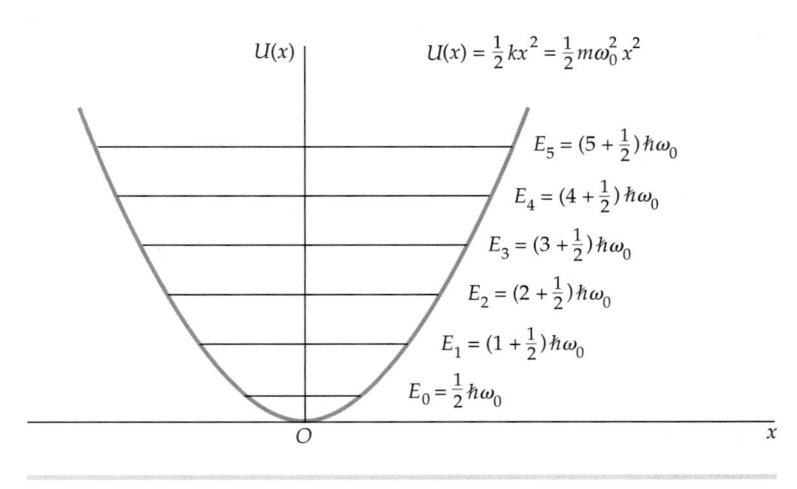

FIGURE 35-8 Energy levels in the harmonic oscillator potential.

$$E_1 = \tfrac{3}{2}\hbar\omega_0$$

In general, the energy of the nth excited state of the harmonic oscillator is

$$E_n = \left(n + \tfrac{1}{2}\right)\hbar\omega_0 \qquad n = 0, 1, 2, \ldots \qquad 35\text{-}26$$

as indicated in Figure 35-8. The fact that the energy levels are evenly spaced by the amount $\hbar\omega_0$ is a peculiarity of the harmonic oscillator potential. As we saw in Chapter 34, the energy levels for a particle in a box, or for the hydrogen atom, are not evenly spaced. The precise spacing of energy levels is closely tied to the particular form of the potential energy function.

35-4 REFLECTION AND TRANSMISSION OF ELECTRON WAVES: BARRIER PENETRATION

In Sections 35-2 and 35-3, we were concerned with bound-state problems in which the potential energy is larger than the total energy for large values of $|x|$. In this section, we consider some simple examples of unbound states for which E is greater than $U(x)$. For these problems, $d^2\psi/dx^2$ and ψ have opposite signs, so $\psi(x)$ curves toward the axis and does not become infinite as x approaches either $+\infty$ or $-\infty$.

STEP POTENTIAL

Consider a particle of energy E moving in a region in which the potential energy is the step function

$$U(x) = \begin{cases} 0 & x < 0 \\ U_0 & x > 0 \end{cases}$$

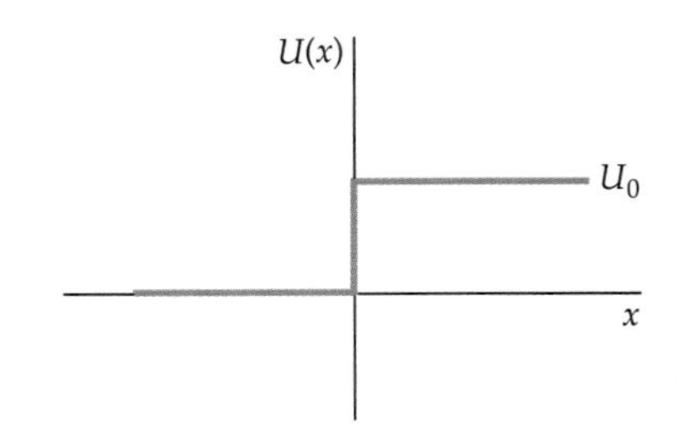

FIGURE 35-9 Step potential. A classical particle that is incident from the left and has total energy $E > U_0$ is always transmitted. The change in potential energy at $x = 0$ merely provides an impulsive force that reduces the speed of the particle. A wave incident from the left is partially transmitted and partially reflected because the wavelength changes abruptly at $x = 0$.

as shown in Figure 35-9. We are interested in what happens when a particle moving from left to right encounters the step.

The classical answer is simple. To the left of the step, the particle moves with a speed $v = \sqrt{2E/m}$. At $x = 0$, an impulsive force acts on the particle. If the initial energy E is less than U_0, the particle will be turned around and will then move to the left at its original speed; that is, the particle will be reflected by the step.

* Each higher-energy state has one additional node in the wave function.

If E is greater than U_0, the particle will continue to move to the right but with reduced speed given by $v = \sqrt{2(E - U_0)/m}$. We can picture this classical problem as a ball rolling along a level surface and coming to a steep hill of height h given by $mgh = U_0$. If the initial kinetic energy of the ball is less than mgh, the ball will roll part way up the hill and then back down and to the left along the lower surface at its original speed. If E is greater than mgh, the ball will roll up the hill and proceed to the right at a lesser speed.

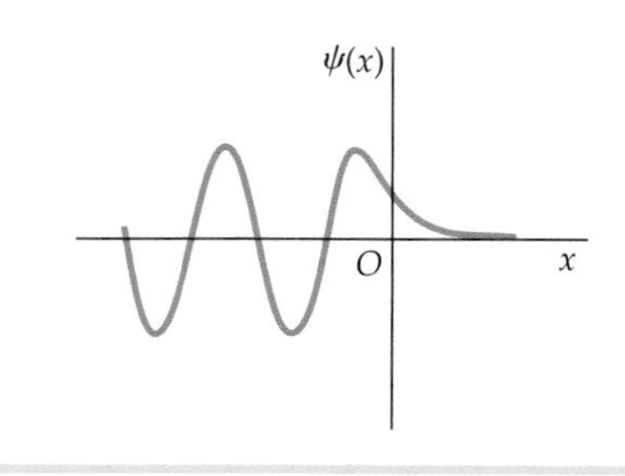

FIGURE 35-10 When the total energy E is less than U_0, the wave function penetrates slightly into the region $x > 0$. However, the probability of reflection for this case is 1, so no energy is transmitted.

The quantum-mechanical result is similar when E is less than U_0. Figure 35-10 shows the wave function for the case $E < U_0$. The wave function does not go to zero at $x = 0$ but rather decays exponentially, like the wave function for the bound state in a finite square-well problem. The wave penetrates slightly into the classically forbidden region $x > 0$, but it is eventually completely reflected. This problem is somewhat similar to that of total internal reflection in optics.

For $E > U_0$, the quantum-mechanical result differs markedly from the classical result. At $x = 0$, the wavelength changes abruptly from $\lambda_1 = h/p_1 = h/\sqrt{2mE}$ to $\lambda_2 = h/p_2 = h/\sqrt{2m(E - U_0)}$. We know from our study of waves that when the wavelength changes suddenly, part of the wave is reflected and part of the wave is transmitted. Because the motion of an electron (or other particle) is governed by a wave equation, the electron sometimes will be transmitted and sometimes will be reflected. The probabilities of reflection and transmission can be calculated by solving the Schrödinger equation in each region of space and comparing the amplitudes of the transmitted waves and reflected waves with the amplitudes of the incident wave. These calculations and their results are similar to finding the fraction of light reflected from an air–glass interface. If R is the probability of reflection, called the reflection coefficient, this calculation gives

$$R = \frac{(k_1 - k_2)^2}{(k_1 + k_2)^2} \qquad 35\text{-}27$$

where k_1 is the wave number for the incident wave and k_2 is the wave number for the transmitted wave. This result is the same as the result in optics for the reflection of light at normal incidence from the boundary between two media having different indexes of refraction n (Equation 31-17). The probability of transmission T, called the **transmission coefficient,** can be calculated from the reflection coefficient, because the probability of transmission plus the probability of reflection must equal 1:

$$T + R = 1 \qquad 35\text{-}28$$

Example 35-2 Reflection and Transmission at a Step Barrier

A particle that has kinetic energy E_0 and is traveling in a region in which the potential energy is zero is incident on a potential-energy barrier of height $U_0 = 0.20E_0$. Find the probability that the particle will be reflected.

PICTURE We need to calculate the wave numbers k_1 and k_2 and use them to calculate the reflection coefficient R from Equation 35-27. The wave numbers are related to the momentum by the de Broglie relation $p = h/\lambda$ (Equation 34-13), where $k = 2\pi/\lambda$. Combining these two equations gives $p = \hbar k$. Thus, the kinetic energy K is related to the wave number by $K = \frac{1}{2}p^2/m = \frac{1}{2}\hbar^2k^2/m$.

SOLVE

1. The probability of reflection is the reflection coefficient:

$$R = \frac{(k_2 - k_1)^2}{(k_1 + k_2)^2}$$

2. Calculate k_1 from the initial kinetic energy E_0:

$$E_0 = \frac{\hbar^2 k_1^2}{2m}$$

$$k_1 = \sqrt{2mE_0/\hbar^2}$$

3. Relate the final kinetic energy K_2 to the initial kinetic energy E_0 and the potential energy U_0 in the region $x > 0$:

$$K_2 = E_0 - U_0 = E_0 - 0.2E_0 = 0.8E_0$$

4. Relate k_2 to the final kinetic energy K_2 and solve for k_2:

$$K_2 = \frac{\hbar k_2^2}{2m}$$

so

$$k_2 = \sqrt{2mK_2/\hbar^2} = \sqrt{2m(0.8E_0)/\hbar^2}$$
$$= \sqrt{0.80}\sqrt{2mE_0/\hbar^2}$$

5. Substitute these values into Equation 35-27 to calculate R:

$$R = \frac{(k_1 - k_2)^2}{(k_1 + k_2)^2} = \left(\frac{1 - \sqrt{0.80}}{1 + \sqrt{0.80}}\right)^2 = \boxed{0.0031}$$

CHECK Classically, the particle would not be reflected by such a low barrier. The step-5 result gives a probability of 0.31 percent that the particle will be reflected. Such a low probability approaches being in agreement with our classical expectations.

TAKING IT FURTHER The probability of reflection is only 0.31 percent. This probability is small because the barrier height reduces the kinetic energy by only 20 percent. Because k is proportional to the square root of the kinetic energy, the wave number and therefore the wavelength is changed by only 10 percent.

PRACTICE PROBLEM 35-1 Express the index of refraction n of light in terms of the wave number k and the frequency ω, and show that the expression $(n_1 - n_2)^2/(n_1 + n_2)^2$ (Equation 31-7) for the reflection coefficient of light at normal incidence is the same as Equation 35-27. *Hint: Express the index of refraction n of light in terms of the wave number k and the angular frequency ω.*

In quantum mechanics, a localized particle is represented by a wave packet, which has a maximum at the most probable position of the particle. Figure 35-11 shows a wave packet representing a particle of energy E incident on a step potential of height U_0, which is less than E. After the encounter, there are two wave packets. The relative heights of the transmitted packet and reflected packet indicate the relative probabilities of transmission and reflection. For the situation shown here, E is much greater than U_0, and the probability of transmission is much greater than that of reflection.

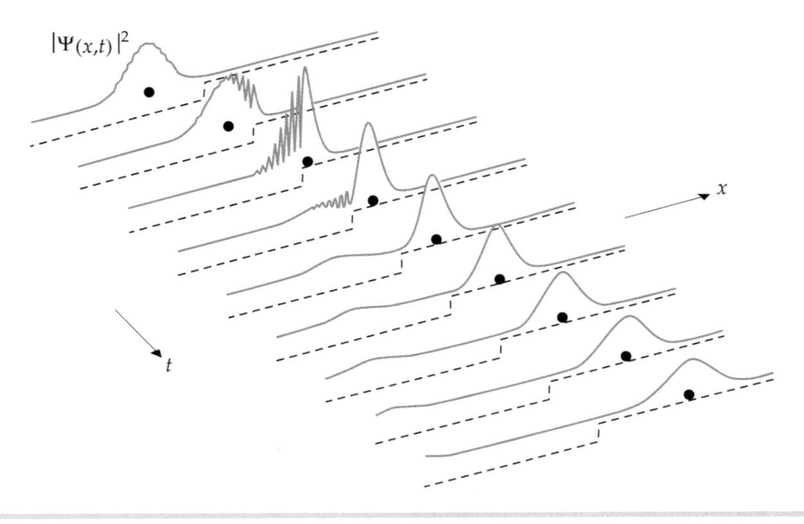

FIGURE 35-11 Time development of a one-dimensional wave packet representing a particle incident on a step potential for $E > U_0$. The position of a classical particle is indicated by the dot. Note that part of the packet is transmitted and part is reflected.

BARRIER PENETRATION

Figure 35-12*a* shows a rectangular potential-energy barrier of height U_0 and width a given by

$$U(x) = \begin{cases} 0 & x < 0 \\ U_0 & 0 < x < a \\ 0 & x > a \end{cases}$$

We consider a particle of energy E, which is slightly less than U_0, that is incident on the barrier from the left. Classically, the particle would always be reflected. However, a wave incident from the left does not decrease immediately to zero at the barrier, but it will instead decay exponentially in the classically forbidden region $0 < x < a$. On reaching the far wall of the barrier ($x = a$), the wave function must join smoothly to a sinusoidal wave function to the right of the barrier, as shown in Figure 35-12*b* This implies that there is some probability of the particle (which is represented by the wave function) being found on the far side of the barrier even though, classically, it should never pass through the barrier. For the case in which the quantity αa [where $\alpha^2 = 2m(U_0 - E)/\hbar^2$] is much greater than 1 the transmission coefficient T is equal to $e^{-2\alpha a}$:

> ! Do not think it might be possible to detect a particle in the classically forbidden region. It cannot. A proof shows this claim to be a consequence of the uncertainty principle.

$$T = e^{-2\alpha a} \qquad \alpha a \gg 1 \qquad \text{35-29}$$

TRANSMISSION THROUGH A BARRIER

The probability of penetration of the barrier thus decreases exponentially with both the barrier thickness a and the square root of the relative barrier height ($U_0 - E$). This phenomenon, the penetrating of a classically forbidden region, is called **quantum tunneling.**

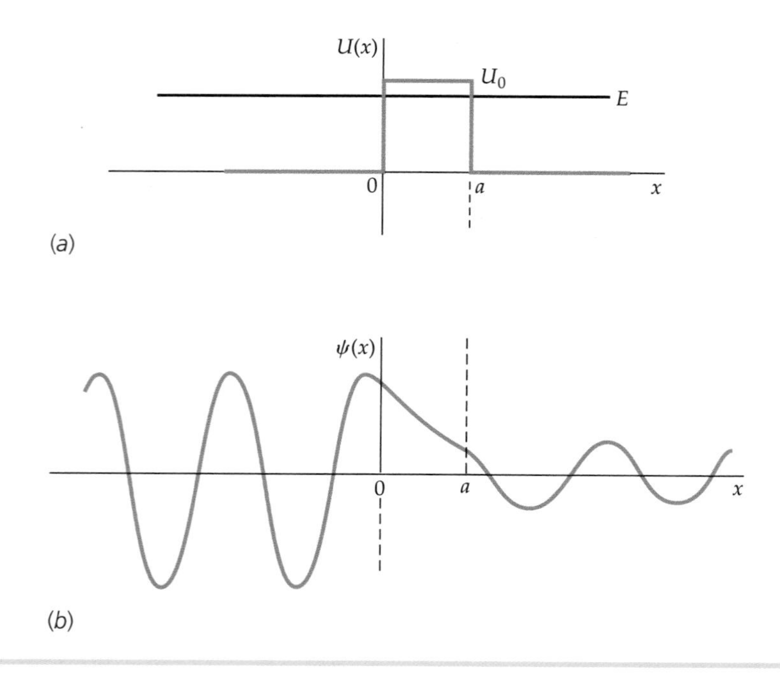

FIGURE 35-12 (*a*) A rectangular potential-energy barrier. (*b*) The penetration of the barrier by a wave that has a total energy less than the barrier energy. Part of the wave is transmitted by the barrier even though, classically, the particle cannot enter the region $0 < x < a$ in which the potential energy is greater than the total energy. To the left of the barrier, there is both an incident wave and a reflected wave. These waves form a resultant wave so that ψ is a superposition of a standing wave and a traveling wave (traveling toward the barrier). Only the transmitted wave exists in the region $x > a$, and it is traveling away from the barrier.

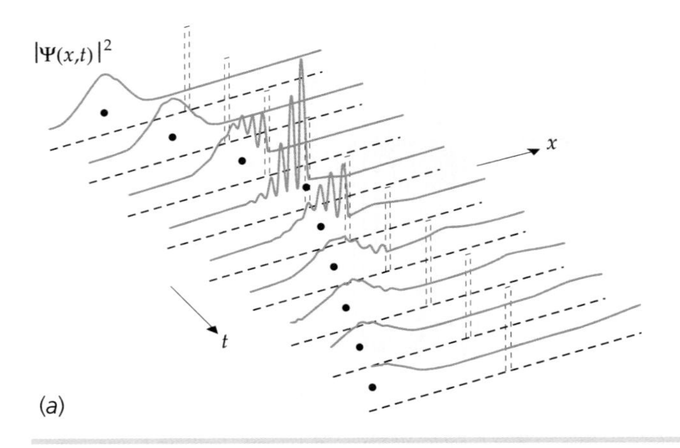

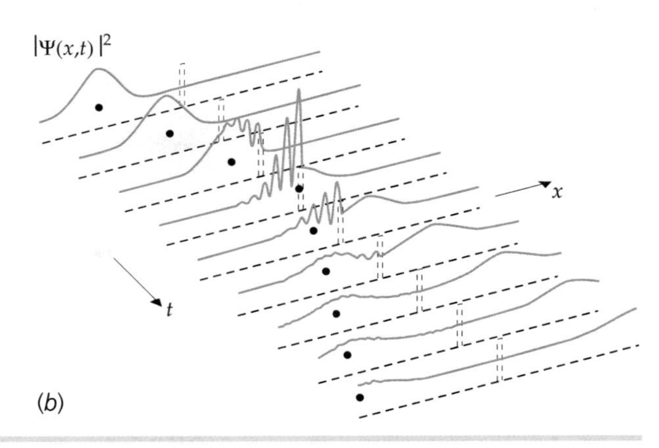

FIGURE 35-13 Barrier penetration. (*a*) The same particle incident on a barrier of height much greater than the energy of the particle. A very small part of the packet tunnels through the barrier. In both drawings, the position of a classical particle is indicated by a dot. (*b*) A wave packet representing a particle incident on a barrier of height just slightly greater than the energy of the particle. For this particular choice of energies, the probability of transmission is approximately equal to the probability of reflection, as indicated by the relative sizes of the transmitted and reflected packets.

Figure 35-13*a* shows a wave packet incident on a potential-energy barrier of height U_0 that is considerably greater than the energy of the particle. The probability of penetration is very small, as indicated by the relative sizes of the reflected and transmitted packets. In Figure 35-13*b*, the barrier is just slightly greater than the energy of the particle. In this case, the probability of penetration is about the same as the probability of reflection. Figure 35-14 shows a particle incident on two potential-energy barriers of height just slightly greater than the energy of the particle.

As we have mentioned, the penetration of a barrier is not unique to quantum mechanics. When light is totally reflected from a glass–air interface, the light wave can penetrate the air barrier if a second piece of glass is brought within a few

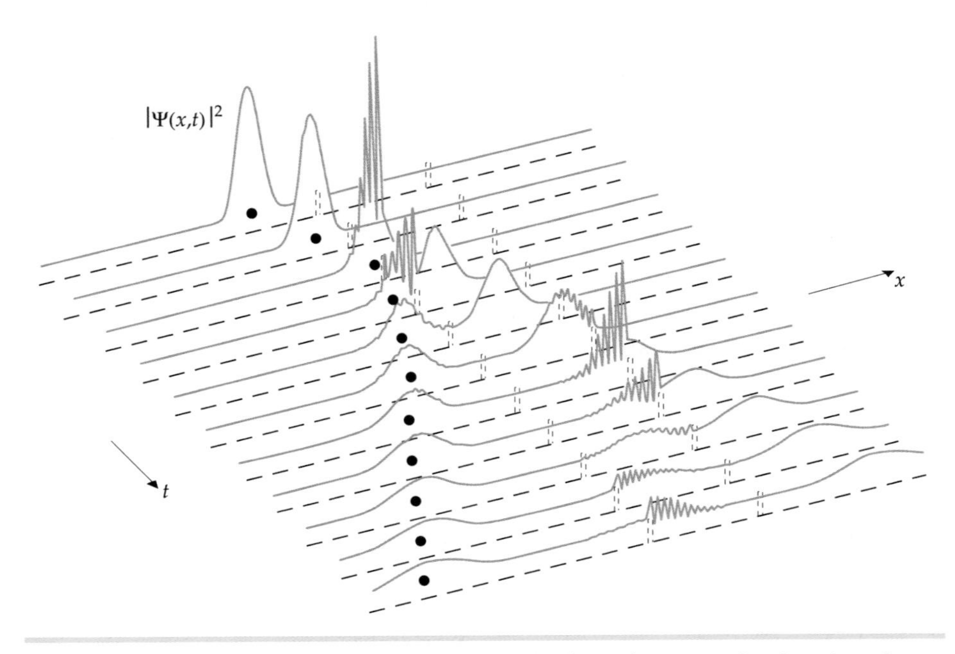

FIGURE 35-14 A wave packet representing a particle incident on two barriers. At each encounter, part of the packet is transmitted and part reflected, resulting in part of the packet being trapped between the barriers for some time.

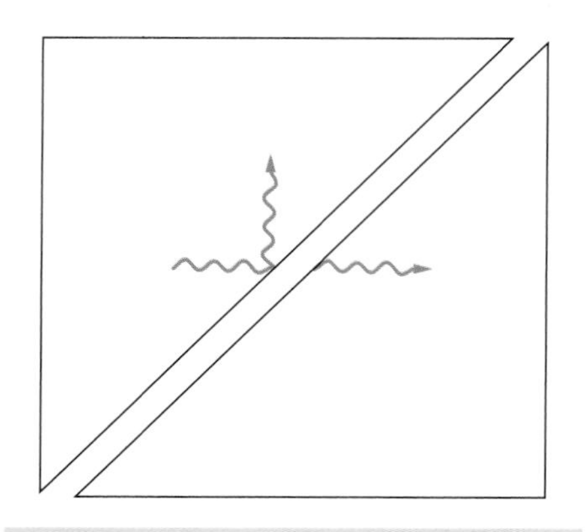

FIGURE 35-15 The penetration of an optical barrier. If the second prism is close enough to the first, part of the wave penetrates the air barrier even when the angle of incidence in the first prism is greater than the critical angle.

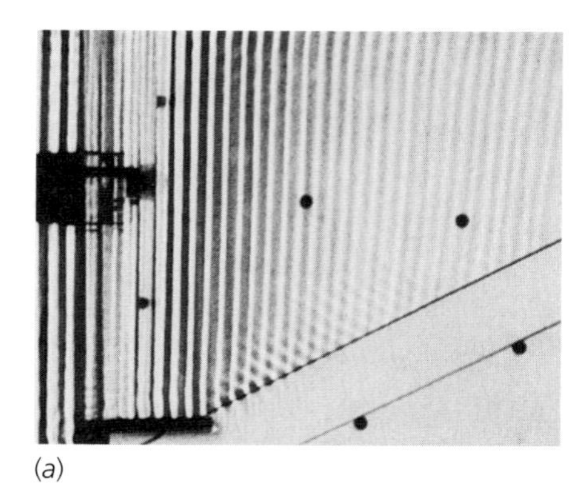

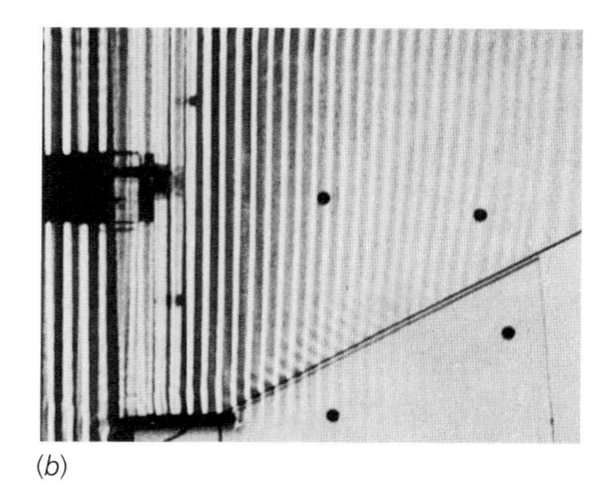

FIGURE 35-16 The penetration of a barrier by water waves in a ripple tank. In Figure 35-16*a*, the waves are totally reflected from a gap of deeper water. When the gap is very narrow, as in Figure 35-16*b*, a transmitted wave appears. The dark circles are spacers that are used to support the prisms from below. *(Education Development Center.)*

wavelengths of the first. This effect can be demonstrated with a laser beam and two 45° prisms (Figure 35-15). Similarly, water waves in a ripple tank can penetrate a gap of deep water (Figure 35-16).

The theory of barrier penetration was used by George Gamow in 1928 to explain the enormous variation in the half-lives for α decay of radioactive nuclei. (Alpha particles are emitted from atoms during radioactive decay and consist of two protons and two neutrons tightly bound together.) In general, the smaller the energy of the emitted α particle, the longer the half-life of the particle is. The energies of α particles from natural radioactive sources range from approximately 4 MeV to 7 MeV, whereas the half-lives range from approximately 10^{-5} seconds to 10^{10} years. Gamow represented a radioactive nucleus by a potential well of finite depth containing an α particle, as shown in Figure 35-17. Without knowing very much about the nuclear force that is exerted by the nucleus on the particle, Gamow represented it by a square well. Just outside the well, the α particle that has a charge of $+2e$ is repelled by the nucleus that has a charge $+Ze$, where Ze is the remaining nuclear charge. This force is represented by the Coulomb potential energy $+k(2e)(Ze)/r$. The energy E is the measured kinetic energy of the emitted α particle, because when it is far from the nucleus its potential energy is zero. After the α particle is formed from the radioactive nucleus, it bounces back and forth inside the nucleus, hitting the barrier at the nuclear radius R. Each time the α particle strikes the barrier, some small probability exists of the particle penetrating the barrier and appearing outside the nucleus. We can see from Figure 35-17 that a small increase in E reduces the relative height of the barrier $U - E$ and also the barrier's thickness. Because the probability of penetration is so sensitive to the barrier thickness and relative height, a small increase in E leads to a large increase in the probability of transmission and therefore to a shorter lifetime. Gamow was able to derive an expression for the half-life as a function of E that is in excellent agreement with experimental results.

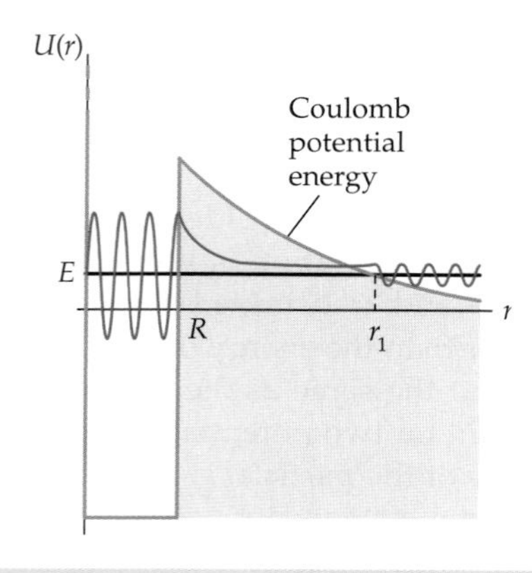

FIGURE 35-17 Model of a potential energy function for an α particle of a radioactive nucleus. The strong attractive nuclear force when r is approximately equal to the nuclear radius R can be approximately described by the potential well shown. The nuclear force is negligible outside the nucleus, and the potential there is given by Coulomb's law, $U(r) = +k(2e)(Ze)/r$, where Ze is the nuclear charge and $2e$ is the charge of the α particle. The wave function of the alpha particle, shown in red, is placed on the graph.

In the **scanning tunneling microscope (STM),** developed in the 1980s, a thin space between the surface of a sample and the tip of a tiny needle-like probe (Figure 35-18) acts as a potential-energy barrier to electrons bound in the sample. (The height of the barrier is the work function of the surface.) A small voltage applied between the probe and the sample causes the electrons to *tunnel* through the vacuum separating the tip of the probe and the surface of the sample if the surfaces are close enough together. The tunneling current is extremely sensitive to the size of the gap between the probe and sample. A constant tunneling current is maintained as the probe scans (travels along) the surface by a feedback

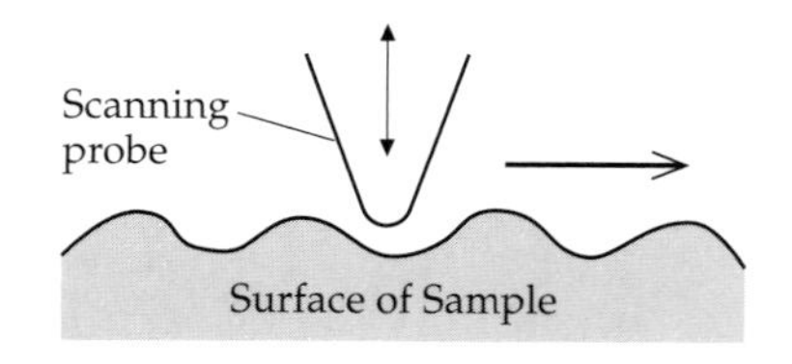

FIGURE 35-18 The tiny probe of a scanning tunneling microscope travels along the surface of a sample. A constant potential difference is maintained between the probe and the sample and electrons tunnel through the potential-energy barrier at the surface. A feedback mechanism maintains a constant tunneling current by moving the probe up and down as it travels along the surface.

mechanism that moves the probe up and down (farther from or closer to the surface). The surface of the sample is mapped out by the tracking of the motions of the probe. In this way, the surface features of the sample can be measured with a resolution of the order of the size of an atom.

35-5 THE SCHRÖDINGER EQUATION IN THREE DIMENSIONS

The one-dimensional time-independent Schrödinger equation is easily extended to three dimensions. In rectangular coordinates, it is

$$-\frac{\hbar^2}{2m}\left(\frac{\partial^2\psi}{\partial x^2}+\frac{\partial^2\psi}{\partial y^2}+\frac{\partial^2\psi}{\partial z^2}\right)+U\psi=E\psi \qquad \text{35-30}$$

where the wave function ψ and the potential energy U are generally functions of all three coordinates, x, y, and z. To illustrate some of the features of problems in three dimensions, we consider a particle in a three-dimensional infinite square well given by $U(x, y, z) = 0$ for $0 < x < L$, $0 < y < L$, and $0 < z < L$. Outside this cubical region, $U(x, y, z) = \infty$. For this problem, the wave function must be zero at the edges of the well.

There are standard methods in partial differential equations for solving Equation 35-30. We can guess the form of the solution from our knowledge of probability. For a one-dimensional box along the x axis, we have found the probability that a particle is in the region between x and $x + dx$ to be $A_1^2 \sin^2(k_1x)\,dx$ (from Equation 35-10), where A_1 is a normalization constant and $k_1 = n\pi/L$ is the wave number. Similarly, for a box along the y axis, the probability of a particle being in a region between y and $y + dy$ is $A_2^2\sin^2(k_2y)dy$. The probability of two independent events occurring is the product of the probabilities of each event occurring.* So the probability of a particle being in region between x and $x + dx$ *and* in region between y and $y + dy$ is $A_1^2\sin^2(k_1x)\,dx\,A_2^2\sin^2(k_2y)\,dy = A_1^2\sin^2(k_1x)\,A_2^2\sin^2(k_2y)\,dx\,dy$. The probability of a particle being in the region between x and $x + dx$, y and $y + dy$, and z and $z + dz$ is $\psi^2(x, y, z)\,dx\,dy\,dz$, where $\psi(x, y, z)$ is the solution of Equation 35-30. This solution is of the form

$$\psi(x, y, z) = A \sin(k_1x) \sin(k_2y) \sin(k_3z) \qquad \text{35-31}$$

where the constant A is determined by normalization. Inserting this solution into Equation 35-30, we obtain for the energy

$$E = \frac{\hbar^2}{2m}(k_1^2 + k_2^2 + k_3^2)$$

which is equivalent to $E = \frac{1}{2}(p_x^2 + p_y^2 + p_z^2)/m$, where $p_x = \hbar k_1$, $p_y = \hbar k_2$, and $p_z = \hbar k_3$. The wave function (Equation 35-31) will be zero at $x = L$ if $k_1 = n_1\pi/L$, where n_1 is an integer. Similarly, the wave function will be zero at $y = L$ if $k_2 = n_2\pi/L$, and the wave function will be zero at $z = L$ if $k_3 = n_3\pi/L$. (It is also zero at $x = 0$, $y = 0$, and $z = 0$.) The energy is thus quantized to the values

$$E_{n_1n_2n_3} = \frac{\hbar^2\pi^2}{2mL^2}(n_1^2 + n_2^2 + n_3^2) = E_1(n_1^2 + n_2^2 + n_3^2) \qquad \text{35-32}$$

where n_1, n_2, and n_3 are positive integers and $E_1 = \hbar^2\pi^2/(2mL^2)$ is the ground-state energy of the one-dimensional well. Note that the energy and wave function are characterized by three quantum numbers, each arising from the boundary conditions for one of the coordinates x, y, and z.

---

* For example, if you throw two dice, the probability of the first die coming up 6 is 1/6 and the probability of the second die coming up an odd number is 1/2. The probability of the first die coming up 6 *and* the second die coming up an odd number is (1/6)(1/2) = 1/12.

The lowest energy state (the ground state) for the cubical well occurs when $n_1 = n_2 = n_3 = 1$ and has the value

$$E_{111} = \frac{3\hbar^2\pi^2}{2mL^2} = 3E_1$$

The first excited energy level can be obtained in three different ways: $n_1 = 2$, $n_2 = n_3 = 1$; $n_2 = 2$, $n_1 = n_3 = 1$; or $n_3 = 2$, $n_1 = n_2 = 1$. Each has a different wave function. For example, the wave function for $n_1 = 2$ and $n_2 = n_3 = 1$ is

$$\psi_{211} = A \sin\frac{2\pi x}{L} \sin\frac{\pi y}{L} \sin\frac{\pi z}{L} \qquad \text{35-33}$$

There are thus three different quantum states as described by the three different wave functions corresponding to the same energy level. An energy level with which more than one wave function is associated is said to be **degenerate.** In this case, there is threefold degeneracy. Degeneracy is related to the spatial symmetry of the system. If, for example, we consider a noncubic well, where $U = 0$ for $0 < x < L_1$, $0 < y < L_2$, and $0 < z < L_3$, the boundary conditions at the edges would lead to the quantum conditions $k_1L_1 = n_1\pi$, $k_2L_2 = n_2\pi$, and $k_3L_3 = n_3\pi$, and the total energy would be

$$E_{n_1n_2n_3} = \frac{\hbar^2\pi^2}{2m}\left(\frac{n_1^2}{L_1^2} + \frac{n_2^2}{L_2^2} + \frac{n_3^2}{L_3^2}\right) \qquad \text{35-34}$$

These energy levels are not degenerate if L_1, L_2, and L_3 are all different. Figure 35-19 shows the energy levels for the ground state and first two excited states for an infinite cubic well in which the excited states are degenerate and for a noncubic infinite well in which L_1, L_2, and L_3 are all slightly different so that the excited levels are slightly split apart and the degeneracy is removed. The ground state is the state where the quantum numbers n_1, n_2, and n_3 all equal 1. None of the three quantum numbers can be zero. If any one of n_1, n_2, and n_3 were zero, the corresponding wave number k would also equal zero and the corresponding wave function (Equation 35-31) would equal zero for all values of x, y, and z.

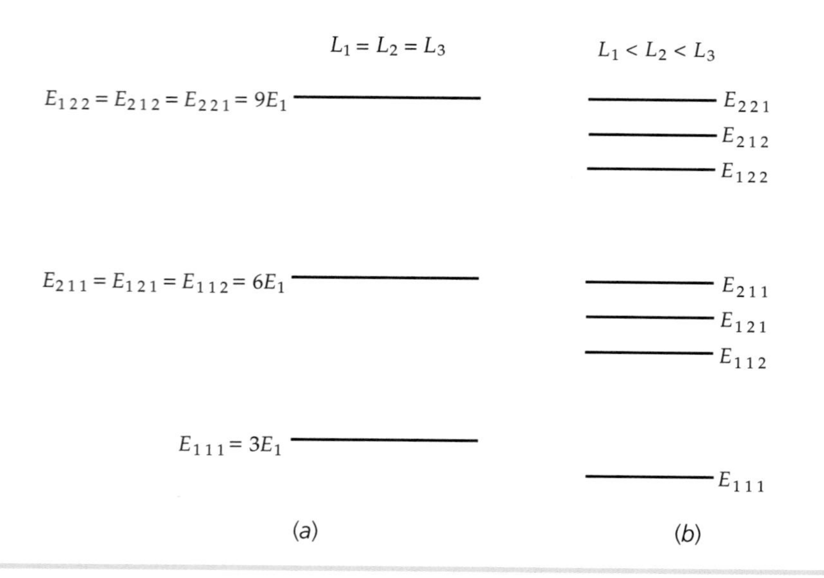

FIGURE 35-19 Energy-level diagrams for (*a*) a cubic infinite well and (*b*) a noncubic infinite well. In Figure 35-19*a* the energy levels are degenerate; that is, there are two or more wave functions having the same energy. The degeneracy is removed when the symmetry of the potential is removed, as in Figure 35-19*b*.

Example 35-3 Energy Levels for a Particle in a Three-Dimensional Box

A particle is in a three-dimensional box where $L_3 = L_2 = 2L_1$. Give the quantum numbers n_1, n_2, and n_3 that correspond to the state(s) in each of the seven lowest energy levels of this box.

PICTURE We can use Equation 35-34 to write the energy in terms of L_1 and the quantum numbers n_1, n_2, and n_3. Then we can find by inspection the values of the quantum numbers that give the lowest energies.

SOLVE

1. The energy of a level is given by Equation 35-34:

$$E_{n_1n_2n_3} = \frac{\hbar^2\pi^2}{2m}\left(\frac{n_1^2}{L_1^2} + \frac{n_2^2}{L_2^2} + \frac{n_3^2}{L_3^2}\right)$$

2. Factor out $1/L_1^2$:

$$E_{n_1n_2n_3} = \frac{\hbar^2\pi^2}{2m}\left(\frac{n_1^2}{L_1^2} + \frac{n_2^2}{4L_1^2} + \frac{n_3^2}{4L_1^2}\right) = \frac{\hbar^2\pi^2}{8mL_1^2}(4n_1^2 + n_2^2 + n_3^2)$$

3. The lowest energy is E_{111}:

$$E_{111} = E_1(4 \cdot 1^2 + 1^2 + 1^2) = \boxed{6E_1} \quad \text{(1st)}$$

where $E_1 = \hbar^2\pi^2/8mL_1^2$.

4. The energy increases the least when we increase n_2 or n_3. Trying various values of the quantum numbers:

$$E_{121} = E_{112} = E_1(4 \cdot 1^2 + 2^2 + 1^2) = \boxed{9E_1} \quad \text{(2nd)}$$

$$E_{122} = E_1(4 \cdot 1^2 + 2^2 + 2^2) = \boxed{12E_1} \quad \text{(3rd)}$$

$$E_{131} = E_{113} = E_1(4 \cdot 1^2 + 3^2 + 1^2) = \boxed{14E_1} \quad \text{(4th)}$$

$$E_{132} = E_{123} = E_1(4 \cdot 1^2 + 3^2 + 2^2) = \boxed{17E_1} \quad \text{(5th)}$$

$$E_{211} = E_1(4 \cdot 2^2 + 1^2 + 1^2) = \boxed{18E_1} \quad \text{(6th)}$$

$$\left.\begin{array}{l} E_{221} = E_{212} = E_1(4 \cdot 2^2 + 2^2 + 1^2) = \boxed{21E_1} \\ E_{141} = E_{114} = E_1(4 \cdot 1^2 + 4^2 + 1^2) = \boxed{21E_1} \end{array}\right\} \text{(7th)}$$

CHECK Because two of the lengths are equal, degenerate energy levels are expected. Our results meet this expectation.

TAKING IT FURTHER Energies E_{221} and E_{212} are exactly equal because L_2 and L_3 are exactly equal. However, energies E_{221} and E_{141} are exactly equal because L_1 is exactly half of L_2.

PRACTICE PROBLEM 35-2 Find the quantum numbers and energies of the next two energy levels in step 4.

Example 35-4 Wave Functions for a Particle in a Three-Dimensional Box

Try It Yourself

Write the degenerate wave functions for the fourth and fifth excited states (the 5th and 6th levels) of the results in step 4 of Example 35-3.

PICTURE Use $\psi(x, y, z) = A\sin(k_1x)\sin(k_2y)\sin(k_3z)$ (a generalized version of Equation 35-31) with $k_i = n_i\pi/L_i$.

SOLVE

Cover the column to the right and try these on your own before looking at the answers.

Steps	**Answers**
Write the wave functions corresponding to the energies E_{131} and E_{113}	$\psi_{131} = A \sin\frac{\pi x}{L_1} \sin\frac{3\pi y}{2L_1} \sin\frac{\pi z}{2L_1}$ $\psi_{113} = A \sin\frac{\pi x}{L_1} \sin\frac{\pi y}{2L_1} \sin\frac{3\pi z}{2L_1}$

35-6 THE SCHRÖDINGER EQUATION FOR TWO IDENTICAL PARTICLES

Our discussion of quantum mechanics has thus far been limited to situations in which a single particle moves in some force field characterized by a potential energy function U. The most important physical problem of this type is the hydrogen atom, in which a single electron moves in the Coulomb potential of a proton. This problem is actually a two-body problem, because the proton also moves in the field of the electron. However, the motion of the much more massive proton requires only a very small correction to the energy of the atom that is easily made in both classical and quantum mechanics. When we consider more complicated problems, such as the helium atom, we must apply quantum mechanics to two or more electrons moving in an external field. Such problems are complicated, not only by the interaction of the electrons with each other, but also by the fact that the electrons are identical.

The interaction of two electrons with each other is electromagnetic and is essentially the same as the classical interaction of two charged particles. The Schrödinger equation for an atom that has two or more electrons cannot be solved exactly, so approximation methods must be used. This situation is not very different from the situation in which a classical expression describes three or more particles. However, the complications arising from the identity of electrons are purely quantum mechanical and have no classical counterpart. They are due to the fact that it is impossible to keep track of which electron is which. Classically, identical particles can be identified by their positions, which in principle can be determined with unlimited accuracy. This is impossible quantum mechanically because of the uncertainty principle. Figure 35-20 offers a schematic illustration of the problem.

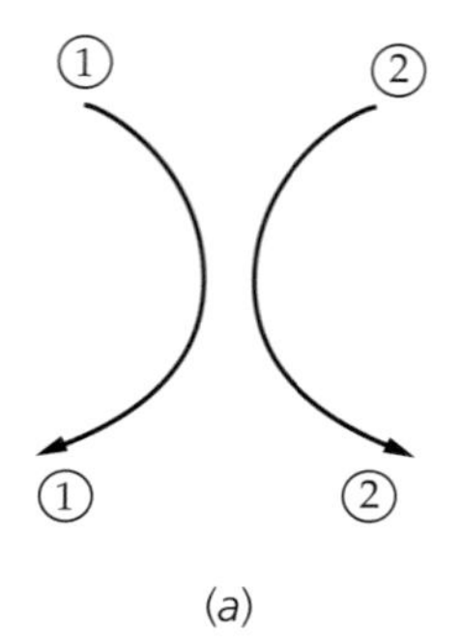

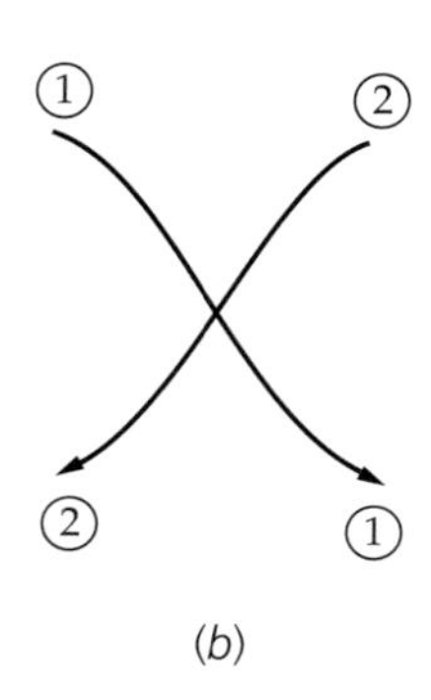

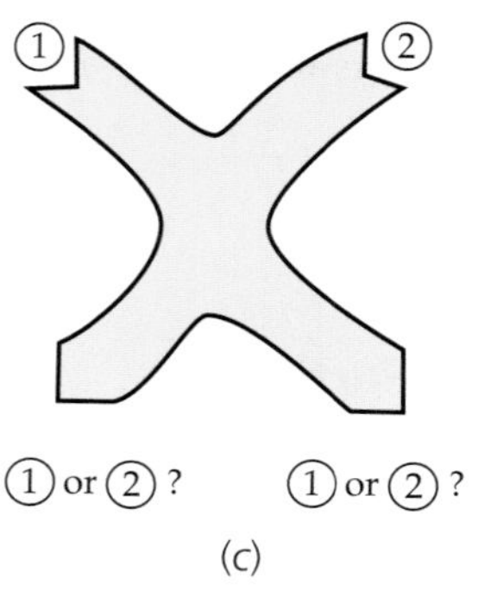

FIGURE 35-20 (*a, b*) Two possible classical electron paths. If electrons were classical particles, they could be distinguished by the paths followed. (*c*) However, because of the quantum-mechanical wave properties of electrons, the paths are spread out, as indicated by the shaded region. It is impossible to distinguish which electron is which after they separate.

The indistinguishability of identical particles has important consequences. For instance, consider the very simple case of two identical, noninteracting particles in a one-dimensional infinite square well. The time-independent Schrödinger equation for two particles, each of mass m, is

$$-\frac{\hbar^2}{2m}\frac{\partial^2\psi(x_1, x_2)}{\partial x_1^2} - \frac{\hbar^2}{2m}\frac{\partial^2\psi(x_1, x_2)}{\partial x_2^2} + U\psi(x_1, x_2) = E\psi(x_1, x_2) \qquad \text{35-35}$$

where x_1 and x_2 are the coordinates of the two particles. If the particles interact, the potential energy U is described by terms that have both x_1 and x_2 and cannot be separated into separate terms having only x_1 or x_2. For example, the electrostatic repulsion of two electrons in one dimension is represented by the potential energy function $ke^2/|x_2 - x_1|$. However, if the particles do not interact (as we are assuming here), we can write $U = U_1(x_1) + U_2(x_2)$. For the infinite square well, we need only solve the Schrödinger equation inside the well where $U = 0$ and require that the wave function be zero at the walls of the well. With $U = 0$, Equation 35-35 looks just like the expression for a particle in a two-dimensional well (Equation 35-30, but no z terms exist and y is replaced by x_2).

Solutions of this equation can be written in the form*

$$\psi_{nm} = \psi_n(x_1)\psi_m(x_2) \qquad \text{35-36}$$

where ψ_n and ψ_m are the single-particle wave functions for a particle in an infinite well and n and m are the quantum numbers of particles 1 and 2, respectively. For example, for $n = 1$ and $m = 2$, the wave function is

$$\psi_{12} = A \sin\frac{\pi x_1}{L} \sin\frac{2\pi x_2}{L} \qquad \text{35-37}$$

The probability of finding particle 1 in the region between $x = x_1$ and $x = x_1 + dx_1$ and particle 2 in the region between $x = x_2$ and $x = x_2 + dx_2$ is $\psi_{nm}^2(x_1, x_2)\,dx_1\,dx_2$, which is just the product of the separate probabilities $\psi_n^2(x_1)\,dx_1$ and $\psi_m^2(x_2)\,dx_2$. However, even though we have labeled the particles 1 and 2, we cannot distinguish which is between x_1 and $x_1 + dx_1$ and which is between x_2 and $x_2 + dx_2$ if they are identical. The mathematical descriptions of identical particles must be the same if we interchange the labels. The probability density $\psi^2(x_1, x_2)$ must therefore be the same as $\psi^2(x_2, x_1)$:

$$\psi^2(x_2, x_1) = \psi^2(x_1, x_2) \qquad \text{35-38}$$

Equation 35-38 is satisfied if $\psi(x_2, x_1)$ is either **symmetric** or **antisymmetric** on the exchange of particles—that is, if either

$$\psi(x_2, x_1) = \psi(x_1, x_2) \quad \text{symmetric} \qquad \text{35-39}$$

or

$$\psi(x_2, x_1) = -\psi(x_1, x_2) \quad \text{antisymmetric} \qquad \text{35-40}$$

Note that the wave functions given by Equations 35-36 and 35-37 are neither symmetric nor antisymmetric. If we interchange x_1 and x_2 in these wave functions, we get a different wave function, which implies that the particles can be distinguished.

* Again, this result can be obtained by solving Equation 35-35, but it also can be understood in terms of our knowledge of probability. The probability of electron 1 being between $x = x_1$ and $x = x_1 + dx_1$ and electron 2 being between $x = x_2$ and $x = x_2 + dx_2$ is the product of the individual probabilities.

We can find symmetric and antisymmetric wave functions that are solutions of the Schrödinger equation by adding or subtracting ψ_{nm} and ψ_{mn}. Adding them, we obtain

$$\psi_S = A'[\psi_n(x_1)\psi_m(x_2) + \psi_n(x_2)\psi_m(x_1)] \quad \text{symmetric} \qquad 35\text{-}41$$

and subtracting them, we obtain

$$\psi_A = A'[\psi_n(x_1)\psi_m(x_2) - \psi_n(x_2)\psi_m(x_1)] \quad \text{antisymmetric} \qquad 35\text{-}42$$

For example, the symmetric and antisymmetric wave functions for the first excited state of two identical particles in an infinite square well would be

$$\psi_S = A'\left(\sin\frac{\pi x_1}{L}\sin\frac{2\pi x_2}{L} + \sin\frac{\pi x_2}{L}\sin\frac{2\pi x_1}{L}\right) \qquad 35\text{-}43$$

and

$$\psi_A = A'\left(\sin\frac{\pi x_1}{L}\sin\frac{2\pi x_2}{L} - \sin\frac{\pi x_2}{L}\sin\frac{2\pi x_1}{L}\right) \qquad 35\text{-}44$$

There is an important difference between antisymmetric and symmetric wave functions. If $n = m$, the antisymmetric wave function is identically zero for all values of x_1 and x_2, whereas the symmetric wave function is not. Thus, if the wave function describing two identical particles is antisymmetric, the quantum numbers n and m of two particles cannot be the same. The idea that no two electrons in an atom can occupy the same quantum state, and thus have the same quantum numbers, was first stated by Wolfgang Pauli in 1925. This idea was soon generalized to include systems other than atoms and particles other than electrons. For example, no two protons of a nucleus can occupy the same quantum state, and no two neutrons of a nucleus can occupy the same quantum state. Electrons, protons, neutrons, neutrinos, and quarks all have a spin quantum number s equal to one-half, and all particles that have half-integer spin are called fermions. The two allowed values for the secondary spin quantum number m_s are plus and minus one half. Pauli's idea is called the **Pauli exclusion principle:**

No two identical fermions can simultaneously occupy the same quantum state.

PAULI EXCLUSION PRINCIPLE

The wave function for two or more identical fermions must be an antisymmetric wave function. Other particles (for example, α particles, deuterons, photons, and mesons) have integer spin and symmetric wave functions. These particles are called **bosons.**

A wave function that is a solution to the multiparticle time-independent wave equation (Equation 35-35) is called a spatial state. A bound system that contains fermions has either one or two identical fermions in each occupied spatial state. However, for a bound system that contain bosons there is no limit to the number of identical bosons in each spatial state.

Summary

1. The Schrödinger equation is a partial differential equation that relates the second space derivative of a wave function to its first time derivative. Wave functions that describe physical situations are solutions of this differential equation.
2. Because a wave function must satisfy the normalization condition, it must be well behaved; this means, among other things, that it must approach zero as $|x|$ approaches infinity. For bound systems such as a particle in a box, a simple harmonic oscillator, or an electron in an atom, this requirement leads to energy quantization.
3. The well-behaved wave functions for bound systems describe standing waves.

TOPIC	RELEVANT EQUATIONS AND REMARKS
1. Time-Independent Schrödinger Equation	$$-\frac{\hbar^2}{2m}\frac{d^2\psi(x)}{dx^2} + U(x)\psi(x) = E\psi(x) \qquad \text{35-4}$$
Allowable solutions	In addition to satisfying the Schrödinger equation, a wave function $\psi(x)$ must be continuous and must have a continuous first derivative $d\psi/dx$.* Because the probability of finding an electron somewhere must be 1, the wave function must obey the normalization condition $$\int_{-\infty}^{\infty} \lvert\psi\rvert^2\,dx = 1$$ This condition implies the boundary condition that ψ must approach 0 as $\lvert x\rvert$ approaches ∞. Such boundary conditions lead to the quantization of energy.
2. Confined Particles	When the total energy E is greater than the potential energy $U(x)$ in some region (the classically allowed region) and less than $U(x)$ outside that region, the wave function ψ oscillates within the classically allowed region and $\lvert\psi\rvert$ decreases exponentially outside that region. The wave function approaches zero as $\lvert x\rvert$ approaches ∞ only for certain values of the total energy E. The energy is thus quantized.
In a finite square well	In a finite well of height U_0, there are only a finite number of allowed energies, and these are slightly less than the corresponding energies in an infinite well.
In the simple harmonic oscillator	For the oscillator with potential energy function $U(x) = \frac{1}{2}m\omega_0^2x^2$, the allowed energies are equally spaced and given by $$E_n = \left(n + \tfrac{1}{2}\right)\hbar\omega_0 \qquad n = 0, 1, 2, \ldots \qquad \text{35-26}$$ The ground-state wave function is given by $$\psi_0(x) = A_0e^{-ax^2} \qquad \text{35-23}$$ where A_0 is the normalization constant and $$a = \tfrac{1}{2}m\omega_0/\hbar.$$
3. Reflection and Barrier Penetration	When the potential changes abruptly over a small distance, a particle may be reflected even though $E > U(x)$. A particle may penetrate a region in which $E < U(x)$. Reflection and penetration of matter waves are similar to those for other kinds of waves.
4. The Schrödinger Equation in Three Dimensions	The wave function for a particle in a three-dimensional box can be written $$\psi(x, y, z) = \psi_1(x)\psi_2(y)\psi_3(z)$$ where ψ_1, ψ_2, and ψ_3 are wave functions for a one-dimensional box.
Degeneracy	When more than one wave function is associated with the same energy level, the energy level is said to be degenerate. Energy-level degeneracy occurs because of spatial symmetry.

* An exception to this claim is for the infinite well potential (where U is equal to zero inside the well and to infinity outside the well). For this potential function $d\psi/dx$ is not continuous at the boundary of the well (see Figure 34-17).

TOPIC	RELEVANT EQUATIONS AND REMARKS
5. The Schrödinger Equation for Two Identical Particles	A wave function that describes two identical particles must be either symmetric or antisymmetric when the coordinates of the particles are exchanged. Fermions (which include electrons, protons, and neutrons) are described by antisymmetric wave functions and obey the Pauli exclusion principle, which states that no two identical particles can simultaneously have the same values for their quantum number. Bosons (which include α particles, deuterons, photons, and mesons) have symmetric wave functions and do not obey the Pauli exclusion principle.

Answers to Concept Checks

35-1 The wave function cannot be normalized.

Answers to Practice Problems

35-2 $E_{133} = 22E_1$, $E_{142} = E_{124} = E_{222} = 24E_1$

Problems

In a few problems, you are given more data than you actually need; in a few other problems, you are required to supply data from your general knowledge, outside sources, or informed estimate.

Interpret as significant all digits in numerical values that have trailing zeros and no decimal points.

- • Single-concept, single-step, relatively easy
- •• Intermediate-level, may require synthesis of concepts
- ••• Challenging
- SSM Solution is in the *Student Solutions Manual*
- Consecutive problems that are shaded are paired problems.

CONCEPTUAL PROBLEMS

1 • Sketch (*a*) the wave function and (*b*) the probability density function for the $n = 5$ state of the finite square-well potential.

2 • Sketch (*a*) the wave function and (*b*) the probability density function for the $n = 4$ state of the finite square-well potential.

THE SCHRÖDINGER EQUATION

3 •• Show that if $\psi_1(x)$ and $\psi_2(x)$ are each solutions to the time-independent Schrödinger equation (Equation 35-4), then $\psi_3(x) = \psi_1(x) + \psi_2(x)$ is also a solution. This result, known as the superposition principle, applies to the solutions of all linear equations.

THE HARMONIC OSCILLATOR

4 •• The harmonic oscillator problem may be used to describe the vibrations of molecules. For example, the hydrogen molecule H_2 is found to have equally spaced vibrational energy levels separated by 8.7×10^{-20} J. What value of the force constant of the spring would be needed to get this energy spacing, assuming that half the molecule can be modeled as a hydrogen atom attached to one end of a spring that has its other end fixed? *Hint: The spacing for the energy levels of this half-molecule would be half of the spacing for the energy levels of the complete molecule. In addition, the force constant of a spring is inversely proportional to its relaxed length, so if half of the spring has force constant k, the entire spring has a force constant that is equal to* $\frac{1}{2}k$.

5 •• Use the procedure of Example 35-1 to verify that the energy of the first excited state of the harmonic oscillator is $E_1 = \frac{3}{2}\hbar\omega_0$. (Note: *Rather than solve for a again, use the step-8 result* $a = \frac{1}{2}m\omega_0/\hbar$ *obtained in* Example 35-1) SSM

6 •• Show that the expectation value $\langle x\rangle = \int_{-\infty}^{\infty} x|\psi|^2\,dx$ is zero for both the ground state and the first excited state of the harmonic oscillator.

7 •• Verify that the normalization constant A_0 in the ground-state harmonic-oscillator wave function $\psi_0(x) = A_0e^{-ax^2}$ (Equation 35-23) is given by $A_0 = (2m\omega_0/h)^{1/4}$.

8 •• Using the result of Problem 7, show that for the ground state of the harmonic oscillator $\langle x^2\rangle = \int x^2|\psi|^2\,dx = \hbar/(2m\omega_0) = 1/(4a)$. Use this result to show that the average potential energy equals half the total energy.

9 •• The quantity $\sqrt{\langle x^2\rangle - \langle x\rangle^2}$ is a measure of the average spread in the location of a particle. (*a*) Consider an electron trapped in a harmonic oscillator potential. Its lowest energy level is found to be 2.1×10^{-4} eV. Calculate $\sqrt{\langle x^2\rangle - \langle x\rangle^2}$ for this electron. (See Problems 6 and 8.) (*b*) Now consider an electron trapped in an infinite square-well potential. If the width of the well is equal to $\sqrt{\langle x^2\rangle - \langle x\rangle^2}$, what would be the lowest energy level for this electron?

10 ••• Classically, the average kinetic energy of the harmonic oscillator equals the average potential energy. Assume that this result is also true for the quantum-mechanical harmonic oscillator, and use this result, along with the result of Problem 8, to determine the expectation value of p_x^2 (where $p_x = mv_x$) for the ground state of the one-dimensional harmonic oscillator.

11 ••• We know that for the classical harmonic oscillator, $p_{x\,\mathrm{av}} = 0$. It can be shown that for the quantum-mechanical harmonic oscillator, $\langle p_x\rangle = 0$. Use the results of Problems 6, 8, and 10 to determine the uncertainty product $\Delta x\,\Delta p_x$ for the ground state of the harmonic oscillator. The uncertainties are defined by $(\Delta x)^2 = \langle(x - \langle x\rangle)^2\rangle$ and $(\Delta p_x)^2 = \langle(p_x - \langle p_x\rangle)^2\rangle$.

REFLECTION AND TRANSMISSION OF ELECTRON WAVES: BARRIER PENETRATION

12 •• A particle of energy E approaches a step barrier of height U_0. What should be the ratio E/U_0 so that the reflection coefficient is $\frac{1}{2}$?

13 •• **SPREADSHEET** A particle that has mass m is traveling in the direction of increasing x. The potential energy of the particle is equal to zero everywhere in the region $x < 0$ and is equal to U_0 everywhere in the region $x > 0$, where $U_0 > 0$. (*a*) Show that if the total energy is $E = \alpha U_0$, where $\alpha \geq 1$, then the wave number k_2 in the region $x > 0$ is given by $k_2 = k_1\sqrt{(\alpha - 1)/\alpha}$, where k_1 is the wave number in the region $x < 0$. (b) Using a spreadsheet program or graphing calculator, graph the reflection coefficient R and the transmission coefficient T as functions of α, for $1 \leq \alpha \leq 5$. SSM

14 •• Suppose that the potential energy in Problem 13 is equal to zero everywhere in the region $x < 0$ and is equal to $-U_0$ everywhere in the region $x > 0$, where $U_0 > 0$. The wave number for the incident particle is again k_1, and the total energy is $2U_0$. (*a*) What is the wave number for the particle in the region where $x > 0$? (*b*) What is the reflection coefficient R? (*c*) What is the transmission coefficient T? (*d*) If each of one million particles that are in the region $x < 0$ are traveling with wave number k_1 in the direction of increasing x and are incident upon the potential energy drop at $x = 0$, how many of these particles are expected to continue along in the direction of increasing x? How does this compare with the classical prediction?

15 •• A 10-eV electron (an electron with a kinetic energy of 10 eV) is incident on a potential-energy barrier that has a height equal to 25 eV and a width equal to 1.0 nm. (*a*) Use Equation 35-29 to calculate the order of magnitude of the probability that the electron will tunnel through the barrier. (*b*) Repeat your calculation for a width of 0.10 nm. SSM

16 •• Use Equation 35-29 to calculate the order of magnitude of the probability that a proton will tunnel out of a nucleus in one collision with the nuclear barrier if the proton has an energy 6.0 MeV below the top of the potential-energy barrier and the barrier thickness is 1.2×10^{-15} m.

17 ••• To understand how a small change in α-particle energy can dramatically change the probability of the α particle tunneling from a nucleus, consider an α particle emitted by a uranium nucleus ($Z = 92$). (*a*) Referring to Figure 35-17, calculate the center-to-center distance of closest approach r_1 that α particles that have kinetic energies of 4.0 MeV and 7.0 MeV could make to the uranium nucleus. (*b*) Use the result from Part (*a*) to calculate the relative transmission coefficient $e^{-2\alpha a}$ for the same α particles. (*Note:* The actual half-lives of uranium nuclei vary over nine orders of magnitude. Your calculation will show a smaller range than this; however, to find half-life, you must also include the frequency with which the α particle strikes the barrier.)

THE SCHRÖDINGER EQUATION IN THREE DIMENSIONS

18 •• (*a*) A particle is confined to a three-dimensional box that has sides L_1, $L_2 = 2L_1$, and $L_3 = 3L_1$. Give the quantum numbers n_1, n_2 and n_3 that correspond to the ten lowest-energy quantum states of this box. *Hint: A spreadsheet can be helpful.* (*b*) What quantum numbers, if any, correspond to degenerate energy levels? (*c*) Give a wave function for the fifth excited state. (There are only five states that have energy levels below the energy level of the fifth excited state.)

19 •• (*a*) A particle is confined to a three-dimensional box that has sides L_1, $L_2 = 2L_1$, and $L_3 = 4L_1$. Give the quantum numbers n_1, n_2 and n_3 that correspond to the ten lowest-energy quantum states of this box. *Hint: A spreadsheet can be helpful.* (*b*) What combinations of these quantum numbers, if any, correspond to degenerate energy levels? (*c*) Give the wave function for the fourth excited energy state. (There are only four states that have energy levels below the energy level of the fourth excited state.)

20 • A particle moves in a potential well given by $U(x, y, z) = 0$ for $-L/2 < x < L/2, 0 < y < L$, and $0 < z < L$; $U = \infty$ outside these ranges. (*a*) Write an expression for the ground-state wave function for the particle. (*b*) How do the allowed energies compare with those for a well having $U = 0$ for $0 < x < L$, rather than for $-L/2 < x < L/2$? Explain your answer.

21 •• A particle is constrained to the two-dimensional region defined by $0 \leq x \leq L$ and $0 \leq y \leq L$ and moves freely throughout that region. (*a*) Find the wave functions that meet these conditions and are solutions of the Schrödinger equation. (*b*) Find the energies that correspond to the wave functions in part (*a*). (*c*) Find the quantum numbers of the two lowest states that have the same energy (that are degenerate). (*d*) Find the quantum numbers of the three lowest states that have the same energy.

THE SCHRÖDINGER EQUATION FOR TWO IDENTICAL PARTICLES

22 • Show that the two-particle wave function $\psi_{12} = A\sin(\pi x_1/L)\sin(2\pi x_2/L)$, $0 < x_1 < L$ and $0 < x_2 < L$, (Equation 35-37), is a solution of

$$-\frac{\hbar^2}{2m}\frac{\partial^2\psi(x_1, x_2)}{\partial x_1^2} - \frac{\hbar^2}{2m}\frac{\partial^2\psi(x_1, x_2)}{\partial x_2^2} + U\psi(x_1, x_2) = E\psi(x_1, x_2)$$

(Equation 35-35), if $U(x_1, x_2) = 0$, and find the energy of the state represented by this wave function.

23 • What is the ground-state energy of ten noninteracting bosons in a one-dimensional box of length L?

24 •• What is the ground-state energy of seven identical noninteracting fermions in a one-dimensional box of length L? (Because the quantum number associated with spin can have two values, each spatial state can be occupied by two fermions.)

ORTHOGONALITY OF WAVE FUNCTIONS

The integral of two functions over some space interval is somewhat analogous to the dot product of two vectors. If this integral is zero, the functions are said to be orthogonal, which is analogous to two vectors being perpendicular. The following problems illustrate the general principle that any two wave functions corresponding to different energy levels in the same potential are orthogonal. A general hint for all these problems is that the integral $\int_{x_1}^{x_2} f(x)\,dx$ *is equal to zero if* x_1 *is equal to* $-x_2$ *and if* $f(x)$ *is equal to* $-f(-x)$.

25 •• Show that the ground-state and the first excited state wave functions of the harmonic oscillator are orthogonal; that is, show that $\int_{-\infty}^{\infty} \psi_0(x)\psi_1(x)\,dx = 0$.

26 •• The wave function for the state $n = 2$ of the harmonic oscillator is $\psi_2(x) = A_2\left(2ax^2 - \frac{1}{2}\right)e^{-ax^2}$, where A_2 is the normalization constant and a is a positive constant. Show that the wave functions for the states $n = 1$ and $n = 2$ of the harmonic oscillator are orthogonal.

27 •• For the wave functions

$$\psi_n(x) = A \sin(n\pi x/L) \qquad n = 1, 2, 3, \ldots$$

corresponding to a particle in an infinite square-well potential from 0 to L, show that $\int_0^L \psi_m(x)\psi_n(x)\,dx = 0$ for all positive integers m and n, where $m \neq n$; that is, show that the wave functions are orthogonal.

GENERAL PROBLEMS

28 •• Consider a particle in an infinite one-dimensional box that has a length L and is centered at the origin. (*a*) What are the values of $\psi_1(0)$ and $\psi_2(0)$? (*b*) What are the values of $\langle x \rangle$ for the $n = 1$ and $n = 2$ states? (*c*) Evaluate $\langle x^2 \rangle$ for the $n = 1$ and $n = 2$ states.

29 •• Eight identical noninteracting fermions are confined to an infinite two-dimensional square box of side length L. Determine the energies of the three lowest energy states. (See Problem 22.) SSM

30 •• A particle is confined to a two-dimensional box defined by the following boundary conditions: $U(x, y) = 0$ for $-L/2 \leq x \leq L/2$ and $-3L/2 \leq y \leq 3L/2$, and $U(x, y) = \infty$ outside these ranges. (*a*) Determine the energies of the three lowest energy states. Are any of these states degenerate? (*b*) Identify the quantum numbers of the two lowest energy degenerate states and determine the energy of these states.

31 ••• The classical probability distribution function for a particle in an infinite one-dimensional well of length L is $P = 1/L$. (See Example 34-5.) (*a*) Show that the classical expectation value of x^2 for a particle in an infinite one-dimensional well of length L that is centered at the origin is $L^2/12$. (*b*) Find the quantum expectation value of x^2 for the nth state of a particle in the one-dimensional box and show that it approaches the classical limit $L^2/12$ as n approaches infinity.

32 •• Show that Equations 35-27 and 35-28 imply that the transmission coefficient for particles of energy E incident on a step barrier $U_0 < E$ is given by

$$T = \frac{4k_1k_2}{(k_1 + k_2)^2} = \frac{4r}{(1 + r)^2}$$

where $r = k_2/k_1$.

33 •• (*a*) Show that for the case of a particle of energy E incident on a step barrier $U_0 < E$, the wave numbers k_1 and k_2 are related by

$$\frac{k_2}{k_1} = r = \sqrt{1 - \frac{U_0}{E}}$$

(*b*) Use this result to show that $R = (1 - r)^2/(1 + r)^2$.

34 •• **SPREADSHEET** (*a*) Using a spreadsheet program or graphing calculator and the results of Problem 32 and Problem 33, graph the transmission coefficient T and reflection coefficient R as a function of incident energy E for values of E ranging from $E = U_0$ to $E = 10.0U_0$. (*b*) What limiting values do your graphs indicate?

35 ••• The wave function for the state $n = 2$ of the harmonic oscillator is $\psi_2(x) = A_2(2ax^2 - \frac{1}{2})e^{-ax^2}$, where $a = \frac{1}{2}m\omega_0/\hbar$. Determine the normalization constant A_2.

36 ••• Consider the time-independent, one-dimensional Schrödinger equation when the potential function is symmetric about the origin, that is, when $U(x)$ is even.* (*a*) Show that if $\psi(x)$ is a solution of the Schrödinger equation and has energy E, then $\psi(-x)$ is also a solution that has the same energy E. Furthermore, $\psi(x)$ and $\psi(-x)$ can differ by only a multiplicative constant. (*b*) Write $\psi(x) = C\psi(-x)$, and show that $C = \pm 1$. Note that $C = +1$ means that $\psi(x)$ is an even function of x, and $C = -1$ means that $\psi(x)$ is an odd function of x.

37 ••• In this problem, you will derive the ground-state energy of the harmonic oscillator using the precise form of the uncertainty principle, $\Delta x\, \Delta p_x \geq \hbar/2$, where Δx and Δp_x are defined to be the standard deviations $(\Delta x)^2 = \langle (x - \langle x \rangle)^2 \rangle$ and $(\Delta p_x)^2 = \langle (p_x - \langle p_x \rangle)^2 \rangle$.

Proceed as follows:

1. Write the total classical energy in terms of the position x and momentum p_x using $U(x) = \frac{1}{2}m\omega_0^2x^2$ and $K = \frac{1}{2}p_x^2/m$.
2. Show that $(\Delta x)^2 = \langle (x - \langle x \rangle)^2 \rangle = \langle x^2 \rangle - \langle x \rangle^2$ and $(\Delta p_x)^2 = \langle (p_x - \langle p_x \rangle)^2 \rangle = \langle p_x^2 \rangle - \langle p_x \rangle^2$. *Hint: See Equations 17-34a and 17-34b.*
3. Use the symmetry of the potential energy function to argue that $\langle x \rangle$ and $\langle p_x \rangle$ must be zero, so that $(\Delta x)^2 = \langle x^2 \rangle$ and $(\Delta p_x)^2 = \langle p_x^2 \rangle$.
4. Assume that $\Delta p_x\, \Delta x = \hbar/2$ to eliminate $\langle p_x^2 \rangle$ from the average energy $\langle E \rangle = \langle \frac{1}{2}p_x^2/m + \frac{1}{2}m\omega_0^2 x^2 \rangle = \frac{1}{2}\langle p_x^2 \rangle/m + \frac{1}{2}m\omega_0^2\langle x^2 \rangle$ and write $\langle E \rangle$ as $\langle E \rangle = \hbar^2/(8mZ) + \frac{1}{2}m\omega^2 Z$, where $Z = \langle x^2 \rangle$.
5. Set $d\langle E \rangle/dZ = 0$ to find the value of Z for which $\langle E \rangle$ is a minimum.
6. Show that the minimum energy is given by $\langle E \rangle_{\min} = +\frac{1}{2}\hbar\omega_0$. SSM

38 ••• A particle that has mass m and is near Earth's surface, at which $z = 0$, can be described by the potential energy function $U = mgz$ in the region $z > 0$, and by $U = \infty$ in the region $z < 0$. Sketch a graph of $U(z)$ versus z. For some positive value of total energy E, indicate the classically allowed region on the graph and plot the classical kinetic energy versus z on the graph. The Schrödinger equation for this problem is quite difficult to solve. Using arguments similar to those in Section 35-2 about the concavity of the wave function as given by the Schrödinger equation, sketch the shape of the wave function for the ground state and for the first two excited states.

* A function $f(x)$ is even if $f(x) = f(-x)$ for all x, and a function $f(x)$ is odd if $f(x) = -f(-x)$ for all x.

AT A DISTANCE OF 6,000 LIGHTYEARS FROM EARTH, THE STAR CLUSTER RCW 38 IS A RELATIVELY CLOSE STAR-FORMING REGION. THIS IMAGE COVERS AN AREA ABOUT 5 LIGHTYEARS ACROSS AND CONTAINS THOUSANDS OF HOT, VERY YOUNG STARS FORMED LESS THAN A MILLION YEARS AGO. X RAYS FROM THE HOT UPPER ATMOSPHERES OF 190 OF THESE STARS WERE DETECTED BY CHANDRA, AN X-RAY OBSERVATORY ORBITING EARTH. THE MECHANISMS GENERATING THESE X RAYS ARE NOT KNOWN. ON EARTH, X-RAY MACHINES PRODUCE X RAYS BY BOMBARDING A TARGET WITH HIGH-ENERGY ELECTRONS. THE ATOMIC NUMBER OF THE ATOMS THAT MAKE UP THE TARGET CAN BE DETERMINED BY ANALYZING THE RESULTING X-RAY SPECTRA. *(NASA/CXC/CfA/S.Wolk et al.)*

? How is the atomic number obtained from the spectral analysis? (See Example 36-8.)

Atoms

One hundred eleven chemical elements have been discovered, and several additional chemical elements recently have been reported but not authenticated. Each element is characterized by an atom that has a number of protons Z and an equal number of electrons. The number of protons Z is called the **atomic number.** The atom that has the fewest protons is called hydrogen (H) and has $Z = 1$. A helium (He) atom has two protons ($Z = 2$), a lithium (Li) atom has three protons ($Z = 3$), and so forth. Nearly all the mass of an atom is concentrated in its tiny nucleus, which is made up of protons and neutrons. An atom's nuclear radius is approximately 1 fm to 10 fm (1 fm $= 10^{-15}$ m) and the radius of an atom is approximately 0.1 nm $=$ 100 000 fm.

The chemical properties and physical properties of an element are determined by the number and arrangement of the electrons in an atom of the element. Because each proton has a positive charge $+e$, the nucleus has a total positive charge $+Ze$. The electrons are negatively charged ($-e$), so they are attracted to the nucleus and repelled by each other. Because electrons and protons have equal but opposite charges and an atom has equal numbers of electrons and protons, atoms are electrically neutral. Atoms that lose or gain one or more electrons are then electrically charged and are called *ions*.

We will begin our study of atoms by discussing the Bohr model, a semiclassical model developed by Niels Bohr in 1913 to explain the electromagnetic spectra produced by hydrogen atoms. Although this pre-quantum mechanics model has many shortcomings, it provides a useful framework for the discussion of atomic phenomena. After discussing the Bohr model, we will then apply our knowledge of quantum mechanics from Chapter 35 to give a much more productive model of the hydrogen atom. We will then discuss the structure of other atoms and the periodic table of the elements. Finally, we will discuss both optical and X-ray spectra.

36-1 THE ATOM

ATOMIC SPECTRA

By the beginning of the twentieth century, a large body of data had been collected on the emission of light by atoms in a gas when the atoms are excited by an electric discharge. Viewed through a spectroscope that has a narrow-slit aperture, light that has been emitted by atoms of a particular element appears as a discrete set of lines of different colors or wavelengths. The spacing and relative intensities of the lines are characteristic of the element. The wavelengths of these spectral lines could be accurately determined, and much effort went into finding regularities in the spectra. Figure 36-1 shows line spectra for hydrogen and for mercury.

In 1885, Johann Balmer determined that the wavelengths of the lines in the visible spectrum of hydrogen can be represented by the formula

$$\lambda = (364.6\text{ nm})\frac{m^2}{m^2 - 4} \qquad m = 3, 4, 5, \ldots \tag{36-1}$$

Balmer suggested that this expression might be a special case of a more general expression that would be applicable to the spectra of other elements. Such an expression, found by Johannes R. Rydberg and Walter Ritz and known as the **Rydberg–Ritz formula,** gives the reciprocal wavelength as

$$\frac{1}{\lambda} = R\left(\frac{1}{n_2^2} - \frac{1}{n_1^2}\right) \tag{36-2}$$

FIGURE 36-1 (*a*) Line spectrum of hydrogen and (*b*) line spectrum of mercury. *((a) and (b) adapted from Eastern Kodak and Wabash Instrument Corporation.)*

where n_1 and n_2 are integers, $n_1 > n_2$, and R is the Rydberg constant. The Rydberg constant does vary slightly, and in a regular way, from element to element. For hydrogen, R has the value

$$R_H = 1.097776 \times 10^7 \text{ m}^{-1}$$

RYDBERG CONSTANT FOR HYDROGEN

The Rydberg–Ritz formula gives the wavelengths for all the lines in the spectra of hydrogen, as well as alkali elements such as lithium and sodium. The hydrogen Balmer series given by Equation 36-1 is also given by Equation 36-2, with $R = R_H$, $n_2 = 2$, and $n_1 = m$.

Many attempts were made to construct a model of the atom that would yield these formulas for an atom's radiation spectrum. The most popular model, created by J. J. Thomson, considered various arrangements of electrons embedded in some kind of fluid that had most of the mass of the atom and had enough positive charge to make the atom electrically neutral. Thomson's model, called the "plum pudding" model, is illustrated in Figure 36-2. Because classical electromagnetic theory predicted that a charge oscillating with frequency f would radiate electromagnetic energy of that frequency, Thomson searched for configurations that were stable and that had normal modes of vibration of frequencies equal to those of the spectrum of the atom. A difficulty of this model and all other models was that, according to classical physics, electric forces alone cannot produce stable equilibrium. Thomson was unsuccessful in finding a model that predicted the observed frequencies for any atom.

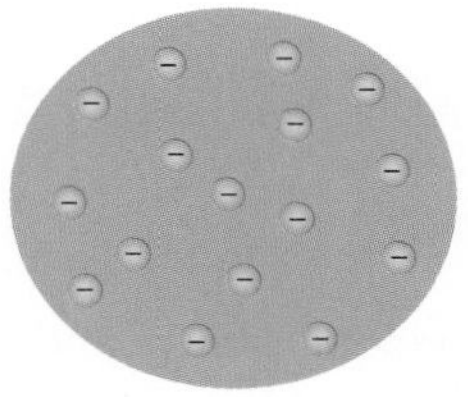

FIGURE 36-2 J. J. Thomson's plum pudding model of the atom. In this model the electrons, which have a negative charge, are embedded in a fluid of positive charge. For a given configuration in such a system, the resonance frequencies of oscillations of the electrons can be calculated. According to classical theory, the atom should radiate light of frequency equal to the frequency of oscillation of the electrons. Thomson could not find any configuration that would give frequencies in agreement with the measured frequencies of the spectrum of any atom.

The Thomson model was essentially ruled out by a set of experiments by H. W. Geiger and E. Marsden, under the supervision of E. Rutherford in approximately 1911, in which alpha particles from radioactive radium were scattered by atoms in a gold foil. Rutherford showed that the number of alpha particles scattered at large angles could not be accounted for by an atom in which the positive charge was distributed throughout the atom (known to be about 0.1 nm in diameter). Instead, the results suggested that the positive charge and most of the mass of an atom is concentrated in a very small region, now called the nucleus, which has a diameter of the order of 10^{-6} nm = 1 fm.

36-2 THE BOHR MODEL OF THE HYDROGEN ATOM

Niels Bohr, working in the Rutherford laboratory in 1912, proposed a model of the hydrogen atom that extended the work of Planck, Einstein, and Rutherford and successfully predicted the observed spectra. According to Bohr's model, the electron of the hydrogen atom moves in either a circular or an elliptical orbit around the positive nucleus according to Coulomb's law and classical mechanics like the planets orbit the Sun. For simplicity, Bohr chose a circular orbit, as shown in Figure 36-3.

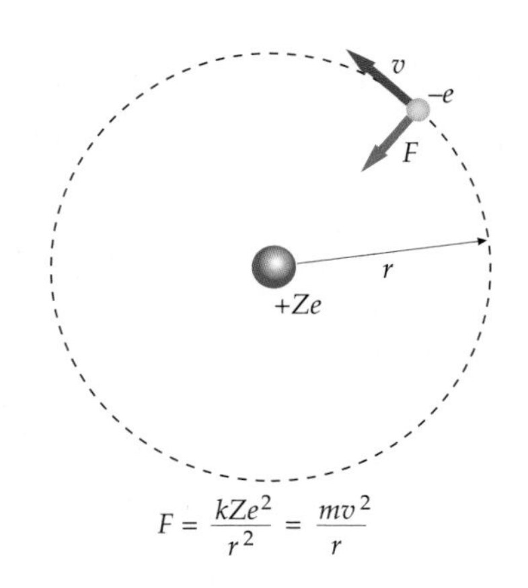

FIGURE 36-3 Electron of charge $-e$ traveling in a circular orbit of radius r around the nuclear charge $+Ze$. The attractive electrical force kZe^2/r^2 keeps the electron in its orbit.

ENERGY FOR A CIRCULAR ORBIT

Consider an electron of charge $-e$ moving in a circular orbit of radius r about a positive charge $+Ze$ such as the nucleus of a hydrogen atom ($Z = 1$) or of a singly ionized helium atom ($Z = 2$). The total energy of the electron can be related to the radius of the orbit. The potential energy of the electron of charge $-e$ at a distance r from a positive charge Ze is

$$U = \frac{kq_1q_2}{r} = \frac{k(Ze)(-e)}{r} = -\frac{kZe^2}{r} \tag{36-3}$$

where k is the Coulomb constant. The kinetic energy K can be obtained as a function of r by using Newton's second law, $F_{net} = ma$. Setting the Coulomb attractive force equal to the mass multiplied by the centripetal acceleration gives

$$\frac{kZe^2}{r^2} = m\frac{v^2}{r} \tag{36-4a}$$

Multiplying both sides by $r/2$ gives

$$K = \frac{1}{2}mv^2 = \frac{1}{2}\frac{kZe^2}{r} \tag{36-4b}$$

Thus, the kinetic energy and the potential energy vary inversely with r. Note that the magnitude of the potential energy is twice that of the kinetic energy:

$$U = -2K \tag{36-5}$$

Equation 36-5 a general result for particles orbiting under the influence of forces that vary inversely with the square of the distance from a fixed point. [It also holds for circular orbits in a gravitational field (see Example 11-6 in Section 11-3)]. The total energy is the sum of the kinetic energy and the potential energy:

$$E = K + U = \frac{1}{2}\frac{kZe^2}{r} - \frac{kZe^2}{r}$$

or

$$E = -\frac{1}{2}\frac{kZe^2}{r} \tag{36-6}$$

ENERGY IN A CIRCULAR ORBIT FOR A $1/r^2$ FORCE

Although mechanical stability is achieved because the Coulomb attractive force provides the centripetal force necessary for the electron to remain in orbit, classical *electromagnetic* theory says that such an atom would be electrically unstable. The atom would be unstable because the electron must accelerate when moving in a circle and therefore radiate electromagnetic energy of frequency equal to that of its motion. According to the classical theory, such an atom would quickly collapse as the electron spiraled into the nucleus and radiated away the energy.

BOHR'S POSTULATES

Bohr circumvented the difficulty of the collapsing atom by *postulating* that only certain orbits, called stationary states, are allowed and that an atom with an electron in one of these orbits does not radiate. An atom radiates only when the electron makes a transition from one allowed orbit (stationary state) to another.

The electron in the hydrogen atom can move only in certain nonradiating, circular orbits called stationary states.

BOHR'S FIRST POSTULATE—NONRADIATING ORBITS

Bohr's second postulate relates the frequency of radiation to the energies of the stationary states. If E_i and E_f are the initial and final energies of the atom, the frequency of the emitted radiation during a transition is given by

$$f = \frac{E_i - E_f}{h} \tag{36-7}$$

BOHR'S SECOND POSTULATE—PHOTON FREQUENCY FROM ENERGY CONSERVATION where h is Planck's constant. This postulate is equivalent to the assumption of conservation of energy when a photon of energy hf is emitted. Combining Equation 36-6 and Equation 36-7, we obtain for the frequency

$$f = \frac{E_i - E_f}{h} = \frac{1}{2}\frac{kZe^2}{h}\left(\frac{1}{r_f} - \frac{1}{r_i}\right) \tag{36-8}$$

where r_i and r_f are the radii of the initial and final orbits.

To obtain the frequencies implied by the Rydberg–Ritz formula, $f = c/\lambda = cR(1/n_2^2 - 1/n_1^2)$, it is evident that the radii of stable orbits must be proportional to the squares of integers. Bohr searched for a quantum condition for the radii of the stable orbits that would yield this result. After much trial and error, Bohr found that he could obtain the desired result if he postulated that the magnitude of the angular momentum of the electron in a stable orbit equals an integer multiplied by $\hbar$. Because the magnitude of the angular momentum of a circular orbit is just mvr, this postulate is

$$mv_n r_n = n\hbar \qquad n = 1, 2, 3, \ldots \tag{36-9}$$

BOHR'S THIRD POSTULATE—QUANTIZED ANGULAR MOMENTUM

where $\hbar = h/2\pi = 1.055 \times 10^{-34}\,\text{J}\cdot\text{s} = 6.582 \times 10^{-16}\,\text{eV}\cdot\text{s}$.

Equation 36-9 relates the speed v_n to the radius r_n of the orbit that has angular momentum $n\hbar$. Equation 36-4*a* gives us another equation relating the speed to the radius:

$$\frac{kZe^2}{r_n^2} = m\frac{v_n^2}{r_n}$$

or

$$v_n^2 = \frac{kZe^2}{mr_n} \tag{36-10}$$

We can determine r_n by first solving for v_n in Equation 36-9. Squaring the result then gives

$$v_n^2 = n^2\frac{\hbar^2}{m^2 r_n^2}$$

Equating this expression for v_n^2 with the expression given by Equation 36-10, we get

$$n^2\frac{\hbar^2}{m^2 r_n^2} = \frac{kZe^2}{mr_n}$$

Solving for r_n, we obtain

$$r_n = n^2\frac{\hbar^2}{mkZe^2} = n^2\frac{a_0}{Z} \tag{36-11}$$

RADII OF THE BOHR ORBITS

where a_0 is called the **first Bohr radius.** According to the Bohr model, a_0 is the orbital radius of the electron in a hydrogen atom that has $n = 1$.

$$a_0 = \frac{\hbar^2}{mke^2} = \frac{\epsilon_0 h^2}{\pi me^2} = 0.0529\text{ nm} \tag{36-12}$$

FIRST BOHR RADIUS

Substituting the expressions for r_n in Equation 36-11 into Equation 36-8 for the frequency gives

$$f = \frac{1}{2}\frac{kZe^2}{h}\left(\frac{1}{r_f} - \frac{1}{r_i}\right) = Z^2\frac{mk^2e^4}{4\pi\hbar^3}\left(\frac{1}{n_f^2} - \frac{1}{n_i^2}\right) \qquad 36\text{-}13$$

If we compare this expression where $Z = 1$ and $f = c/\lambda$ with the empirical Rydberg–Ritz formula (Equation 36-2), we obtain for the Rydberg constant

$$R = \frac{mk^2e^4}{4\pi c\hbar^3} = \frac{me^4}{8\epsilon_0^2ch^3} \qquad 36\text{-}14$$

Using the values of m, e, c, k, and $\hbar$ known in 1913, Bohr calculated R and found his result to agree (within the limits of the uncertainties of the constants) with the value obtained from spectroscopy.

Example 36-1 Standing-Wave Condition Implies Quantization of Angular Momentum

For waves in a circle, the standing-wave condition is that there is an integral number of wavelengths in the circumference. That is, $n\lambda_n = 2\pi r_n$, where $n = 1, 2, 3$, and so on. Show that this condition for electron waves implies quantization of angular momentum.

PICTURE The wavelength and the momentum are related by the de Broglie relation $p = h/\lambda$ (Equation 34-13). Using this relation and the standing wave condition $n\lambda_n = 2\pi r_n$ to relate the momentum to the radius.

SOLVE

1. Write the standing-wave condition: $n\lambda_n = 2\pi r_n$
2. Use the de Broglie relation (Equation 34-13) to relate the momentum p to λ_n: $p = \frac{h}{\lambda_n} = \frac{nh}{2\pi r_n} = n\frac{\hbar}{r_n}$
3. Solve for pr_n. The angular momentum of an electron in a circular orbit is $mvr_n = pr_n$, where $p = mv$: $pr_n = \boxed{mvr_n = n\hbar}$

CHECK The step-3 result is the check. It is what the problem statement asks us to show.

ENERGY LEVELS

The total mechanical energy of the electron in the hydrogen atom is related to the radius of the circular orbit by Equation 36-6. If we substitute the quantized values of r as given by Equation 36-11, we obtain

$$E_n = -\frac{1}{2}\frac{kZe^2}{r_n} = -\frac{1}{2}\frac{kZ^2e^2}{n^2a_0} = -\frac{1}{2}\frac{mk^2Z^2e^4}{n^2\hbar^2}$$

or

$$E_n = -Z^2\frac{E_0}{n^2} \qquad 36\text{-}15$$

ENERGY LEVELS IN THE HYDROGEN ATOM

where

$$E_0 = \frac{mk^2e^4}{2\hbar^2} = \frac{1}{2}\frac{ke^2}{a_0} = 13.6\text{ eV} \qquad 36\text{-}16$$

The energies E_n corresponding to $Z = 1$ are the quantized allowed energies for the hydrogen atom.

Transitions between these allowed energies result in the emission or absorption of a photon whose frequency is given by $f = (E_i - E_f)/h$, and whose wavelength is

$$\lambda = \frac{c}{f} = \frac{hc}{E_i - E_f} \qquad 36\text{-}17$$

As we found in Chapter 34, it is convenient to have the value of hc in electron-volt nanometers:

$$hc = 1240 \text{ eV} \cdot \text{nm} \qquad 36\text{-}18$$

Because the energies are quantized, the frequencies and wavelengths of the radiation emitted by the hydrogen atom are quantized in agreement with the observed line spectrum.

Figure 36-4 shows the energy-level diagram for hydrogen. The energy of the hydrogen atom in the ground state is $E_1 = -13.6$ eV. As n approaches infinity the energy approaches zero. The process of removing an electron from an atom is called **ionization,** and the minimum amount of energy required to remove the electron is the **ionization energy.** The ionization energy of the ground-state hydrogen atom, which is also its **binding energy,** is 13.6 eV. A few transitions from a higher state to a lower state are indicated in Figure 36-4. When Bohr published his model of the hydrogen atom, the Balmer series (corresponding to $n_f = 2$ and $n_i = 3, 4, 5, \ldots$) and the Paschen series (corresponding to $n_f = 3$ and $n_i = 4, 5, 6, \ldots$) were known. In 1916, T. Lyman found the series corresponding to $n_f = 1$. F. Brackett in 1922 and H. A. Pfund in 1924 found the series corresponding to $n_f = 4$ and $n_f = 5$, respectively. Only the Balmer series lies in the visible portion of the electromagnetic spectrum.

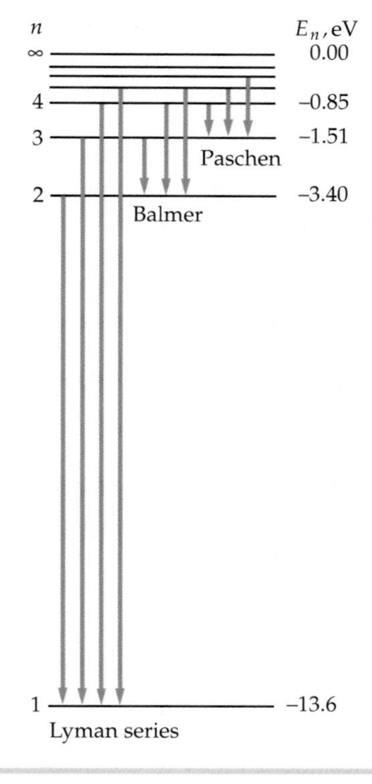

FIGURE 36-4 Energy-level diagram for hydrogen showing the first few transitions in each of the Lyman, Balmer, and Paschen series. The energies of the levels are given by Equation 36-15.

! Do not think the λ in Equation 36-17 is the wavelength of the electron. It is not. It is the wavelength of the emitted or absorbed photon.

Example 36-2 Longest Wavelength in the Lyman Series

Find (*a*) the energy and (*b*) the wavelength of the spectral line that has the longest wavelength in the Lyman series.

PICTURE For the Lyman series $n_f = 1$. From Figure 36-4, we can see that the Lyman series corresponds to transitions ending at the ground-state energy, $E_f = E_1 = -13.6$ eV. Because the photon wavelength λ varies inversely with energy, the transition that has the longest wavelength is the transition that has the lowest energy, which is from the first excited state $n = 2$ to the ground state $n = 1$.

SOLVE

1. The energy of the photon is the difference in the energies of the initial and final atomic states:

$$E_{\text{photon}} = \Delta E_{\text{atom}} = E_i - E_f$$
$$= E_2 - E_1 = \frac{-13.6 \text{ eV}}{2^2} - \frac{-13.6 \text{ eV}}{1^2}$$
$$= -3.40 \text{ eV} + 13.6 \text{ eV} = \boxed{10.2 \text{ eV}}$$

2. The wavelength of the photon is

$$\lambda = \frac{hc}{E_2 - E_1} = \frac{1240 \text{ eV} \cdot \text{nm}}{10.2 \text{ eV}} = \boxed{122 \text{ nm}}$$

CHECK The step-1 result of 10.2 eV is less than 13.6 eV (the binding energy of ground-state hydrogen). This result is expected.

TAKING IT FURTHER This photon has a wavelength that corresponds to the ultraviolet region of the electromagnetic spectrum. Because all the other lines in the Lyman series have even greater energies and shorter wavelengths, the Lyman series is completely in the ultraviolet region.

PRACTICE PROBLEM 36-1 Find the shortest wavelength for a line in the Lyman series.

Despite its spectacular successes, the Bohr model of the hydrogen atom had many shortcomings. There was no justification for the postulates of stationary states or for the quantization of angular momentum other than the fact that these postulates led to energy levels that agreed with spectroscopic data. Furthermore, attempts to apply the model to atoms that have more electrons and protons had little success. The quantum-mechanical model resolves these difficulties. The stationary states of the Bohr model are replaced by the standing-wave solutions of the Schrödinger equation analogous to the standing electron waves for a particle in a box discussed in Chapter 34 and Chapter 35. Energy quantization is a direct consequence of the standing-wave solutions of the Schrödinger equation. For hydrogen, these quantized energies agree with those obtained from the Bohr model and with experiment. The quantization of angular momentum that had to be postulated in the Bohr model is predicted by the quantum theory.

36-3 QUANTUM THEORY OF ATOMS

THE SCHRÖDINGER EQUATION IN SPHERICAL COORDINATES

In quantum theory, an electron in an atom is described by its wave function ψ. The probability of finding the electron in some volume dV of space equals the square of the absolute value of the wave function $|\psi|^2$ multiplied by dV. Boundary conditions on the wave function lead to the quantization of the wavelengths and frequencies and thereby to the quantization of the electron energy.

Consider a single electron of mass m moving in three dimensions in a region in which the potential energy is U. The time-independent Schrödinger equation for such a particle is given by Equation 35-30:

$$-\frac{\hbar^2}{2m}\left(\frac{\partial^2\psi}{\partial x^2}+\frac{\partial^2\psi}{\partial y^2}+\frac{\partial^2\psi}{\partial z^2}\right)+U(x,y,z)\psi=E\psi \qquad 36\text{-}19$$

For a single isolated atom, the potential energy U depends only on the radial distance $r=\sqrt{x^2+y^2+z^2}$ of the electron from the center of the nucleus. The problem is then most conveniently treated using the spherical coordinates r, θ, and ϕ, which are related to the rectangular coordinates x, y, and z by

$$\begin{aligned} z &= r\cos\theta \\ x &= r\sin\theta\cos\phi \\ y &= r\sin\theta\sin\phi \end{aligned} \qquad 36\text{-}20$$

These relations are shown in Figure 36-5. The transformation of Equation 36-19 from rectangular to spherical coordinates is a straightforward, but tedious, calculation, which we will omit. The result of this transformation can be found in Problem 42. We will discuss qualitatively some of the interesting features of the wave functions that satisfy this equation.

The transformed version of Equation 36-19 can be solved using the technique called separation of variables. This is accomplished by expressing the wave function $\psi(r,\theta,\phi)$ as a product of three functions, each of which is a function of only one of the three spherical coordinates:

$$\psi(r,\theta,\phi)=R(r)f(\theta)g(\phi) \qquad 36\text{-}21$$

where R is a function that depends only on the radial coordinate r, f is a function that depends only on the polar coordinate θ, and g is a function that depends only on the azimuth coordinate ϕ. When this form of $\psi(r,\theta,\phi)$ is substituted into the Schrödinger equation, the Schrödinger equation can be transformed into three ordinary differential equations, one for $R(r)$, one for $f(\theta)$, and one for $g(\phi)$. The potential energy $U(r)$ appears only in the equation for $R(r)$, which is called the **radial equation.**

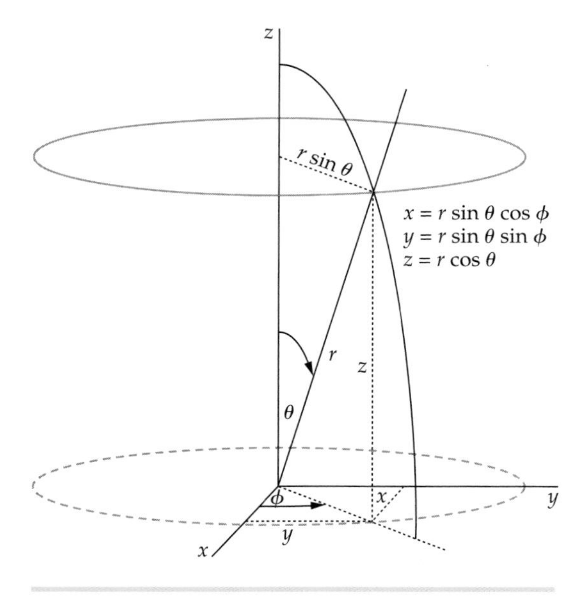

FIGURE 36-5 Geometric relations between spherical coordinates and rectangular coordinates.

Because the potential energy depends only on the coordinate r, the potential energy has no effect on the solutions of the equations for $f(\theta)$ and $g(\phi)$, and therefore has no effect on the angular dependence of the wave function $\psi(r, \theta, \phi)$. These solutions are applicable to *any* problem in which the potential energy depends only on r.

QUANTUM NUMBERS IN SPHERICAL COORDINATES

In three dimensions, the requirement that the wave function be continuous and normalizable introduces three quantum numbers, one associated with each spatial dimension. In spherical coordinates, the quantum number associated with r is labeled n, that associated with θ is labeled ℓ, and that associated with ϕ is labeled m_ℓ.* The quantum numbers n_1, n_2, and n_3 that we found in Chapter 35 for a particle in a three-dimensional square well in rectangular coordinates x, y, and z were independent of one another, but the quantum numbers associated with wave functions in spherical coordinates are dependent on each other. The possible values of these quantum numbers are

$$n = 1, 2, 3, \ldots$$
$$\ell = 0, 1, 2, 3, \ldots, n - 1$$
$$m_\ell = -\ell, -\ell + 1, -\ell + 2, \ldots, 0, \ldots, \ell - 2, \ell - 1, \ell \qquad \text{36-22}$$

QUANTUM NUMBERS IN SPHERICAL COORDINATES

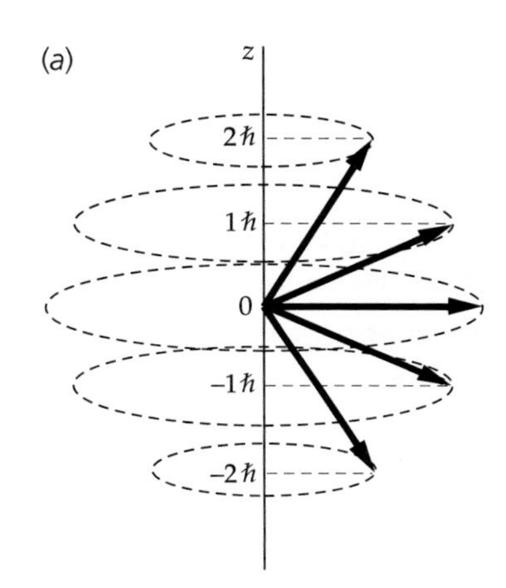

That is, n can be any positive integer; ℓ can be 0 or any positive integer up to $n - 1$; and m_ℓ can have $2\ell + 1$ possible values, ranging from $-\ell$ to $+\ell$ in integral steps.

The number n is called the **principal quantum number.** It is associated with the dependence of the wave function on the distance r and therefore with the probability of finding the electron at various distances from the nucleus. The quantum numbers ℓ and m_ℓ are associated with the orbital angular momentum of the electron and with the angular dependence of the electron wave function. The quantum number ℓ is called the **orbital quantum number.** The magnitude L of the orbital angular momentum $\vec{L}$ is related to the orbital quantum number ℓ by

$$L = \sqrt{\ell(\ell + 1)}\hbar \qquad \text{36-23}$$

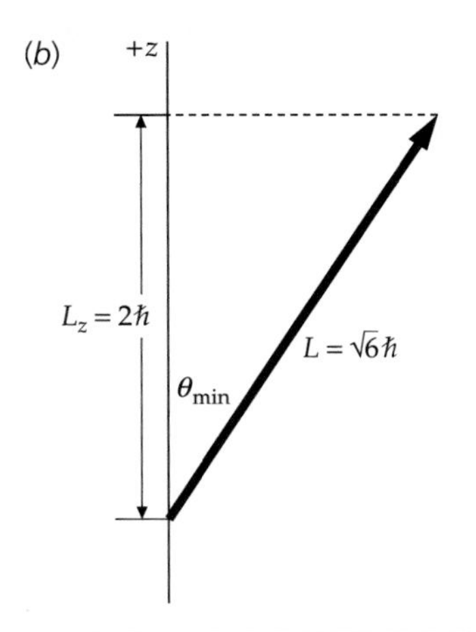

FIGURE 36-6 (*a*) Vector-model diagram illustrating the possible values of the z component of the orbital angular-momentum vector for the case $\ell = 2$. The magnitude of the orbital angular momentum is $L = \hbar\sqrt{\ell(\ell + 1)} = \hbar\sqrt{2(2 + 1)} = \hbar\sqrt{6}$. (*b*) The values of the z component of the orbital angular-momentum vector for the case $\ell = 2$ and $m_\ell = 2$. The value of the z component of the orbital angular momentum is $L_z = 2\hbar$.

The quantum number m_ℓ is called the **magnetic quantum number.** It is related to the component of the orbital angular momentum along some direction in space. All spatial directions are equivalent for an isolated atom, but placing the atom in a magnetic field results in the direction of the magnetic field being separated out from the other directions. The convention is that the $+z$ direction is chosen for the magnetic-field direction. Then the z component of the orbital angular momentum of the electron is given by the quantum condition

$$L_z = m_\ell \hbar \qquad \text{36-24}$$

This quantum condition arises from the boundary condition on the azimuth coordinate ϕ that the probability of finding the electron at some arbitrary angle ϕ_1 must be the same as that of finding the electron at angle $\phi_2 = \phi_1 + 2\pi$, because these two values of ϕ represent the same point in space.

If we measure the angular momentum of the electron in units of $\hbar$, we see that the magnitude of the orbital angular momentum is quantized to the value $\sqrt{\ell(\ell + 1)}$ units, and that its component along any direction can have only the $2\ell + 1$ values ranging from $-\ell$ to $+\ell$ units. Figure 36-6 shows a vector-model diagram illustrating the possible orientations of the angular-momentum vector for $\ell = 2$. Note that only specific values of θ are allowed; that is, the directions in space are quantized.

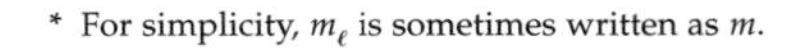

* For simplicity, m_ℓ is sometimes written as m.

Example 36-3 The Directions of the Angular Momentum

If the orbital angular momentum is characterized by the quantum number $\ell = 2$, what are the possible values of L_z, and what is the smallest possible angle between $\vec{L}$ and the direction of increasing z?

PICTURE The possible orientations of $\vec{L}$ and the z axis are shown in Figure 36-6. The direction of increasing z is the direction of the external magnetic field in the vicinity of the atom.

SOLVE

1. Write the possible values of L_z:

$$\boxed{L_z = m_\ell \hbar \quad \text{where} \quad m_\ell = -2, -1, 0, 1, 2}$$

2. Express the angle θ between $\vec{L}$ and the $+z$ direction in terms of L and L_z:

$$\cos\theta = \frac{L_z}{L} = \frac{m_\ell \hbar}{\sqrt{\ell(\ell+1)}\hbar} = \frac{m_\ell}{\sqrt{\ell(\ell+1)}}$$

3. The smallest angle occurs when $\ell = 2$ and $m_\ell = 2$:

$$\cos\theta_{\min} = \frac{2}{\sqrt{2(2+1)}} = \frac{2}{\sqrt{6}} = 0.816$$

$$\theta_{\min} = \boxed{35.3^\circ}$$

CHECK The angle in Figure 36-6*b* looks to be between about 30° and 40°, so the step-3 result of 35.3° is plausible.

TAKING IT FURTHER We note the somewhat strange result that the orbital angular-momentum vector cannot be parallel to the z axis.

PRACTICE PROBLEM 36-2 An atom is in a region that has a magnetic field. An electron in the atom has an angular momentum characterized by the quantum number $\ell = 4$. What are the possible values of m_ℓ for this electron?

36-4 QUANTUM THEORY OF THE HYDROGEN ATOM

We can treat the simplest atom, the hydrogen atom, as a stationary nucleus (a proton) and a single moving particle, an electron, which has linear momentum p and kinetic energy $p^2/2m$. The potential energy $U(r)$ due to the electrostatic attraction between the electron and the proton* is

$$U(r) = \frac{kZe^2}{r} \qquad 36\text{-}25$$

For this potential-energy function, the Schrödinger equation can be solved exactly. In the lowest energy state, which is the ground state, the principal quantum number n has the value 1, ℓ is 0, and m_ℓ is 0.

ENERGY LEVELS

The allowed energies of the hydrogen atom that result from the solution of the Schrödinger equation are

$$E_n = -Z^2\frac{E_0}{n^2} \qquad n = 1, 2, 3, \ldots \qquad 36\text{-}26$$

ENERGY LEVELS FOR HYDROGEN

* We include the factor Z, which is 1 for hydrogen, so that we can apply our results to other one-electron "atoms," such as singly ionized helium He^+, for which $Z = 2$.

where

$$E_0 = \frac{mk^2e^4}{2\hbar^2} = 13.6 \text{ eV} \qquad 36\text{-}27$$

These energies are the same as those obtained using the Bohr model. Note that the energies E_n are negative, indicating that the electron is bound to the nucleus (thus the term *bound state*), and that the energies depend only on the principal quantum number n. The fact that the energy does not depend on the orbital quantum number ℓ is a peculiarity of the inverse-square force and holds only for an inverse r potential such as Equation 36-25. For atoms having multiple electrons, the interaction of the electrons leads to a dependence of the energy on ℓ. In general, the lower the value of ℓ, the lower the energy for such atoms. Because there is usually no preferred direction in space, the energy for any atom does not ordinarily depend on the magnetic quantum number m_ℓ, which is related to the z component of the angular momentum. However, the energy does depend on m_ℓ if the atom is in a magnetic field.

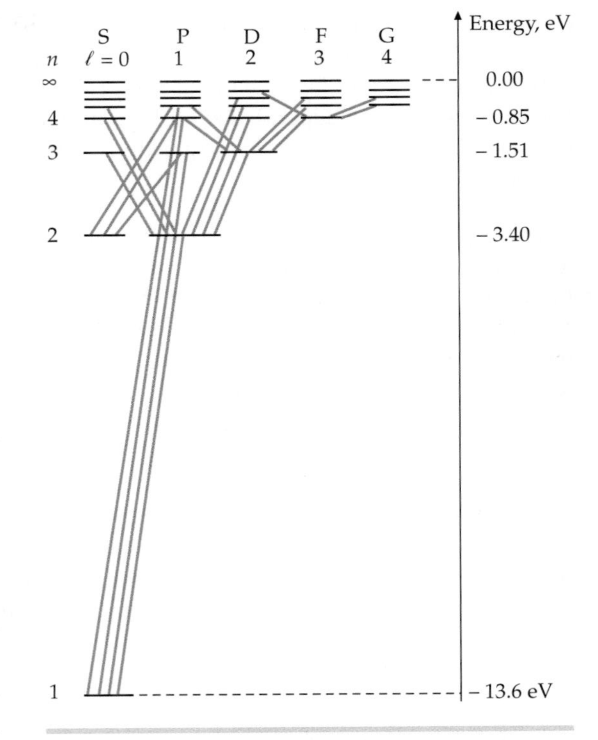

FIGURE 36-7 Energy-level diagram for hydrogen. The diagonal lines show transitions that involve emission or absorption of radiation and obey the selection rule $\Delta\ell = \pm 1$. States that have the same value of n but with different values of ℓ have the same energy $-E_0/n^2$, where $E_0 = 13.6$ eV as in the Bohr model.

Figure 36-7 shows an energy-level diagram for hydrogen. This diagram is similar to Figure 36-4, except that the states which have the same value of n but different values of ℓ are shown separately. These states (called *terms*) are referred to by giving the value of n along with a code letter: s for $\ell = 0$, p for $\ell = 1$, d for $\ell = 2$, and f for $\ell = 3$.* (Lowercase letters s, p, d, f, and so on, are used to specify the orbital angular momentum of an individual electron, whereas uppercase letters S, P, D, F, and so on, are used to identify the orbital angular momentum for the entire multielectron atom. For a single-electron atom, like hydrogen, either uppercase or lowercase letters will suffice.) When an atom makes a transition from one allowed energy state to another, electromagnetic radiation in the form of a photon is emitted or absorbed. Such transitions result in spectral lines that are characteristic of the atom. The transitions obey the **selection rules:**

$$\Delta m_\ell = -1, 0, \text{ or } +1$$
$$\Delta\ell = -1 \text{ or } +1 \qquad 36\text{-}28$$

These selection rules are related to the conservation of angular momentum and to the fact that the photon itself has an intrinsic angular momentum that has a maximum angular-momentum component of $1\hbar$ in any direction. The wavelengths of the light emitted by hydrogen (and by other atoms) are related to the energy levels by

$$hf = \frac{hc}{\lambda} = E_\text{i} - E_\text{f} \qquad 36\text{-}29$$

where E_i and E_f are the energies of the initial and final states.

WAVE FUNCTIONS AND PROBABILITY DENSITIES

The solutions of the Schrödinger equation in spherical coordinates are characterized by the quantum numbers n, ℓ, and m_ℓ, and are written $\psi_{n\ell m_\ell}$. The principal quantum number n can take on any of the values 1, 2, 3, ... In addition, for each value of n, ℓ can take on any of the values $0, 1, \ldots, n-1$, and for each value of ℓ, m_ℓ can take on any of the values $-\ell, -\ell+1, -\ell+2, \ldots, +\ell$. Thus, for any given value of n, there are n possible values of ℓ, and for any given value of ℓ, there are $2\ell+1$ possible values of m_ℓ. For hydrogen, the energy depends only on n, so there are generally many different wave functions that correspond to the same

* These code letters for the values of ℓ are remnants of spectroscopists' descriptions of various spectral lines as *sharp*, *principal*, *diffuse*, and *fundamental*. For values greater than 3, the letters follow alphabetically, starting with g for $\ell = 4$.

energy (except at the lowest energy level, for which $n = 1$ and therefore ℓ and m_ℓ must both equal 0). These energy levels are therefore degenerate (see Section 35-5). The origins of this degeneracy are the $1/r$ dependence of the potential energy and the fact that, in the absence of any external fields, there is no preferred direction in space.*

The ground state In the lowest energy state, the ground state of hydrogen, the principal quantum number n has the value 1, ℓ is 0, and m_ℓ is 0. The energy is −13.6 eV, and the angular momentum is zero. (In the Bohr model of the atom the angular momentum in the ground state is equal to $\hbar$, not zero.) The wave function for the ground state is

$$\psi_{100} = C_{100}e^{-Zr/a_0} \tag{36-30}$$

where

$$a_0 = \frac{\hbar^2}{mke^2} = 0.0529 \text{ nm}$$

is the first Bohr radius and C_{100} is a constant that is determined by normalization. In three dimensions, the normalization condition is

$$\int |\psi|^2 dV = 1$$

where dV is a volume element and the integration is performed over all space. In spherical coordinates, the volume element (Figure 36-8) is

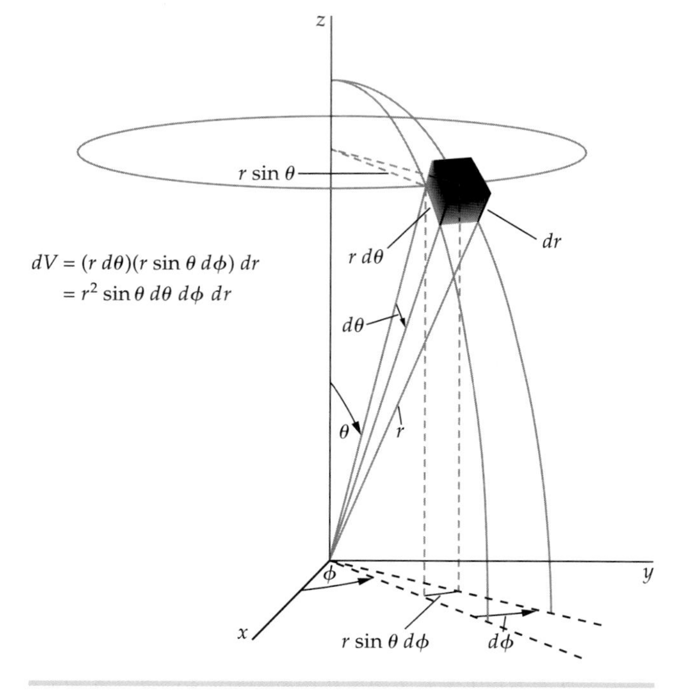

FIGURE 36-8 Volume element in spherical coordinates.

$$dV = (r\sin\theta\, d\phi)(r\, d\theta)dr = r^2 \sin\theta\, d\theta\, d\phi\, dr$$

We integrate over all space by integrating over ϕ from $\phi = 0$ to $\phi = 2\pi$, over θ from $\theta = 0$ to $\theta = \pi$, and over r from $r = 0$ to $r = \infty$. The normalization condition is thus

$$\int |\psi|^2\, dV = \int_0^\infty \left[\int_0^\pi \left(\int_0^{2\pi} |\psi|^2 r^2 \sin\theta\, d\phi\right) d\theta\right] dr$$

$$= \int_0^\infty \left[\int_0^\pi \left(\int_0^{2\pi} C_{100}^2 e^{-2Zr/a_0} r^2 \sin\theta\, d\phi\right) d\theta\right] dr = 1$$

Because there is no θ or ϕ dependence in ψ_{100}, the triple integral can be factored into the product of three integrals. This gives

$$\int |\psi|^2\, dV = \left(\int_0^{2\pi} d\phi\right)\left(\int_0^\pi \sin\theta\, d\theta\right)\left(\int_0^\infty C_{100}^2 e^{-2Zr/a_0} r^2\, dr\right)$$

$$= 2\pi \cdot 2 \cdot C_{100}^2 \left(\int_0^\infty e^{-2Zr/a_0} r^2\, dr\right) = 1$$

The remaining integral is of the form $\int_0^\infty x^n e^{-ax}\, dx$, where n a positive integer and $a > 0$. Using successive integration-by-parts operations† yields the result

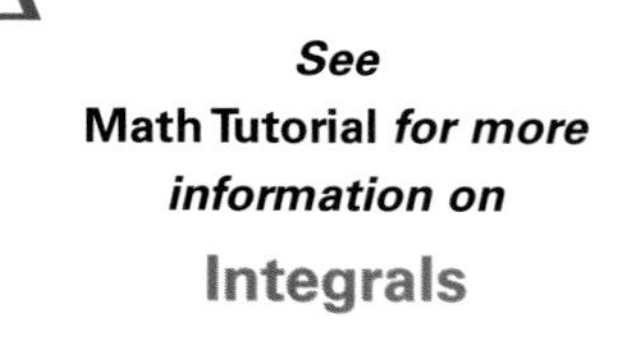

$$\int_0^\infty x^n e^{-ax}\, dx = \frac{n!}{a^{n+1}}$$

so

$$\int_0^\infty r^2 e^{-2Zr/a_0}\, dr = \frac{a_0^3}{4Z^3}$$

* If spin, relativistic effects, the spin of the nucleus, and quantum electrodynamics are considered, the degeneracy is broken.

† This integral can also be looked up in a table of integrals.

Then

$$4\pi C_{100}^2\left(\frac{a_0^3}{4Z^3}\right) = 1$$

so

$$C_{100} = \frac{1}{\sqrt{\pi}}\left(\frac{Z}{a_0}\right)^{3/2} \qquad 36\text{-}31$$

The normalized ground-state wave function is thus

$$\psi_{100} = \frac{1}{\sqrt{\pi}}\left(\frac{Z}{a_0}\right)^{3/2} e^{-Zr/a_0} \qquad 36\text{-}32$$

The probability of finding the electron in a volume dV is $|\psi|^2\,dV$. The probability density $|\psi|^2$ is shown in Figure 36-9. Note that this probability density is spherically symmetric; that is, the probability density depends only on r, and is independent of θ or ϕ. The probability density is maximum at the origin.

We are more often interested in the probability of finding the electron at some radial distance r between r and $r + dr$. This radial probability $P(r)\,dr$ is the probability density $|\psi|^2$ multiplied by the volume of the spherical shell of thickness dr, which is $dV = 4\pi r^2\,dr$. The probability of finding the electron in the range from r to $r + dr$ is thus $P(r)\,dr = |\psi|^2\,4\pi r^2\,dr$, and the **radial probability density** is

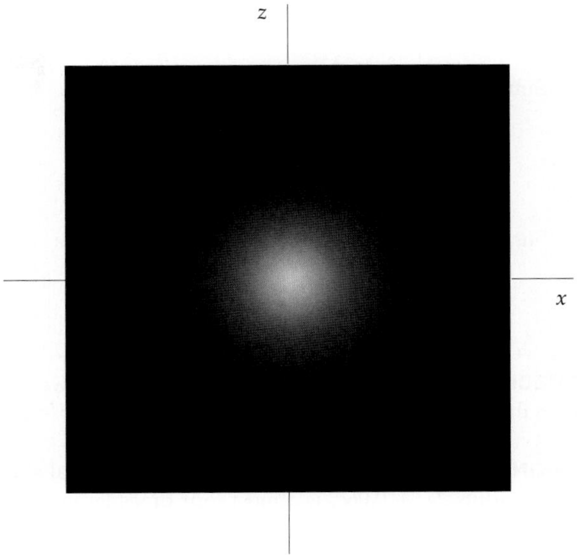

FIGURE 36-9 Computer-generated picture of the probability density $|\psi|^2$ for the ground state of hydrogen. The quantity $-e|\psi|^2$ can be thought of as the electron charge density in the atom. The density is spherically symmetric, is greatest at the origin, and decreases exponentially with r.

$$P(r) = 4\pi r^2|\psi|^2 \qquad 36\text{-}33$$

RADIAL PROBABILITY DENSITY

For the hydrogen atom in the ground state, the radial probability density is

$$P(r) = 4\pi r^2|\psi|^2 = 4\pi C_{100}^2 r^2 e^{-2Zr/a_0} = 4\left(\frac{Z}{a_0}\right)^3 r^2 e^{-2Zr/a_0} \qquad 36\text{-}34$$

Figure 36-10 shows the radial probability density $P(r)$ as a function of r. The maximum value of $P(r)$ occurs at $r = a_0/Z$, which for $Z = 1$ is the first Bohr radius. In contrast to the Bohr model, in which the electron stays in a well-defined orbit at $r = a_0$, we see that it is possible for the electron to be found at any distance from the nucleus. However, the most probable distance is a_0 (assuming $Z = 1$), and the chance of finding the electron at a much different distance is small. It is often useful to think of the electron in an atom as a charged cloud of charge density $\rho = -e|\psi|^2$, but we should remember that when it interacts with matter, an electron is always observed as a single charge.

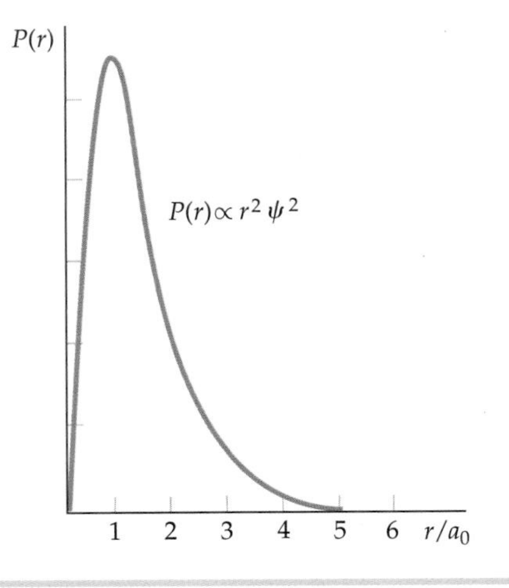

FIGURE 36-10 Radial probability density $P(r)$ versus r/a_0 for the ground state of the hydrogen atom. $P(r)$ is proportional to $r^2\psi^2$. The value of r for which $P(r)$ is maximum is the most probable distance $r = a_0$.

Example 36-4 Probability That the Electron Is in a Thin Spherical Shell

Consider a hydrogen atom that is in the ground state. Estimate the probability of finding the electron in a thin spherical shell of inner radius r and outer radius $r + \Delta r$, where $\Delta r = 0.06a_0$ at (*a*) $r = a_0$ and (*b*) $r = 2a_0$.

PICTURE Because Δr is so small compared to r, the variation in the radial probability density $P(r)$ in the shell can be neglected. The probability of finding the electron in some small range Δr is then $P(r)\,\Delta r$.

SOLVE

1. Substitute $Z = 1$ and $r = a_0$ into Equation 36-34:

$$P(r)\Delta r = \left[4\left(\frac{1}{a_0}\right)^3 r^2 e^{-2r/a_0}\right]\Delta r$$

$$P(a_0)(0.06a_0) = \left[4\left(\frac{1}{a_0}\right)^3 a_0^2 e^{-2}\right](0.06a_0) = \boxed{0.0325}$$

2. Substitute $Z = 1$ and $r = 2a_0$ into Equation 36-34:

$$P(r)\Delta r = \left[4\left(\frac{1}{a_0}\right)^3 r^2 e^{-2r/a_0}\right]\Delta r$$

$$P(2a_0)(0.06a_0) = \left[4\left(\frac{1}{a_0}\right)^3 4a_0^2 e^{-4}\right](0.06a_0) = \boxed{0.0176}$$

CHECK The probability of finding the electron between $r = a_0$ and $r = a_0 + 0.06a_0$ is larger than the probability of finding the particle between $r = 2a_0$ and $r = 2a_0 + 0.06a_0$, as expected.

TAKING IT FURTHER The volume of the spherical shell that has an inner radius $2a_0$ and outer radius $2a_0 + 0.06a_0$ is almost four times larger than the volume of the spherical shell that has an inner radius a_0 and outer radius $a_0 + 0.06a_0$. In spite of this there is approximately a 3 percent chance of finding the electron in this range at $r = a_0$, but at $r = 2a_0$ the chance is slightly less than 2 percent.

The first excited state In the first excited state of a hydrogen atom, n is equal to 2 and ℓ can equal either 0 or 1. For $\ell = 0$, $m_\ell = 0$, and we again have a spherically symmetric wave function, this time given by

$$\psi_{200} = C_{200}\left(2 - \frac{Zr}{a_0}\right)e^{-Zr/(2a_0)} \qquad 36\text{-}35$$

For $\ell = 1$, m_ℓ can be $+1$, 0, or -1. The corresponding wave functions are

$$\psi_{210} = C_{210}\frac{Zr}{a_0}e^{-Zr/(2a_0)}\cos\theta \qquad 36\text{-}36$$

$$\psi_{21\pm1} = C_{211}\frac{Zr}{a_0}e^{-Zr/(2a_0)}\sin\theta\, e^{\pm i\phi} \qquad 36\text{-}37$$

where C_{200}, C_{210}, and C_{211} are normalization constants. The probability densities are given by

$$\psi_{200}^2 = C_{200}^2\left(2 - \frac{Zr}{a_0}\right)^2 e^{-Zr/a_0} \qquad 36\text{-}38$$

$$\psi_{210}^2 = C_{210}^2\left(\frac{Zr}{a_0}\right)^2 e^{-Zr/a_0}\cos^2\theta \qquad 36\text{-}39$$

$$|\psi_{21\pm1}|^2 = C_{211}^2\left(\frac{Zr}{a_0}\right)^2 e^{-Zr/a_0}\sin^2\theta \qquad 36\text{-}40$$

The wave functions and probability densities for $\ell \neq 0$ are not spherically symmetric, but instead depend on the angle θ. The probability densities do not depend on ϕ. Figure 36-11 shows the probability density $|\psi|^2$ for $n = 2$, $\ell = 0$, and $m_\ell = 0$ (Figure 36-11*a*); for $n = 2$, $\ell = 1$, and $m_\ell = 0$ (Figure 36-11*b*); and for $n = 2$, $\ell = 1$, and $m_\ell = \pm1$ (Figure 36-11*c*). An important feature of these plots is that the electron cloud is spherically symmetric for $\ell = 0$ and is not spherically symmetric for $\ell \neq 0$. These angular distributions of the electron charge density depend only on the values of ℓ and m_ℓ and not on the radial part of the wave function. Similar charge distributions for the valence electrons of more complicated atoms play an important role in the chemistry of molecular bonding. (Electrons in the outermost shell are called **valence electrons.**)

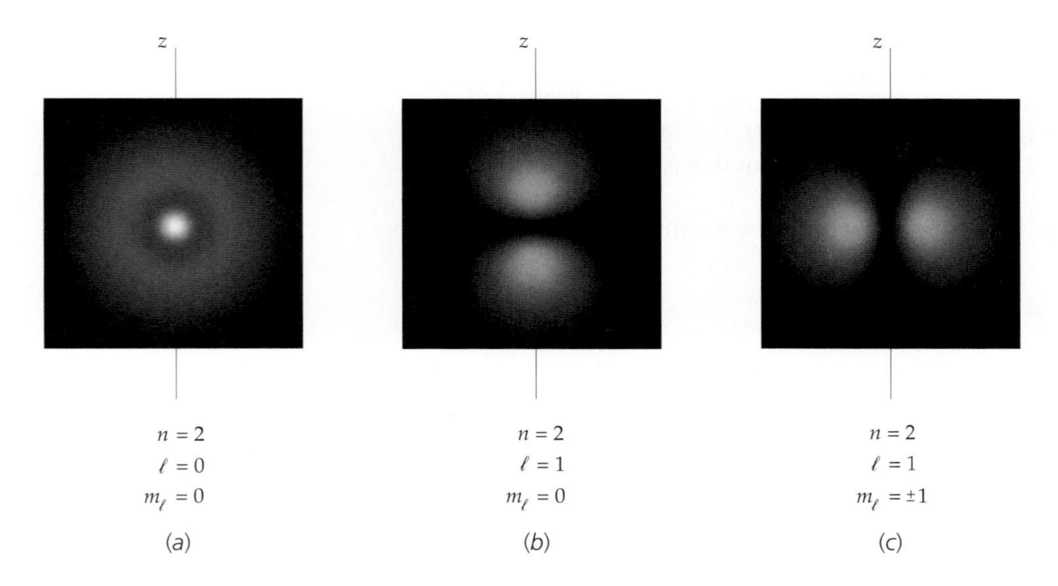

FIGURE 36-11 Computer-generated picture of the probability densities $|\psi|^2$ for the electron in the $n = 2$ states of hydrogen. All three images represent figures of revolution about the z axis. (*a*) For $\ell = 0$, $|\psi|^2$ is spherically symmetric. (*b*) For $\ell = 1$ and $m_\ell = 0$, $|\psi|^2$ is proportional to $\cos^2\theta$. (*c*) For $\ell = 1$ and $m_\ell = +1$ or -1, $|\psi|^2$ is proportional to $\sin^2\theta$.

Figure 36-12 shows the probability density for finding the electron at a distance r from the nucleus for $n = 2$, when $\ell = 1$ and when $\ell = 0$. We can see from the figure that this radial probability density depends on ℓ as well as on n.

For $n = 1$, we found that the most likely distance between the electron and the nucleus is a_0, which is the first Bohr radius, whereas for $n = 2$ and $\ell = 1$, the most probable distance between the electron and the nucleus is $4a_0$. These are the orbital radii for the first and second Bohr orbits (Equation 36-11). For $n = 3$ (and $\ell = 2$),* the most likely distance between the electron and nucleus is $9a_0$, which is the radius of the third Bohr orbit.

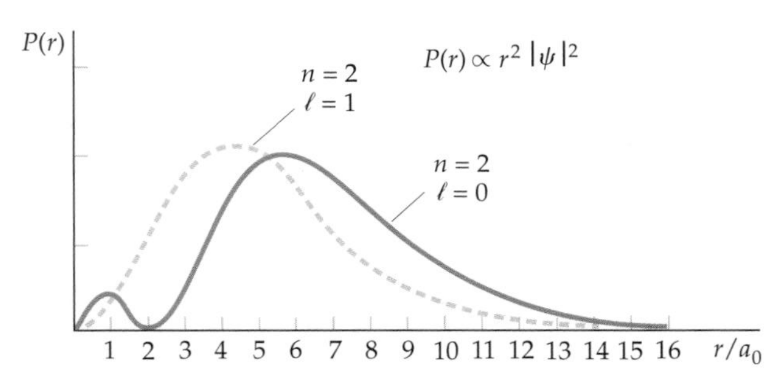

FIGURE 36-12 Radial probability density $P(r)$ versus r/a_0 for the $n = 2$ states of hydrogen. For $\ell = 1$, $P(r)$ is maximum at the Bohr value $r_2 = 2^2 a_0$. For $\ell = 0$, there is a maximum near this value and a much smaller maximum near the origin.

36-5 THE SPIN–ORBIT EFFECT AND FINE STRUCTURE

The orbital magnetic moment of an electron in an atom can be derived semiclassically, even though it is quantum mechanical in origin.† Consider a particle of mass m and charge q moving with speed v in a circle of radius r. The magnitude of the orbital angular momentum of the particle is $L = mvr$, and the magnitude of the magnetic moment is the product of the current and the area of the circle $\mu = IA = I\pi r^2$. If T is the time for the charge to complete one revolution, the current (charge passing a point per unit time) is q/T. Because the period T is the distance $2\pi r$ divided by the velocity v, the current is $I = q/T = qv/(2\pi r)$. The magnetic moment is then

$$\mu = IA = \frac{qv}{2\pi r}\pi r^2 = \frac{1}{2}qvr = \frac{q}{2m}L$$

where we have substituted L/m for vr. If the charge q is positive, the orbital angular momentum and orbital magnetic moment vectors are in the same direction. We can therefore write

$$\vec{\mu} = \frac{q}{2m}\vec{L} \qquad 36\text{-}41$$

* The correspondence with the Bohr model is closest for the maximum value of ℓ, which is $n - 1$.

† This topic was first presented in Section 27-5.

Equation 36-41 is the general classical relation between magnetic moment and angular momentum. It also holds in the quantum theory of the atom for orbital angular momentum, but not for the intrinsic spin angular momentum of the electron. For electron spin, the magnetic moment is twice that predicted by Equation 36-41.* The extra factor of 2 is a result from quantum theory that has no analog in classical mechanics.

The quantum of angular momentum is $\hbar$, so we express the magnetic moment in terms of $\vec{L}/\hbar$:

$$\vec{\mu} = \frac{q\hbar}{2m}\frac{\vec{L}}{\hbar}$$

For an electron, $m = m_e$ and $q = -e$, so the magnetic moment of the electron due to its orbital motion is

$$\vec{\mu}_\ell = -\frac{e\hbar}{2m_e}\frac{\vec{L}}{\hbar} = -\mu_B\frac{\vec{L}}{\hbar}$$

where $\mu_B = e\hbar/(2m_e) = 5.79 \times 10^{-5}$ eV/T is the quantum unit of magnetic moment called a Bohr magneton. The magnetic moment of an electron due to its intrinsic spin angular momentum $\vec{S}$ is

$$\vec{\mu}_S = -2\frac{e\hbar}{2m_e}\frac{\vec{S}}{\hbar} = -2\mu_B\frac{\vec{S}}{\hbar}$$

In general, an electron in an atom has both orbital angular momentum characterized by the quantum number ℓ and spin angular momentum characterized by the quantum number s. Analogous classical systems that have two kinds of angular momentum are Earth, which is spinning about its axis of rotation in addition to revolving about the Sun, and a precessing gyroscope that has angular momentum of precession in addition to its spin. The total angular momentum $\vec{J}$ is the sum of the orbital angular momentum $\vec{L}$ and the spin angular momentum $\vec{S}$, where

$$\vec{J} = \vec{L} + \vec{S} \qquad \text{36-42}$$

Classically $\vec{J}$ is an important quantity because the resultant torque on a system equals the rate of change of the total angular momentum, and in the case of only central forces, the total angular momentum is conserved. For a classical system, the direction of the total angular momentum $\vec{J}$ is without restrictions and the magnitude of $\vec{J}$ can take on any value between $J_{max} = L + S$ and $J_{min} = |L - S|$. In quantum mechanics, however, the directions of both $\vec{L}$ and $\vec{S}$ are more restricted and the magnitudes L and S are both quantized. Furthermore, like $\vec{L}$ and $\vec{S}$, the direction of the total angular momentum $\vec{J}$ is restricted and the magnitude of $\vec{J}$ is quantized. For an electron that has an orbital angular momentum characterized by the quantum number ℓ and spin $s = \frac{1}{2}$, the total angular-momentum magnitude J is equal to $\sqrt{j(j+1)}\hbar$, where the quantum number j is given by

$$j = +\tfrac{1}{2} \qquad \text{if} \qquad \ell = 0$$

and either

$$j = \ell + \tfrac{1}{2} \quad \text{or} \quad j = \ell - \tfrac{1}{2} \qquad \text{if} \qquad \ell > 0 \qquad \text{36-43}$$

Figure 36-13 is a vector model illustrating the two possible combinations $j = \frac{3}{2}$ and $j = \frac{1}{2}$ for the case of $\ell = 1$. The lengths of the vectors are proportional to $\sqrt{\ell(\ell+1)}\hbar$, $\sqrt{s(s+1)}\hbar$, and $\sqrt{j(j+1)}\hbar$. The spin angular momentum and the orbital angular momentum are said to be *parallel* when $j = \ell + s$ and *antiparallel* when $j = \ell - s$.

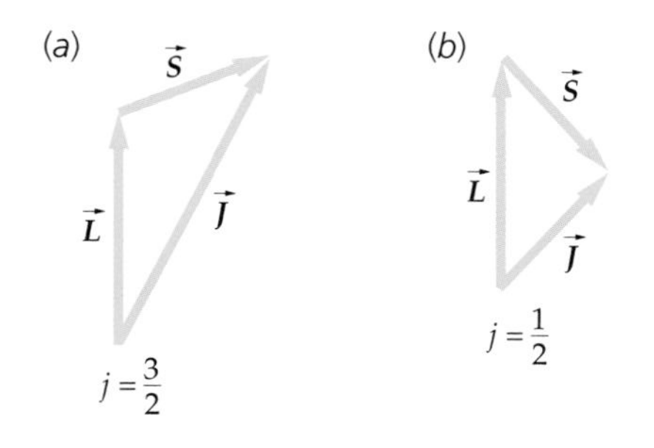

FIGURE 36-13 Vector diagrams illustrating the addition of orbital angular momentum and spin angular momentum for the case $\ell = 1$ and $s = \frac{1}{2}$. There are two possible values of the quantum number for the total angular momentum: $j = \ell + s = \frac{3}{2}$ and $j = \ell - s = \frac{1}{2}$.

* This result and the phenomenon of electron spin itself, was predicted in 1927 by Paul Dirac, who combined special relativity and quantum mechanics into a relativistic wave equation called the *Dirac equation*. Precise measurements indicate that the magnetic moment of the electron due to its spin is 2.00232 times that predicted by Equation 36-42. The fact that the intrinsic magnetic moment of the electron is approximately twice what we would expect makes it clear that the simple model of the electron as a spinning ball is not to be taken literally.

Atomic states that have the same n and ℓ values but with different j values have slightly different energies because of the interaction of the spin of the electron with its orbital motion. This effect is called the **spin–orbit effect.** The resulting splitting of spectral lines is called **fine-structure splitting.**

In the notation $n\ell_j$, the ground state of the hydrogen atom is written $1s_{1/2}$, where the 1 indicates that $n = 1$, the s indicates that $\ell = 0$, and the 1/2 indicates that $j = \frac{1}{2}$. The $n = 2$ states can have either $\ell = 0$ or $\ell = 1$, and the $\ell = 1$ state can have either $j = \frac{3}{2}$ or $j = \frac{1}{2}$. These states are thus denoted by $2s_{1/2}$, $2p_{3/2}$, and $2p_{1/2}$. Because of the spin–orbit effect, the $2p_{3/2}$ and $2p_{1/2}$ states have slightly different energies resulting in the fine-structure splitting of the transitions $2p_{3/2} \rightarrow 2p_{1/2}$ and $2p_{1/2} \rightarrow 2s_{1/2}$.

We can understand the spin–orbit effect qualitatively from a simple Bohr-model picture, as shown in Figure 36-14. In this figure, the electron moves in a circular orbit around a fixed proton. In Figure 36-14*a*, the orbital angular momentum $\vec{L}$ is up. In an inertial reference frame in which the electron is momentarily at rest (see Figure 36-14*b*), the proton is moving at right angles to the line connecting the proton and the electron. The moving proton produces a magnetic field $\vec{B}$ at the position of the electron. The direction of $\vec{B}$ is up, parallel to $\vec{L}$. The energy of the electron depends on its spin because of the magnetic moment $\vec{\mu}_s$ associated with the electron's spin. The energy is lowest when $\vec{\mu}_s$ is parallel to $\vec{B}$ and the energy is highest when it is antiparallel. This energy is given by (Equation 36-16)

$$U = -\vec{\mu}_s \cdot \vec{B} = -\mu_{s_z} B \approx -\mu_B B \qquad 36\text{-}44^*$$

Because $\vec{\mu}_s$ is directed opposite to its spin (because the electron has a negative charge), the energy is lowest when the spin $\vec{S}$ is antiparallel to $\vec{B}$ and thus to $\vec{L}$. The energy of the $2p_{1/2}$ state in hydrogen, in which $\vec{L}$ and $\vec{S}$ are antiparallel (Figure 36-15), is therefore slightly lower than that of the $2p_{3/2}$ state, in which $\vec{L}$ and $\vec{S}$ are parallel.

* Transferring the energy of the dipole to the frame of the proton gives a factor of 2, which is included in this result.

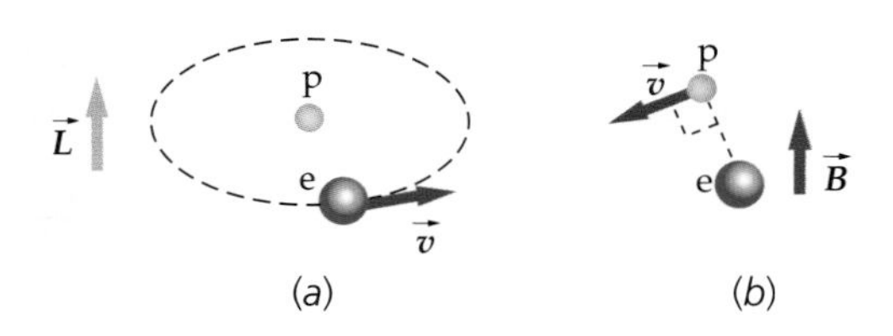

FIGURE 36-14 (*a*) An electron moving about a proton in a circular orbit in the horizontal plane with angular momentum $\vec{L}$ up. (*b*) In an inertial reference frame in which the electron is momentarily at rest there is, at the location of the electron, a magnetic field $\vec{B}$ due to the motion of the proton that is also directed up. When the electron spin $\vec{S}$ is parallel to $\vec{L}$, its magnetic moment $\vec{\mu}_s$ is antiparallel to $\vec{L}$ and $\vec{B}$, so the spin–orbit energy is at its greatest.

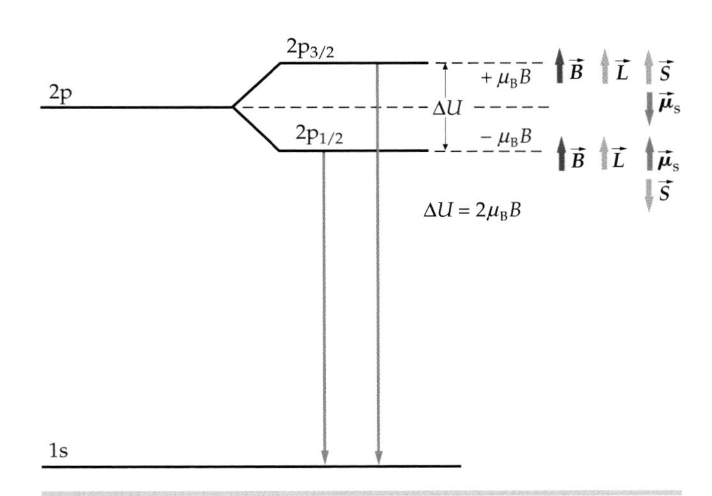

FIGURE 36-15 Fine-structure energy-level diagram. On the left, the levels in the absence of a magnetic field are shown. The effect of the field is shown on the right. Because of the spin-orbit interaction, the magnetic field splits the 2p level into two energy levels, with the $j = \frac{3}{2}$ level having slightly greater energy than the $j = \frac{1}{2}$ level. The spectral line due to the transition $2p \rightarrow 1s$ is therefore split into two lines of slightly different wavelengths.

Example 36-5 Determining *B* by Fine-Structure Splitting

As a consequence of fine-structure splitting, the energies of the $2p_{3/2}$ and $2p_{1/2}$ levels in hydrogen differ by 4.5×10^{-5} eV. If the 2p electron sees an internal magnetic field of magnitude B, the spin–orbit energy splitting will be of the order of $\Delta E = 2\mu_B B$, where μ_B is the Bohr magneton. From this, estimate the magnetic field that the 2p electron in hydrogen experiences.

PICTURE Use the equation $\Delta E = 2\mu_B B$ along with the given value of the energy difference and known value of μ_B.

SOLVE

1. Write the spin–orbit energy splitting in terms of the magnetic moment:

$$\Delta E = 2\mu_B B$$

where

$$\Delta E = 4.5 \times 10^{-5}\ \text{eV}$$

2. Solve for the magnetic field B:

$$B = \frac{\Delta E}{2\mu_B} = \frac{4.5 \times 10^{-5}\ \text{eV}}{2(5.79 \times 10^{-5}\ \text{eV/T})} = \boxed{0.389\ \text{T}}$$

36-6 THE PERIODIC TABLE

For atoms that have more than one electron, the Schrödinger equation cannot be solved exactly. However, powerful approximation methods allow us to calculate the energy levels of the atoms and wave functions of the electrons to a high degree of accuracy. As a first approximation, the Z electrons in an atom are assumed to be noninteracting. The Schrödinger equation can then be solved, and the resulting wave functions used to calculate the interaction of the electrons, which in turn can be used to better approximate the wave functions. Because the spin of an electron can have two possible components along an axis, there is an additional quantum number m_s, which can have the possible values $+\frac{1}{2}$ or $-\frac{1}{2}$. The state of each electron is thus described by the four quantum numbers n, ℓ, m_ℓ, and m_s, and such states are called **stationary states.** The energy of the electron is determined mainly by the principal quantum number n (which is related to the radial dependence of the wave function) and by the orbital angular-momentum quantum number ℓ. Generally, the lower the values of n, the lower the energy; and for a given value of n, the lower the value of ℓ, the lower the energy. The dependence of the energy on ℓ is due to the interaction of the electrons in the atom with each other. In hydrogen, of course, there is only one electron, and the energy is independent of ℓ. The specification of n, ℓ, m_ℓ, and m_s for each electron in an atom is called the **electron configuration.** Customarily, ℓ is specified according to the same code used to label the states of the hydrogen atom rather than by its numerical value. The code is

	s	p	d	f	g	h
ℓ value	0	1	2	3	4	5

The n values are sometimes referred to as shells, which are identified by another letter code: $n = 1$ denotes the K shell*; $n = 2$, the L shell; and so on.

The electron configuration of atoms is constrained by the Pauli exclusion principle—no two electrons in an atom can be in the same quantum state. That is, no two electrons can have the same set of values for the quantum numbers n, ℓ, m_ℓ, and m_s. Using the exclusion principle and the restrictions on the quantum numbers discussed in the previous sections (n is a positive integer, ℓ is an integer that ranges from 0 to $n - 1$, m_ℓ can have $2\ell + 1$ values from $-\ell$ to ℓ in integral steps, and m_s can be either $+\frac{1}{2}$ or $-\frac{1}{2}$), we can understand much of the structure of the periodic table.

We have already discussed the lightest element, hydrogen, which has just one electron. In the ground (lowest energy) state, the electron has $n = 1$ and $\ell = 0$, with $m_\ell = 0$ and $m_s = +\frac{1}{2}$ or $-\frac{1}{2}$. We call this a 1s electron. The 1 signifies that $n = 1$, and the s signifies that $\ell = 0$.

Electrons of atoms whose atomic numbers are greater than 1 will have states that will give the lowest total energy consistent with the Pauli exclusion principle.

CONCEPT CHECK 36-1

The following table lists candidates for the quantum numbers of an electron in an atom. Which of these candidates are not found in nature?

	n	ℓ	m_ℓ	m_s
(*a*)	2	2	-1	$+\frac{1}{2}$
(*b*)	3	2	-1	$+\frac{1}{2}$
(*c*)	2	-1	-1	$-\frac{1}{2}$
(*d*)	3	0	1	$+\frac{1}{2}$
(*e*)	3	1	1	$+\frac{1}{2}$

HELIUM ($Z = 2$)

The next element after hydrogen in the periodic table is helium ($Z = 2$); a helium atom has two electrons. In its ground state, both electrons are in the K shell, where $n = 1$, $\ell = 0$, and $m_\ell = 0$; one electron has $m_s = +\frac{1}{2}$ and the other has $m_s = -\frac{1}{2}$. This configuration is lower in energy than any other two-electron configuration. The resultant spin of the two electrons is zero. Because the orbital angular momentum is also zero, the total angular momentum is zero. The electron configuration for helium is written $1s^2$. The 1 signifies that $n = 1$, the s signifies that $\ell = 0$, and the superscript 2 signifies that there are two electrons in this state. Because ℓ

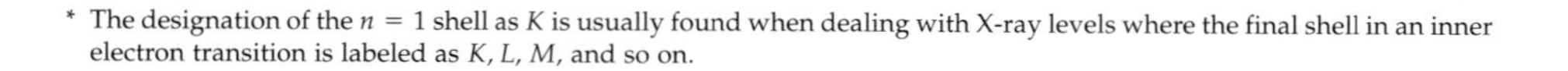

* The designation of the $n = 1$ shell as K is usually found when dealing with X-ray levels where the final shell in an inner electron transition is labeled as K, L, M, and so on.

can be only 0 for $n = 1$, these two electrons fill the K ($n = 1$) shell. The energy required to remove the most loosely bound electron from an atom in the ground state is called the **first ionization energy.** For a helium atom, the first ionization energy is 24.6 eV, which is relatively large. Helium is therefore basically inert.

Example 36-6 Electron Interaction Energy in Helium

(*a*) Use the measured first ionization energy to calculate the energy of interaction of the two electrons in the ground state of the helium atom. (*b*) Use your result to estimate the average separation of the two electrons.

PICTURE The energy of one electron in the ground state of helium is given by $E_n = -Z^2E_0/n^2$ (Equation 36-26), where $n = 1$ and $Z = 2$. If the electrons did not interact, the energy of the second electron would also be E_1, the same as that of the first electron. Thus, for an atom that has noninteracting electrons, the ionization energy of the first electron removed would be $|E_1|$ and the ground-state energy would be $E_{non} = 2E_1$. This is represented by the lowest level in Figure 36-16. Because of the interaction energy, the ground-state energy is greater than $2E_1$, which is represented by the higher level labeled E_g in the figure. The measured first ionization energy of helium is 24.6 eV. When we add $E_{ion} = 24.6$ eV to ionize He, we obtain ionized helium, written He^+, which has just one electron and therefore energy E_1.

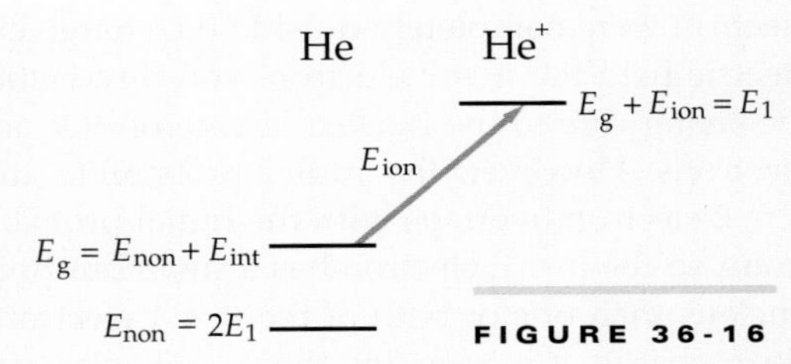

FIGURE 36-16

SOLVE

(*a*) 1. The sum of the energy of interaction E_{int} and the energy of two noninteracting electrons E_{non} equals the ground-state energy of helium:

$$E_{int} + E_{non} = E_g$$

2. Solve for E_{int} and substitute $E_{non} = 2E_1$:

$$E_{int} = E_g - E_{non} = E_g - 2E_1$$

3. Use Equation 36-26 to calculate the energy E_1 of one electron in the ground state:

$$E_n = -Z^2\frac{E_0}{n^2}$$

so

$$E_1 = -(2)^2\frac{13.6\text{ eV}}{1^2} = -54.4\text{ eV}$$

4. Substitute this value for E_1:

$$E_{int} = E_g - 2E_1 = E_g - 2(-54.4\text{ eV})$$
$$= E_g + 108.8\text{ eV}$$

5. The sum of the ground-state energy of He E_g and the ionization energy E_{ion} equals the ground-state energy of He^+, which is E_1:

$$E_g + E_{ion} = E_1 = -54.4\text{ eV}$$

6. Substitute $E_{ion} = 24.6$ eV to calculate E_g:

$$E_g = E_1 - E_{ion} = -54.4\text{ eV} - 24.6\text{ eV}$$
$$= -79.0\text{ eV}$$

7. Substitute this result for E_g to obtain E_{int}:

$$E_{int} = E_g + 108.8\text{ eV} = -79.0\text{ eV} + 108.8\text{ eV}$$
$$= \boxed{29.8\text{ eV}}$$

(*b*) 1. The energy of interaction of two electrons separated by distance r_s is the potential energy:

$$U = +\frac{ke^2}{r_s}$$

2. Set U equal to 29.8 eV, and solve for r. It is convenient to express r in terms of a_0, the radius of the first Bohr orbit in hydrogen, and to use $E_0 = ke^2/(2a_0) = 13.6$ eV (Equation 36-16):

$$r_s = \frac{ke^2}{U} = \frac{ke^2}{a_0}\frac{a_0}{U} = 2\frac{ke^2}{2a_0}\frac{a_0}{U} = 2\frac{E_0}{U}a_0$$
$$= 2\frac{13.6\text{ eV}}{29.8\text{ eV}}a_0 = \boxed{0.913a_0}$$

CHECK This separation is approximately the size of the diameter d_1 of the first Bohr orbit for an electron in helium, which is $d_1 = 2r_1 = 2a_0/Z = a_0$.

LITHIUM ($Z = 3$)

The next element, lithium, has an atom that has three electrons. Because the K shell ($n = 1$) of a ground-state lithium atom is completely filled with two electrons, the third electron occupies a higher energy shell. The next lowest energy shell after $n = 1$ is the $n = 2$ or L shell. This $n = 2$ electron has a greater probability of being much farther from the nucleus than the two $n = 1$ electrons. It is most likely to be found at a radius near that of the second Bohr orbit, which is four times the radius of the first Bohr orbit.

The nuclear charge is partially screened from the $n = 2$ electron by the two $n = 1$ electrons. Recall that the electric field outside a spherically symmetric charge density is the same as if all the charge were at the center of the sphere. If the $n = 2$ electron were completely outside the charge cloud of the two $n = 1$ electrons, the electric field the $n = 2$ electron would see would be that of a single charge $+e$ at the center due to the nuclear charge of $+3e$ and the charge $-2e$ of the two $n = 1$ electrons. However, the radial probability distribution (Equation 36-33) of the $n = 2$ electron overlaps with the radial probability distributions of the $n = 1$ electrons, so the $n = 2$ electron has a small but finite probability of being closer to the nucleus than one or both of the $n = 1$ electrons. Because of this, the effective nuclear charge $Z'e$ seen by the $n = 2$ electron is somewhat greater than $+1e$. Consequently, the energy of the $n = 2$ electron at a distance r from a point charge $+Z'e$ is given by Equation 36-6, where the nuclear charge $+Z$ replaced by $+Z'$.

$$E = -\frac{1}{2}\frac{kZ'e^2}{r} \qquad \text{36-45}$$

The greater the overlap of the radial probability distributions of a higher energy electron with lower energy electrons, the greater is the effective nuclear charge $Z'e$ seen by the higher energy electron and the lower is the energy of the higher energy electron. Because the overlap is greater for ℓ values closer to zero (see Figure 36-12), the energy of the $n = 2$ electron in lithium is lower for the s state ($\ell = 0$) than for the p state ($\ell = 1$). The electron configuration of a lithium atom in the ground state is therefore $1s^22s$. The first ionization energy of a lithium atom is only 5.39 eV. Because its $n = 2$ electron is so loosely bound to the atom, lithium is very active chemically. It behaves like a one-electron atom, similar to hydrogen.

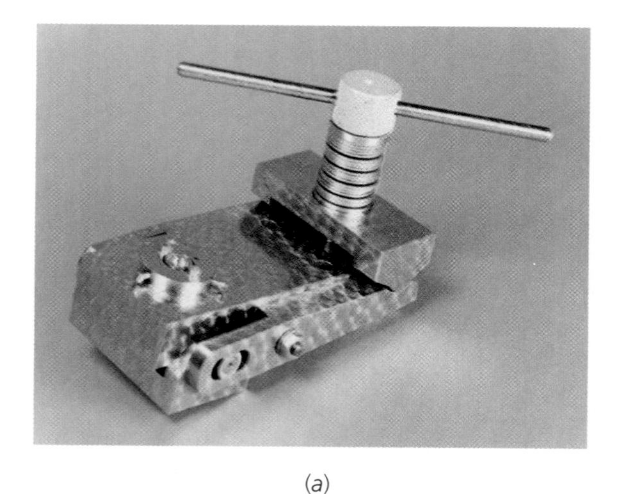

(a)

(b)

(*a*) A diamond anvil cell, in which the facets of two diamonds (approximately 1 mm^2 each) are used to compress a sample substance, subjecting it to very high pressure. (*b*) Samarium monosulfide (SmS) is normally a black, dull-looking semiconductor. When it is subjected to pressure above 7000 atm, an electron from the 4f state moves into the 5d state. The resulting compound glitters like gold and behaves like a metal. *((a) and (b) A. Jayaraman/AT&T Bell Labs.)*

Example 36-7 Effective Nuclear Charge for an Electron of a Lithium Atom

Suppose the radial probability distribution of the $n = 2$ electron in the lithium atom in the ground state did not overlap the probability distribution of the two $n = 1$ electrons; the nuclear charge would be shielded by the two $n = 1$ electrons and the effective nuclear charge would be $Z'e$, where $Z' = 1$. Then the energy of the $n = 2$ electron would be $-(13.6 \text{ eV})/2^2 = -3.4$ eV. However, the first ionization energy of lithium is 5.39 eV, not 3.4 eV. Use this fact to calculate the effective nuclear charge Z' seen by the $n = 2$ electron in lithium.

PICTURE Because the $n = 2$ electron is in the $n = 2$ shell, we will take $r = 4a_0$ for its average distance from the nucleus. We can then calculate Z' from Equation 36-45. Because r is given in terms of a_0, it will be convenient to use the fact that $E_0 = ke^2/(2a_0) = 13.6$ eV (Equation 36-16).

SOLVE

1. Equation 36-45 relates the energy of the $n = 2$ electron to its average distance r from the nucleus and the effective nuclear charge Z':

$$E = -\frac{1}{2}\frac{kZ'e^2}{r}$$

2. Substitute the given values $r = 4a_0$ and $E = -5.39$ eV:

$$-5.39 \text{ eV} = -\frac{1}{2}\frac{kZ'e^2}{4a_0}$$

3. Use $ke^2/(2a_0) = E_0 = 13.6$ eV and solve for Z':

$$-5.39 \text{ eV} = -\frac{Z'}{4}\frac{ke^2}{2a_0} = -\frac{Z'}{4}(13.6 \text{ eV})$$

so

$$Z' = 4\frac{5.39 \text{ eV}}{13.6 \text{ eV}} = \boxed{1.59}$$

CHECK We expected Z' to be larger than one and certainly less than 3. Our step-3 result meets these expectations.

TAKING IT FURTHER This calculation is interesting but not very rigorous. We essentially used the radius ($r = 4a_0$) for the circular orbit from the semiclassical Bohr model and the measured first ionization energy to calculate the effective nuclear charge seen by the $n = 2$ electron. We know, of course, that this $n = 2$ electron does not move in a circular orbit of constant radius, but is better represented by a probability density $|\psi|^2$ that overlaps the probability distributions of the $n = 1$ electrons.

BERYLLIUM ($Z = 4$)

The energy of the beryllium atom is a minimum if both $n = 2$ electrons are in the 2s state. There can be two electrons that have $n = 2$, $\ell = 0$, and $m_\ell = 0$ because of the two possible values for the spin quantum number m_s. The configuration of a beryllium atom is thus $1s^2 2s^2$.

BORON TO NEON ($Z = 5$ TO $Z = 10$)

If the 2s subshell of a ground-state boron atom is filled, the fifth electron must be in the next available (lowest energy) subshell, which is the 2p subshell, where $n = 2$ and $\ell = 1$. Because there are three possible values of m_ℓ (+1, 0, and −1) and two values of m_s for each value of m_ℓ, there can be six electrons in this subshell.

Hydrogen

Carbon

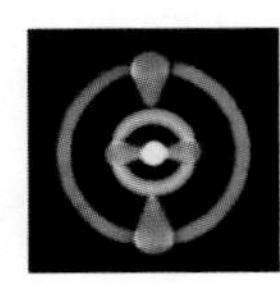

Silicon

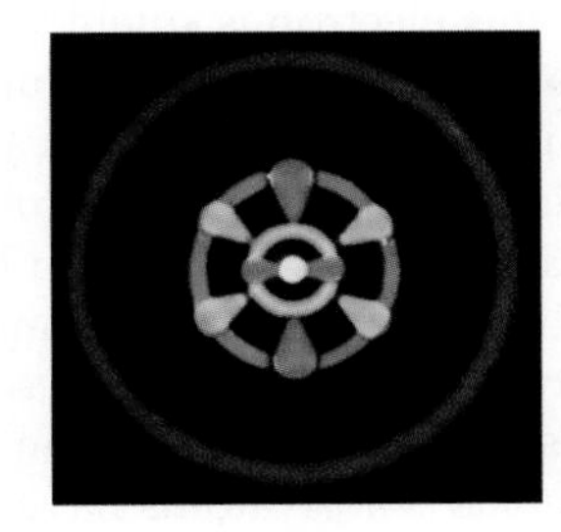

Iron

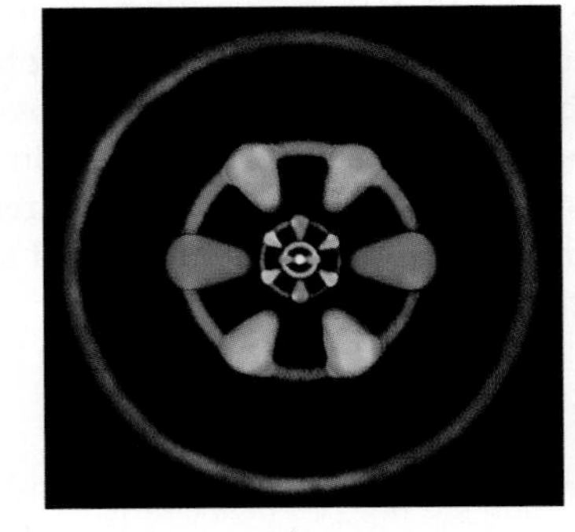

Silver

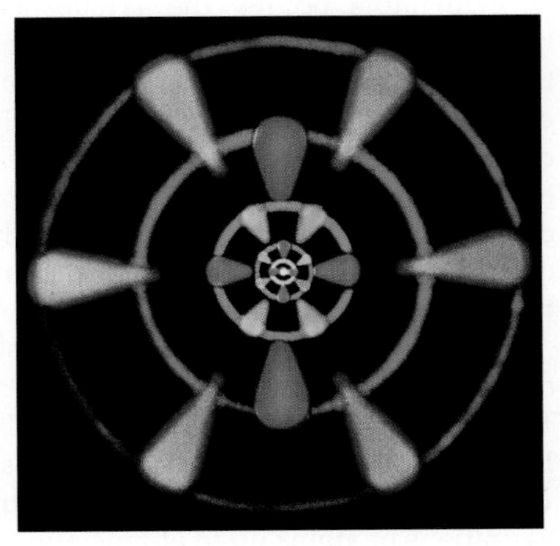

Europium

A schematic depiction of the electron configurations in atoms. The spherically symmetric s states can have 2 electrons and are shown as white and blue. The dumbbell-shaped p states can have up to 6 electrons and are shown as orange. The d states can have up to 10 electrons and are shown as yellow-green. The f states can have up to 14 electrons and are shown as purple. *(David Parker/Photo Researchers.)*

The electron configuration for a boron atom is $1s^22s^22p$. The electron configurations for carbon atoms ($Z = 6$) to neon atoms ($Z = 10$) differ from that for boron atoms only in the number of electrons in the 2p subshell. The first ionization energy increases with Z for these elements, reaching the value of 21.6 eV for the last element in the group, neon. A neon atom has the maximum number of electrons allowed in the $n = 2$ shell and its electron configuration is $1s^22s^22p^6$. Because of its very large first ionization energy, neon, like helium, basically is chemically inert. The atom whose atomic number is one less than neon's atomic number is fluorine, which has an unoccupied electron state in the 2p subshell; that is, a fluorine atom can have one more electron in its 2p subshell. Fluorine readily combines with elements such as lithium that have one electron in its highest energy shell (that is, an electron in an unfilled highest energy shell of an atom in the ground state). Lithium, for example, will donate its single valence electron to the fluorine atom to make an F^- ion and a Li^+ ion. These ions then bond together to form lithium fluoride.

SODIUM TO ARGON ($Z = 11$ TO $Z = 18$)

The eleventh electron of a ground-state sodium atom must be the $n = 3$ shell. Because this electron is shielded from the nucleus by $n = 2$ and $n = 1$ electrons, it is weakly bound in the sodium ($Z = 11$) atom. The first ionization energy of sodium is only 5.14 eV. Sodium atoms therefore combine readily with atoms such as fluorine. With $n = 3$, the value of ℓ can be 0, 1, or 2. A 3s electron has a lower energy than a 3p or 3d electron because its probability density overlap with the probability densities of $n = 2$ and $n = 1$ electrons is greatest. This energy difference between subshells of the same n value becomes greater as the number of electrons increases. The electron configuration of a sodium atom is $1s^22s^22p^63s^1$. For elements whose atoms have larger values of Z, the 3s subshell and then the 3p subshell are occupied. These two subshells can accommodate $2 + 6 = 8$ electrons. The configuration of an argon ($Z = 18$) atom is $1s^22s^22p^63s^23p^6$. One might expect the nineteenth electron of potassium would occupy the third subshell (the d subshell where $\ell = 2$), but the overlap effect is now so strong that the energy of the nineteenth electron is lower in the 4s subshell than in the 3d subshell. There is thus another large energy difference between the eighteenth and nineteenth electrons of a potassium atom, and so an argon atom, with its full 3p subshell, is basically stable and inert.

ELEMENTS WITH $Z > 18$

The nineteenth electron in a potassium ($Z = 19$) atom and the twentieth electron in a calcium ($Z = 20$) atom occupy the 4s subshell rather than the 3d subshell. The electron configurations of the atoms of the next ten elements, scandium ($Z = 21$) through zinc ($Z = 30$), differ only in the number of electrons in the 3d shell, except for a chromium ($Z = 24$) atom and a copper ($Z = 29$) atom, each of which has only one 4s electron. These ten elements are called **transition elements.**

Figure 36-17 shows a plot of the first ionization energies versus Z for $Z = 1$ to $Z = 60$. The peaks in first ionization energy at $Z = 2$, 10, 18, 36, and 54 mark a filled shell or subshell. Table 36-1 gives the ground-state electron configurations of atoms up to atomic number 111.

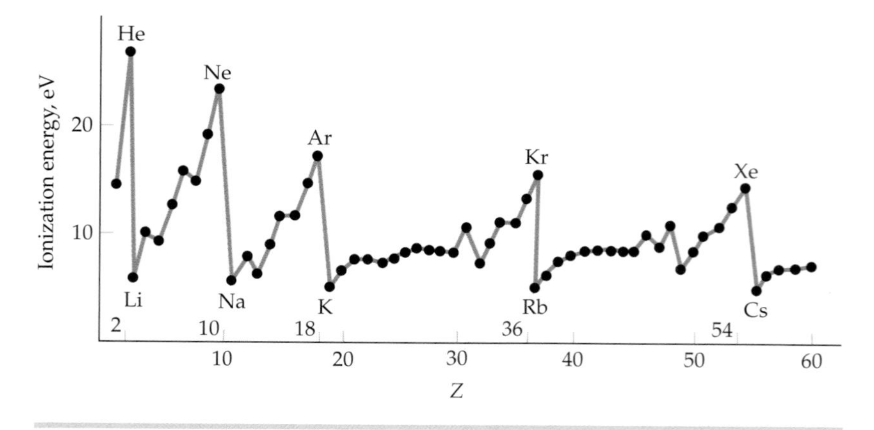

FIGURE 36-17 First ionization energies versus Z for $Z = 1$ to $Z = 60$. The first ionization energy increases with Z until a shell is filled at $Z = 2$, 10, 18, 36, and 54. An atom that has a filled shell and a single valence electron, such as sodium ($Z = 11$), has a very low ionization energy because the valence electron is shielded by the core electrons.

Table 36-1 Electron Configurations of the Atoms in Their Ground States
For some of the rare-earth elements ($Z = 57$ to 71) and the heavy elements ($Z > 89$) the configurations are not firmly established.

		Shell (n):	K (1)	L (2)		M (3)			N (4)				O (5)				P (6)			Q (7)
			s	s	p	s	p	d	s	p	d	f	s	p	d	f	s	p	d	s
Z	Element	Subshell (ℓ):	(0)	(0)	(1)	(0)	(1)	(2)	(0)	(1)	(2)	(3)	(0)	(1)	(2)	(3)	(0)	(1)	(2)	(1)
1	H	hydrogen	1																	
2	He	helium	2																	
3	Li	lithium	2	1																
4	Be	beryllium	2	2																
5	B	boron	2	2	1															
6	C	carbon	2	2	2															
7	N	nitrogen	2	2	3															
8	O	oxygen	2	2	4															
9	F	fluorine	2	2	5															
10	Ne	neon	2	2	6															
11	Na	sodium	2	2	6	1														
12	Mg	magnesium	2	2	6	2														
13	Al	aluminum	2	2	6	2	1													
14	Si	silicon	2	2	6	2	2													
15	P	phosphorus	2	2	6	2	3													
16	S	sulfur	2	2	6	2	4													
17	Cl	chlorine	2	2	6	2	5													
18	Ar	argon	2	2	6	2	6													
19	K	potassium	2	2	6	2	6	.	1											
20	Ca	calcium	2	2	6	2	6	.	2											
21	Sc	scandium	2	2	6	2	6	1	2											
22	Ti	titanium	2	2	6	2	6	2	2											
23	V	vanadium	2	2	6	2	6	3	2											
24	Cr	chromium	2	2	6	2	6	5	1											
25	Mn	manganese	2	2	6	2	6	5	2											
26	Fe	iron	2	2	6	2	6	6	2											
27	Co	cobalt	2	2	6	2	6	7	2											
28	Ni	nickel	2	2	6	2	6	8	2											
29	Cu	copper	2	2	6	2	6	10	1											
30	Zn	zinc	2	2	6	2	6	10	2											
31	Ga	gallium	2	2	6	2	6	10	2	1										
32	Ge	germanium	2	2	6	2	6	10	2	2										
33	As	arsenic	2	2	6	2	6	10	2	3										
34	Se	selenium	2	2	6	2	6	10	2	4										
35	Br	bromine	2	2	6	2	6	10	2	5										
36	Kr	krypton	2	2	6	2	6	10	2	6										
37	Rb	rubidium	2	2	6	2	6	10	2	6	.	.	1							
38	Sr	strontium	2	2	6	2	6	10	2	6	.	.	2							

Continued on next page

Table 36-1 Continued

		Shell (n):	K (1)	L (2)		M (3)			N (4)				O (5)				P (6)			Q (7)
			s	s	p	s	p	d	s	p	d	f	s	p	d	f	s	p	d	s
Z	Element	Subshell (ℓ):	(0)	(0)	(1)	(0)	(1)	(2)	(0)	(1)	(2)	(3)	(0)	(1)	(2)	(3)	(0)	(1)	(2)	(1)
39	Y	yttrium	2	2	6	2	6	10	2	6	1	.	2							
40	Zr	zirconium	2	2	6	2	6	10	2	6	2	.	2							
41	Nb	niobium	2	2	6	2	6	10	2	6	4	.	1							
42	Mo	molybdenum	2	2	6	2	6	10	2	6	5	.	1							
43	Tc	technetium	2	2	6	2	6	10	2	6	6	.	1							
44	Ru	ruthenium	2	2	6	2	6	10	2	6	7	.	1							
45	Rh	rhodium	2	2	6	2	6	10	2	6	8	.	1							
46	Pd	palladium	2	2	6	2	6	10	2	6	10	.	.							
47	Ag	silver	2	2	6	2	6	10	2	6	10	.	1							
48	Cd	cadmium	2	2	6	2	6	10	2	6	10	.	2							
49	In	indium	2	2	6	2	6	10	2	6	10	.	2	1						
50	Sn	tin	2	2	6	2	6	10	2	6	10	.	2	2						
51	Sb	antimony	2	2	6	2	6	10	2	6	10	.	2	3						
52	Te	tellurium	2	2	6	2	6	10	2	6	10	.	2	4						
53	I	iodine	2	2	6	2	6	10	2	6	10	.	2	5						
54	Xe	xenon	2	2	6	2	6	10	2	6	10	.	2	6						
55	Cs	cesium	2	2	6	2	6	10	2	6	10	.	2	6	.	.	1			
56	Ba	barium	2	2	6	2	6	10	2	6	10	.	2	6	.	.	2			
57	La	lanthanum	2	2	6	2	6	10	2	6	10	.	2	6	1	.	2			
58	Ce	cerium	2	2	6	2	6	10	2	6	10	1	2	6	1	.	2			
59	Pr	praseodymium	2	2	6	2	6	10	2	6	10	3	2	6	.	.	2			
60	Nd	neodymium	2	2	6	2	6	10	2	6	10	4	2	6	.	.	2			
61	Pm	promethium	2	2	6	2	6	10	2	6	10	5	2	6	.	.	2			
62	Sm	samarium	2	2	6	2	6	10	2	6	10	6	2	6	.	.	2			
63	Eu	europium	2	2	6	2	6	10	2	6	10	7	2	6	.	.	2			
64	Gd	gadolinium	2	2	6	2	6	10	2	6	10	7	2	6	1	.	2			
65	Tb	terbium	2	2	6	2	6	10	2	6	10	9	2	6	.	.	2			
66	Dy	dysprosium	2	2	6	2	6	10	2	6	10	10	2	6	.	.	2			
67	Ho	holmium	2	2	6	2	6	10	2	6	10	11	2	6	.	.	2			
68	Er	erbium	2	2	6	2	6	10	2	6	10	12	2	6	.	.	2			
69	Tm	thulium	2	2	6	2	6	10	2	6	10	13	2	6	.	.	2			
70	Yb	ytterbium	2	2	6	2	6	10	2	6	10	14	2	6	.	.	2			
71	Lu	lutetium	2	2	6	2	6	10	2	6	10	14	2	6	1	.	2			
72	Hf	hafnium	2	2	6	2	6	10	2	6	10	14	2	6	2	.	2			
73	Ta	tantalum	2	2	6	2	6	10	2	6	10	14	2	6	3	.	2			
74	W	tungsten (wolfram)	2	2	6	2	6	10	2	6	10	14	2	6	4	.	2			
75	Re	rhenium	2	2	6	2	6	10	2	6	10	14	2	6	5	.	2			
76	Os	osmium	2	2	6	2	6	10	2	6	10	14	2	6	6	.	2			
77	Ir	iridium	2	2	6	2	6	10	2	6	10	14	2	6	7	.	2			
78	Pt	platinum	2	2	6	2	6	10	2	6	10	14	2	6	9	.	1			
79	Au	gold	2	2	6	2	6	10	2	6	10	14	2	6	10	.	1			

Continued on next page

Table 36-1 Continued

		Shell (n):	K (1)	L (2)		M (3)			N (4)				O (5)				P (6)			Q (7)
			s	s	p	s	p	d	s	p	d	f	s	p	d	f	s	p	d	s
Z	Element	Subshell (ℓ):	(0)	(0)	(1)	(0)	(1)	(2)	(0)	(1)	(2)	(3)	(0)	(1)	(2)	(3)	(0)	(1)	(2)	(1)
80	Hg	mercury	2	2	6	2	6	10	2	6	10	14	2	6	10	.	2			
81	Tl	thallium	2	2	6	2	6	10	2	6	10	14	2	6	10	.	2	1		
82	Pb	lead	2	2	6	2	6	10	2	6	10	14	2	6	10	.	2	2		
83	Bi	bismuth	2	2	6	2	6	10	2	6	10	14	2	6	10	.	2	3		
84	Po	polonium	2	2	6	2	6	10	2	6	10	14	2	6	10	.	2	4		
85	At	astatine	2	2	6	2	6	10	2	6	10	14	2	6	10	.	2	5		
86	Rn	radon	2	2	6	2	6	10	2	6	10	14	2	6	10	.	2	6		
87	Fr	francium	2	2	6	2	6	10	2	6	10	14	2	6	10	.	2	6	.	1
88	Ra	radium	2	2	6	2	6	10	2	6	10	14	2	6	10	.	2	6	.	2
89	Ac	actinium	2	2	6	2	6	10	2	6	10	14	2	6	10	.	2	6	1	2
90	Th	thorium	2	2	6	2	6	10	2	6	10	14	2	6	10	.	2	6	2	2
91	Pa	protactinium	2	2	6	2	6	10	2	6	10	14	2	6	10	2	2	6	1	2
92	U	uranium	2	2	6	2	6	10	2	6	10	14	2	6	10	3	2	6	1	2
93	Np	neptunium	2	2	6	2	6	10	2	6	10	14	2	6	10	4	2	6	1	2
94	Pu	plutonium	2	2	6	2	6	10	2	6	10	14	2	6	10	6	2	6	.	2
95	Am	americium	2	2	6	2	6	10	2	6	10	14	2	6	10	7	2	6	.	2
96	Cm	curium	2	2	6	2	6	10	2	6	10	14	2	6	10	7	2	6	1	2
97	Bk	berkelium	2	2	6	2	6	10	2	6	10	14	2	6	10	9	2	6	.	2
98	Cf	californium	2	2	6	2	6	10	2	6	10	14	2	6	10	10	2	6	.	2
99	Es	einsteinium	2	2	6	2	6	10	2	6	10	14	2	6	10	11	2	6	.	2
100	Fm	fermium	2	2	6	2	6	10	2	6	10	14	2	6	10	12	2	6	.	2
101	Md	mendelevium	2	2	6	2	6	10	2	6	10	14	2	6	10	13	2	6	.	2
102	No	nobelium	2	2	6	2	6	10	2	6	10	14	2	6	10	14	2	6	.	2
103	Lr	lawrencium	2	2	6	2	6	10	2	6	10	14	2	6	10	14	2	6	1	2
104	Rf	rutherfordium	2	2	6	2	6	10	2	6	10	14	2	6	10	14	2	6	2	2
105	Db	dubnium	2	2	6	2	6	10	2	6	10	14	2	6	10	14	2	6	3	2
106	Sg	seaborgium	2	2	6	2	6	10	2	6	10	14	2	6	10	14	2	6	4	2
107	Bh	bohrium	2	2	6	2	6	10	2	6	10	14	2	6	10	14	2	6	5	2
108	Hs	hassium	2	2	6	2	6	10	2	6	10	14	2	6	10	14	2	6	6	2
109	Mt	meitnerium	2	2	6	2	6	10	2	6	10	14	2	6	10	14	2	6	7	2
110	Ds	darmstadtium	2	2	6	2	6	10	2	6	10	14	2	6	10	14	2	6	9	1
111	Rg	roentgenium	2	2	6	2	6	10	2	6	10	14	2	6	10	14	2	6	10	1

36-7 OPTICAL SPECTRA AND X-RAY SPECTRA

When an atom is in an excited state (when it is in an energy state above the ground state), it makes transitions to lower energy states, and in doing so emits electromagnetic radiation. The wavelength of the electromagnetic radiation emitted is related to the initial and final states by the Bohr formula (Equation 36-17), $\lambda = hc/(E_i - E_f)$, where E_i and E_f are the initial and final energies and h is Planck's constant. The atom can be excited to a higher energy state by bombarding the atom with a beam of electrons, as in a spectral tube that has a high voltage across it. Because the excited energy states of an atom form a discrete (rather than continuous) set, only certain wavelengths are emitted. These wavelengths of the emitted radiation constitute the emission spectrum of the atom.

OPTICAL SPECTRA

To understand atomic spectra we need to understand the excited states of the atom. The situation for an atom that has many electrons is, in general, much more complicated than that of a hydrogen atom that has just one electron. An excited state of the atom may involve a change in the state occupied by any one of the electrons, or even two or more electrons. Fortunately, in most cases, an excited state of an atom involves the excitation of just one of the electrons in the atom. The energies of excitation of the valence electrons of an atom are of the order of a few electron volts. Transitions involving these electrons result in photons in or near the visible or **optical spectrum.** (Recall that the energies of visible photons range from approximately 1.5 eV to 3 eV.) The excitation energies can often be calculated from a simple model in which the atom is pictured as a single electron plus a stable core consisting of the nucleus plus the other electrons. This model works particularly well for the alkali metals: Li, Na, K, Rb, and Cs. These elements are in the first column of the periodic table. The optical spectra of these elements are similar to the optical spectra of hydrogen.

Figure 36-18 shows an energy-level diagram for the optical transitions of a sodium atom, whose electrons form a neon core plus one electron. Because the total spin angular momentum of the core adds up to zero, the spin of each state of the sodium atom is $\frac{1}{2}$ (the spin of the valence electron). Because of the spin–orbit effect, the atomic states for which $J = L - \frac{1}{2}$ have a slightly different energy than those for which $J = L + \frac{1}{2}$ (except for states with $L = 0$). Each state (except for the $L = 0$ states) is therefore split into two states, called a doublet. The doublet splitting is very small and not evident on the energy scale of this diagram. The usual spectroscopic notation is that the states of these atoms are labeled with a superscript given by $2S + 1$, followed by a letter denoting the orbital angular momentum, followed by a subscript denoting the total angular momentum J. For states that have a total spin angular momentum $S = \frac{1}{2}$ the superscript is 2, indicating the state is a doublet. Thus, $^2P_{3/2}$, read as "doublet P three halves," denotes a state in which $L = 1$ and $J = \frac{3}{2}$. (The $L = 0$, or S, states are customarily labeled as if they were doublets even though they are not.) For a sodium atom in the first excited state, the electron is excited from the 3s level to the 3p level, which is approximately 2.1 eV above the ground state. The energy difference between the $P_{3/2}$ and $P_{1/2}$ states due to the spin–orbit effect is about 0.002 eV. Transitions from these states to the ground state give the familiar sodium yellow doublet:

$$3p(^2P_{1/2}) \rightarrow 3s(^2S_{1/2}) \qquad \lambda = 589.6 \text{ nm}$$
$$3p(^2P_{3/2}) \rightarrow 3s(^2S_{1/2}) \qquad \lambda = 589.0 \text{ nm}$$

The energy levels and spectra of other alkali metal atoms are similar to those for sodium. The optical spectrum for atoms such as helium, beryllium, and magnesium that have two valence electrons is considerably more complex because of the interaction of the two electrons.

A neon sign outside a Chinatown restaurant in Paris. Neon atoms in the tube are excited by an electron current passing through the tube. The excited neon atoms emit light in the visible range as they decay toward their ground states. The colors of neon signs result from the characteristic red-orange spectrum of neon plus the color of the glass tube itself. *(Robert Landau/Westlight.)*

X-RAY SPECTRA

X rays are usually produced in the laboratory by bombarding a target element with a high-energy beam of electrons in an X-ray tube. The result (Figure 36-19) consists of a continuous spectrum that depends only on the energy of the bombarding electrons and a line spectrum that is characteristic of the target element. The characteristic spectrum results from excitation of the core electrons in the target element.

The energy needed to excite a core electron—for example, an electron in the $n = 1$ state (K shell)—is much greater than the energy required to excite a valence

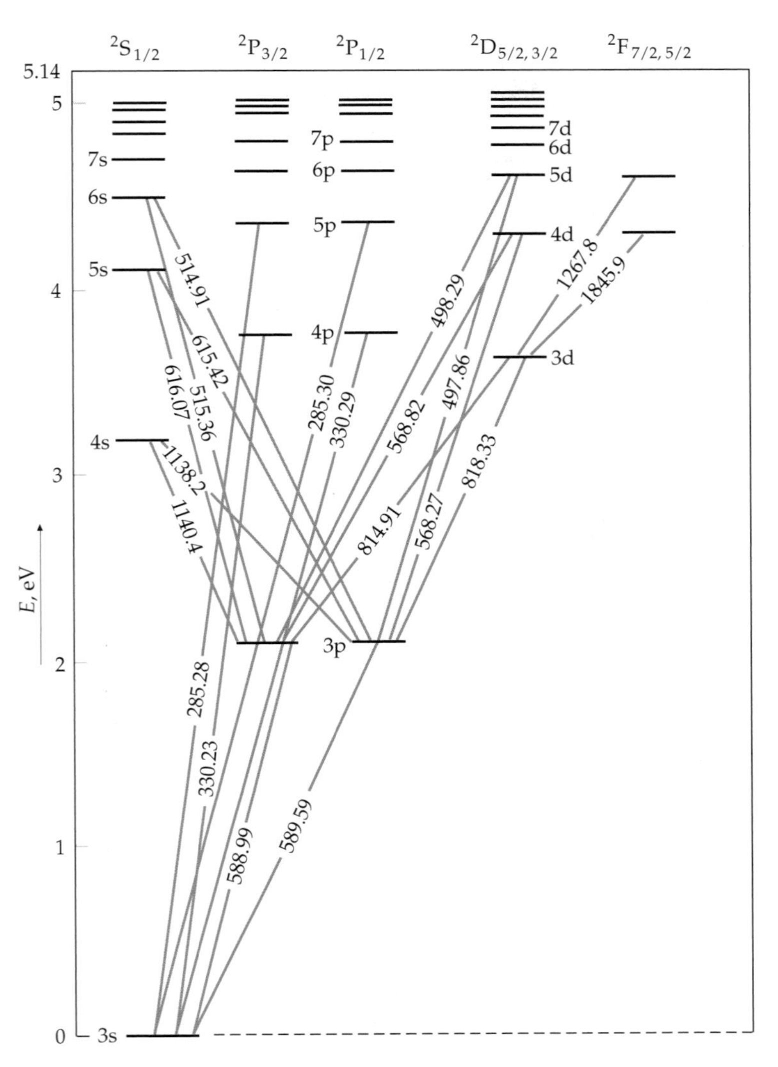

FIGURE 36-18 Energy-level diagram for sodium. The diagonal lines show observed optical transitions, where wavelengths are given in nanometers. The energy of the ground state has been chosen as the zero point for the scale on the left.

electron. A core electron cannot be excited to any of the filled states (for example, the $n = 2$ states in an atom with $Z \geq 10$) because of the exclusion principle. The energy required to excite a core electron to an unoccupied state is typically of the order of several thousand electron volts. If an electron is knocked out of the $n = 1$ shell (the K shell), there is a vacancy left in that shell. The vacancy can be filled if an electron in a higher energy shell makes a transition into the K shell. The photons emitted by electrons making such transitions also have energies of the order of several thousand electron volts and produce the sharp peaks in the X-ray spectrum, as shown in Figure 36-19. The K_α spectral line arises from transitions from the $n = 2$ shell (the L shell) to the $n = 1$ shell (the K shell). The K_β spectral line arises from transitions from the $n = 3$ shell to the $n = 1$ shell. These and other lines arising from transitions ending at the $n = 1$ shell make up the K series of the characteristic X-ray spectrum of the target element. Similarly, a second series, the L series, is produced by transitions from higher energy states to a vacated place in the $n = 2$ (L) shell. The letters K, L, M, and so on, designate the final shell of the electron making the transition and the series α, β, and so on, designates the number of shells above the final shell for the initial state of the electron.

In 1913, the English physicist Henry Moseley measured the wavelengths of the characteristic K_α X-ray spectra for approximately forty elements. Using this data, Moseley showed that a plot of $\lambda^{-1/2}$ versus the order in which the elements appeared in the periodic table resulted in a straight line (with a few gaps and a few outliers). From his data, Moseley was able to accurately determine the atomic

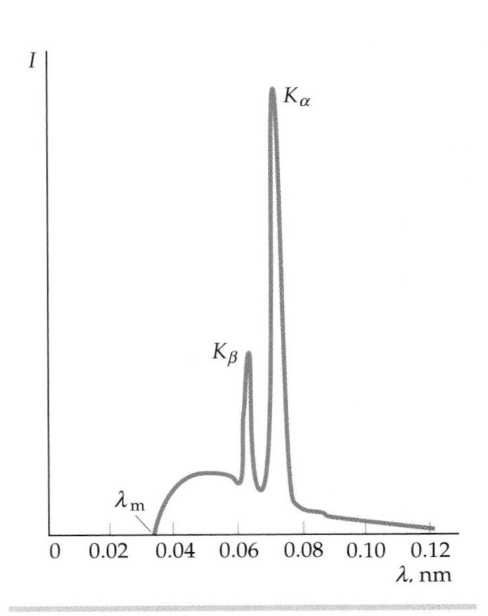

FIGURE 36-19 X-ray spectrum of molybdenum. The sharp peaks labeled K_α and K_β are characteristic of the element. The cutoff wavelength λ_m is independent of the target element and is related to the voltage V of the X-ray tube by $\lambda_m = hc/eV$.

number Z for each known element, and to predict the existence of some elements that were later discovered. The equation of the straight line of his plot is given by

$$\frac{1}{\sqrt{\lambda_{K_\alpha}}} = a(Z - 1)$$

The work of Bohr and Moseley can be combined to obtain an equation relating the wavelength of the emitted photon and the atomic number. According to the Bohr model of a single-electron atom (see Equation 36-13), the wavelength of the emitted photon when the electron makes the transition from $n = 2$ to $n = 1$ is given by

$$\frac{1}{\lambda} = Z^2 \frac{E_0}{hc}\left(1 - \frac{1}{2^2}\right)$$

where $E_0 = 13.6$ eV is the binding energy of the ground-state hydrogen atom. Taking the square root of both sides gives

$$\frac{1}{\sqrt{\lambda_{K_\alpha}}} = \left[\frac{E_0}{hc}\left(1 - \frac{1}{2^2}\right)\right]^{1/2} Z$$

Moseley's equation and this equation are in agreement if $Z - 1$ is substituted for Z in Bohr's equation and if $a = 3E_0/(4hc)$. This result raises the question, why a factor of $Z - 1$ instead of a factor of Z? Part of the explanation is that the formula from the Bohr theory ignores the shielding of the nuclear charge. In a multielectron atom, electrons in the $n = 2$ states are electrically shielded from the nuclear charge by the two electrons in the $n = 1$ state, so the $n = 2$ state electrons are attracted by an effective nuclear charge of about $(Z - 2)e$. However, when there is only one electron in the K shell, the $n = 2$ electrons are attracted by an effective nuclear charge of about $(Z - 1)e$. When an electron from state n drops into the vacated state in the $n = 1$ shell, a photon of energy $E_n - E_1$ is emitted. For $n = 2$, the wavelength of this photon is

$$\lambda_{K_\alpha} = \frac{hc}{(Z - 1)^2 E_0\left(1 - \frac{1}{2^2}\right)} \qquad \text{36-46}$$

which is obtained from the previous equation with $Z - 1$ substituted for Z.

Example 36-8 Identifying the Element from the K_α X-Ray Line

The wavelength of the K_α X-ray line for a certain element is $\lambda = 0.0721$ nm. What is the element?

PICTURE The K_α line corresponds to a transition from $n = 2$ to $n = 1$. The wavelength is related to the atomic number Z by Equation 36-46.

SOLVE

1. Solve Equation 36-46 for $(Z - 1)^2$:

$$\lambda_{K_\alpha} = \frac{hc}{(Z - 1)^2 E_0\left(1 - \frac{1}{2^2}\right)}$$

so

$$(Z - 1)^2 = \frac{4hc}{3\lambda_{K_\alpha} E_0}$$

2. Substitute the given data and solve for Z:

$$(Z - 1)^2 = \frac{4(1240\ \text{eV}\cdot\text{nm})}{3(0.0721\ \text{nm})(13.6\ \text{eV})} = 1686$$

so

$$Z = 1 + \sqrt{1686} = 42.06$$

3. Because Z is an integer, we round to the nearest integer:

$$Z = 42$$

The element is molybdenum.

CHECK The naturally occurring atom that has the largest atomic number is uranium, which has an atomic number $Z = 92$. That our step-3 result is greater than 0 and less than 93 is as expected.

Summary

1. The Bohr model is important because it was the first model to succeed at explaining the discrete optical spectrum of atoms in terms of the quantization of energy. It has been superceded by the quantum-mechanical model.
2. The quantum theory of atoms results from the application of the Schrödinger equation to a bound system consisting of nucleus of charge $+Ze$ and Z electrons of charge $-e$.
3. For the hydrogen atom, an atom that consists of one proton and one electron, the time-independent Schrödinger equation can be solved exactly to obtain the wave functions ψ, which depend on the quantum numbers n, ℓ, m_ℓ, and m_s.
4. The electron configuration of atoms is governed by the Pauli exclusion principle—no two electrons in an atom can have the same set of values for the quantum numbers n, ℓ, m_ℓ, and m_s. Using the exclusion principle and the restrictions on the quantum numbers, we can understand much of the structure of the periodic table.

TOPIC	RELEVANT EQUATIONS AND REMARKS	
1. The Bohr Model of the Hydrogen Atom		
Postulates for the hydrogen atom		
Nonradiating orbits	The idea that an electron moves in a circular nonradiating orbit around the proton.	
Photon frequency from energy conservation	$f = \frac{E_i - E_f}{h}$	36-7
Quantized angular momentum	$L_n = mv_nr_n = n\hbar \qquad n = 1, 2, 3, \ldots$	36-9
First Bohr radius	$a_0 = \frac{\hbar^2}{mke^2} = 0.0529\text{ nm}$	36-12
Radii of the Bohr orbits	$r_n = n^2\frac{a_0}{Z}$	36-11
Energy levels in hydrogen-like atoms	$E_n = -Z^2\frac{E_0}{n^2}$	36-15
	where $E_0 = -\frac{mk^2e^4}{2\hbar^2} = \frac{1}{2}\frac{ke^2}{a_0} = 13.6\text{ eV}$	36-16
Wavelengths emitted by the hydrogen atom	$\lambda = \frac{c}{f} = \frac{hc}{E_i - E_f} = \frac{1240\text{ eV}\cdot\text{nm}}{E_i - E_f}$	36-17, 36-18
2. Quantum Theory of Atoms	The electron is described by a wave function ψ that is a solution of the Schrödinger equation. Energy quantization arises from standing-wave conditions. ψ is described by the principal, orbital, and magnetic quantum numbers n, ℓ, and m_ℓ, and the spin quantum number $m_s = \pm\frac{1}{2}$.	
Time-independent Schrödinger equation	$-\frac{\hbar^2}{2m}\left(\frac{\partial^2\psi}{\partial x^2} + \frac{\partial^2\psi}{\partial y^2} + \frac{\partial^2\psi}{\partial z^2}\right) + U\psi = E\psi$	36-19
For an isolated atom, the solutions can be written as products of functions of r, θ, and ϕ separately	$\psi(r, \theta, \phi) = R(r)f(\theta)g(\phi)$	36-21
Quantum numbers in spherical coordinates		
Principal quantum number	$n = 1, 2, 3, \ldots$	
Orbital quantum number	$\ell = 0, 1, 2, 3, \ldots, n - 1$	
Magnetic quantum number	$m_\ell = -\ell, (-\ell + 1), \ldots, 0, \ldots, (\ell + 1), \ell$	36-22
Orbital angular momentum	$L = \sqrt{\ell(\ell + 1)}\hbar$	36-23

TOPIC	RELEVANT EQUATIONS AND REMARKS	
z component of orbital angular momentum	$L_z = m_\ell \hbar$	36-24
3. Quantum Theory of the Hydrogen Atom		
Energy levels for hydrogen-like atoms (same as for the Bohr model)	$E_n = -Z^2 \dfrac{E_0}{n^2} \quad n = 1, 2, 3, \ldots$	36-26
	where $E_0 = -\dfrac{mk^2e^4}{2\hbar^2} = 13.6 \text{ eV}$	36-27
Wavelengths emitted by the hydrogen atom (same as for Bohr model)	$\lambda = \dfrac{c}{f} = \dfrac{hc}{E_i - E_f} = \dfrac{1240 \text{ eV}\cdot\text{nm}}{E_i - E_f}$	36-17, 36-18
Wave functions		
The ground state	$\psi_{100} = C_{100}e^{-Zr/a_0} = \dfrac{1}{\sqrt{\pi}}\left(\dfrac{Z}{a_0}\right)^{3/2} e^{-Zr/a_0}$	36-30, 36-32
The first excited state	$\psi_{200} = C_{200}\left(2 - \dfrac{Zr}{a_0}\right)e^{-Zr/2a_0}$	36-35
	$\psi_{210} = C_{210}\dfrac{Zr}{a_0}e^{-Zr/2a_0}\cos\theta$	36-36
	$\psi_{21\pm1} = C_{21\pm1}\dfrac{Zr}{a_0}e^{-Zr/2a_0}\sin\theta\, e^{\pm i\phi}$	36-37
Probability densities	For $\ell = 0$, $\lvert\psi\rvert^2$ is spherically symmetric. For $\ell > 0$, $\lvert\psi\rvert^2$ depends on the angle θ.	
Radial probability density	$P(r) = 4\pi r^2 \lvert\psi\rvert^2$	36-33
	The radial probability density is maximum at the distances corresponding roughly to the Bohr orbits.	
4. The Spin–Orbit Effect and Fine Structure	The total angular momentum of an electron in an atom is a combination of the orbital angular momentum and spin angular momentum. It is characterized by the quantum number j, which can be either $\lvert\ell - \frac{1}{2}\rvert$ or $\ell + \frac{1}{2}$. Because of the interaction of the orbital and spin magnetic moments, the state $j = \lvert\ell - \frac{1}{2}\rvert$ has lower energy than the state $j = \ell + \frac{1}{2}$, for $\ell > 0$. This small splitting of the energy states gives rise to a small splitting of the spectral lines called fine structure.	
5. The Periodic Table	An atom of an element has Z electrons, where Z is the atomic number of the element. For an atom in the ground state, the electrons are in those states that will give the lowest energy consistent with the Pauli exclusion principle. The state of an atom is described by its electron configuration, which gives the values of n and ℓ for each electron. The ℓ values are specified by a code: s p d f g h; ℓ value 0 1 2 3 4 5	
Pauli exclusion principle	No two electrons in an atom can have the same set of values for the quantum numbers n, ℓ, m_ℓ, and m_s.	
6. Atomic Spectra	Atomic spectra include optical spectra and X-ray spectra. Optical spectra result from transitions between energy levels of a single valence electron moving in the field of the nucleus and core electrons of the atom. Characteristic X-ray spectra result from the excitation of a core electron and the subsequent filling of the vacancy by other electrons in the atom.	
Selection rules	Transitions between energy states with the emission of a photon are governed by the following selection rules	
	$\Delta m_\ell = 0 \quad \text{or} \quad \Delta m_\ell = \pm 1$ $\Delta\ell = \pm 1$	36-28

Answers to Concept Checks

36-1 (*a*), (*c*), and (*d*)

Answers to Practice Problems

36-1 91.2 nm

36-2 −4, −3, −2, −1, 0, 1, 2, 3, 4

Problems

In a few problems, you are given more data than you actually need; in a few other problems, you are required to supply data from your general knowledge, outside sources, or informed estimate.

Interpret as significant all digits in numerical values that have trailing zeros and no decimal points.

• Single-concept, single-step, relatively easy

•• Intermediate-level, may require synthesis of concepts

••• Challenging

SSM Solution is in the *Student Solutions Manual*

Consecutive problems that are shaded are paired problems.

CONCEPTUAL PROBLEMS

1 • For the hydrogen atom, as n increases, does the spacing of adjacent energy levels on an energy-level diagram increase or decrease? SSM

2 • The energy of the ground state of doubly ionized lithium ($Z = 3$) is ______, where $E_0 = 13.6$ eV. (*a*) $-9E_0$, (*b*) $-3E_0$, (*c*) $-E_0/3$, (*d*) $-E_0/9$

3 • Bohr's quantum condition on electron orbits requires (*a*) that the orbital angular momentum of the electron about the hydrogen nucleus equals $n\hbar$, where n is an integer, (*b*) that no more than one electron occupy a given state, (*c*) that the electrons spiral into the nucleus while radiating electromagnetic waves, (*d*) that the energies of an electron in a hydrogen atom be equal to nE_0, where E_0 is a constant and n is an integer, (*e*) none of the above.

4 • According to the Bohr model, if an electron moves to a larger orbit, does the electron's total energy increase or decrease? Does the electron's kinetic energy increase or decrease?

5 • According to the Bohr model, the kinetic energy of the electron in the ground state of hydrogen is E_0, where $E_0 = 13.6$ eV. The kinetic energy of the electron in the state $n = 2$ is (*a*) $4E_0$, (*b*) $2E_0$, (*c*) $E_0/2$, (*d*) $E_0/4$.

6 • According to the Bohr model, the radius of the $n = 1$ orbit in the hydrogen atom is $a_0 = 0.053$ nm. What is the radius of the $n = 5$ orbit? (*a*) $25a_0$, (*b*) $5a_0$, (*c*) a_0, (*d*) $a_0/5$, (*e*) $a_0/25$.

7 • For the principal quantum number $n = 4$, how many different values can the orbital quantum number ℓ have? (*a*) 4, (*b*) 3, (*c*) 7, (*d*) 16, (*e*) 25 SSM

8 • For the principal quantum number $n = 4$, how many different combinations of ℓ and m_ℓ can occur? (*a*) 4, (*b*) 3, (*c*) 7, (*d*) 16, (*e*) 25

9 •• Why is the energy of the 3s state considerably lower than the energy of the 3p state for sodium, whereas in hydrogen 3s and 3p states have essentially the same energy? SSM

10 • The d state of an electron configuration corresponds to (*a*) $n = 2$, (*b*) $\ell = 3$, (*c*) $\ell = 2$, (*d*) $n = 3$, (*e*) $\ell = 0$.

11 •• Why are three quantum numbers inadequate to describe the states of the electrons in atoms that have more than one electron?

12 •• Group the following six atoms—potassium, calcium, titanium, chromium, manganese, and copper—according to their ground-state electron configurations for the $n = 4$ shell.

13 • What element has the electron configuration (*a*) $1s^22s^22p^63s^23p^3$ and (*b*) $1s^22s^22p^63s^23p^63d^54s^1$?

14 • For the principal quantum number $n = 3$, what are the possible combinations of the quantum numbers ℓ and m_ℓ?

15 • An electron in the L shell means that the electron is represented by (*a*) $\ell = 0$, (*b*) $\ell = 1$, (*c*) $n = 1$, (*d*) $n = 2$, or (*e*) $m_\ell = 2$.

16 •• The Bohr model and the quantum-mechanical model of the hydrogen atom give the same results for the energy levels. Discuss the advantages and disadvantages of each model.

17 •• The Sommerfeld–Hosser displacement theorem states that the optical spectrum of any atom is very similar to the spectrum of the singly charged positive ion of the element immediately following it in the periodic table. Discuss why this theorem is accurate.

18 • Using the triplet of numbers (n, ℓ, m_ℓ) to represent an electron that has the principal quantum number n, orbital quantum number ℓ, and magnetic quantum number m_ℓ, which of the following transitions is allowed? (*a*) $(5, 2, 2) \rightarrow (3, 1, -2)$, (*b*) $(2, 1, 0) \rightarrow (3, 0, 0)$, (*c*) $(4, 3, -2) \rightarrow (3, 2, -1)$, (*d*) $(1, 0, 0) \rightarrow (2, 1, -1)$, (*e*) $(2, 1, 0) \rightarrow (3, 1, 0)$.

19 •• The Ritz combination principle states that for any atom, one can find different spectral lines λ_1, λ_2, λ_3, and λ_4, so that $1/\lambda_1 + 1/\lambda_2 = 1/\lambda_3 + 1/\lambda_4$. Show why this is true using an energy-level diagram. SSM

ESTIMATION AND APPROXIMATION

20 •• (*a*) We can define a thermal de Broglie wavelength λ_T for an atom in a gas at temperature T as being the de Broglie wavelength for an atom moving at the rms speed appropriate to that temperature. (The average kinetic energy of an atom is equal to $\frac{3}{2}kT$, where k is the Boltzmann constant. Use this value to calculate the rms speed of the atoms.) Show that $\lambda_T = h/\sqrt{3mkT}$ where m is the mass of the atom. (*b*) Cooled atoms can form a Bose *condensate* (a new state of matter) when their thermal de Broglie wavelength becomes larger than the average interatomic spacing. From this criterion, estimate the temperature needed to create a Bose condensate in a gas of ^{85}Rb atoms whose number density is 10^{12} atoms/cm^3.

21 •• In laser cooling and trapping, a beam of atoms traveling in one direction is slowed by interaction with an intense laser beam in the opposite direction. The photons scatter off the atoms by resonance absorption, a process by which the incident photon is

absorbed by the atom, and a short time later a photon of equal energy is emitted in a random direction. The net result of a single such scattering event is a transfer of momentum to the atom in a direction opposite to the motion of the atom, followed by a second transfer of momentum to the atom in a random direction. Thus, during photon absorption the atom loses speed, but during photon emission the change in speed of the atom is, on average, zero (because the directions of the emitted photons are random). An analogy often made to this process is that of slowing down a bowling ball by bouncing ping-pong balls off of it. (*a*) Given that the typical photon energy used in these experiments is about 1 eV, and that the typical kinetic energy of an atom in the beam is the typical kinetic energy of the atoms in a gas that has a temperature of about 500 K (a typical temperature for an oven that produces an atomic beam), estimate the number of photon–atom collisions that are required to bring an atom to rest. (The average kinetic energy of an atom is equal to $\frac{3}{2}kT$, where k is the Boltzmann constant and T is the temperature. Use this to estimate the speed of the atoms.) (*b*) Compare the Part (*a*) result with the number of ping-pong ball–bowling ball collisions that are required to bring the bowling ball to rest. (Assume the typical speed of the incident ping-pong balls are all equal to the initial speed of the bowling ball.) (*c*) ^{85}Rb is a type of atom often used during cooling experiments. The wavelength of the light resonant with the cooling transition of the atoms is $\lambda = 780.24$ nm. Estimate the number of photons needed to slow down an ^{85}Rb atom from a typical thermal velocity of 300 m/s to a stop. **SSM**

THE BOHR MODEL OF THE HYDROGEN ATOM

22 • The first Bohr radius is given by $a_0 = \hbar^2/(mke^2) = 0.0529$ nm (Equation 36-12). Use the known values of the constants in the equation to show that a_0 is equal to 0.0529 nm.

23 • The longest wavelength in the Lyman series for the hydrogen atom was calculated in Example 36-2. Find the wavelengths for the transitions (*a*) $n_i = 3$ to $n_f = 1$ and (*b*) $n_i = 4$ to $n_f = 1$.

24 • Find the photon energies for the three longest wavelengths in the Balmer series for the hydrogen atom, and calculate the three wavelengths.

25 •• Find the photon energy and wavelength for the series limit (shortest wavelength) in the Paschen series ($n_f = 3$) for the hydrogen atom. (*b*) Calculate the wavelength for the three longest wavelengths in Paschen series.

26 •• (*a*) Find the photon energy and wavelength for the series limit (shortest wavelength) in the Brackett series ($n_f = 4$) for the hydrogen atom. (*b*) Calculate the wavelength for the three longest wavelengths in Brackett series.

27 ••• In the center-of-mass reference frame of a hydrogen atom, the electron and nucleus have momenta that have equal magnitudes p and opposite directions. (*a*) Using the Bohr model, show that the total kinetic energy of the electron and nucleus can be written $K = p^2/(2\mu)$ where $\mu = m_eM/(M + m_e)$ is called the reduced mass, m_e is the mass of the electron, and M is the mass of the nucleus. (*b*) For the equations for the Bohr model of the atom, the motion of the nucleus can be taken into account by replacing the mass of the electron with the reduced mass. Use Equation 36-14 to calculate the Rydberg constant for a hydrogen atom that has a nucleus of mass $M = m_p$. Find the approximate value of the Rydberg constant by letting M go to infinity in the reduced mass formula. To how many figures does this approximate value agree with the actual value? (*c*) Find the percentage correction for the ground-state energy of the hydrogen atom by using the reduced mass in Equation 36-16.

Note: In general, the reduced mass for a two-body problem with masses m_1 and m_2 is given by

$$\mu = \frac{m_1 m_2}{m_1 + m_2}$$ **SSM**

28 •• The Pickering series of the spectrum of He^+ (singly ionized helium) consists of spectral lines due to transitions to the $n = 4$ state of He^+. Every other line of the Pickering series is very close to a spectral line in the Balmer series for hydrogen transitions to $n = 2$. (*a*) Show that this statement is accurate. (*b*) Calculate the wavelength of the photon during a transition from the $n = 6$ level to the $n = 4$ level of He^+, and show that it corresponds to one of the Balmer series lines.

QUANTUM NUMBERS IN SPHERICAL COORDINATES

29 • For an electron in an atom that has an orbital quantum number $\ell = 1$, find (*a*) the magnitude of the angular momentum L and (*b*) the possible values of the magnetic quantum number m_ℓ. (*c*) Draw a vector diagram to scale showing the possible orientations of $\vec{L}$ relative to the $+z$ direction.

30 • For an electron in an atom that has an orbital quantum number $\ell = 3$, find (*a*) the magnitude of the angular momentum L and (*b*) the possible values of m_ℓ. (*c*) Draw a vector diagram to scale showing the possible orientations of $\vec{L}$ relative to the $+z$ direction.

31 • An electron in an atom has principal quantum number $n = 3$. (*a*) What are the possible values of ℓ? (*b*) What are the possible combinations of ℓ and m_ℓ? (*c*) Using the fact that there are two quantum states for each combination of ℓ and m_ℓ because of electron spin, find the total number of electron states for $n = 3$.

32 • In an atom, find the total number of electron states that have (*a*) $n = 4$ and (*b*) $n = 2$. (See Problem 31.)

33 •• Find the minimum value of the angle θ between $\vec{L}$ and the $+z$ direction for an electron in an atom that has (*a*) $\ell = 1$, (*b*) $\ell = 4$, and (*c*) $\ell = 50$. **SSM**

34 •• What are the possible values of n and m_ℓ for an electron in an atom that has (*a*) $\ell = 3$, (*b*) $\ell = 4$, and (*c*) $\ell = 0$?

35 •• For an electron in an atom that is in an $\ell = 2$ state, find (*a*) the magnitude of the angular momentum squared L^2, (*b*) the maximum value of L_z^2, and (*c*) the smallest value of $L_x^2 + L_y^2$.

QUANTUM THEORY OF THE HYDROGEN ATOM

36 • For the ground state of the hydrogen atom, find the values of (*a*) $\psi(r)$ at $r = a_0$, (*b*) $\psi^2(r)$ at $r = a_0$, and (*c*) the radial probability density $P(r)$ at $r = a_0$. Give your answers in terms of a_0.

37 • (*a*) If electron spin is not included, how many different wave functions are there corresponding to the first excited energy level $n = 2$ for a hydrogen atom? (*b*) Specify the quantum numbers for each of these wave functions. **SSM**

38 •• For the ground state of the hydrogen atom, calculate the probability of finding the electron in the region between r and $r + \Delta r$, where $\Delta r = 0.03a_0$ and (*a*) $r = a_0$ and (*b*) $r = 2a_0$.

39 •• The value of the constant C_{200} in the equation

$$\psi_{200} = C_{200}\left(2 - \frac{Zr}{a_0}\right)e^{-Zr/(2a_0)}$$

(Equation 36-35) is given by

$$C_{200} = \frac{1}{4\sqrt{2\pi}}\left(\frac{Z}{a_0}\right)^{3/2}$$

Find the values of (*a*) $\psi(r)$ at $r = a_0$, (*b*) $\psi^2(r)$ at $r = a_0$, and (*c*) the radial probability density $P(r)$ at $r = a_0$ for the state $n = 2$, $\ell = 0$, and $m_\ell = 0$ of a hydrogen atom. Give your answers in terms of a_0.

40 ••• Show that the radial probability density for the $n = 2$, $\ell = 1$, and $m_\ell = 0$ state of a one-electron atom can be written as $P(r) = A\cos^2\theta\, r^4 e^{-Zr/a_0}$, where A is a constant.

41 ••• Calculate the probability of finding the electron in the region between r and $r + \Delta r$, where $\Delta r = 0.02a_0$ and (*a*) $r = a_0$ and (*b*) $r = 2a_0$ for the state $n = 2$, $\ell = 0$, and $m_\ell = 0$ in hydrogen. (See Problem 39 for the value of C_{200}.)

42 •• Show that the ground-state hydrogen atom wave function $\psi_{100} = \pi^{-1/2}(Z/a_0)^{3/2}e^{-Zr/a_0}$ (Equation 36-32) is a solution to Schrödinger's equation in spherical coordinates:

$$\frac{-\hbar}{2mr^2}\left\{\frac{\partial}{\partial r}\left(r^2\frac{\partial\psi}{\partial r}\right) + \left[\frac{1}{\sin\theta}\frac{\partial}{\partial\theta}\left(\sin\theta\frac{\partial\psi}{\partial\theta}\right) + \frac{1}{\sin^2\theta}\frac{\partial^2\psi}{\partial\phi^2}\right]\right\} + U(r)\psi = E\psi$$

where $U(r) = kZe^2/r$ (Equation 36-25).

43 •• Show by dimensional analysis that the expression for the hydrogen atom ground-state energy given by $E_0 = \frac{1}{2}mk^2e^4/\hbar^2$ (Equation 36-27) has the dimensions of energy.

44 •• By dimensional analysis, show that the expression for the first Bohr radius given by $a_0 = \hbar^2/(mke^2)$ (Equation 36-12) has the dimensions of length.

45 •• The radial probability distribution function for a one-electron atom in its ground state can be written $P(r) = Cr^2e^{-2Zr/a_0}$, where C is a constant. Show that $P(r)$ has its maximum value at $r = a_0/Z$.

46 ••• Show that the number of states in the hydrogen atom for a given n is $2n^2$.

47 ••• Calculate the probability that the electron in the ground state of a hydrogen atom is in the region $0 < r < a_0$.

THE SPIN–ORBIT EFFECT AND FINE STRUCTURE

48 • The potential energy of a magnetic moment in an external magnetic field is given by $U = -\vec{\mu}\cdot\vec{B}$. (*a*) Calculate the difference in energy between the two possible orientations of an electron in a magnetic field $\vec{B} = 1.50\,\mathrm{T}\hat{k}$. (*b*) If the electrons are bombarded with photons of energy equal to that energy difference, "spin flip" transitions can be induced. Find the wavelength of the photons needed for such transitions. This phenomenon is called *electron spin resonance*.

49 • The total angular momentum of a hydrogen atom in a certain excited state has the quantum number $j = \frac{1}{2}$. What can you say about the value of the orbital angular-momentum quantum number ℓ?

50 • A hydrogen atom is in the state $n = 3$, $\ell = 2$. What are the possible values of j?

51 • Using a scaled vector diagram, show how the orbital angular momentum $\vec{L}$ combines with the spin angular momentum $\vec{S}$ to produce the two possible values of total angular momentum $\vec{J}$ for the $\ell = 3$ state of the hydrogen atom.

THE PERIODIC TABLE

52 • The total number of states of a hydrogen atom that has principal quantum number $n = 4$ is (*a*) 4, (*b*) 16, (*c*) 32, (*d*) 36, (*e*) 48

53 • How many of the eight electrons in an oxygen atom in the ground state are in a p state? (*a*) 0, (*b*) 2, (*c*) 4, (*d*) 6, (*e*) 8

54 • Write the ground-state electron configuration of (*a*) an atom of carbon and (*b*) an atom of oxygen.

55 • Give the possible values of the *z* component of the orbital angular momentum of (*a*) a d electron and (*b*) an f electron.

OPTICAL SPECTRA AND X-RAY SPECTRA

56 • The optical spectra of atoms that have two electrons in the same highest energy shell are similar, but they are quite different from the spectra of atoms that have just one electron in the highest energy shell because of the interaction of the two electrons. Group the elements according to similar spectra: lithium, beryllium, sodium, magnesium, potassium, calcium, chromium, nickel, cesium, and barium.

57 • Write down the possible electron configurations for the first excited state of (*a*) a hydrogen atom, (*b*) a sodium atom, and (*c*) a helium atom.

58 • Indicate which of the following atoms should have optical spectra similar to a hydrogen atom and which of the following atoms should have optical spectra similar to a helium atom: Li, Ca, Ti, Rb, Hg, Ag, Cd, Ba, Fr, and Ra.

59 • (*a*) Calculate the next two longest wavelengths in the K series (after the K_α line) of molybdenum. (*b*) What is the wavelength of the shortest wavelength in this series?

60 • The wavelength of the K_α line for a certain element is 0.3368 nm. What is the element?

61 • Calculate the wavelength of the K_α line in (*a*) a magnesium ($Z = 12$) atom and (*b*) a copper ($Z = 29$) atom.

GENERAL PROBLEMS

62 • What is the energy of the shortest wavelength photon emitted by the hydrogen atom?

63 • The wavelength of a spectral line of hydrogen is 97.254 nm. Identify the transition that results in this line, assuming that the transition is to the ground state.

64 •• The wavelength of a spectral line of hydrogen is 1093.8 nm. Identify the transition that results in this line.

65 •• Spectral lines of the following wavelengths are emitted by a singly ionized helium atom: 164 nm, 230.6 nm, and 541 nm. Identify the transitions that result in those spectral lines.

66 •• The combination of physical constants $\alpha = e^2k/\hbar c$, where k is the Coulomb constant, is known as the *fine-structure constant*. It appears in numerous relations in atomic physics. (*a*) Show that α is dimensionless. (*b*) Show that in the Bohr model of the hydrogen atom $v_n = c\alpha/n$, where v_n is the speed of the electron in the state of quantum number n.

67 •• The wavelengths of the photons emitted by a potassium atom corresponding to transitions from the $4P_{3/2}$ and $4P_{1/2}$ states to the ground state are 766.41 nm and 769.90 nm. (*a*) Calculate the energies of the photons in electron volts. (*b*) The difference in the energies of the photons equals the difference in energy ΔE between the $4P_{3/2}$ and $4P_{1/2}$ states in potassium. Calculate ΔE. (*c*) Estimate the magnetic field that the 4p electron in potassium experiences.

68 •• To observe the characteristic K lines of the X-ray spectrum, one of the $n = 1$ electrons must be ejected from the atom. This is generally accomplished by bombarding the target material with electrons of sufficient energy to eject this tightly bound electron. What is the minimum energy required to observe the K lines of (*a*) a tungsten atom, (*b*) a molybdenum atom, and (*c*) a copper atom?

69 •• We are often interested in finding the quantity ke^2/r in electron volts when r is given in nanometers. Show that $ke^2 = 1.44\ \text{eV}\cdot\text{nm}$. SSM

70 •• The *positron* is a particle that has the same mass as the electron and carries a charge equal to $+e$. *Positronium* is a bound state of an electron–positron combination. (*a*) Calculate the energies of the five lowest energy states of positronium using the reduced mass, as given in Problem 27. (*b*) Do transitions between any of the levels found in Part (*a*) fall in the visible range of wavelengths? If so, which transitions are they?

71 • In 1947, Lamb and Retherford showed that there is a very small energy difference between the $2S_{1/2}$ and the $2P_{1/2}$ states of the hydrogen atom. They measured this difference essentially by causing transitions between the two states using very long wavelength electromagnetic radiation. The energy difference (the Lamb shift) is 4.372×10^{-6} eV and is explained by quantum electrodynamics as being due to fluctuations in the energy level of the vacuum. (*a*) What is the frequency of a photon whose energy is equal to the Lamb shift energy? (*b*) What is the wavelength of that photon? In what spectral region does it belong?

72 • A Rydberg atom is one in which an electron is in a *very* high excited state ($n \approx 40$ or higher). Such atoms are useful for experiments that probe the transition from quantum-mechanical behavior to classical. Furthermore, these excited states have extremely long lifetimes (i.e., the electron will stay in this high excited state for a very long time). A hydrogen atom is in the $n = 45$ state. (*a*) What is the ionization energy of the atom when it is in that state? (*b*) What is the energy level separation (in electron volts) between that state and the $n = 44$ state? (*c*) What is the wavelength of a photon resonant with the transition between these two states? (*d*) What is the radius of the atom when it is in the $n = 45$ state?

73 •• The deuteron, the nucleus of deuterium (heavy hydrogen), was first recognized from the spectrum of hydrogen. The deuteron has a mass that is approximately twice the mass of the proton. (*a*) Calculate the Rydberg constant for hydrogen and for deuterium using the reduced mass as given in Problem 27. (*b*) Using the result obtained in Part (*a*), determine the difference between the longest wavelength Balmer line of hydrogen (protium) and the longest wavelength Balmer line of deuterium.

74 •• The muonium atom is a hydrogen atom that has the electron replaced by a μ^- particle. The μ^- has a mass 207 times as great as the electron. (*a*) Calculate the energies of the five lowest energy levels of muonium using the reduced mass as given in Problem 27. (*b*) Do transitions between any of the levels found in Part (*a*) fall in the visible range of wavelengths (for example, between $\lambda = 700$ nm and $\lambda = 400$ nm)? If so, which transitions are they?

75 •• The triton, a nucleus consisting of a proton and two neutrons, is unstable and has a half-life of approximately 12 years. Tritium is an atom consisting of an electron and a triton. (*a*) Calculate the Rydberg constant of tritium using the reduced mass as given in Problem 27. (*b*) Determine the difference between the longest wavelength of the Balmer lines of tritium and the longest wavelength of the Balmer lines of deuterium (see Problem 73). In addition, (*c*) determine the difference between the longest wavelength of the Balmer lines of tritium and the longest wavelength of the Balmer lines of hydrogen (protium).

A MICROGRAPH OF SODIUM FLUORIDE CRYSTALS. SODIUM FLUORIDE IS OFTEN ADDED TO PUBLIC WATER SUPPLIES AS A TOOTH-DECAY PREVENTATIVE. *(National Institutes of Health/Photo Researchers.)*

Molecules

? How much energy is needed to form sodium fluoride? (See Example 37-1).

Most atoms bond together to form molecules or solids. Molecules may exist as separate entities, as in gaseous O_2 or N_2, or they may bond together to form liquids or solids. A molecule is the smallest constituent of a substance that retains its chemical properties.

In this chapter, we use our understanding of quantum mechanics to discuss bonding and the energy levels and spectra of diatomic molecules. Much of our discussion will be qualitative because, as in atomic physics, the quantum-mechanical calculations are very difficult.

37-1 BONDING

Consider a hydrogen molecule (H_2). We can think of H_2 either as two H atoms joined together or as a quantum-mechanical system of two protons and two electrons. The latter picture is more useful in this case because neither of the electrons in the H_2 molecule is confined to the region surrounding either one of the two protons. Instead, each electron is equally shared by both protons. For more complicated molecules, however, an intermediate picture is useful. For example, the fluorine molecule F_2 consists of 18 protons and 18 electrons, but only two of the electrons take part in the bonding. We therefore can consider this molecule as two F^+ ions and two electrons that belong to the molecule as a whole. The molecular wave functions for the bonding electrons are called **molecular orbitals.** In many cases, these molecular wave functions can be constructed from combinations of the atomic wave functions with which we are familiar.

The two principal types of bonds responsible for the formation of solids and molecules are the ionic bond and the covalent bond. Other types of bonds that are important in the bonding of liquids and solids are van der Waals bonds, metallic bonds, and hydrogen bonds. In many cases, bonding is a mixture of these mechanisms.

THE IONIC BOND

The simplest type of bond is the **ionic bond,** which is found in salts such as sodium chloride (NaCl). The sodium atom has one 3s electron outside a stable ten-electron core. The first ionization energy of sodium is the energy needed to remove the 3s electron from an isolated sodium atom. This energy is just 5.14 eV (see Figure 36-18). The removal of this electron results in an isolated positive ion that has its $n = 1$ and $n = 2$ electron shells filled. A chlorine atom has 17 electrons, and so is one electron short of having its first three shells filled. A measure of the energy released when an isolated atom gains one electron is called its **electron affinity;** a chlorine atom releases 3.62 eV of energy when it acquires an electron to form a Cl^- ion. Thus, the chlorine atom is said to have an electron affinity of -3.62 eV. The acquisition of one electron by chlorine results in a negative ion that has a filled outer electron shell. Thus, the formation of a Na^+ ion and a Cl^- ion by the donation of one electron of sodium to chlorine requires only 5.14 eV $-$ 3.62 eV $=$ 1.52 eV at infinite separation. The electrostatic potential energy U_e of the two ions when they are a distance r apart is $-ke^2/r$. When the separation of the ions is less than approximately 0.95 nm, the negative potential energy of attraction is of greater magnitude than the 1.52 eV of energy needed to create the ions. Thus, at separation distances less than 0.95 nm, it is energetically favorable (the total energy of the system is reduced) for the sodium atom to donate an electron to the chlorine atom to form NaCl.

Because the electrostatic attraction increases as the ions get closer together, it might seem that equilibrium could not exist. However, when the separation of the ions is very small, there is a strong repulsion that can be described by quantum mechanics and the exclusion principle. This repulsion is also responsible for the repulsion of the atoms in all molecules and ions (except H_2).* We can understand it qualitatively as follows. When the ions are very far apart, the probability distribution for a core electron in one of the ions does not overlap the probability distribution of any electron in the other ion. We can distinguish the electrons by the ion to which they belong. This means that electrons in the two ions can have the same quantum numbers because they occupy different regions of space. However, as the distance between the ions decreases, the probability distributions of the core electrons begin to overlap; that is, the electrons in the two ions begin to occupy the same region of space. Some of these electrons must go into higher energy quantum states as described by the exclusion principle.† But energy is required to shift the electrons into higher energy quantum states. This increase in energy when the ions are pushed closer together is equivalent to the repulsion energy of the ions. It is not a sudden process. The energy states of the electrons change gradually as the ions are brought together. A sketch of the potential energy $U(r)$ of the Na^+ and Cl^- ions versus separation distance r is shown in Figure 37-1. The energy is lowest at an

---

* In H_2, the repulsion is simply that of the two positively charged protons.

† Recall from our discussion in Chapter 35 that the exclusion principle is related to the fact that the wave function for two identical electrons is antisymmetric on the exchange of the electrons and that an antisymmetric wave function for two electrons with the same quantum numbers is zero if the space coordinates of the electrons are the same.

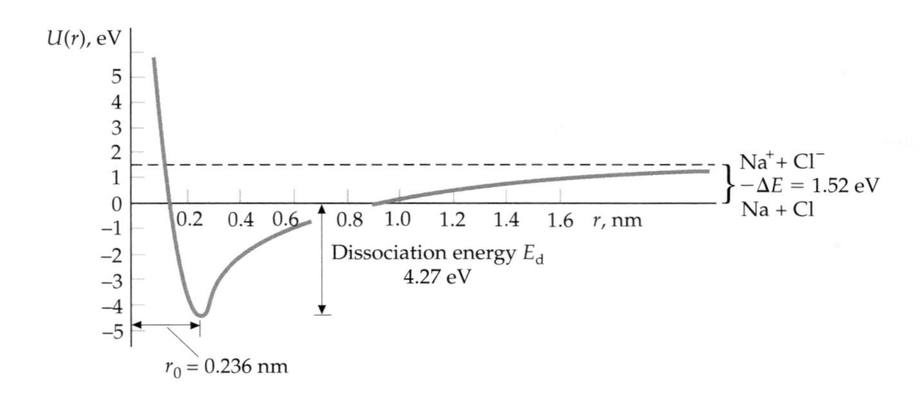

FIGURE 37-1 Potential energy for Na^+ and Cl^- ions as a function of separation distance r. The energy at infinite separation is chosen to be 1.52 eV, corresponding to the energy $-\Delta E$ needed to form the ions from atoms. The minimum energy is at the equilibrium separation $r_0 = 0.236$ nm for the ions.

equilibrium separation r_0 of approximately 0.236 nm. At smaller separations, the energy increases steeply. The energy required to separate the ions and form sodium and chlorine atoms is called the **dissociation energy** E_d, which is approximately 4.27 eV for NaCl.

The equilibrium separation distance of the gaseous NaCl, which can be obtained by evaporating solid NaCl, is 0.236 nm. Normally, NaCl exists as a solid in a cubic crystal structure, where the Na^+ and Cl^- ions are at the alternate corners of a cube. The separation of the Na^+ and Cl^- ions in a crystal is approximately 0.28 nm, which is somewhat larger than the 0.236 nm separation for the gaseous NaCl. Because of the presence of neighboring ions of opposite charge, the electrostatic energy per ion pair is lower when the ions are in a crystal.

Example 37-1 The Energy of Sodium Fluoride

The electron affinity of fluorine is −3.40 eV and the equilibrium separation of sodium fluoride (NaF) is 0.193 nm. (*a*) How much energy is needed to form Na^+ and F^- ions from sodium and fluorine atoms? (*b*) What is the electrostatic potential energy of the Na^+ and F^- ions at their equilibrium separation? (*c*) The dissociation energy of NaF is 5.38 eV. What is the energy due to repulsion of the ions at the equilibrium separation?

PICTURE (*a*) The energy ΔE needed to form Na^+ and F^- ions from the sodium and fluorine atoms is the sum of the first ionization energy of sodium (5.14 eV) and the electron affinity of fluorine. (*b*) The electrostatic potential energy, where $U = 0$ at infinity, is $U_e = -ke^2/r$. (*c*) If we choose the potential energy at infinity to be ΔE, the total potential energy is $U_{tot} = U_e + \Delta E + U_{rep}$, where U_{rep} is the energy of repulsion, which is found by setting the dissociation energy equal to $-U_{tot}$.

SOLVE

(*a*) Calculate the energy needed to form Na^+ and F^- ions from the sodium and fluorine atoms (see the Picture section):

$$\Delta E = 5.14\text{ eV} - 3.40\text{ eV} = \boxed{1.74\text{ eV}}$$

(*b*) Calculate the electrostatic potential energy at the equilibrium separation of $r = 0.193$ nm:

$$U_e = -\frac{ke^2}{r}$$
$$= -\frac{(8.99 \times 10^9\text{ N}\cdot\text{m}^2/\text{C}^2)(1.60 \times 10^{-19}\text{ C})^2}{1.93 \times 10^{-10}\text{ m}}$$
$$= -1.19 \times 10^{-18}\text{ J} = \boxed{-7.45\text{ eV}}$$

(*c*) The dissociation energy equals the negative of the total potential energy:

$$E_d = -U_{tot} = -(U_e + \Delta E + U_{rep})$$

so

$$U_{rep} = -(E_d + \Delta E + U_e)$$
$$= -(5.38\text{ eV} + 1.74\text{ eV} - 7.45\text{ eV}) = \boxed{0.33\text{ eV}}$$

CHECK The Part (*c*) result is greater than zero as expected.

THE COVALENT BOND

A completely different mechanism, the **covalent bond**, is responsible for the bonding of identical or similar atoms to form molecules such as gaseous hydrogen (H_2), nitrogen (N_2), and carbon monoxide (CO). If we calculate the energy needed to form H^+ and H^- ions by the transfer of an electron from one atom to the other and then add this energy to the electrostatic potential energy, we find that there is no separation distance for which the total energy is negative. The bond thus cannot be ionic. Instead, the attraction of two hydrogen atoms can only be explained quantum-mechanically. The decrease in energy when two hydrogen atoms approach each other is due to the sharing of the two electrons by both atoms, which can be explained using the symmetry properties of the wave functions of electrons.

We can gain some insight into covalent bonding by considering a simple, one-dimensional quantum-mechanics problem of two identical finite square wells. We first consider a single electron that is equally likely to be in either well. Because the wells are identical, the probability distribution, which is proportional to $|\psi^2|$, must be symmetric about the midpoint between the wells. Then ψ must be either symmetric or antisymmetric with respect to the two wells. The two possibilities for the ground state are shown in Figure 37-2*a* for the case in which the wells are far apart and in Figure 37-2*b* for the case in which the wells are close together. An important feature of Figure 37-2*b* is that in the region between the wells the symmetric wave function is large and the antisymmetric wave function is small.

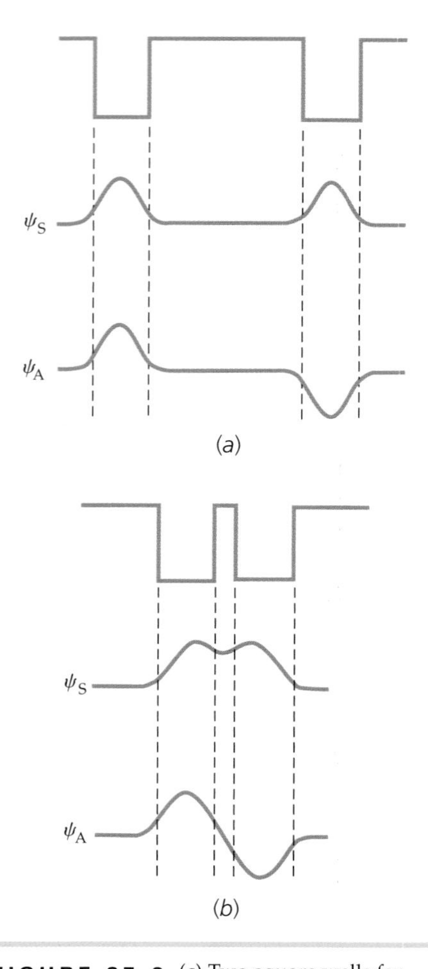

FIGURE 37-2 (*a*) Two square wells far apart. The electron wave function can be either symmetric (ψ_S) or antisymmetric (ψ_A) in space. The probability distributions and energies are the same for the two wave functions when the wells are far apart. (*b*) Two square wells that are close together. Between the wells, the antisymmetric space wave function is approximately zero, whereas the symmetric space wave function is quite large.

Now consider adding a second electron to the two wells. We saw in Section 6 of Chapter 35 that the wave functions for particles that obey the exclusion principle are antisymmetric on exchange of the particles. Thus, the total wave function for the two electrons must be antisymmetric on exchange of the electrons. Note that exchanging the electrons while keeping the wells in place is equivalent to keeping the electrons in place and exchanging the wells. The total wave function for two electrons can be written as a spatial expression and an expression for spin. So, an antisymmetric wave function can be the product of a symmetric spatial expression and an antisymmetric expression for spin or of a symmetric expression for spin and an antisymmetric spatial expression.

To understand the symmetry of the total wave function, we must therefore understand the symmetry of the expression for spin of the wave function. The spin of a single electron can have two possible values for its quantum number m_S: $m_S = +\frac{1}{2}$, which we call spin up, or $m_S = -\frac{1}{2}$, which we call spin down. We will use arrows to designate the spin wave function for a single electron: $\uparrow_1$ or $\uparrow_2$ for electron 1 or electron 2 that both are spin up and $\downarrow_1$ or $\downarrow_2$ for electron 1 or electron 2 that are both spin down. The total spin quantum number for two electrons can be $S = 1$, where $m_S = +1, 0,$ or -1; or $S = 0$, where $m_S = 0$. We use $\phi_{S\,m_S}$ to denote the spin wave function for two electrons. The spin state $\phi_{1\,+1}$, corresponding to $S = 1$ and $m_S = +1$, can be written

$$\phi_{1\,+1} = \uparrow_1\uparrow_2 \qquad S = 1, m_S = +1 \qquad \text{37-1}$$

Similarly, the spin state for $S = 1, m_S = -1$ is

$$\phi_{1\,-1} = \downarrow_1\downarrow_2 \qquad S = 1, m_S = -1 \qquad \text{37-2}$$

Note that both of these states are symmetric upon exchange of the electrons. The spin state corresponding to $S = 1$ and $m_S = 0$ is not quite so obvious. It turns out to be proportional to

$$\phi_{10} = \uparrow_1\downarrow_2 + \uparrow_2\downarrow_1 \qquad S = 1, m_S = 0 \qquad \text{37-3}$$

This spin state is also symmetric upon exchange of the electrons. The spin state for two electrons with antiparallel spins ($S = 0$) is

$$\phi_{00} = \uparrow_1\downarrow_2 - \uparrow_2\downarrow_1 \qquad S = 0, m_S = 0 \qquad 37\text{-}4$$

This spin state is antisymmetric upon exchange of electrons.

We thus have the important result that the spin part of the wave function is symmetric for parallel spins ($S = 1$) and antisymmetric for antiparallel spins ($S = 0$). Because the total wave function is the product of the spatial expression and the expression for spin, we have the following important result:

> For the total wave function of two electrons to be antisymmetric, the spatial part of the wave function must be antisymmetric for parallel spins ($S = 1$) and symmetric for antiparallel spins ($S = 0$).
>
> SPIN ALIGNMENT AND WAVE-FUNCTION SYMMETRY

We can now consider the problem of two hydrogen atoms. Figure 37-3*a* shows a spatially symmetric wave function ψ_S and a spatially antisymmetric wave function ψ_A for two hydrogen atoms that are far apart, and Figure 37-3*b* shows the same two wave functions for two hydrogen atoms that are close together. The squares of these two wave functions are shown in Figure 37-3*c*. Note that the probability distribution $|\psi|^2$ in the region between the protons is large for the symmetric wave function and small for the antisymmetric wave function. Thus, when the spatial part of the wave function is symmetric ($S = 0$), the electrons are often found in the region between the protons. The electron cloud, as shown in the upper part of Figure 37-3*c*, is concentrated in the space between the protons and the protons are bound together by this negatively charged cloud. Conversely, when the spatial part of the wave function is antisymmetric ($S = 1$), the electrons spend little time between the protons and the atoms do not bond together to form a molecule. In this case, the electron cloud is not concentrated in the space between the protons, as shown in the lower part of Figure 37-3*c*.

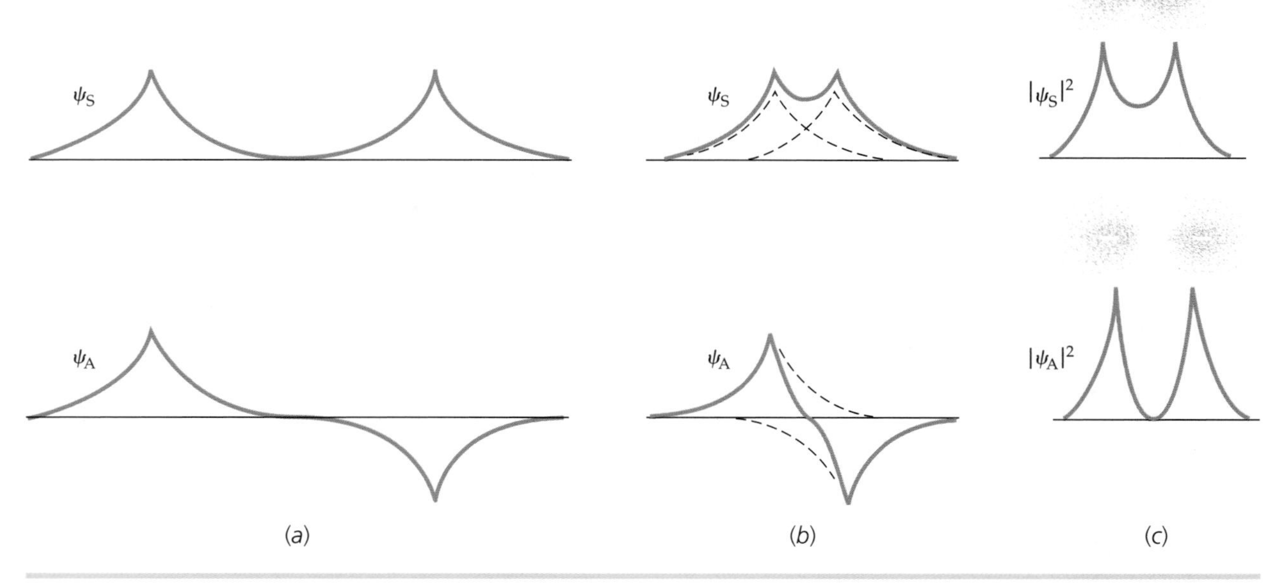

FIGURE 37-3 One-dimensional symmetric and antisymmetric wave functions for two hydrogen atoms (*a*) far apart and (*b*) close together. (*c*) Electron probability distributions ($|\psi|^2$) for the wave functions in Figure 37-3*b*. For the symmetric wave function, the electron charge density is large between the protons. This negative charge density holds the protons together in the hydrogen molecule H_2. For the antisymmetric wave function, the electron charge density is not large between the protons.

The total electrostatic potential energy for the H_2 molecule consists of the positive energy of repulsion of the two electrons and the negative potential energy of attraction of each electron for each proton. Figure 37-4 shows the electrostatic potential energy function U_S for two hydrogen atoms versus separation for the case in which the spatial part of the electron wave function is symmetric, and the electrostatic potential energy function U_A for the case in which the spatial part of the wave function is antisymmetric. We can see that the potential energy for the symmetric state is lower than the potential energy for the antisymmetric state and that the shape of the potential energy curve for the symmetric state is similar to the shape of the potential energy curve for ionic bonding (Figure 37-1). The equilibrium separation for H_2 is $r_0 = 0.074$ nm, and the binding energy is 4.52 eV. For the antisymmetric state, the potential energy is never negative and there is no bonding.

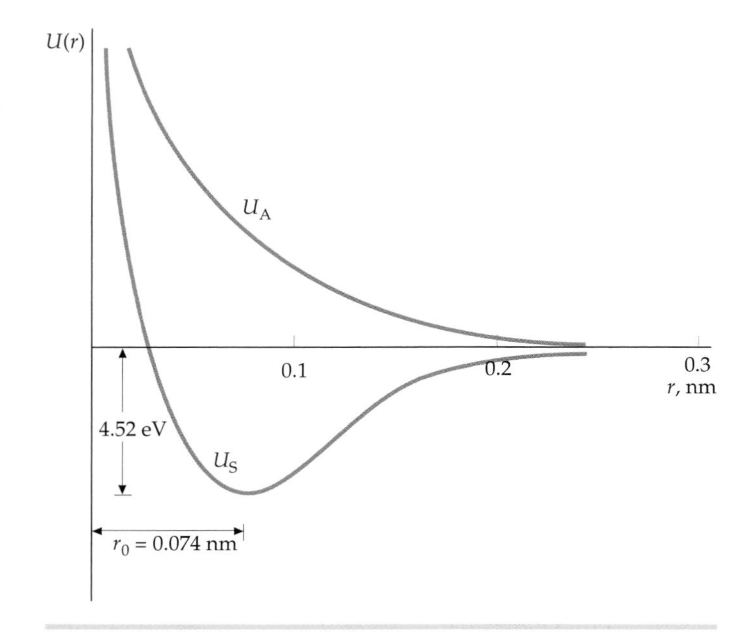

FIGURE 37-4 Potential energy versus separation for two hydrogen atoms. The curve labeled U_S is for a wave function that has a symmetric expression for the spatial part and the curve labeled U_A is for a wave function that has an antisymmetric expression for the spatial part.

We can now see why three hydrogen atoms do not bond to form H_3. If a third hydrogen atom is brought near an H_2 molecule, the third electron cannot be in a 1s state and have its spin antiparallel to the spin of both of the other electrons. If that electron is in an antisymmetric spatial state with respect to exchange with one of the electrons, the repulsion of this atom is greater than the attraction of the other. As the three atoms are pushed together, the third electron is, in effect, forced into a higher quantum-energy state according to the exclusion principle. The bond between two hydrogen atoms is called a **saturated bond** because there is no room for another electron. The two shared electrons essentially fill the 1s states of both atoms.

We can also see why two helium atoms do not normally bond together to form the He_2 molecule. There are no valence electrons that can be shared. The electrons in the filled shells are forced into higher energy states when the two atoms are brought together. At low temperatures or high pressures, helium atoms do bond together due to van der Waals forces, which we will discuss next. This bonding is so weak that at atmospheric pressure helium boils at 4 K, and it does not form a solid at any temperature unless the pressure is greater than about 20 atm.

> ! Don't think all bonds between atoms are partly ionic. They are not. Bonds between two identical atoms are always 100 percent covalent.

When two identical atoms bond, as in O_2 or N_2, the bonding is purely covalent. However, the bonding of two dissimilar atoms is often a mixture of covalent and ionic bonding. Even in NaCl, the electron donated by sodium to chlorine has some probability of being at the sodium atom because its wave function in the vicinity of the sodium atom, while small, is not zero. Thus, this electron is partially shared in a covalent bond, although this bonding is only a small part of the total bond, which is mainly ionic.

A measure of the degree to which a bond is ionic or covalent can be obtained from the electric dipole moment of the molecule or ionic unit. For example, if the bonding in NaCl were purely ionic, the center of positive charge would be at the Na^+ ion and the center of negative charge would be at the Cl^- ion. The electric dipole moment would have the magnitude

$$p_{\text{ionic}} = er_0 \tag{37-5}$$

where $r_0 = 2.36 \times 10^{-10}$ m is the equilibrium separation of the ions. Thus, the dipole moment of NaCl would be (from Figure 37-1)

$$\begin{aligned} p_{\text{ionic}} &= er_0 \\ &= (1.60 \times 10^{-19}\ \text{C})(2.36 \times 10^{-10}\ \text{m}) = 3.78 \times 10^{-29}\ \text{C}\cdot\text{m} \end{aligned}$$

The actual measured electric dipole moment of NaCl is

$$p_{\text{measured}} = 3.00 \times 10^{-29}\ \text{C}\cdot\text{m}$$

We can define the ratio of p_{measured} to p_{ionic} as the fractional amount of ionic bonding. For NaCl, this ratio is $3.00/3.78 = 0.79$. Thus, the bonding in NaCl is about 79 percent ionic.

> **PRACTICE PROBLEM 37-1**
> The equilibrium separation of HCl is 0.128 nm and its measured electric dipole moment is 3.60×10^{-30} C · m. What is the percentage of ionic bonding in HCl?

OTHER BONDING TYPES

The van der Waals Bond Any two separated molecules will be attracted to one another by electrostatic forces called *van der Waals forces.* So will any two atoms that do not form ionic or covalent bonds. The **van der Waals bonds** due to these forces are much weaker than the bonds already discussed. At high enough temperatures, these forces are not strong enough to overcome the motion of the atoms or molecules due to thermal energy. At sufficiently low temperatures, these motions become negligible and the van der Waals forces will cause virtually all substances to condense into a liquid and then a solid form.* The van der Waals forces arise from the interaction of the instantaneous electric dipole moments of the molecules.

Figure 37-5 shows how two polar molecules—molecules that have *permanent* electric dipole moments, such as H_2O—can bond. The electric field due to the dipole moment of one molecule orients the other molecule so that the two dipole moments attract. Nonpolar molecules also attract other nonpolar molecules by the van der Waals forces. Although nonpolar molecules have zero electric dipole moments on the average, they have instantaneous dipole moments that are generally not zero because of fluctuations in the positions of the charges. When two nonpolar molecules are near each other, the fluctuations in the instantaneous dipole moments tend to become correlated so as to produce attraction. This is illustrated in Figure 37-6.

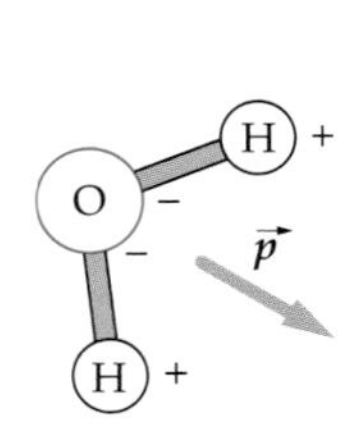

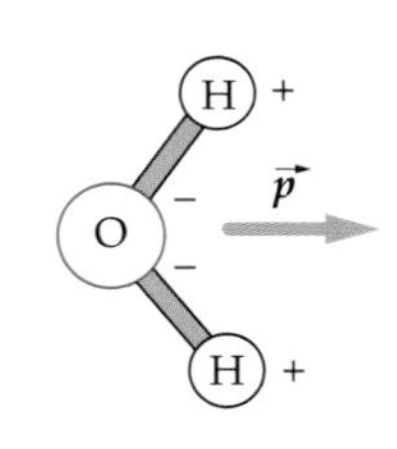

FIGURE 37-5 Bonding of H_2O molecules due to the attraction of the electric dipoles. The dipole moment of each molecule is indicated by $\vec{p}$. The electric field of one dipole orients the other dipole so the two dipole moments tend to be parallel. When the dipole moments are approximately parallel, the center of negative charge of one molecule is closer to the center of positive charge of the other molecule than it is to the center of the negative charge, so the molecules attract.

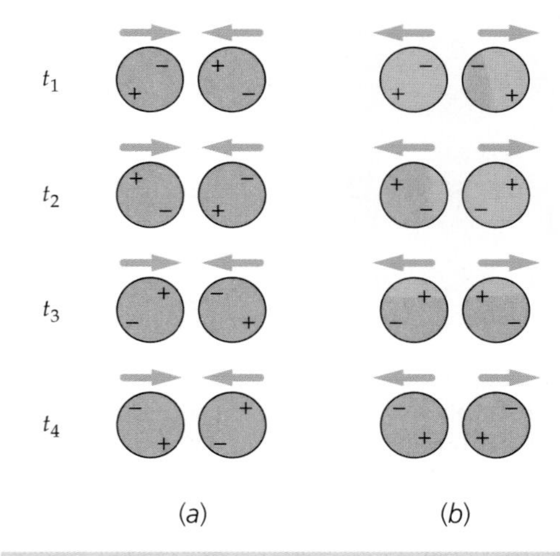

FIGURE 37-6 van der Waals attraction of molecules that have zero permanent dipole moments. (*a*) Possible orientations of instantaneous dipole moments at different times leading to attraction. (*b*) Possible orientations leading to repulsion. The electric field of the instantaneous dipole moment of one molecule tends to polarize the other molecule; thus the orientations leading to attraction (Figure 37-6*a*) are much more likely than those leading to repulsion (Figure 37-6*b*).

* Helium is the only element that does not solidify at any temperature at atmospheric pressure.

The hydrogen bond Another bonding mechanism of great importance is the hydrogen bond, which is formed by the sharing of a proton (the nucleus of the hydrogen atom) between two atoms, frequently two oxygen atoms. This sharing of a proton is similar to the sharing of electrons responsible for the covalent bond already discussed. It is facilitated by the small mass of the proton and by the absence of core electrons in hydrogen. The hydrogen bond often holds groups of molecules together and is responsible for the cross-linking that allows giant biological molecules and polymers to hold their fixed shapes. The well-known helical structure of DNA is due to hydrogen-bond linkages across turns of the helix (Figure 37-7).

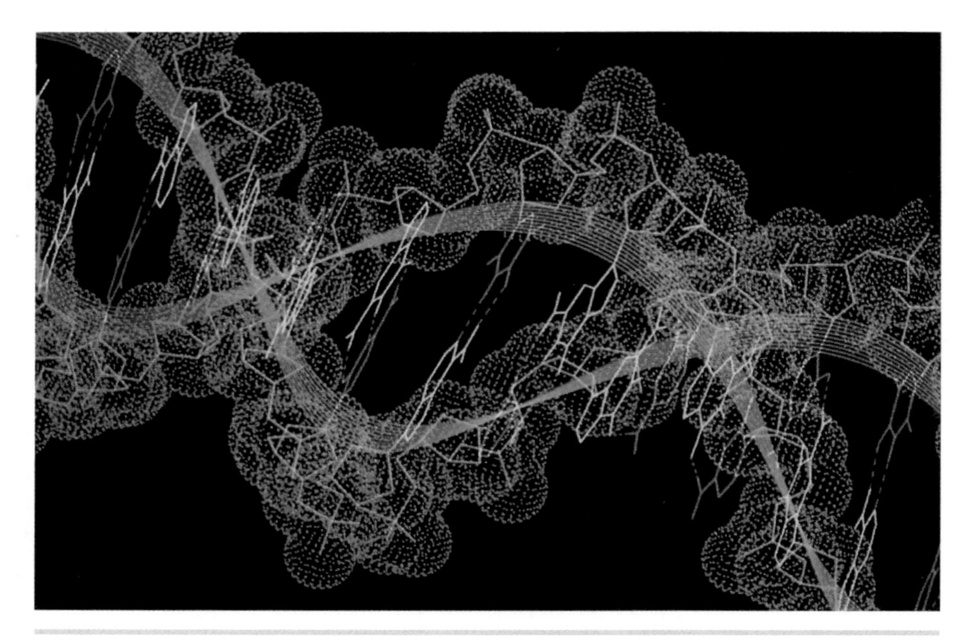

FIGURE 37-7 The DNA molecule. *(© Will and Demi McIntire/Photo Researchers.)*

(*a*) The discoverers of the structure of DNA. James Watson at left and Francis Crick are shown with their model of part of a DNA molecule in 1953. Crick and Watson met at the Cavendish Laboratory, Cambridge, in 1951. Their work on the structure of DNA was performed with a knowledge of Chargaff's ratios of the bases in DNA and some access to the X-ray crystallography of Maurice Wilkins and Rosalind Franklin at King's College London. Combining all of this work led to the deduction that DNA exists as a double helix, thus to its structure. Crick, Watson, and Wilkins shared the 1962 Nobel Prize for Physiology or Medicine; Franklin died from cancer in 1958. *((a) Norman Collection for the History of Molecular Biology.)*

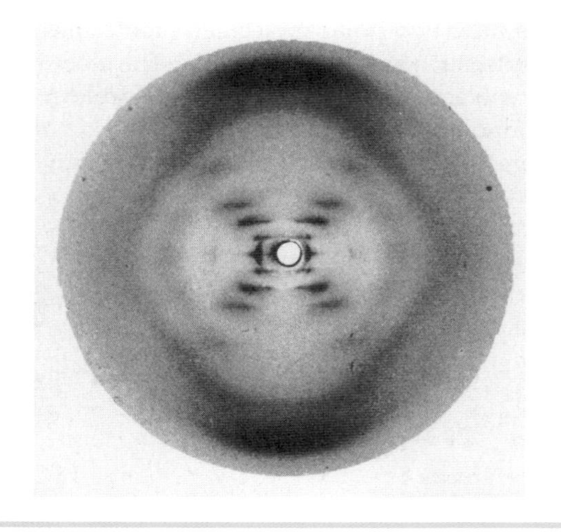

(*b*) X-ray diffraction pattern of the B form of DNA. Rosalind Franklin's colleague Maurice Wilkins, without obtaining her permission, made available to Watson and Crick her then unpublished X-ray diffraction pattern of the B form of DNA, which was crucial evidence for the helical structure. In his account of this discovery, Watson wrote: "The instant I saw the picture, my mouth fell open and pulse began to race. . . . The black cross of reflections which dominated the picture could arise only from a helical structure. . . . Mere inspection of the X-ray picture gave several of the vital helical parameters" (from Stent, Gunther, *The Double Helix,* New York: Norton, 1980). *((b) © A. Barrington Brown/Photo Researchers, NY.)*

The metallic bond In a metal, two atoms do not bond together by exchanging or sharing an electron to form a molecule. Instead, each valence electron is shared by many atoms. The bonding is thus distributed throughout the entire metal. A metal can be thought of as a lattice of positive ions held together by essentially free electrons that roam throughout the solid. In the quantum-mechanical picture, these free electrons form a cloud of negative charge density between the positively charged lattice ions that holds the ions together. In this respect, the metallic bond is somewhat similar to the covalent bond. However, with the metallic bond, there are far more than just two atoms involved, and the negative charge is distributed uniformly throughout the volume of the metal. The number of free electrons per lattice ion varies from metal to metal but is of the order of one free electron per ion.

* 37-2 POLYATOMIC MOLECULES

Molecules that have more than two atoms range from relatively simple molecules such as water, which has a molecular mass number of 18, to such giants as proteins and DNA, which can have molecular mass numbers of hundreds of thousands up to many millions. As with diatomic molecules, the structure of polyatomic molecules can be understood by applying basic quantum mechanics to the bonding of individual atoms. The bonding mechanisms for most polyatomic molecules are the covalent bond and the hydrogen bond. We will discuss only some of the simplest polyatomic molecules—H_2O, NH_3, and CH_4—to illustrate both the simplicity and complexity of the application of quantum mechanics to molecular bonding.

The basic requirement for the sharing of electrons in a covalent bond is that the wave functions of the valence electrons in the individual atoms must overlap as much as possible. As our first example, we will consider the water molecule. The ground-state configuration of the oxygen atom is $1s^22s^22p^4$. The 1s and 2s electrons are in filled shells and do not contribute to the bonding. The 2p shell has room for six electrons, two in each of the three spatial states (orbitals) corresponding to $\ell = 1$. In an isolated atom, we describe these spatial states by the hydrogen-like wave functions corresponding to $\ell = 1$ and $m_\ell = +1, 0,$ and -1. Because the energy is the same for these three spatial states, we could equally well use any linear combination of these wave functions. When an atom participates in molecular bonding, certain combinations of these atomic wave functions are important. These combinations are called **p_x, p_y, and p_z atomic orbitals.** The angular dependence of these orbitals is

$$p_x \propto \sin\theta \cos\phi \qquad 37\text{-}6$$

$$p_y \propto \cos\theta \cos\phi \qquad 37\text{-}7$$

$$p_z \propto \cos\phi \qquad 37\text{-}8$$

The electron charge distribution for these orbitals is maximum along the x, y, or z axis, respectively, as shown in Figure 37-8.

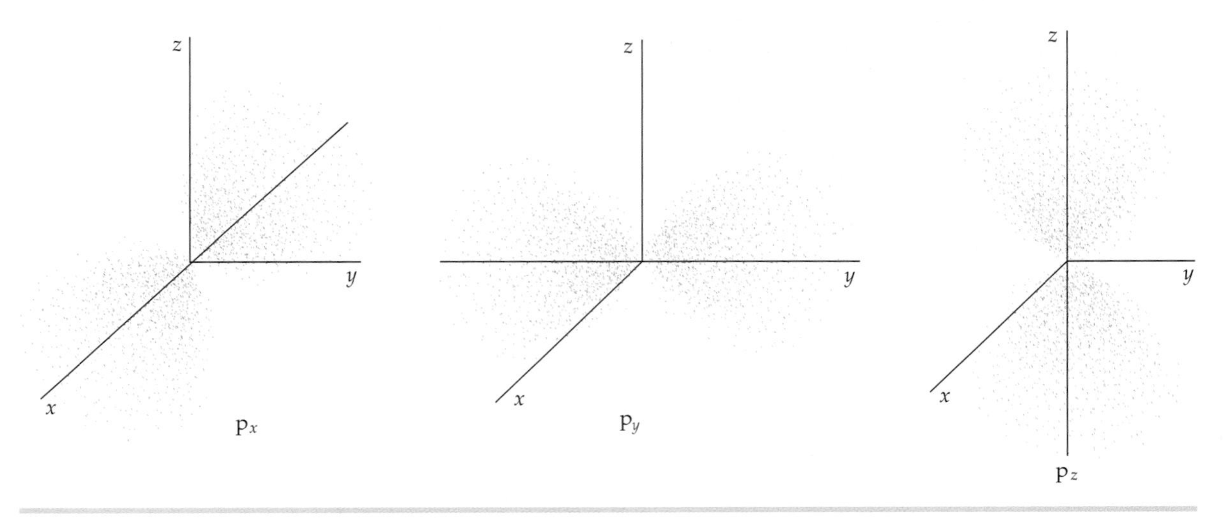

FIGURE 37-8 Computer-generated dot plot illustrating the spatial dependence of the electron charge distribution in the p_x, p_y, and p_z atomic orbitals.

For the oxygen in an H_2O molecule, maximum overlap of the electron wave functions occurs when two of the four 2p electrons are in one of the atomic orbitals (for this example, assume the p_z orbital) with their spins antiparallel, the third 2p electron is in a second orbital (the p_x orbital), and the fourth 2p electron is in the third orbital (the p_y orbital). Each of the unpaired electrons (in the p_x and p_y orbitals for this example) forms a bond with the electron of a hydrogen atom, as shown in Figure 37-9. Because of the repulsion of the two hydrogen atoms, the angle between the O—H bonds is actually greater than 90°. The effect of this repulsion can be calculated, and the result is in agreement with the measured angle of 104.5°.

Similar reasoning leads to an understanding of the bonding in NH_3 (not shown). In the ground state, nitrogen has three electrons in the 2p state. When these three electrons are in the p_x, p_y, and p_z atomic orbitals, they bond to the electrons of hydrogen atoms. Again, because of the repulsion of the hydrogen atoms, the angles between the bonds are somewhat larger than 90°.

The bonding of carbon atoms is somewhat more complicated. Carbon forms single, double and triple bonds, leading to a great diversity in the kinds of organic molecules. The ground-state configuration of carbon is $1s^2 2s^2 2p^2$. From our previous discussion, we might expect carbon to be divalent–that is, bonding only through its two 2p electrons–with the two bonds forming at approximately 90°. However, one of the most important features of the chemistry of carbon is that tetravalent carbon compounds, such as CH_4, are overwhelmingly favored.

The observed valence of 4 for carbon comes about in an interesting way. One of the first excited states of carbon occurs when a 2s electron is excited to a 2p state, giving a configuration of $1s^2 2s^1 2p^3$. In this excited state, we can have four unpaired electrons, one each in the 2s, $2p_x$, $2p_y$, and $2p_z$ atomic orbitals. We might expect there to be three similar bonds corresponding to the three p orbitals and one different bond corresponding to the s orbital. However, when carbon forms tetravalent bonds, these four atomic orbitals become mixed and form four new *equivalent* molecular orbitals called **hybrid orbitals.** This mixing of atomic orbitals, called hybridization, is among the most important features involved in the physics of complex molecular bonds. Figure 37-10 shows the tetrahedral structure of the methane molecule (CH_4), and Figure 37-11 shows the structure of the ethane molecule (CH_3—CH_3), which is similar to two joined methane molecules in which one of the C—H bonds is replaced with a C—C bond.

Carbon orbitals can also hybridize such that the s, p_x, and p_y orbitals combine to form three hybrid orbitals that are in the *xy* plane and form bonds that are 120° apart (the p_z orbital does not participate in bonding). An example of this configuration is graphite, in which the bonds in the *xy* plane provide the strongly layered structure characteristic of the material.

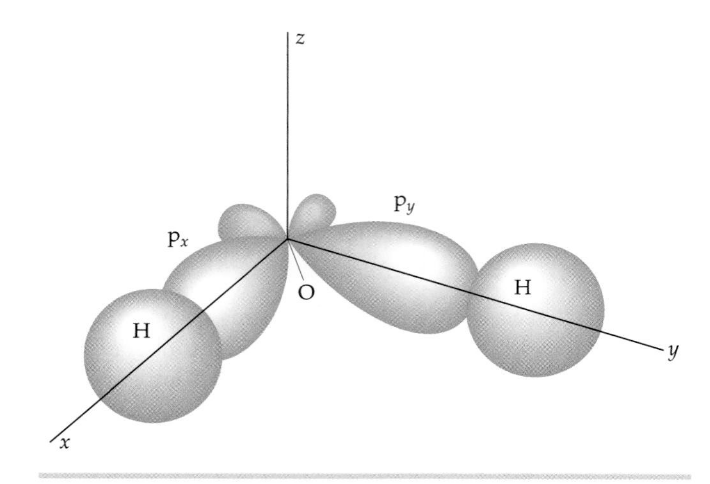

FIGURE 37-9 Electron charge distribution in the H_2O molecule.

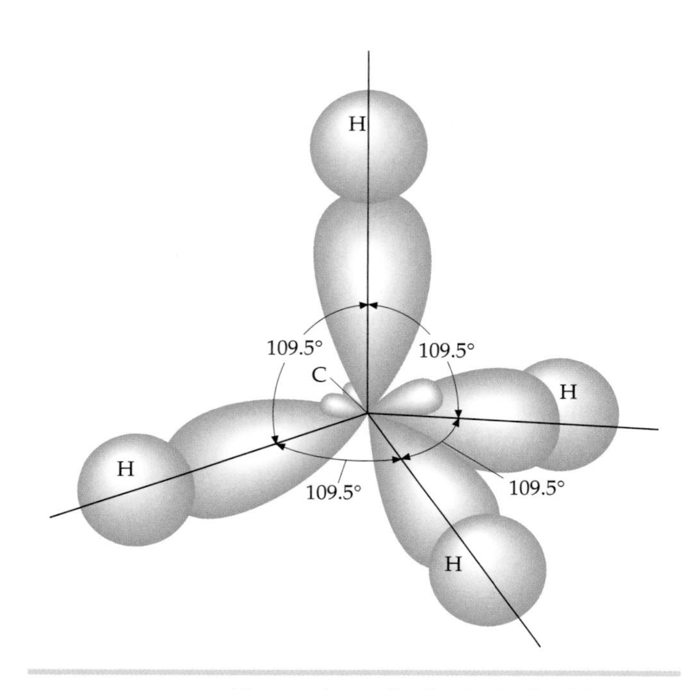

FIGURE 37-10 Electron charge distribution in the CH_4 (methane) molecule.

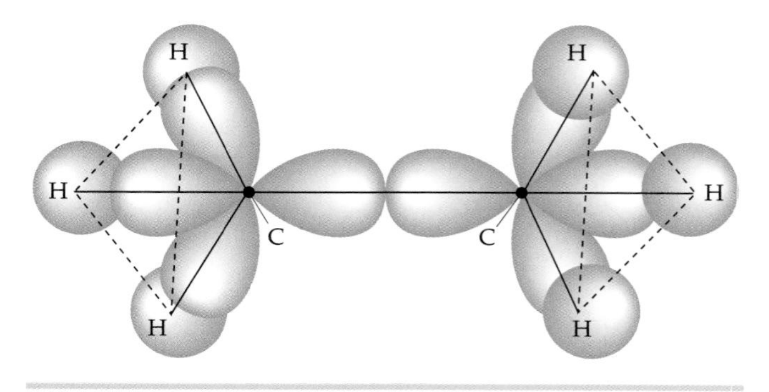

FIGURE 37-11 Electron charge distribution in the CH_3—CH_3 (ethane) molecule.

37-3 ENERGY LEVELS AND SPECTRA OF DIATOMIC MOLECULES

As is the case with an atom, a molecule often emits electromagnetic radiation when it makes a transition from an excited energy state to a state of lower energy. Conversely, a molecule can absorb radiation and make a transition from a lower energy state to a higher energy state. The study of molecular emission and absorption spectra thus provides us with information about the energy states of molecules. For simplicity, we will consider only diatomic molecules here.

The internal energy of a molecule can be conveniently separated into three parts: electronic, due to the excitation of the electrons of the molecule; vibrational, due to the oscillations of the atoms of the molecule; and rotational, due to the rotation of the molecule about its center of mass. The magnitudes of these energies are sufficiently different that they can be treated separately. The energies due to the electronic excitations of a molecule are typically of the order of magnitude of 1 eV, the same as for the electronic excitations of an atom. The energies of the vibrations of the atoms and of the rotation of the molecule are much smaller than the electronic excitation energy.

ROTATIONAL ENERGY LEVELS

Figure 37-12 shows a simple schematic model of a diatomic molecule consisting of particles that have masses of m_1 and m_2, are separated by a distance r and are rotating about the center of mass. Classically, the kinetic energy of rotation (Equation 9-11) is

$$E = \frac{1}{2}I\omega^2 \quad \text{37-9}$$

where I is the moment of inertia and ω is the angular speed of the rotation motion. If we write this in terms of the angular momentum $L = I\omega$, we have

$$E = \frac{(I\omega)^2}{2I} = \frac{L^2}{2I} \quad \text{37-10}$$

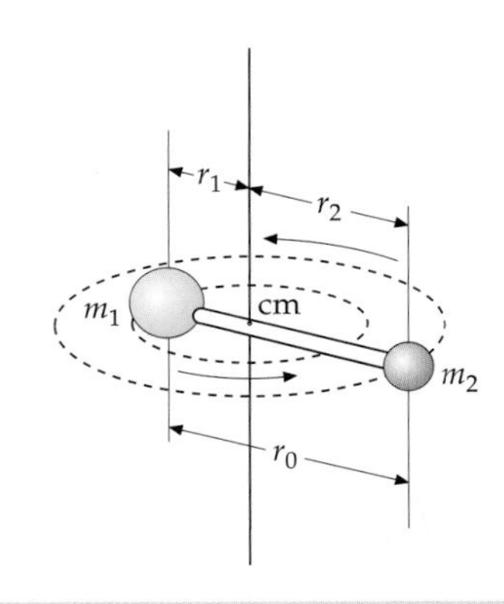

FIGURE 37-12 Diatomic molecule rotating about an axis through its center of mass.

The solution of the Schrödinger equation for rotation leads to quantization of the angular momentum with values given by

$$L^2 = \ell(\ell + 1)\hbar^2 \qquad \ell = 0, 1, 2, \ldots \quad \text{37-11}$$

where ℓ is the **rotational quantum number.** This is the same quantum condition on angular momentum that holds for the orbital angular momentum of an electron in an atom. Note, however, that L in Equation 37-10 refers to the angular momentum of the entire molecule rotating about its center of mass. The energy levels of a rotating molecule are therefore given by

$$E_\ell = \frac{\ell(\ell + 1)\hbar^2}{2I} = \ell(\ell + 1)E_{0r} \qquad \ell = 0, 1, 2, \ldots \quad \text{37-12}$$

ROTATIONAL ENERGY LEVELS

where E_{0r} is the characteristic rotational energy of a particular molecule, which is inversely proportional to its moment of inertia:

$$E_{0r} = \frac{\hbar^2}{2I} \quad \text{37-13}$$

CHARACTERISTIC ROTATIONAL ENERGY

A measurement of the rotational energy of a molecule from its rotational spectrum can be used to determine the moment of inertia of the molecule, which can then be used to find the separation of the atoms in the molecule. The moment of inertia about an axis through the center of mass of a diatomic molecule (see Figure 37-12) is

$$I = m_1 r_1^2 + m_2 r_2^2$$

Using $m_1 r_1 = m_2 r_2$, where r_1 is the distance of atom 1 from the center of mass, r_2 is the distance of atom 2 from the center of mass, and $r_0 = r_1 + r_2$, we can write the moment of inertia (see Problem 26) as

$$I = \mu r_0^2 \qquad \text{37-14}$$

where μ, called the **reduced mass,** is

$$\mu = \frac{m_1 m_2}{m_1 + m_2} \qquad \text{37-15}$$

DEFINITION—REDUCED MASS

If the masses are equal ($m_1 = m_2 = m$), as in H_2 and O_2, the reduced mass is $\mu = \frac{1}{2}m$ and

$$I = \tfrac{1}{2} m r_0^2 \qquad \text{37-16}$$

A unit of mass convenient for discussing atomic and molecular masses is the **unified atomic mass unit,** u, which is defined as one-twelfth the mass of the carbon-12 (^{12}C) atom. The mass of one ^{12}C atom is thus 12 u. The mass of an atom in unified mass units is therefore numerically equal to the molar mass of the atom in grams. The unified mass unit is related to the gram and kilogram by

$$1\text{ u} = \frac{1\text{ g}}{N_A} = \frac{10^{-3}\text{ kg}}{6.0221 \times 10^{23}} = 1.6606 \times 10^{-27}\text{ kg} \qquad \text{37-17}$$

where N_A is Avogadro's number.

Example 37-2 The Reduced Mass of a Diatomic Molecule

Find the reduced mass of the HCl molecule.

PICTURE We find the masses of the hydrogen and chlorine atoms in Appendix C* and use the definition of reduced mass (Equation 37-15).

SOLVE

1. The reduced mass μ is related to the individual masses m_H and m_{Cl}: $\mu = \frac{m_H m_{Cl}}{m_H + m_{Cl}}$
2. Find the masses in the periodic table: $m_H = 1.01\text{ u}, \quad m_{Cl} = 35.5\text{ u}$
3. Substitute to calculate the reduced mass: $\mu = \frac{m_H m_{Cl}}{m_H + m_{Cl}} = \frac{(1.01\text{ u})(35.5\text{ u})}{1.01\text{ u} + 35.5\text{ u}} = \boxed{0.982\text{ u}}$

CHECK The formula for reduced mass is identical to the formula for the equivalent resistance for two resistors in parallel. As expected, the reduced mass is less than either mass.

TAKING IT FURTHER When one atom of a diatomic molecule is much more massive than the other, the center of mass of the molecule is approximately at the center of the more massive atom and the reduced mass is approximately equal to the mass of the lighter atom.

* The masses in the tables are weighted according to the natural isotopic distribution. Thus, the mass of carbon is given as 12.011 rather than 12.000 because natural carbon consists of about 98.9 percent ^{12}C and 1.1 percent ^{13}C. Similarly, natural chlorine consists of about 76 percent ^{35}Cl and 24 percent ^{37}Cl.

Example 37-3 Rotational Kinetic Energy of a Molecule

Estimate the characteristic rotational kinetic energy of an O_2 molecule, assuming that the separation of the atoms is 0.100 nm.

PICTURE The characteristic rotational kinetic energy is given by $E_{0r} = \hbar/(2I)$ (Equation 37-13), where I is the moment of inertia. The moment of inertia is given by $I = \mu r_0^2$ (Equation 37-14), where μ is the reduced mass and r_0 is the average center-to-center separation of the atomic nuclei.

SOLVE

1. The characteristic rotational energy is inversely proportional to the moment of inertia:
$$E_{0r} = \frac{\hbar^2}{2I}$$
2. Calculate the moment of inertia:
$$I = \mu r_0^2 = \tfrac{1}{2} m r_0^2$$
3. Substitute this expression for I into the expression for E_{0r}:
$$E_{0r} = \frac{\hbar^2}{m r_0^2}$$
4. Use $m = 16$ u for the mass of oxygen to calculate E_{0r}:
$$E_{0r} = \frac{\hbar^2}{m r_0^2} = \frac{(1.055 \times 10^{-34}\,\text{J}\cdot\text{s})^2}{(16\,\text{u})(10^{-10}\,\text{m})^2} \times \left(\frac{1\,\text{u}}{1.66 \times 10^{-27}\,\text{kg}}\right)$$
$$= 4.19 \times 10^{-23}\,\text{J} = \boxed{2.62 \times 10^{-4}\,\text{eV}}$$

CHECK As expected, the characteristic rotational kinetic energy is small compared with 1 eV (a typical electronic excitation energy).

We can see from Example 37-3 that the rotational energy levels are several orders of magnitude smaller than energy levels due to electron excitation. Transitions within a given set of rotational energy levels yield photons in the microwave region of the electromagnetic spectrum. The rotational energies are also small compared with the typical thermal energy kT at normal temperatures. For $T = 300$ K, for example, kT is about 2.6×10^{-2} eV, which is approximately 100 times the characteristic rotational energy as calculated in Example 37-3 and approximately 1 percent of the typical electronic energy. Thus, at ordinary temperatures, a molecule can be easily excited to the lower rotational energy levels by collisions with other molecules. But such collisions cannot excite the molecule to its electronic energy levels above the ground state.

CONCEPT CHECK 37-1

At room temperature, the molecules of a diatomic gas undergo transitions between rotational states, but the atoms of a monatomic gas do not. Why?

VIBRATIONAL ENERGY LEVELS

The quantization of energy in a simple harmonic oscillator was one of the first problems solved by Schrödinger in his paper proposing his wave equation. Solving the Schrödinger equation for a simple harmonic oscillator gives

$$E_\nu = (\nu + \tfrac{1}{2})hf \qquad \nu = 0, 1, 2, \ldots \qquad \text{37-18}$$

VIBRATIONAL ENERGY LEVELS

where f is the frequency of the oscillator and ν (lowercase Greek nu) is the **vibrational quantum number.*** An interesting feature of this result is that the energy levels are equally spaced with intervals equal to hf. The frequency of vibration of a diatomic molecule can be related to the force exerted by one atom on the other. Consider two objects of mass m_1 and m_2 connected by a spring of force constant k_F.

* We use ν here rather than n so as not to confuse the vibrational quantum number with the principal quantum number n for electronic energy levels.

The frequency of oscillation of this system (see Problem 32) can be shown to be

$$f = \frac{1}{2\pi}\sqrt{\frac{k_F}{\mu}} \qquad 37\text{-}19$$

where μ is the reduced mass given by Equation 37-15. The effective force constant k_F of a diatomic molecule can thus be determined from a measurement of the frequency of oscillation of the molecule.

A selection rule on transitions between vibrational states (of the same electronic state) requires that the vibrational quantum number ν can change only by ± 1, so the energy of a photon emitted by such a transition is hf and the frequency of the photon is f, the same as the frequency of vibration. There is a similar selection rule that ℓ must change by ± 1 for transitions between rotational states.

A typical measured frequency of a transition between vibrational states is 5×10^{13} Hz, which gives

$$E \approx hf = (4.14 \times 10^{-15}\ \text{eV}\cdot\text{s})(5.0 \times 10^{13}\ \text{s}^{-1}) = 0.2\ \text{eV}$$

and is an estimate for the order of magnitude of vibrational energies. This typical vibrational energy is approximately 1000 times greater than the typical rotational energy E_{0r} of the O_2 molecule we found in Example 37-3 and about 8 times greater than the typical thermal energy $kT = 0.026$ eV at $T = 300$ K. Thus, the vibrational levels are almost never excited by molecular collisions at ordinary temperatures.

Example 37-4 Determining the Force Constant

The frequency of vibration of the CO molecule is 6.42×10^{13} Hz. What is the effective force constant for this molecule?

PICTURE We use $2\pi f = \sqrt{k_F/\mu}$ (Equation 37-19) to relate k_F to the frequency and the reduced mass, and calculate μ from its definition.

SOLVE

1. The effective force constant is related to the frequency and reduced mass by Equation 37-19:

$$f = \frac{1}{2\pi}\sqrt{\frac{k_F}{\mu}}$$

$$k_F = (2\pi f)^2\mu$$

2. Calculate the reduced mass using 12 u for the mass of the carbon atom and 16 u for the mass of the oxygen atom:

$$\mu = \frac{m_1 m_2}{m_1 + m_2} = \frac{(12\ \text{u})(16\ \text{u})}{12\ \text{u} + 16\ \text{u}} = 6.86\ \text{u}$$

3. Substitute this value of μ into the equation for k_F in step 1 and convert to SI units:

$$k_F = (2\pi f)^2\mu$$
$$= 4\pi^2(6.42 \times 10^{13}\ \text{Hz})^2(6.86\ \text{u})$$
$$= 1.12 \times 10^{30}\ \text{u/s}^2 \times \left(\frac{1.66 \times 10^{-27}\ \text{kg}}{1\ \text{u}}\right)$$
$$= \boxed{1.85 \times 10^3\ \text{N/m}}$$

CHECK From Newton's second law we know that 1 kg m/s² = 1 N, so the units of kg/s² that remain after canceling out the u's in step 3 are equal to N/m, which is what is expected for the force constant of a "spring."

EMISSION SPECTRA

Figure 37-13 shows schematically some electronic, vibrational, and rotational energy levels of a diatomic molecule. The vibrational levels are labeled with the quantum number ν and the rotational levels are labeled with ℓ. The lower vibrational levels are evenly spaced, with $\Delta E = hf$. For higher vibrational levels, the approximation that the vibration is simple harmonic is not valid and the levels are not quite evenly spaced.

Note that the potential energy curves representing the force between the two atoms in the molecule do not have exactly the same shape for the electronic ground and excited states. This implies that the fundamental frequency of vibration f is different for different electronic states. For transitions between vibrational states of different electronic states, the selection rule $\Delta\nu = \pm 1$ does not hold. Such transitions result in the emission of photons of wavelengths in or near the visible spectrum, so the emission spectrum of a molecule for electronic transitions is also sometimes called the optical spectrum.

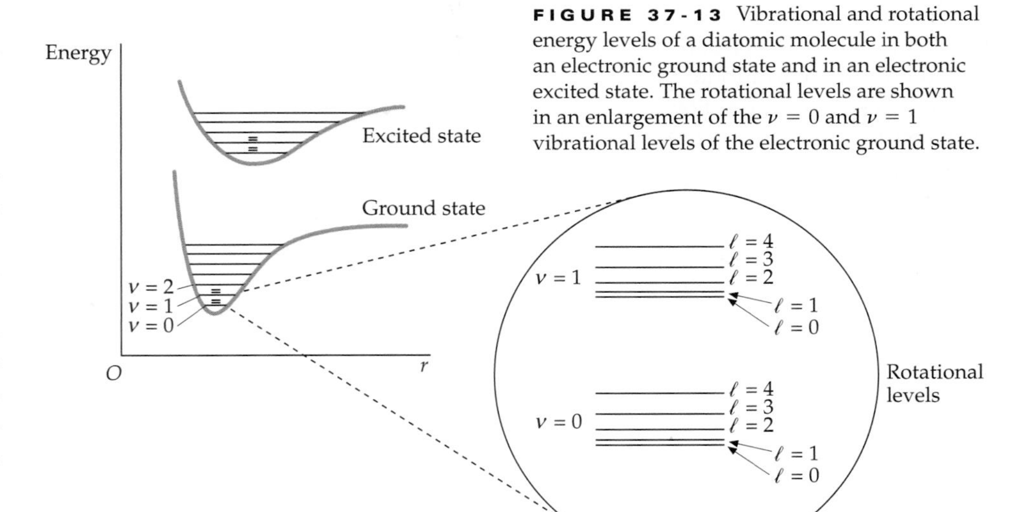

FIGURE 37-13 Vibrational and rotational energy levels of a diatomic molecule in both an electronic ground state and in an electronic excited state. The rotational levels are shown in an enlargement of the $\nu = 0$ and $\nu = 1$ vibrational levels of the electronic ground state.

The spacing of the rotational levels increases with increasing values of ℓ. Because the energies of rotation are so much smaller than those of vibrational excitation or electronic excitation of a molecule, molecular rotation shows up in optical spectra as a fine splitting of the spectral lines. When this fine structure is not resolved, the spectrum appears as bands, as shown in Figure 37-14*a*. Close inspection of these bands reveals that they have a fine structure due to the rotational energy levels, as shown in the enlargement in Figure 37-14*c*.

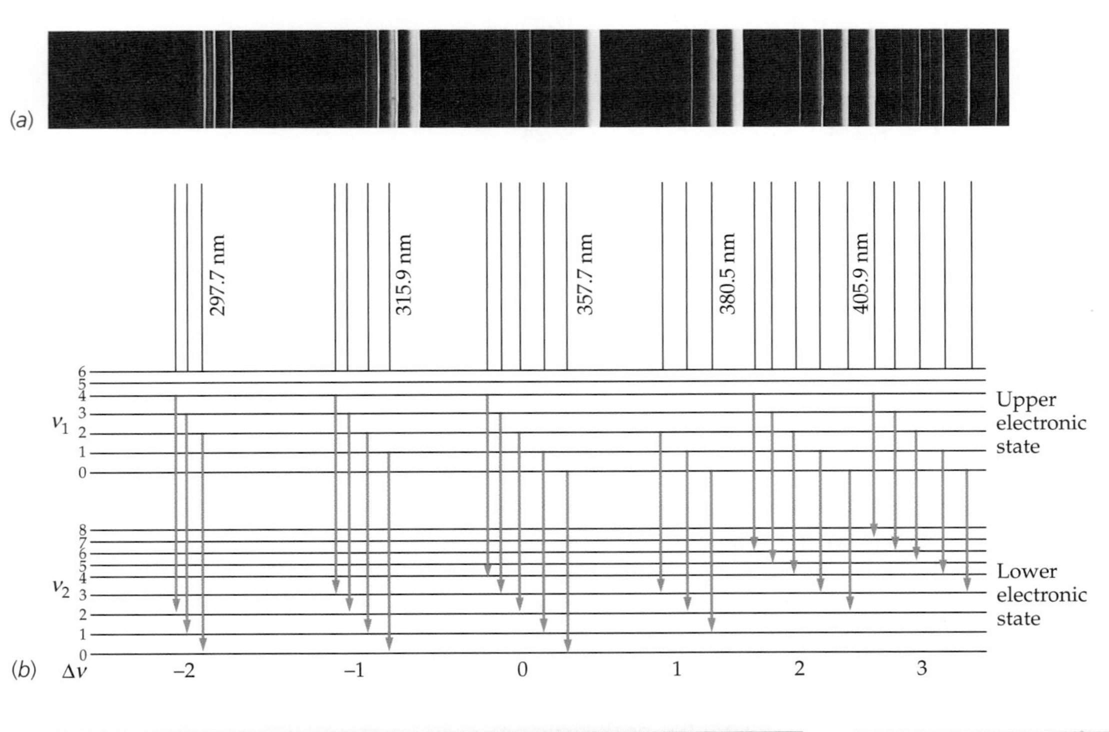

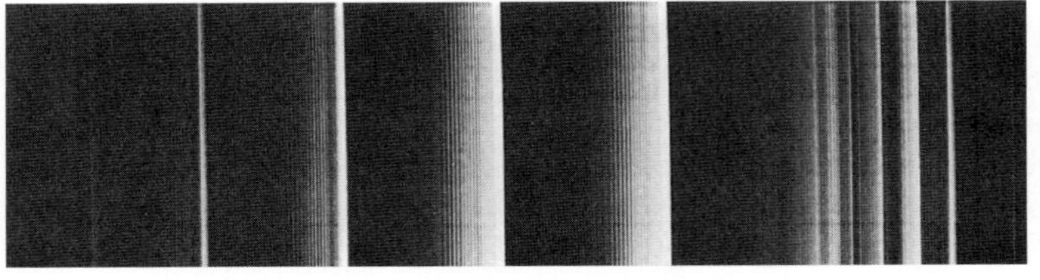

FIGURE 37-14 (*a*) Part of the emission spectrum of N_2. The spectral lines are due to transitions between the vibrational levels of two electronic states, as indicated in the energy level diagram (*b*). (*c*) An enlargement of part of Figure 37-14*a* shows that the apparent lines are in fact bands with structure caused by rotational levels. *(Courtesy of Dr. J. A. Marquissee.)*

ABSORPTION SPECTRA

Much molecular spectroscopy is done using infrared absorption techniques in which only the vibrational and rotational energy levels of the ground-state electronic level are excited. For ordinary temperatures, the vibrational energies are sufficiently large in comparison with the thermal energy kT that most of the molecules are in the lowest vibrational state $\nu = 0$, for which the energy is $E_0 = \frac{1}{2}hf$. The transition from $\nu = 0$ to $\nu = 1$ is the predominant transition in absorption. However, at room temperature the rotational energies are much less than the thermal energy kT. Thus, a number of the rotational energy states are occupied. If the molecule is originally in a vibrational state characterized by $\nu = 0$ and a rotational state characterized by the quantum number ℓ, the molecule's initial energy is

$$E_\ell = \tfrac{1}{2}hf + \ell(\ell + 1)E_{0r} \qquad 37\text{-}20$$

where E_{0r} is given by Equation 37-13. From this state, two transitions are permitted by the selection rules. For a transition to the next higher vibrational state $\nu = 1$ and a rotational state characterized by $\ell + 1$, the final energy is

$$E_{\ell+1} = \tfrac{3}{2}hf + (\ell + 1)(\ell + 2)E_{0r} \qquad 37\text{-}21$$

For a transition to the next higher vibrational state and to a rotational state characterized by $\ell - 1$, the final energy is

$$E_{\ell-1} = \tfrac{3}{2}hf + (\ell - 1)\ell E_{0r} \qquad 37\text{-}22$$

The energy differences therefore are

$$\Delta E_{\ell\to\ell+1} = E_{\ell+1} - E_\ell = hf + 2(\ell + 1)E_{0r} \qquad \ell = 0, 1, 2, \ldots \qquad 37\text{-}23$$

and

$$\Delta E_{\ell\to\ell-1} = E_{\ell-1} - E_\ell = hf - 2\ell E_{0r} \qquad \ell = 1, 2, 3, \ldots \qquad 37\text{-}24$$

(In Equation 37-24, ℓ begins at $\ell = 1$ rather than at $\ell = 0$ because from $\ell = 0$ only the transition $\ell \to \ell + 1$ can occur.) Figure 37-15 illustrates these transitions. The frequencies of these transitions are given by

$$f_{\ell\to\ell+1} = \frac{\Delta E_{\ell\to\ell+1}}{h} = f + \frac{2(\ell + 1)E_{0r}}{h} \qquad \ell = 0, 1, 2, \ldots \qquad 37\text{-}25$$

and

$$f_{\ell\to\ell-1} = \frac{\Delta E_{\ell\to\ell-1}}{h} = f - \frac{2\ell E_{0r}}{h} \qquad \ell = 1, 2, 3, \ldots \qquad 37\text{-}26$$

The frequencies for the transitions $\ell \to \ell + 1$ are thus $f + 2(E_{0r}/h)$, $f + 4(E_{0r}/h)$, $f + 6(E_{0r}/h)$, and so forth; those corresponding to the transition $\ell \to \ell - 1$ are $f - 2(E_{0r}/h)$, $f - 4(E_{0r}/h)$, $f - 6(E_{0r}/h)$, and so forth. We thus expect the absorption spectrum to contain frequencies equally spaced by $2E_{0r}/h$ except for a gap of $4E_{0r}/h$ at the vibrational frequency f, as shown in Figure 37-16. A measurement of the position of the gap gives f and a measurement of the spacing of the absorption peaks gives E_{0r}, which is inversely proportional to the moment of inertia of the molecule.

Figure 37-17 shows the absorption spectrum of HCl. The double-peak structure results from the fact that chlorine occurs naturally in two isotopes, ^{35}Cl and ^{37}Cl, which gives HCl with two different moments of inertia. If all the rotational levels were equally populated initially, we would expect the intensities of each absorption line to be equal. However, the population of a rotational level is proportional to the degeneracy of the level, that is, to the number of states with the same value of ℓ, which is $2\ell + 1$, and to the Boltzmann factor $e^{-E/kT}$, where E is the energy of

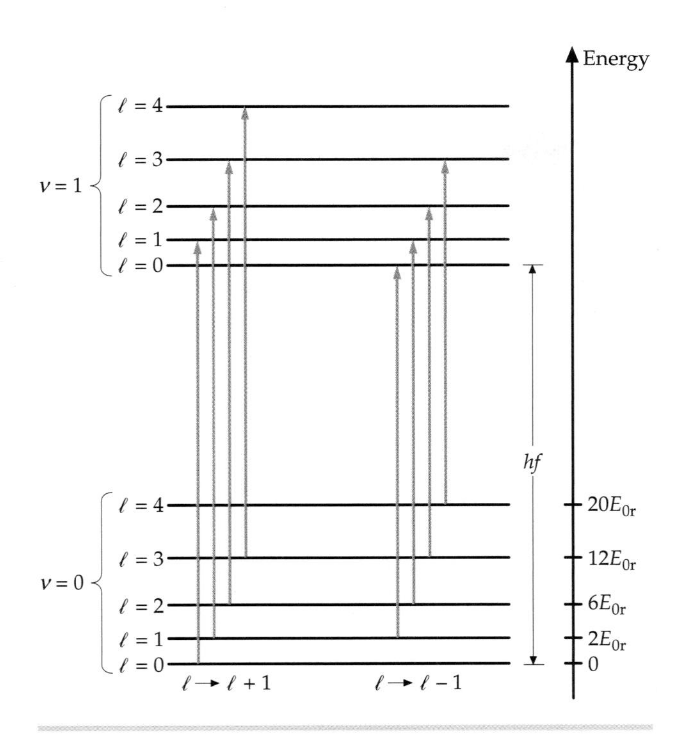

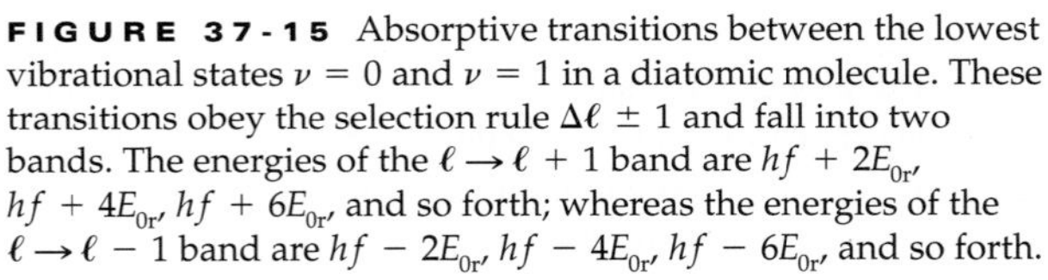

FIGURE 37-15 Absorptive transitions between the lowest vibrational states $\nu = 0$ and $\nu = 1$ in a diatomic molecule. These transitions obey the selection rule $\Delta\ell \pm 1$ and fall into two bands. The energies of the $\ell \rightarrow \ell + 1$ band are $hf + 2E_{0r}$, $hf + 4E_{0r}$, $hf + 6E_{0r}$, and so forth; whereas the energies of the $\ell \rightarrow \ell - 1$ band are $hf - 2E_{0r}$, $hf - 4E_{0r}$, $hf - 6E_{0r}$, and so forth.

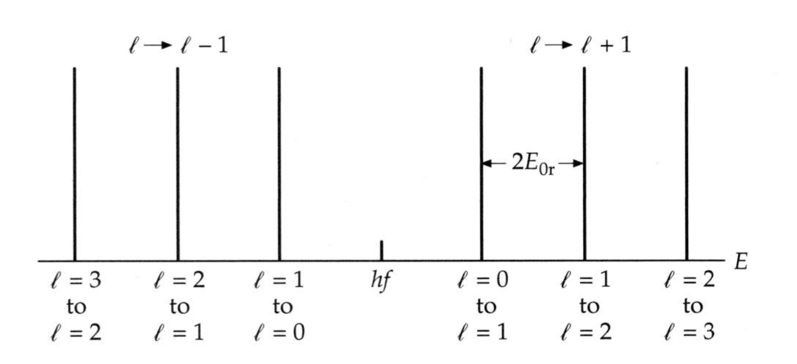

FIGURE 37-16 Expected absorption spectrum of a diatomic molecule. The right branch corresponds to transitions $\ell \rightarrow \ell + 1$ and the left branch corresponds to the transitions $\ell \rightarrow \ell - 1$. The lines are equally spaced by $2E_{0r}$. The energy midway between the branches is hf, where f is the frequency of vibration of the molecule.

the state. (The Boltzmann factor is presented in Chapter 17.) For low values of ℓ, the population increases slightly because of the degeneracy factor, whereas for higher values of ℓ, the population decreases because of the Boltzmann factor. The intensities of the absorption lines therefore increase with ℓ for low values of ℓ and then decrease with ℓ for high values of ℓ, as can be seen from the figure.

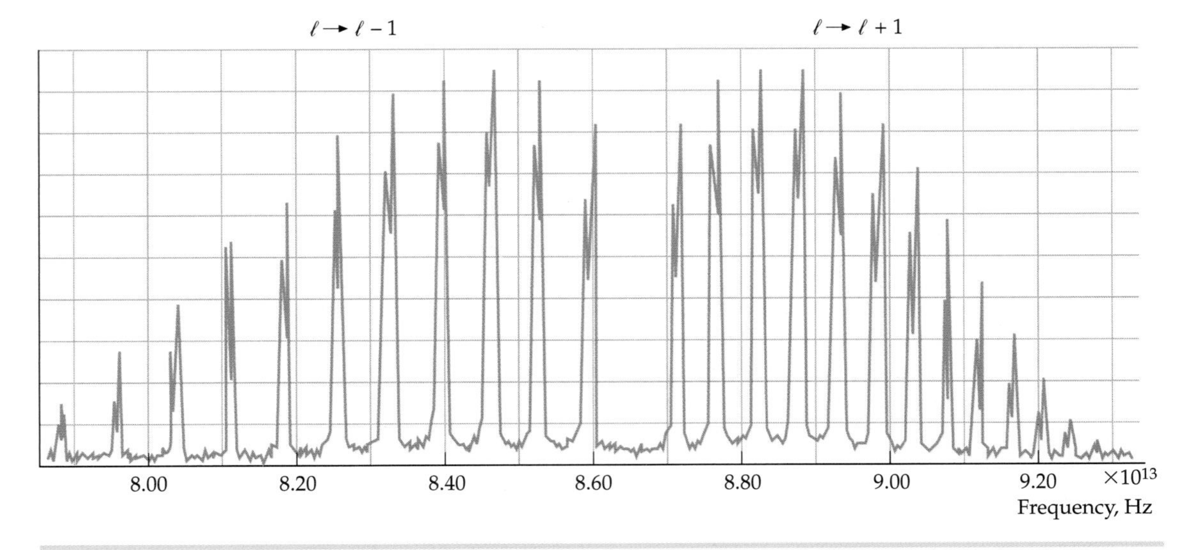

FIGURE 37-17 Absorption spectrum of the diatomic molecule HCl. The double-peak structure results from the two isotopes of chlorine, ^{35}Cl (abundance 75.5 percent) and ^{37}Cl (abundance 24.5 percent). The intensities of the peaks vary because the population of the initial state depends on ℓ.

Summary

1. Atoms are usually found in nature bonded to form molecules or in the lattices of crystalline solids.
2. Ionic bonnds and covalent bonds are the principal mechanisms responsible for forming molecules. Metallic bonds and van der Waals bonds are important in the formation of solids and liquids. Hydrogen bonds enable large biological molecules to maintain their shape.
3. Like atoms, molecules emit electromagnetic radiation when making a transition from a higher energy state to a lower energy state. The internal energy of a molecule can be separated into three parts: electronic, vibrational, and rotational energy.

TOPIC	RELEVANT EQUATIONS AND REMARKS	
1. Molecular Bonding		
Ionic	Ionic bonds result when an electron is transferred from one atom to another, resulting in a positive ion and a negative ion that bond together.	
Covalent	The covalent bond is the sharing of one or more electrons by atoms.	
van der Waals	The van der Waals bonds are weak bonds that result from the interaction of the instantaneous electric dipole moments of molecules.	
Hydrogen	The hydrogen bond results from the sharing of a proton of the hydrogen atom by other atoms.	
Metallic	In the metallic bond, the positive lattice ions of the metal are held together by a cloud of negative charge composed of free electrons.	
Mixed	A diatomic molecule formed from two identical atoms, such as O_2, must bond by covalent bonding. The bonding of two nonidentical atoms is often a mixture of covalent and ionic bonding. The percentage of ionic bonding can be found from the ratio of the magnitude of the measured electric dipole moment to the magnitude of the ionic electric dipole moment defined by	
	$p_{\text{ionic}} = er_0$	37-5
	where r_0 is the equilibrium separation of the ions.	
2. *Polyatomic Molecules	The shapes of such polyatomic molecules as H_2O and NH_3 can be understood from the spatial distribution of the atomic-orbital or molecular-orbital wave functions. The tetravalent nature of the carbon atom is a result of the hybridization of the 2s and 2p atomic orbitals.	
3. Diatomic Molecules		
Moment of inertia	$I = \mu r_0^2$	37-14
	where $\mu = \dfrac{m_1 m_2}{m_1 + m_2}$	37-15
	r_0 is the equilibrium separation, and μ is the reduced mass.	
Rotational energy levels	$E_\ell = \ell(\ell + 1)E_{0r}$ where $E_{0r} = \hbar/2\pi$ and $\ell = 0, 1, 2, \ldots$	37-12
Vibrational energy levels	$E_\nu = \left(\nu + \frac{1}{2}\right)hf \quad \nu = 0, 1, 2, \ldots$	37-18
Effective force constant k_F	$f = \dfrac{1}{2\pi}\sqrt{\dfrac{k_F}{\mu}}$	37-19
4. Molecular Spectra	The optical spectra of molecules have a band structure due to transitions between rotational levels. Information about the structure and bonding of a molecule can be found from its rotational and vibrational absorption spectrum involving transitions from one vibrational-rotational level to another. These transitions obey the selection rules $\Delta\nu = \pm 1 \quad \Delta\ell = \pm 1$	

Answers to Practice Problems

37-1 17.6 percent

Answers to Concept Check

37-1 The moment of inertia of an atom is much much less than the moment of inertia of a diatomic molecule, so the amount of energy needed to change the rotational state of a single atom is much much larger than the amount needed for a diatomic molecule. At 300 K, the required energy is not available by way of collisions between atoms.

Problems

In a few problems, you are given more data than you actually need; in a few other problems, you are required to supply data from your general knowledge, outside sources, or informed estimate.

Interpret as significant all digits in numerical values that have trailing zeros and no decimal points.

- • Single-concept, single-step, relatively easy
- •• Intermediate-level, may require synthesis of concepts
- ••• Challenging
- SSM Solution is in the *Student Solutions Manual*
- Consecutive problems that are shaded are paired problems.

CONCEPTUAL PROBLEMS

1 • Would you expect NaCl to be polar or nonpolar? SSM

2 • Would you expect N_2 to be polar or nonpolar?

3 • Does neon naturally occur as Ne or Ne_2? Explain your answer.

4 • What type of bonding mechanism would you expect for atoms of (*a*) HF, (*b*) KBr, (*c*) N_2, (*d*) Ag in solid silver?

5 •• The elements in the far right column of the periodic table are sometimes called noble gases, both because they are gases under a wide range of conditions and because atoms of these elements almost never react with other atoms to form molecules or ionic compounds. However, atoms of noble gases can react if the resulting molecule is formed in an electronic excited state. An example is ArF. When it is formed in the excited state, it is written ArF* and is called an excimer (for *ex*cited di*mer*). Refer to Figure 37-13 and discuss how a diagram for the electronic, vibrational, and rotation energy levels of ArF and ArF* would look in which the ArF ground state is unstable and the ArF* excited state is stable. (Note: Excimers are used in certain kinds of lasers.) SSM

6 • Find other atoms that have the same subshell electron configurations in their two highest energy orbitals as carbon atoms do. Would you expect the same type of hybridization for these orbitals as for carbon?

7 • How does the value of the effective force constant calculated for a CO molecule in Example 37-4 compare with the value of the force constant of the suspension springs on a typical automobile, which is about 1.5 kN/m?

8 • Explain why the moment of inertia of a diatomic molecule increases slightly with increasing angular momentum.

9 • Why would you expect the separation distance between the two protons to be larger in a H_2^+ ion than in a H_2 molecule?

10 • At room temperature an atom typically absorb radiation only from the ground state, whereas a diatomic molecule typically absorbs radiation from many different rotational states. Why?

11 •• The vibrational energy levels of diatomic molecules are described by a single vibrational frequency f that is the frequency of vibration of the two atoms of the molecule along the line through their centers. Would you expect to see one or more than one vibrational frequency in molecules that have three or more atoms? Consider in particular a water molecule H_2O (Figure 37-9).

ESTIMATION AND APPROXIMATION

12 •• The potential energy for a diatomic molecule has a minimum as shown in Figure 37-13. Near this minimum, the graph for the energy as a function of distance between the atoms may be approximated as a parabola, leading to the harmonic oscillator model for the vibrating molecule. An improved approximation is called the anharmonic oscillator and leads to a modification of the expression for the energy $E_\nu = (\nu + \frac{1}{2})hf$, where $\nu = 0, 1, 2, \ldots$ (Equation 37-18). The modified expression for energy is $E_\nu = (\nu + \frac{1}{2})hf - (\nu + \frac{1}{2})^2 hf\alpha$, where $\nu = 0, 1, 2, \ldots$. For an O_2 molecule, the constants have the values $f = 4.74 \times 10^{13}\ \mathrm{s^{-1}}$ and $\alpha = 7.6 \times 10^{-3}$. Use this formula to estimate the smallest value of the quantum number ν for which the modified expression differs from the original expression by 10 percent.

13 •• To understand why quantum mechanics is not needed to describe many macroscopic systems, estimate the rotational energy quantum number ℓ and spacing between adjacent energy levels for a baseball ($m \sim 300$ g, $r \sim 3$ cm) spinning about its own axis at 20 rev/min. *Hint: Pick ℓ so the quantum energy formula $E_\ell = \ell(\ell + 1)\hbar^2/(2I)$, where $\ell = 0, 1, 2, \ldots$ (Equation 37-12) gives the correct energy for the given system. Then find the energy increase for the next highest energy level.*

14 •• Estimate the quantum number ν and spacing between adjacent energy levels for a 1.0-kg mass attached to spring. The spring has a force constant equal to 1200 N/m and the mass-spring system is vibrating with an amplitude of 3.0 cm. *Hint: Pick ν so that the quantum energy formula $E_\nu = (\nu + \frac{1}{2})hf$, where $\nu = 0, 1, 2, \ldots$ (Equation 37-18) gives the correct energy for the given system. Then find the energy increase for the next highest energy level.*

MOLECULAR BONDING

15 • Calculate the separation of Na^+ and Cl^- ions for which the potential energy of a single ionic unit (one Na^+ ion and one Cl^- ion) is -1.52 eV.

16 • The equilibrium separation of the atoms in a HF molecule is 0.0917 nm and the measured electric dipole moment of the molecule is $6.40 \times 10^{-30}\ \mathrm{C \cdot m}$. What percentage of the HF bond is ionic?

17 •• The dissociation energy of RbF is 5.12 eV, and the equilibrium separation of RbF is 0.227 nm. The electron affinity of a fluorine atom is -3.40 eV and the ionization energy of rubidium is 4.18 eV. Determine the core-repulsion energy of RbF.

18 •• The equilibrium separation of the K^+ and Cl^- ions in KCl is about 0.267 nm. (*a*) Calculate the potential energy of attraction of the ions. Assume that the ions are point charges at this separation. (*b*) The ionization energy of potassium is 4.34 eV and the electron affinity of chlorine is -3.62 eV. Calculate a value for the dissociation energy using the assumption that the energy of repulsion is negligible. (See Figure 37-1.) (*c*) The measured dissociation energy is 4.49 eV. What is the energy due to repulsion of the ions at the equilibrium separation?

19 •• Indicate an approximate value for the average value of the separation distance r for two vibrational levels on the potential energy curve for a diatomic molecule (one of the curves in Figure 37-13). Your teacher claims that the increase in r_{av} with increases in vibration energy explains why solids expand when heated. Do you agree? If so, give an argument supporting this claim. If not, give an argument opposing this claim.

20 •• Calculate the potential energy of attraction between the Na^+ and Cl^- ions at the equilibrium separation $r_0 = 0.236$ nm. Compare this result with the dissociation energy given in Figure 37-1. What is the energy due to repulsion of the ions at the equilibrium separation?

21 •• The equilibrium separation of the K^+ and F^- ions in KF is about 0.217 nm. (*a*) Calculate the potential energy of attraction of these ions. Assume that the ions are point charges at this separation. (*b*) The ionization energy of potassium is 4.34 eV and the electron affinity of fluorine is -3.40 eV. Find the dissociation energy by neglecting any energy of repulsion. (*c*) The measured dissociation energy is 5.07 eV. Calculate the energy due to repulsion of the ions at the equilibrium separation.

ENERGY LEVELS OF SPECTRA OF DIATOMIC MOLECULES

22 • The characteristic rotational energy E_{0r} for the rotation of a N_2 molecule is 2.48×10^{-4} eV. Using this value, find the separation distance of the 2 nitrogen atoms.

23 • The separation of the two oxygen atoms in a molecule of O_2 is actually slightly greater than the 0.100 nm used in Example 37-3. Furthermore, the characteristic energy of rotation E_{0r} for O_2 is 1.78×10^{-4} eV rather than the result obtained in that example. Use this value to calculate the separation distance of the two oxygen atoms. SSM

24 •• Show that the reduced mass of a diatomic molecule is always smaller than the mass of the molecule. Calculate the reduced mass for (*a*) H_2, (*b*) N_2, (*c*) CO, and (*d*) HCl. Express your answers in unified atomic mass units.

25 •• A CO molecule has a binding energy of approximately 11 eV. Find the vibrational quantum number ν that corresponds to 11 eV. (If a CO molecule actually had this much vibrational energy, it would "shake" apart.)

26 •• Derive the equation $I = \mu r_0^2$ (Equation 37-14) for the moment of inertia in terms of the reduced mass of a diatomic molecule.

27 •• The equilibrium separation between the atoms of a LiH molecule is 0.16 nm. Determine the energy separation between the $\ell = 3$ and $\ell = 2$ rotational levels of the diatomic molecule. SSM

28 •• The equilibrium separation of the K^+ and Cl^- ions in KCl is about 0.267 nm. Use this value together with the reduced mass of KCl to calculate the characteristic rotational energy E_{0r} (Equation 37-13) of KCl.

29 •• The central frequency for the absorption band of HCl shown in Figure 37-17 is at 8.66×10^{13} Hz, and the absorption peaks to either side of the central frequency are separated by about 6×10^{11} Hz. Use this information to find (*a*) the lowest (zero-point) vibrational energy for HCl, (*b*) the moment of inertia of HCl, and (*c*) the equilibrium separation of the two atoms.

30 •• Calculate the effective force constant for HCl from its reduced mass and from the fundamental vibrational frequency obtained from Figure 37-17.

31 •• The equilibrium separation between the atoms of a CO molecule is 0.113 nm. For a molecule, such as CO, that has a permanent electric dipole moment, radiative transitions obeying the selection rule $\Delta\ell = \pm 1$ between two rotational energy levels of the same vibrational level are allowed. (That is, the selection rule $\Delta\nu = \pm 1$ does not hold.) (*a*) Find the moment of inertia of CO and calculate the characteristic rotational energy E_{0r} (in eV). (*b*) Make an energy-level diagram for the rotational levels from $\ell = 0$ to $\ell = 5$ for some vibrational level. Label the energies in electron volts, starting with $E = 0$ for $\ell = 0$. Indicate on your diagram the transitions that obey $\Delta\ell = -1$, and calculate the energies of the photons emitted. (*c*) Find the wavelength of the photons emitted during each transition in (*b*). In what region of the electromagnetic spectrum are those photons? SSM

32 •• Two objects, one of mass m_1 and the other of mass m_2, are connected to opposite ends of a spring of force constant k_F. The objects are released from rest with the spring compressed. (*a*) Show that when the spring is extended and the object of mass m_1 is a distance Δr_1 from its equilibrium position in the center-of-mass reference frame, the force exerted by the spring is given by $F = -k_F(m_1/\mu)\Delta r_1$, where μ is the reduced mass. (*b*) Show that the frequency of oscillation f is related to k_F and μ by $2\pi f = \sqrt{k_F/\mu}$.

33 ••• Calculate the reduced masses μ for $H^{35}Cl$ and $H^{37}Cl$ molecules and the fractional difference $\Delta\mu/\mu$. Show that the mixture of isotopes in HCl leads to a fractional difference in the frequency of a transition from one rotational state to another given by $\Delta f/f = -\Delta\mu/\mu$. Compute $\Delta f/f$ and compare your result with Figure 37-17.

GENERAL PROBLEMS

34 • Show that when one atom of a diatomic molecule is much more massive than the other the reduced mass is approximately equal to the mass of the lighter atom.

35 •• The equilibrium separation between the nuclei of a CO molecule is 0.113 nm. Determine the energy difference between the $\ell = 2$ and $\ell = 1$ rotational energy levels of the molecule.

36 •• The effective force constant for a HF molecule is 970 N/m. Find the frequency of vibration for the molecule.

37 •• The frequency of vibration of a NO molecule is 5.63×10^{13} Hz. Find the effective force constant for NO.

38 •• The effective force constant of the hydrogen bond in a H_2 molecule is 580 N/m. Obtain the energies of the four lowest vibrational levels of the H_2, HD, and D_2 molecules, where H is protium and D is deuterium, and find the wavelengths of photons resulting from transitions between adjacent vibrational levels for each of the molecules.

39 •• The potential energy between two atoms in a molecule separated by a distance r can often be described rather well by the Lenard-Jones (or 6-12) potential function, which can be written as

$$U = U_0\left[\left(\frac{a}{r}\right)^{12} - 2\left(\frac{a}{r}\right)^{6}\right]$$

where U_0 and a are constants. Find the equilibrium separation r_0 in terms of a. *Hint: At the equilibrium separation the potential energy is a minimum.* Find U_{min}, the value of U when $r = r_0$. Use Figure 37-4 to obtain numerical values of r_0 and U_0 for a H_2 molecule, and express your answers in nanometers and electron volts.

40 •• In this problem, you are to determine how the van der Waals force between a polar molecule and a nonpolar molecule depends on the distance between the molecules. Let the polar molecule be at the origin and let its dipole moment be in the $+x$ direction. In addition, let the nonpolar molecule be on the x axis a distance x away. (*a*) How does the electric field strength due to an electric dipole vary with the distance from the dipole in a given direction? (*b*) Use (1) that the potential energy U of an electric dipole of dipole moment $\vec{p}$ in an electric field $\vec{E}$ can be expressed as $U = -\vec{p} \cdot \vec{E}$, and (2) that the induced dipole moment $\vec{p}'$ of the nonpolar molecule is in the direction of $\vec{E}$, and that p' is proportional to E, to determine how the potential energy of interaction of the two molecules depends on the separation distance x. (*c*) Using $F_x = -dU/dx$, determine how the force between the two molecules depends on distance.

41 •• Find the dependence of the force on separation distance between the two polar molecules described in Problem 40.

42 •• Use the infrared absorption spectrum of HCl in Figure 37-17 to obtain (*a*) the characteristic rotational energy E_{0r} (in eV) and (*b*) the vibrational frequency f and the vibrational energy hf (in eV).

43 • The dissociation energy is sometimes expressed in kilocalories per mole (kcal/mol). (*a*) Find the relation between the units eV/molecule and kcal/mol. (*b*) Find the dissociation energy of NaCl in kcal/mol.

IT IS WELL KNOWN THAT ARSENIC IS A POISON. IT IS LESS WELL KNOWN THAT SILICON CRYSTALS THAT HAVE SMALL CONCENTRATIONS OF ARSENIC ATOMS HAVE A MUCH LOWER RESISTIVITY THAN DO CRYSTALS THAT ARE 100 PERCENT SILICON. *(The Natural History Museum/Alamy.)*

? Do you know how many atoms of arsenic it takes to increase the charge-carrier density by a factor of 5 million? (See Example 38-7.)

Solids

The first microscopic model of electric conduction in metals was proposed by Paul K. Drude in 1900 and developed by Hendrik A. Lorentz about 1909. This model successfully predicts that the current is proportional to the potential drop (Ohm's law) and relates the resistivity of conductors to the mean speed and the mean free path* of the free electrons within the conductor. However, when mean speed and mean free path are interpreted classically, there is a disagreement between the calculated values and the measured values of the resistivity, and a similar disagreement between the predicted temperature dependence and the observed temperature dependence that resistivity values have. Thus, the classical theory fails to adequately describe the resistivity of metals. Furthermore, the classical theory says nothing about the most striking

* The mean free path is the average distance traveled between collisions.

property of solids, namely, that some substances are conductors, others are insulators, and still others are semiconductors, which are substances whose resistivity falls between that of conductors and insulators.

When mean speed and mean free path are interpreted using quantum theory, both the magnitude and the temperature dependence of the resistivity are correctly predicted. In addition, quantum theory allows us to determine if a substance will be a conductor, an insulator, or a semiconductor.

In this chapter, we use our understanding of quantum mechanics to discuss the structure of solids and solid-state semiconducting devices. Much of our discussion will be qualitative because, as in atomic physics, the quantum-mechanical calculations are mathematically sophisticated.

38-1 THE STRUCTURE OF SOLIDS

The three phases of matter we observe everyday—gas, liquid, and solid—result from the relative strengths of the attractive forces between atoms and molecules and the thermal energies of the particles. Molecules and atoms in the gas phase have relatively large thermal kinetic energies, and such particles have little influence on one another except during their frequent but brief collisions. (By using the term thermal kinetic energies, we mean the kinetic energies of the molecules and atoms in the center-of-mass reference frame of the gas.) At sufficiently low temperatures, van der Waals forces will cause practically every substance to condense into a liquid and then into a solid. In liquids, the molecules or atoms are close enough—and their thermal kinetic energies are low enough—that they can develop a temporary **short-range order.** As their thermal kinetic energies are further reduced, the molecules or atoms form solids, which are characterized by a lasting order.

If a liquid is cooled slowly so that the kinetic energy of its molecules is reduced slowly, the molecules (or atoms or ions) may arrange themselves in a regular crystalline array, producing the maximum number of bonds and leading to a minimum potential energy. However, if the liquid is cooled rapidly so that its internal energy is removed before the molecules have a chance to arrange themselves, the solid formed is often not crystalline or the arrangement is not regular. Such a solid is called an **amorphous solid.** It displays short-range order but not the long-range order (the order over many molecular, atomic, or ionic diameters) that is characteristic of a crystal. Glass is a typical amorphous solid. A characteristic result of the long-range ordering of a crystal is that it has a well-defined melting point, whereas an amorphous solid merely softens as its temperature is increased. Many substances may solidify into either an amorphous state or a crystalline state depending on how the substances are prepared; others exist only in one such state or the other.

Most common solids are polycrystalline; that is, they consist of many single crystals that meet at *grain boundaries.* The size of a single crystal is typically a fraction of a millimeter. However, large single crystals do occur naturally and can be produced artificially. The most important property of a single crystal is the symmetry and regularity of its structure. It can be thought of as having a single unit structure that is repeated throughout the crystal. This smallest unit of a crystal is called the **unit cell;** its structure depends on the type of bonding—ionic, covalent, metallic, hydrogen, van der Waals—between the atoms, ions, or molecules. If more than one kind of atom is present, the structure will also depend on the relative sizes of the atoms.

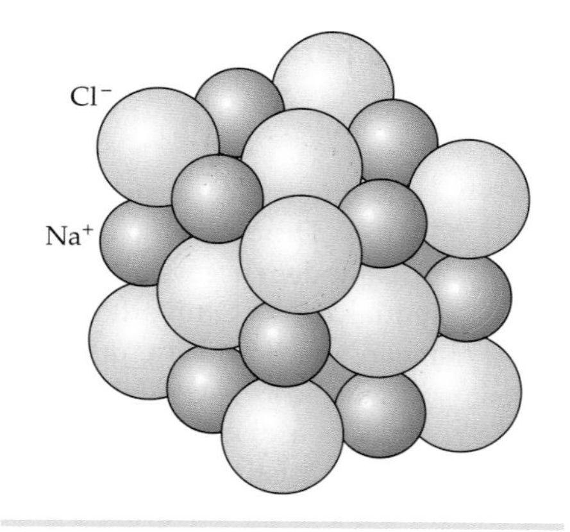

FIGURE 38-1 Face-centered-cubic structure of the NaCl crystal.

Figure 38-1 shows the structure unit cell of crystalline sodium chloride (NaCl). The Na^+ and Cl^- ions are spherically symmetric, and the Cl^- ion is approximately twice as large as the Na^+ ion. The minimum potential energy for this crystal occurs when an ion of either kind has six nearest neighbors of the other kind. This structure is called *face-centered-cubic* (fcc). Note that the Na^+ and Cl^- ions in solid NaCl are *not* paired into NaCl molecules.

The net attractive part of the potential energy of an ion in a crystal can be written

$$U_{\text{att}} = -\alpha\frac{ke^2}{r} \qquad 38\text{-}1$$

where r is the (center-to-center) separation distance between neighboring ions (0.281 nm for the Na^+ and Cl^- ions in crystalline NaCl) and α, called the **Madelung constant,** depends on the geometry of the crystal. If only the six nearest neighbors of each ion in a face-centered-cubic crystalline structure were important, α would be six. However, in addition to the six neighbors of the opposite charge at a distance r, there are twelve ions of the same charge at a distance $\sqrt{2}r$, eight ions of opposite charge at a distance $\sqrt{3}r$, and so on. The Madelung constant is thus an infinite sum:

$$\alpha = 6 - \frac{12}{\sqrt{2}} + \frac{8}{\sqrt{3}} - \cdots \qquad 38\text{-}2$$

The value of the Madelung constant for face-centered-cubic structures is $\alpha = 1.7476$.*

* A large number of terms are needed to calculate the Madelung constant accurately because the sum converges very slowly.

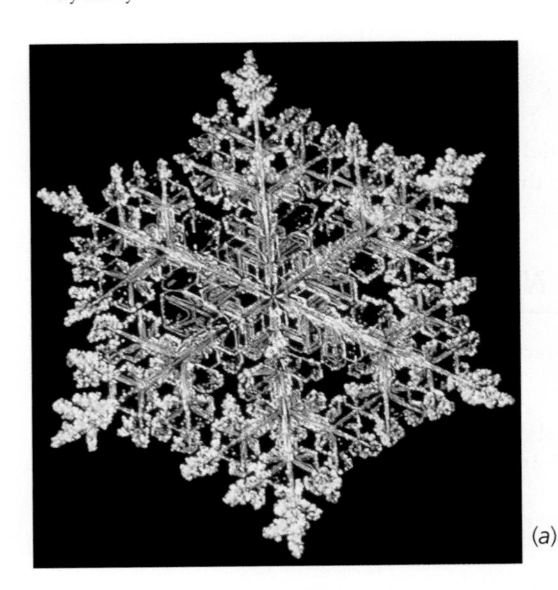
(a)

Crystal structure. (*a*) The hexagonal symmetry of a snowflake arises from a hexagonal symmetry in its lattice of hydrogen atoms and oxygen atoms. (*b*) NaCl (salt) crystals, magnified approximately thirty times. The crystals are built up from a cubic lattice of sodium and chloride ions. In the absence of impurities, an exact cubic crystal is formed. This (false-color) scanning electron micrograph shows that in practice the basic cube is often disrupted by dislocations, giving rise to crystals that have a wide variety of shapes. The underlying cubic symmetry, though, remains evident. (*c*) A crystal of quartz (SiO_2, silicon dioxide), the most abundant and widespread mineral on Earth. If molten quartz solidifies without crystallizing, glass is formed. (*d*) A soldering iron tip, ground down to reveal the copper core within its iron sheath. Visible in the iron is its underlying microcrystalline structure.
((a) Richard Waters 2/89 p. 52 Discover. (b) © Dr. Jeremy Burgess/Science Photo Library/Photo Researchers. (c) © Thomas R. Taylor/Photo Researchers. (d) Courtesy the AT&T Archives.)

(b)

(c)

(d)

When Na^+ and Cl^- ions are very close together, they repel each other because of the overlap of their electron orbitals and the exclusion-principle repulsion discussed in Section 37-1. A simple empirical expression for the potential energy associated with this repulsion that works fairly well is

$$U_{\text{rep}} = \frac{A}{r^n}$$

where A and n are constants. The total potential energy of an ion is then

$$U = -\alpha\frac{ke^2}{r} + \frac{A}{r^n} \qquad \text{38-3}$$

The equilibrium separation $r = r_0$ is that at which the force $F = -dU/dr$ is zero. Differentiating and setting $dU/dr = 0$ at $r = r_0$, we obtain

$$A = \frac{\alpha ke^2 r_0^{n-1}}{n} \qquad \text{38-4}$$

Substituting for A in Equation 38-3 gives

$$U = -\alpha\frac{ke^2}{r_0}\left[\frac{r_0}{r} - \frac{1}{n}\left(\frac{r_0}{r}\right)^n\right] \qquad \text{38-5}$$

At $r = r_0$, we have

$$U(r_0) = -\alpha\frac{ke^2}{r_0}\left(1 - \frac{1}{n}\right) \qquad \text{38-6}$$

If we know the equilibrium separation r_0, the value of n can be found approximately from the *dissociation energy* of the crystal, which is the energy needed to break up the crystal into atoms.

Example 38-1 Separation Distance between Na^+ and Cl^- in NaCl

Calculate the equilibrium separation r_0 for NaCl from the measured density of NaCl, which is $\rho = 2.16\ \text{g/cm}^3$.

PICTURE We consider each ion to occupy a cubic volume of side r_0. The mass of 1 mol of NaCl is 58.4 g, which is the sum of the molar masses of sodium and chlorine. There are $2N_A$ ions in 1 mol of NaCl, where $N_A = 6.02 \times 10^{23}$ is Avogadro's number.

SOLVE

1. We consider each ion to occupy a cubic volume of side r_0. The volume v of one mole of NaCl equals the number of ions multiplied by the volume per ion:

$$v = 2N_A r_0^3$$

2. Relate r_0 to the density ρ and the molar mass M of NaCl:

$$\rho = \frac{M}{v} = \frac{M}{2N_A r_0^3}$$

3. Solve for r_0^3 and substitute the known values:

$$r_0^3 = \frac{M}{2N_A\rho} = \frac{58.4\ \text{g}}{2(6.02 \times 10^{23})(2.16\ \text{g/cm}^3)} = 2.25 \times 10^{-23}\ \text{cm}^3$$

so

$$r_0 = 2.82 \times 10^{-8}\ \text{cm} = \boxed{0.282\ \text{nm}}$$

CHECK In Chapter 36, we found the diameter of the hydrogen atom in the ground state to be about 0.11 nm. Our step 3 result is less than three times larger. Thus, $r_0 = 0.282$ nm is plausible.

The measured dissociation energy of NaCl is 770 kJ/mol. Using 1 eV = 1.602×10^{-19} J and the fact that 1 mol of NaCl has N_A pairs of ions, we can express the dissociation energy in electron volts per ion pair. The conversion between electron volts per ion pair and kilojoules per mole is

$$1\frac{\text{eV}}{\text{ion pair}} \times \frac{6.022 \times 10^{23}\ \text{ion pairs}}{1\ \text{mol}} \times \frac{1.602 \times 10^{-19}\ \text{J}}{1\ \text{eV}}$$

The result is

$$1\,\frac{\text{eV}}{\text{ion pair}} = 96.47\,\frac{\text{kJ}}{\text{mol}} \qquad 38\text{-}7$$

Thus, 770 kJ/mol = 7.98 eV per ion pair. Substituting −7.98 eV for $U(r_0)$, 0.282 nm for r_0, and 1.75 for α in Equation 38-6, we can solve for n. The result is $n = 9.35 \approx 9$.

Most ionic crystals, such as LiF, KF, KCl, KI, and AgCl, have a face-centered-cubic structure. Some elemental solids that have fcc structure are silver, aluminum, gold, calcium, copper, nickel, and lead.

Figure 38-2 shows the structure of CsCl, which is called the *body-centered-cubic* (bcc) structure. In this structure, each ion has eight nearest neighbor ions of the opposite charge. The Madelung constant for these crystals is 1.7627. Elemental solids that have bcc structure include barium, cesium, iron, potassium, lithium, molybdenum, and sodium.

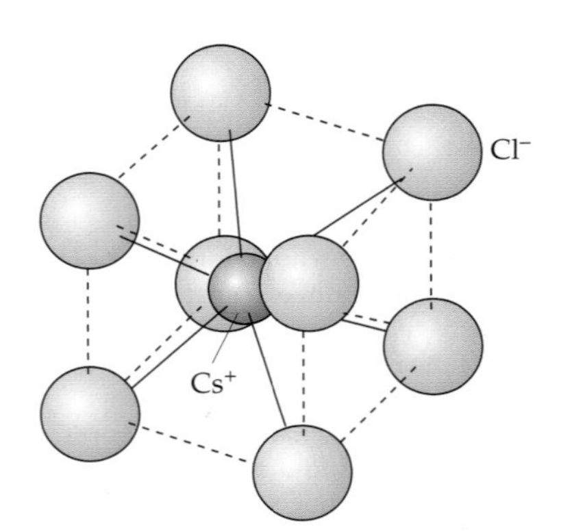

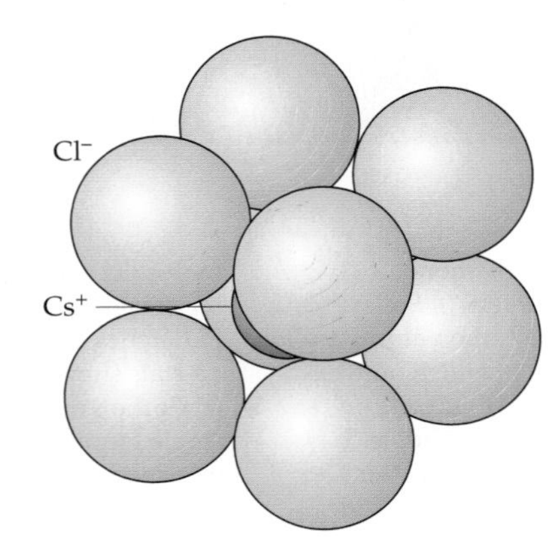

FIGURE 38-2 Body-centered-cubic structure of the CsCl crystal.

Figure 38-3 shows another important crystal structure: the *hexagonal close-packed* (hcp) structure. This structure is obtained by stacking identical spheres, such as bowling balls. In the first layer, each ball touches six others; thus, the name *hexagonal.* In the next layer, each ball fits into a triangular depression of the first layer. In the third layer, each ball fits into a triangular depression of the second layer, so it lies directly over a ball in the first layer. Elemental solids that have hcp structure include beryllium, cadmium, cerium, magnesium, osmium, and zinc.

For solids that have covalent bonding, the crystal structure is determined by the configuration of the bonds. Figure 38-4 illustrates the diamond structure of carbon, in which each atom is bonded to four other atoms as a result of hybridization, which is discussed in Section 37-2. This configuration is also the structure of germanium and silicon.

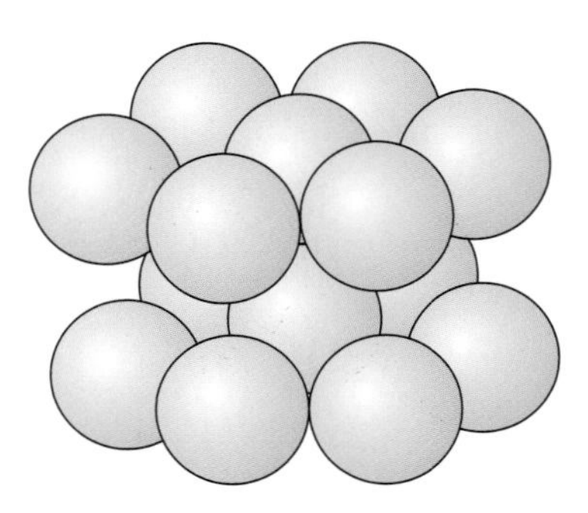

FIGURE 38-3 Hexagonal close-packed crystal structure.

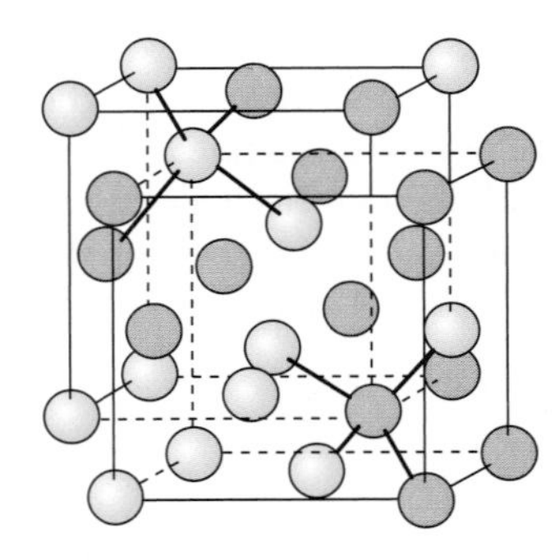

FIGURE 38-4 Diamond crystal structure. This structure can be considered to be a combination of two interpenetrating face-centered-cubic structures.

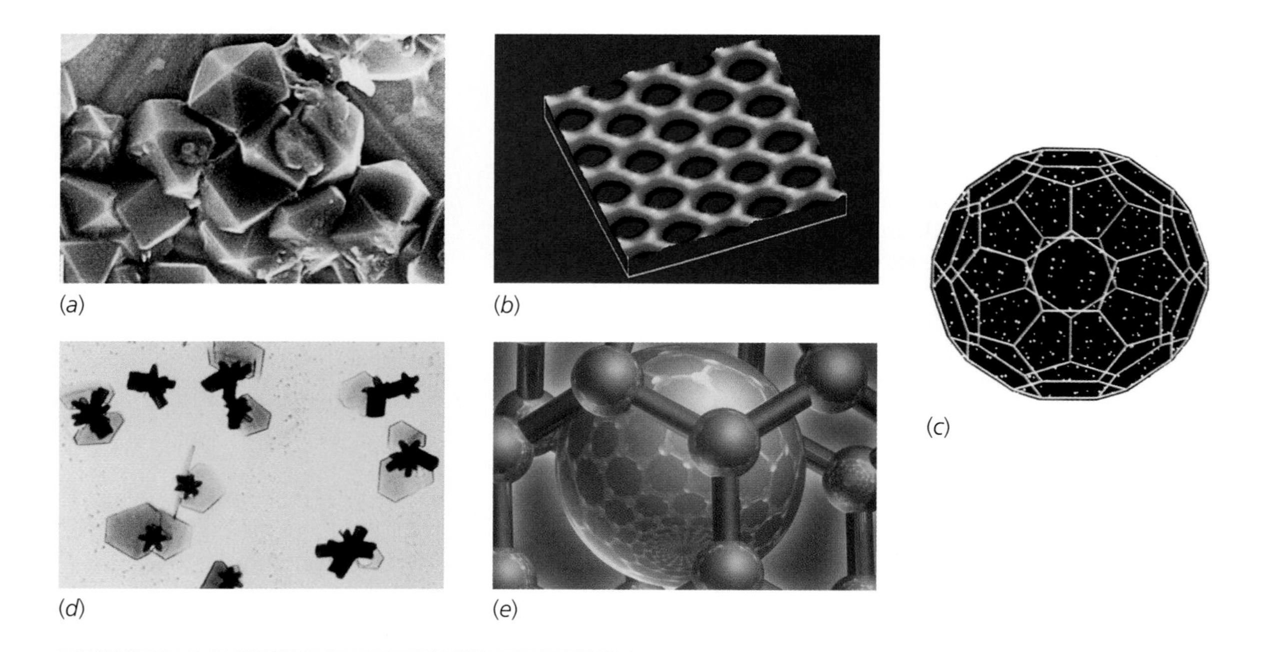

Carbon exists in three well-defined crystalline forms: diamond, graphite, and fullerenes (short for "buckminsterfullerenes"). Fullerenes were discovered in 1985. The forms differ in how the carbon atoms are packed together in a lattice. A fourth form of carbon, in which no well-defined crystalline form exists, is common charcoal. (*a*) Synthetic diamonds, magnified approximately 75,000 times. In diamond, each carbon atom is centered in a tetrahedron of four other carbon atoms. The strength of these bonds accounts for the hardness of a diamond. (*b*) An atomic-force micrograph of graphite. In graphite, carbon atoms are arranged in sheets, where each sheet is made up of atoms in hexagonal rings. The sheets slide easily across one another, a property that allows graphite to function as a lubricant. (*c*) A single sheet of carbon rings can be closed on itself if certain rings are allowed to be pentagonal, instead of hexagonal. A computer-generated image of the smallest such structure, C_{60}, is shown here. Each of the sixty vertices corresponds to a carbon atom; twenty of the faces are hexagons and twelve of the faces are pentagons. The same geometric pattern is encountered in a soccer ball. (*d*) Fullerene crystals, in which C_{60} molecules are close-packed. The smaller crystals tend to form thin brownish platelets; larger crystals are usually rodlike in shape. Fullerenes exist in which more than sixty carbon atoms appear. In the crystals shown here, about one-sixth of the molecules are C_{70}. (*e*) Carbon nanotubes have very interesting electrical properties. A single graphite sheet is a semimetal, which means that it has properties intermediate between those of semiconductors and those of metals. When a graphite sheet is rolled into a nanotube, not only do the carbon atoms have to line up around the circumference of the tube, but the wave functions of the electrons must also match up. This boundary-matching requirement places restrictions on these wave functions, which affects the motion of the electrons. Depending on exactly how the tube is rolled up, the nanotube can be either a semiconductor or a metal. *((a) Chris Kovach 3/91 p. 69 Discover. (b) Srinivas Manne, University of California, Santa Barbara. (c) Dr. F. A. Quiocho and J. S. Spurlino/Howard Hughes Medical Institute, Baylor College of Medicine. (d) W. Krätschmer/ Max-Planck-Institute for Nuclear Physics. (e) © Kenneth Weard/BioGrafx/ Science Source/Photo Researchers.)*

38-2 A MICROSCOPIC PICTURE OF CONDUCTION

We consider a metal as a regular three-dimensional lattice of ions filling some volume V and having a large number N of electrons that are free to move throughout the whole metal. The number of free electrons in a metal is approximately one to four electrons per atom. In the absence of an electric field, the free electrons move about the metal randomly, much the way gas molecules move about in a container.

The current in a conducting wire segment is proportional to the voltage drop across the segment:

$$I = \frac{V}{R} \qquad (\text{or } V = IR)$$

The resistance R is proportional to the length L of the wire segment and inversely proportional to the cross-sectional area A:

$$R = \rho\frac{L}{A}$$

where ρ is the resistivity. Substituting $\rho L/A$ for R, and EL for V, we can write the current in terms of the electric field strength E and the resistivity. We have

$$I = \frac{V}{R} = \frac{EL}{\rho L/A} = \frac{1}{\rho}EA$$

Dividing both sides by the area A gives $I/A = (1/\rho)E$, or $J = (1/\rho)E$, where $J = I/A$ is the magnitude of the **current density** vector $\vec{J}$. The current density vector is defined as

$$\vec{J} = qn\vec{v}_{\mathrm{d}} \qquad 38\text{-}8$$

DEFINITION—CURRENT DENSITY

where q, n, and $\vec{v}_{\mathrm{d}}$ are the charge, the number density, and the drift velocity of the charge carrier. (This follows from Equation 25-3.) In vector form, the relation between the current density and the electric field is

$$\vec{J} = \frac{1}{\rho}\vec{E} \qquad 38\text{-}9$$

This relation is the point form of Ohm's law. The reciprocal of the resistivity is called the **conductivity.**

According to Ohm's law, the resistivity is independent of both the current density and the electric field $\vec{E}$. Combining Equations 38-8 and 38-9 gives

$$-en_{\mathrm{e}}\vec{v}_{\mathrm{d}} = \frac{1}{\rho}\vec{E} \qquad 38\text{-}10$$

where $-e$ and n_{e} have been substituted for q and n, respectively. According to Equation 38-10, the drift velocity $\vec{v}_{\mathrm{d}}$ is proportional to $\vec{E}$.

In the presence of an electric field, a free electron experiences a force $-e\vec{E}$. If this were the only force acting, the electron would have a constant acceleration $-e\vec{E}/m_{\mathrm{e}}$. However, Equation 38-10 implies a steady-state situation with a constant drift velocity that is proportional to the field $\vec{E}$. In the microscopic model, it is assumed that a free electron is accelerated for a short time and then makes a collision with a lattice ion. The velocity of the electron immediately after the collision is completely unrelated to the drift velocity. The justification for this assumption is that the magnitude of the drift velocity is extremely small compared with the speeds associated with the thermal kinetic energies of the free electrons.

For a typical free electron, its velocity a time t after its last collision is $\vec{v}_0 - (-e\vec{E}/m_{\mathrm{e}})t$, where $\vec{v}_0$ is its velocity immediately after that collision. Because the direction of $\vec{v}_0$ is random, it does not contribute to the average velocity of the electrons. Thus, the average velocity or drift velocity of the electrons is

$$\vec{v}_{\mathrm{d}} = -\frac{e\vec{E}}{m_{\mathrm{e}}}\tau \qquad 38\text{-}11$$

where τ is the average time since the last collision. Substituting for $\vec{v}_{\mathrm{d}}$ in Equation 38-10, we obtain

$$-n_{\mathrm{e}}e\left(\frac{e\vec{E}}{m_{\mathrm{e}}}\tau\right) = \frac{1}{\rho}\vec{E}$$

so

$$\rho = \frac{m_{\mathrm{e}}}{n_{\mathrm{e}}e^2\tau} \qquad 38\text{-}12$$

The time τ, called the **collision time,** is also the average time between collisions.*

* It is tempting but incorrect to think that if τ is the average time between collisions, the average time since its last collision is $\frac{1}{2}\tau$ rather than τ. If you find this confusing, you may take comfort in the fact that Drude used the incorrect result $\frac{1}{2}\tau$ in his original work.

The average distance an electron travels between collisions is $v_{av}\tau$, which is called the mean free path λ:

$$\lambda = v_{av}\tau \qquad \text{38-13}$$

where v_{av} is the mean speed of the electrons. (The mean speed is many orders of magnitude greater than the drift speed.) In terms of the mean free path and the mean speed, the resistivity is

$$\rho = \frac{m_e v_{av}}{n_e e^2 \lambda} \qquad \text{38-14}$$

RESISTIVITY IN TERMS OF v_{AV} AND λ

According to Ohm's law, the resistivity ρ is independent of the electric field $\vec{E}$. Because m_e, n_e, and e are constants, the only quantities that could possibly depend on $\vec{E}$ are the mean speed v_{av} and the mean free path λ. Let us examine these quantities to see if they can possibly depend on the applied field $\vec{E}$.

CLASSICAL INTERPRETATION OF v_{av} AND λ

Classically, at $T = 0$ all the free electrons in a conductor should have zero kinetic energy. As the conductor is heated, the lattice ions acquire an average kinetic energy of $\frac{3}{2}kT$, which is imparted to the free electrons by the collisions between the electrons and the ions. (This is a result of the equipartition theorem studied in Chapters 17 and 18.) The free electrons would then have a Maxwell–Boltzmann distribution just like a gas of molecules. In equilibrium, the electrons would be expected to have a mean kinetic energy of $\frac{3}{2}kT$, which at ordinary temperatures (~300 K) is approximately 0.04 eV. At $T = 300$ K, their root-mean-square (rms) speed,* which is slightly greater than the mean speed, is

$$v_{av} \approx v_{rms} = \sqrt{\frac{3kT}{m_e}} = \sqrt{\frac{3(1.38 \times 10^{-23}\,\text{J/K})(300\,\text{K})}{9.11 \times 10^{-31}\text{kg}}} \qquad \text{38-15}$$

$$= 1.17 \times 10^5\ \text{m/s}$$

Note that this is about nine orders of magnitude greater than the typical drift speed of 3.5×10^{-5} m/s, which was calculated in Example 25-1. The very small drift speed caused by the electric field therefore has essentially no effect on the very large mean speed of the electrons, so v_{av} in Equation 38-14 cannot depend on the electric field $\vec{E}$.

The mean free path is related classically to the size of the lattice ions in the conductor and to the number of ions per unit volume. Consider one electron moving with speed v through a region of stationary ions that are assumed to be hard spheres (Figure 38-5). Assume the size of the electron is negligible. The electron will collide with an ion if it comes within a distance r from the center of the ion, where r is the radius of the ion. During some time interval Δt_1, the electron moves a distance vt_1. If there is an ion whose center is in the cylindrical volume $\pi r^2 v\Delta t_1$, the electron will collide with the ion. The electron will then change directions and collide with another ion in time Δt_2 if the center of the ion is in the volume $\pi r^2 vt_2$. Thus, in the total time $\Delta t = \Delta t_1 + \Delta t_2 + \ldots$, the electron will collide with the number of ions whose centers are in the volume $\pi r^2 v\Delta t$. The number of ions in this volume is $n_{ion}\pi r^2 v\Delta t$, where n_{ion} is the number of ions per unit volume.

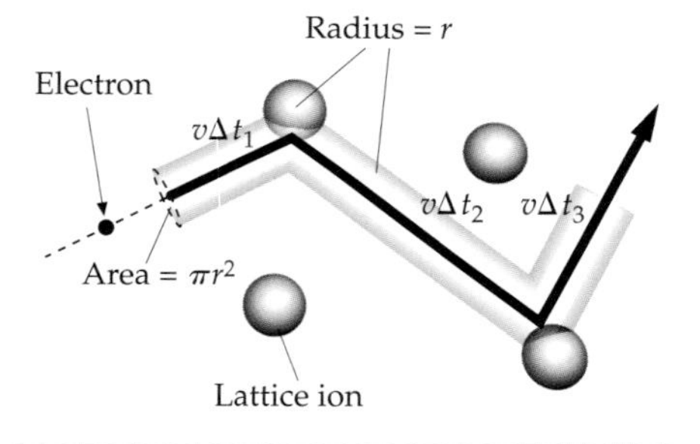

FIGURE 38-5 Model of an electron moving through the lattice ions of a conductor. The electron, which is considered to be a point particle, collides with an ion if it comes within a distance r of the center of the ion, where r is the radius of the ion. If the electron speed is v, it collides in time Δt with all the ions whose centers are in the volume $\pi r^2 v\Delta t$. While this picture is in accord with the classical Drude model for conduction in metals, it is in conflict with the current quantum-mechanical model presented later in this chapter.

* See Equation 17–21.

The total path length divided by the number of collisions is the mean free path:

$$\lambda = \frac{v\Delta t}{n_{\text{ion}}\pi r^2 v\Delta t} = \frac{1}{n_{\text{ion}}\pi r^2} = \frac{1}{n_{\text{ion}}A} \qquad \text{38-16}$$

where $A = \pi r^2$ is the cross-sectional area of a lattice ion.

SUCCESSES AND FAILURES OF THE CLASSICAL MODEL

Neither n_{ion} nor r depends on the electric field $\vec{E}$, so λ also does not depend on $\vec{E}$. v_{av} and λ do not depend on $\vec{E}$ according to their classical interpretations, so the resistivity ρ does not depend on $\vec{E}$ in accordance with Ohm's law. However, the classical theory gives an incorrect temperature dependence for the resistivity. Because λ depends only on the radius and the number density of the lattice ions, the only quantity in Equation 38-14 that depends on temperature in the classical theory is v_{av}, which is proportional to $\sqrt{T}$. But experiments show that ρ varies linearly with temperature. Furthermore, when ρ is calculated at $T = 300$ K using the Maxwell–Boltzmann distribution for v_{av} and Equation 38-16 for λ, the calculated result is about six times greater than the measured value.

The classical theory of conduction fails because electrons are not classical particles. The wave nature of the electrons must be considered. Because of the wave properties of electrons and the constraints described by the exclusion principle (to be discussed in the following section), the energy distribution of the free electrons in a metal is not even approximately given by the Maxwell–Boltzmann distribution. Furthermore, the collision of an electron with a lattice ion is not similar to the collision of a baseball with a tree. Instead, it involves the scattering of electron waves by the lattice. To understand the quantum theory of conduction, we need a qualitative understanding of the energy distribution of free electrons in a metal. This will also help us understand the origin of contact potentials between two dissimilar metals in contact and the contribution of free electrons to the heat capacity of metals.

38-3 FREE ELECTRONS IN A SOLID

One may want to consider free electrons in a metal to be an *electron gas* in a metal. However, molecules in an ordinary gas, such as air, obey the classical Maxwell–Boltzmann energy distribution, but the free electrons in a metal do not. Instead, they obey a quantum energy distribution called the *Fermi–Dirac distribution.* The main features of a free electron can be understood by considering the electron in a metal to be a particle in a box, a problem whose one-dimensional version we studied extensively in Chapter 34. We discuss the main features of a free electron semiquantitatively in this section and leave the details of the Fermi–Dirac distribution to Section 38-9.

ENERGY QUANTIZATION IN A BOX

In Chapter 34, we found that the wavelength associated with an electron of momentum p is given by the de Broglie relation:

$$\lambda = \frac{h}{p} \qquad \text{38-17}$$

where h is Planck's constant. When a particle is confined to a finite region of space, such as a box, only certain wavelengths λ_n, where $n = 1, 2, \ldots,$ that are specified by standing-wave conditions are allowed. For a one-dimensional box of length L, the standing-wave condition is

$$n\frac{\lambda_n}{2} = L \qquad n = 1, 2, \ldots \qquad \text{38-18}$$

This results in the quantization of energy:

$$E_n = \frac{p_n^2}{2m} = \frac{(h/\lambda_n)^2}{2m} = \frac{h^2}{2m}\frac{1}{\lambda_n^2} = \frac{h^2}{2m}\frac{1}{(2L/n)^2}$$

or

$$E_n = n^2 E_1 \qquad \text{38-19}$$

where $E_1 = h^2/(8mL^2)$. The spatial wave function for the nth state is given by

$$\psi_n(x) = \sqrt{\frac{2}{L}}\sin(n\pi x/L) \qquad \text{38-20}$$

The quantum number n characterizes the wave function for a particular state and the energy of that state. In three-dimensional problems, three quantum numbers arise, one associated with each dimension.

THE EXCLUSION PRINCIPLE

The distribution of electrons among the possible energy states is described by the exclusion principle, which states that no two electrons in an atom can be in the same quantum state; that is, they cannot have the same set of values for their quantum numbers. The exclusion principle applies to all "spin one-half" particles (fermions), which include electrons, protons, and neutrons. These particles have a *spin* quantum number m_s which has two possible values, $+\frac{1}{2}$ and $-\frac{1}{2}$. The quantum state of a particle is characterized by the spin quantum number m_s and the quantum numbers associated with the spatial part of the wave function. Because the spin quantum numbers have just two possible values, the exclusion principle can be stated in terms of the spatial states:

> There can be at most two electrons with the same set of values for their *spatial* quantum numbers.
>
> EXCLUSION PRINCIPLE IN TERMS OF SPATIAL STATES

When there are more than two electrons in a system, such as an atom, only two can be in the lowest energy state. The third and fourth electrons must go into the second-lowest state, and so on.

Example 38-2 Boson-System Energy versus Fermion-System Energy

Compare the total energy of the ground state of five identical bosons of mass m in a one-dimensional box with the total energy of the ground state of five identical fermions of mass m in the same box.

PICTURE The ground state is the lowest possible energy state. The energy levels in a one-dimensional box are given by $E_n = n^2E_1$, where $E_1 = h^2/(8mL^2)$. (This is in accord with Equation 38-19.) The lowest energy for five bosons occurs when all the bosons are in the state $n = 1$, as shown in Figure 38-6a. For fermions, the lowest state occurs when two fermions are in the state $n = 1$, two fermions are in the state $n = 2$, and one fermion is in the state $n = 3$, as shown in Figure 38-6b.

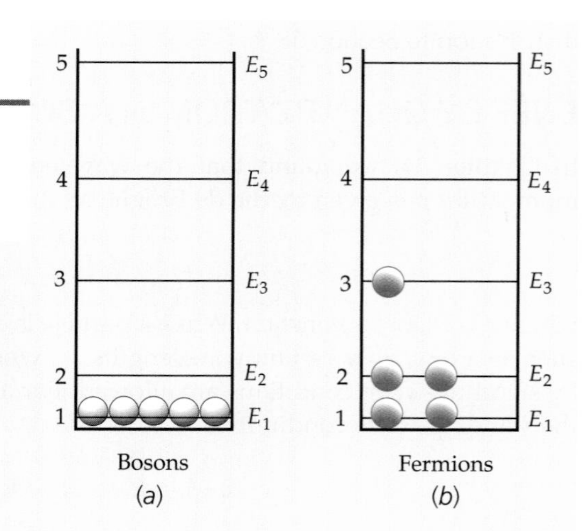

FIGURE 38-6

SOLVE

1. The energy of five bosons in the state $n = 1$ is:

$$E = 5E_1$$

2. The energy of two fermions in the state $n = 1$, two fermions in the state $n = 2$, and one fermion in the state $n = 3$ is:

$$E = 2E_1 + 2E_2 + 1E_3 = 2E_1 + 2(2)^2E_1 + 1(3)^2E_1$$
$$= 2E_1 + 8E_1 + 9E_1 = 19E_1$$

3. Compare the total energies:

The five identical fermions have 3.8 times the total energy of the five identical bosons.

CHECK The fact that fermions must have different quantum states has a large effect on the total energy of a multiple-particle system, as expected.

THE FERMI ENERGY

When there are many electrons in a box, at $T = 0$ the electrons will occupy the lowest energy states consistent with the exclusion principle. If we have N electrons, we can put two electrons in the lowest energy level, two electrons in the next lowest energy level, and so on. The N electrons thus fill the lowest $N/2$ energy levels (Figure 38-7). The energy of the last filled (or half-filled) level at $T = 0$ is called the Fermi energy E_F. If the electrons moved in a one-dimensional box, the Fermi energy would be given by Equation 38-19, with $n = N/2$:

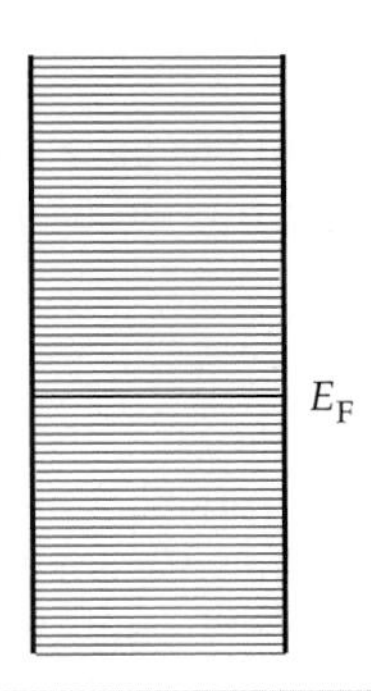

FIGURE 38-7 At $T = 0$ the electrons fill up the allowed energy states to the Fermi energy E_F. The levels are so closely spaced that they can be assumed to be continuous.

$$E_F = \left(\frac{N}{2}\right)^2 \frac{h^2}{8m_eL^2} = \frac{h^2}{32m_e}\left(\frac{N}{L}\right)^2 \qquad 38\text{-}21$$

FERMI ENERGY AT $T = 0$ IN ONE DIMENSION

In a one-dimensional box, the Fermi energy depends on the number of free electrons per unit length of the box.

PRACTICE PROBLEM 38-1

Suppose there is an ion, and therefore a free electron, every 0.100 nm in a one-dimensional box. Calculate the Fermi energy. *Hint: Write Equation* 38-21 *as*

$$E_F = \frac{(hc)^2}{32m_ec^2}\left(\frac{N}{L}\right)^2 = \frac{(1240\ \text{eV}\cdot\text{nm})^2}{32(0.511\ \text{MeV})}\left(\frac{N}{L}\right)^2$$

In our model of conduction, the free electrons move in a *three-dimensional* box of volume V. The derivation of the Fermi energy in three dimensions is somewhat difficult, so we will just give the result. In three dimensions, the Fermi energy at $T = 0$ is

$$E_F = \frac{h^2}{8m_e}\left(\frac{3N}{\pi V}\right)^{2/3} \qquad 38\text{-}22a$$

FERMI ENERGY AT $T = 0$ IN THREE DIMENSIONS

The Fermi energy depends on the number density of free electrons N/V. Substituting numerical values for the constants gives

$$E_F = (0.3646\ \text{eV}\cdot\text{nm}^2)\left(\frac{N}{V}\right)^{2/3} \qquad 38\text{-}22b$$

FERMI ENERGY AT $T = 0$ IN THREE DIMENSIONS

Example 38-3 The Fermi Energy for Copper

The number density for electrons in copper was calculated in Example 25-1 and found to be $84.7/\text{nm}^3$. Calculate the Fermi energy at $T = 0$ for copper.

PICTURE The Fermi energy is given by Equations 38-22.

SOLVE

1. The Fermi energy is given by Equation 38-22*b*: $E_F = (0.3646\ \text{eV}\cdot\text{nm}^2)\left(\frac{N}{V}\right)^{2/3}$

2. Substitute the given number density for copper: $E_F = (0.3646\ \text{eV}\cdot\text{nm}^2)(84.7/\text{nm}^3)^{2/3} = \boxed{7.03\ \text{eV}}$

CHECK The Fermi energy (the step-2 result) is much greater than kT at room temperatures as expected. For example, at $T = 300$ K, kT is only about 0.026 eV.

PRACTICE PROBLEM 38-2 Use Equation 38-22*b* to calculate the Fermi energy at $T = 0$ for gold, which has a free-electron number density of $59.0/\text{nm}^3$.

Table 38-1 lists the free-electron number densities and Fermi energies at $T = 0$ for several metals.

The free electrons in a metal are sometimes referred to as a Fermi gas. (They constitute a gas of fermions.) The average energy of a free electron can be calculated from the complete energy distribution of the electrons, which is discussed in Section 38-9. At $T = 0$, the average energy turns out to be

$$E_{av} = \tfrac{3}{5}E_F \tag{38-23}$$

AVERAGE ENERGY OF ELECTRONS IN A FERMI GAS AT $T = 0$

Table 38-1 Free-Electron Number Densities* and Fermi Energies at $T = 0$ for Selected Elements

	Element	N/V, electrons/nm³	E_F, eV
Al	Aluminum	181	11.7
Ag	Silver	58.6	5.50
Au	Gold	59.0	5.53
Cu	Copper	84.7	7.03
Fe	Iron	170	11.2
K	Potassium	14.0	2.11
Li	Lithium	47.0	4.75
Mg	Magnesium	86.0	7.11
Mn	Manganese	165	11.0
Na	Sodium	26.5	3.24
Sn	Tin	148	10.2
Zn	Zinc	132	9.46

* Number densities are measured using the Hall effect, discussed in Section 26-4.

For copper, E_{av} is approximately 4 eV. This average energy is huge compared with thermal energies of about $kT \approx 0.026$ eV at a temperature of $T = 300$ K. This result is very different from the classical Maxwell–Boltzmann distribution result that at $T = 0$, $E = 0$, and that at some temperature T, E is of the same order as kT.

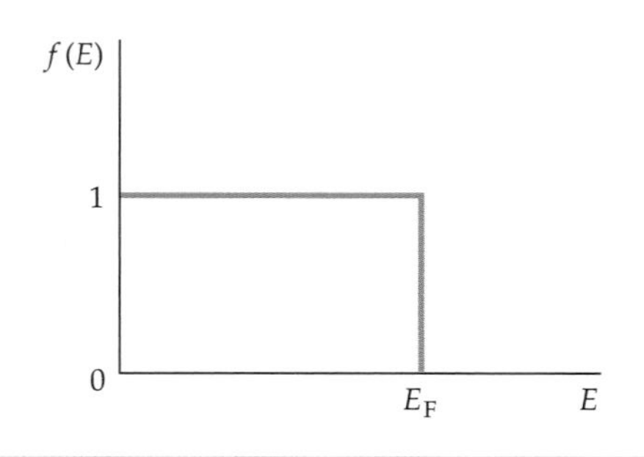

FIGURE 38-8 Fermi factor versus energy at $T = 0$.

THE FERMI FACTOR AT $T = 0$

The probability of an energy state being occupied is called the **Fermi factor,** $f(E)$. At $T = 0$ all the states below E_F are filled, whereas all those above that energy are empty, as shown in Figure 38-8. Thus, at $T = 0$ the Fermi factor is simply

$$f(E) = \begin{cases} 1 & E < E_F \\ 0 & E > E_F \end{cases} \quad \text{38-24}$$

THE FERMI FACTOR FOR $T > 0$

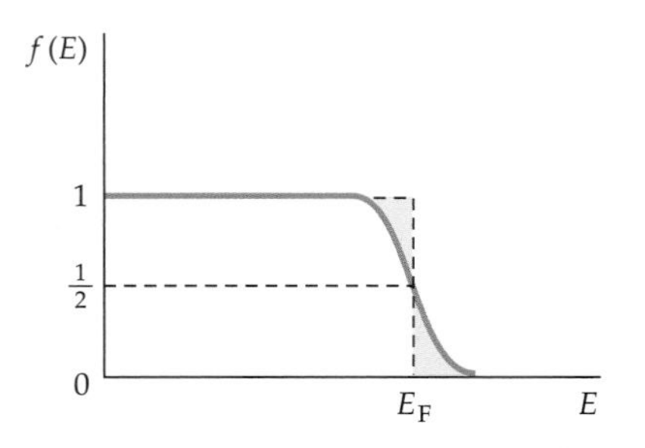

FIGURE 38-9 The Fermi factor for some temperature T. Some electrons that have energies near the Fermi energy are excited, as indicated by the shaded regions. The Fermi energy E_F is that value of E for which $f(E) = \frac{1}{2}$.

At temperatures greater than $T = 0$, some electrons will occupy higher energy states because of thermal energy gained during collisions with the lattice. However, an electron cannot move to a higher or lower state unless it is unoccupied. Because the kinetic energy of the lattice ions is of the order of kT, electrons cannot gain much more energy than kT in collisions with the lattice ions. Therefore, only those electrons that have energies within about kT of the Fermi energy can gain energy as the temperature is increased. At 300 K, kT is only 0.026 eV, so the exclusion principle prevents all but a very few electrons near the top of the energy distribution from gaining energy through random collisions with the lattice ions. Figure 38-9 shows a plot of the Fermi factor for some temperature T. Because for $T > 0$ there is no distinct energy that separates filled levels from unfilled levels, the definition of the Fermi energy must be slightly modified. At temperature T, the Fermi energy is defined to be the energy of the energy state for which the probability of being occupied is $\frac{1}{2}$. For all but extremely high temperatures, the difference between the Fermi energy at temperature T and the Fermi energy at temperature $T = 0$ is very small.

The **Fermi temperature** T_F is defined by

$$kT_F = E_F \quad \text{38-25}$$

For temperatures much lower than the Fermi temperature, the average energy of the lattice ions will be much less than the Fermi energy, and the electron energy distribution will not differ greatly from that at $T = 0$.

Example 38-4 The Fermi Temperature for Copper

Find the Fermi temperature for copper.

PICTURE We use Equation 38-25 to find the Fermi temperature. The Fermi energy for copper at $T = 0$, calculated in Example 38-3, is 7.03 eV.

SOLVE

Use $E_F = 7.03$ eV and $k = 8.617 \times 10^{-5}$ eV/K in Equation 38-25:

$$T_F = \frac{E_F}{k} = \frac{7.03 \text{ eV}}{8.617 \times 10^{-5} \text{ eV/K}} = \boxed{81\,600 \text{ K}}$$

CHECK The Fermi temperature is very high, as expected.

TAKING IT FURTHER We can see from this example that the Fermi temperature of copper is much greater than any temperature T for which copper remains a solid.

Because an electric field in a conductor accelerates all of the conduction electrons together, the exclusion principle does not prevent the free electrons in filled states from participating in conduction. Figure 38-10 shows the Fermi factor in one dimension versus *velocity* for an ordinary temperature. The factor is approximately 1 for velocities v_x in the range $-u_F < v_x < u_F$, where the Fermi speed u_F is related to the Fermi energy by $E_F = \frac{1}{2}mu_F^2$. Then

$$u_F = \sqrt{\frac{2E_F}{m_e}} \qquad 38\text{-}26$$

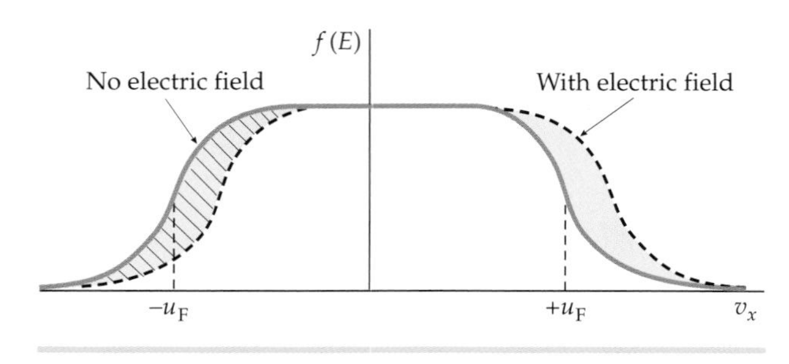

FIGURE 38-10 Fermi factor versus velocity in one dimension with no electric field (solid) and with an electric field in the $-x$ direction (dashed). The difference is greatly exaggerated.

Example 38-5 The Fermi Speed for Copper

Calculate the Fermi speed for copper.

PICTURE We use Equation 38-26 to find the Fermi speed. The Fermi energy for copper at $T = 0$, calculated in Example 38-3, is 7.03 eV.

SOLVE

Use Equation 38-26 with $E_F = 7.03$ eV:

$$u_F = \sqrt{\frac{2(7.03 \text{ eV})}{9.11 \times 10^{-31} \text{ kg}}\left(\frac{1.60 \times 10^{-19} \text{ J}}{1 \text{ eV}}\right)} = \boxed{1.57 \times 10^6 \text{ m/s.}}$$

CHECK As expected, the result (the Fermi speed for copper) is high, but less than the speed of light.

The dashed curve in Figure 38-10 shows the Fermi factor after the electric field has been acting for some time t. Although all of the free electrons have their velocities shifted in the direction opposite to the electric field, the net effect is equivalent to shifting only the electrons near the Fermi energy.

CONTACT POTENTIAL

When two different metals are placed in contact, a potential difference V_{contact} called the **contact potential** develops between them. The contact potential depends on both the work functions of the two metals, ϕ_1 and ϕ_2 (we encountered work functions when the photoelectric effect was introduced in Chapter 34), and the Fermi energies of the two metals. When the metals are in contact, the total energy of the system is lowered if electrons near the boundary move from the metal that has the higher Fermi energy into the metal that has the lower Fermi energy until the Fermi energies of the two metals are the same, as shown in Figure 38-11. When equilibrium is established, the metal that has the lower initial Fermi energy is negatively charged and the other metal is positively charged, so that between them there is a potential difference V_{contact} given by

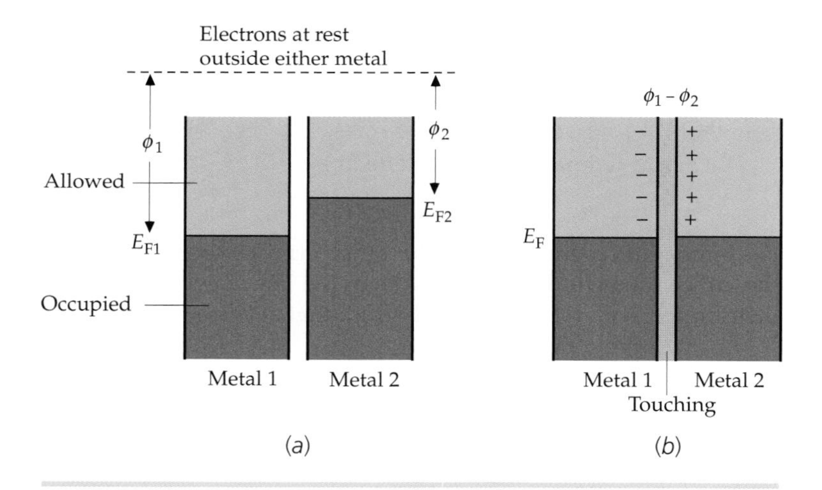

FIGURE 38-11 (*a*) Energy levels for two different metals that have different Fermi energies E_F and work functions ϕ. The work function is the difference between the energy of an electron at rest outside the metal and the Fermi energy within the metal. (*b*) When the metals are in contact, electrons flow from the metal that initially has the higher Fermi energy to the metal that initially has the lower Fermi energy until the Fermi energies are equal.

$$V_{\text{contact}} = \frac{\phi_1 - \phi_2}{e} \qquad 38\text{-}27$$

Table 38-2 lists the work functions for several metals.

Table 38-2 Work Functions for Some Metals

	Metal	ϕ, eV		Metal	ϕ, eV
Ag	Silver	4.7	K	Potassium	2.1
Au	Gold	4.8	Mn	Manganese	3.8
Ca	Calcium	3.2	Na	Sodium	2.3
Cu	Copper	4.1	Ni	Nickel	5.2

Example 38-6 Contact Potential between Silver and Tungsten

The threshold wavelength for the photoelectric effect is 271 nm for tungsten and 262 nm for silver. What is the contact potential developed when silver and tungsten are placed in contact?

PICTURE The contact potential is proportional to the difference in the work functions for the two metals (Equation 38-27). The work function ϕ can be found from the given threshold wavelengths using $\phi = hc/\lambda_t$ (Equation 34-4).

SOLVE

1. The contact potential is given by Equation 38-27:
$$V_{\text{contact}} = \frac{\phi_1 - \phi_2}{e}$$

2. The work function is related to the threshold wavelength (Equation 34-4):
$$\phi = \frac{hc}{\lambda_t}$$

3. Substitute $\lambda_t = 271$ nm for tungsten (the symbol for tungsten is W):
$$\phi_W = \frac{hc}{\lambda_t} = \frac{1240\ \text{eV}\cdot\text{nm}}{271\ \text{nm}} = 4.58\ \text{eV}$$

4. Substitute $\lambda_t = 262$ nm for silver:
$$\phi_{Ag} = \frac{1240\ \text{eV}\cdot\text{nm}}{262\ \text{nm}} = 4.73\ \text{eV}$$

5. The contact potential is thus:
$$V_{\text{contact}} = \frac{\phi_{Ag} - \phi_W}{e} = 4.73\ \text{V} - 4.58\ \text{V} = \boxed{0.15\ \text{V}}$$

CHECK As expected, the contact potential is small (less than one volt). You do not get large potential differences just by putting two metals in contact.

HEAT CAPACITY DUE TO ELECTRONS IN A METAL

The quantum-mechanical description of the electron distribution in metals allows us to understand why the contribution of the free electrons to the heat capacity of a metal is much less that of the ions. According to the classical equipartition theorem, the energy of the lattice ions in n moles of a solid is $3nRT$, and thus the molar specific heat is $c' = 3R$, where R is the universal gas constant (see Section 18-7). In a metal, the number of free electrons is approximately equal to the number of lattice ions. If these electrons obey the classical equipartition theorem, they should have an energy of $\frac{3}{2}nRT$ and contribute an additional $\frac{3}{2}R$ to the molar specific heat. But measured heat capacities of metals are just slightly greater than those of insulators. We can understand this result because at some temperature T, only those electrons that have energies near the Fermi energy can be excited by random collisions with the lattice ions. The number of those electrons is of the order of $(kT/E_F)N$, where N is the total number of free electrons. The energy of those electrons is increased from that at $T = 0$ by an amount that is of the order of kT. So the total increase in thermal energy is of the order of $(kT/E_F)N \times kT$.

We can thus express the energy of N electrons at temperature T as

$$E = NE_{\text{av}}(0) + \alpha N \frac{kT}{E_{\text{F}}} kT \qquad \text{38-28}$$

where $E_{\text{av}}(0)$ is the average energy at $T = 0$ and α is a constant that we expect to be of the order of 1 if our reasoning is correct. The calculation of α is quite challenging. The result is $\alpha = \pi^2/4$. Using this result and writing E_{F} in terms of the Fermi temperature, $E_{\text{F}} = kT_{\text{F}}$, we obtain the following for the contribution of the free electrons to the heat capacity at constant volume:

$$C_{\text{V}} = \frac{dE}{dT} = 2\alpha Nk\frac{kT}{E_{\text{F}}} = \frac{1}{2}\pi^2 nR\frac{T}{T_{\text{F}}}$$

where we have written Nk in terms of the gas constant R ($R = Nk/n$). The molar specific heat at constant volume is then

$$c'_{\text{V}} = \frac{1}{2}\pi^2 R\frac{T}{T_{\text{F}}} \qquad \text{38-29}$$

We can see that because of the large value of T_{F}, the contribution of the free electrons is a small fraction of R at ordinary temperatures. Because $T_{\text{F}} = 81\,600$ K for copper, the molar specific heat of the free electrons at $T = 300$ K is

$$c'_{\text{V}} = \frac{1}{2}\pi^2 \frac{300\text{ K}}{81\,600\text{ K}} R \approx 0.02R$$

which is in good agreement with the experiment.

38-4 QUANTUM THEORY OF ELECTRICAL CONDUCTION

We can use Equation 38-14 for the resistivity if we use the Fermi speed u_{F} (Equation 38-26) in place of v_{av}:

$$\rho = \frac{m_{\text{e}} u_{\text{F}}}{n_{\text{e}} e^2 \lambda} \qquad \text{38-30}$$

We now have two problems. First, because the Fermi speed u_{F} is approximately independent of temperature, the resistivity given by Equation 38-30 is independent of temperature unless the mean free path should depend on the temperature. The second problem concerns magnitudes. As mentioned earlier, the classical expression for resistivity using v_{av} calculated from the Maxwell–Boltzmann distribution gives values that are about 6 times too large at $T = 300$ K. Because the Fermi speed u_{F} is about 16 times the Maxwell-Boltzmann value of v_{av}, the magnitude of ρ predicted by Equation 38-30 will be approximately 100 times greater than the experimentally determined value. The resolution of both of these problems lies in the calculation of the mean free path λ.

THE SCATTERING OF ELECTRON WAVES

In Equation 38-16 for the classical mean free path $\lambda = 1/(n_{\text{ion}}A)$, the quantity $A = \pi r^2$ is the cross-sectional area of the lattice ion as seen by an electron. In the quantum calculation, the mean free path is related to the scattering of electron waves by the crystal lattice. Detailed calculations show that, for a *perfectly* ordered crystal, $\lambda = \infty$; that is, there is no scattering of the electron waves. The scattering of electron waves arises because of *imperfections* in the crystal lattice, which have nothing to do with the actual cross-sectional area A of the lattice ions.

According to the quantum theory of electron scattering, A depends merely on *deviations* of the lattice ions from a perfectly ordered array and not on the size of the ions. The most common causes of such deviations are thermal vibrations of the lattice ions or impurities.

We can use $\lambda = 1/(n_{\text{ion}}A)$ for the mean free path if we reinterpret the area A. Figure 38-12 compares the classical picture and the quantum picture of this area. In the quantum picture, the lattice ions are points that have no size but present an area $A = \pi r_0^2$, where r_0 is the amplitude of thermal vibrations. In Chapter 14, we saw that the energy of vibration in simple harmonic motion is proportional to the square of the amplitude, which is πr_0^2. Thus, the effective area A is proportional to the energy of vibration of the lattice ions. From the equipartition theorem,* we know that the average energy of vibration is proportional to kT. Thus, A is proportional to T, and λ is proportional to $1/T$. Then the resistivity given by Equation 38-14 is proportional to T, in agreement with experiment.

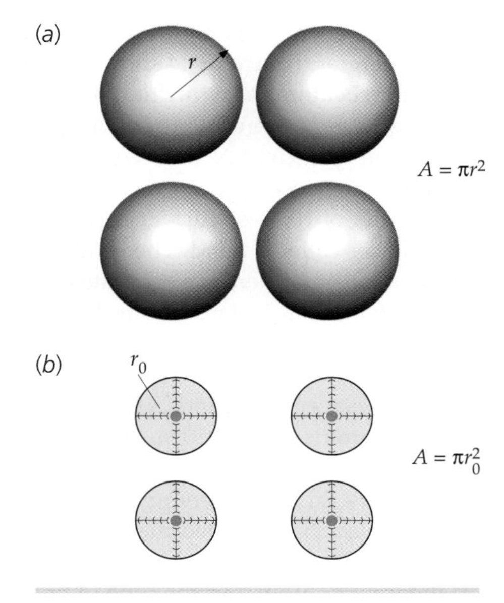

FIGURE 38-12 (*a*) Classical picture of the lattice ions as spherical balls of radius r that each present an area πr^2 to the electrons. (*b*) Quantum-mechanical picture of the lattice ions as points that are vibrating in three dimensions. The area presented to the electrons is πr_0^2, where r_0 is the amplitude of oscillation of the ions.

The effective area A due to thermal vibrations can be calculated, and the results give values for the resistivity that are in agreement with experiments. At $T = 300$ K, for example, the effective area turns out to be about 100 times smaller than the actual cross-sectional area of a lattice ion. We see, therefore, that the free-electron model of metals gives a good account of electrical conduction if the classical mean speed v_{av} is replaced by the Fermi speed u_F and if the collisions between electrons and the lattice ions are interpreted in terms of the scattering of electron waves, for which only deviations from a perfectly ordered lattice are important.

The presence of impurities in a metal also causes deviations from perfect regularity in the crystal lattice. The effects of impurities on resistivity are approximately independent of temperature. The resistivity of a metal containing impurities can be written $\rho = \rho_t + \rho_i$, where ρ_t is the resistivity due to the thermal motion of the lattice ions and ρ_i is the resistivity due to impurities. Figure 38-13 shows typical resistance versus temperature curves for metals with impurities. As the absolute temperature approaches zero, the resistivity due to thermal motion approaches zero, and the total resistivity approaches the resistivity due to impurities, which is constant.

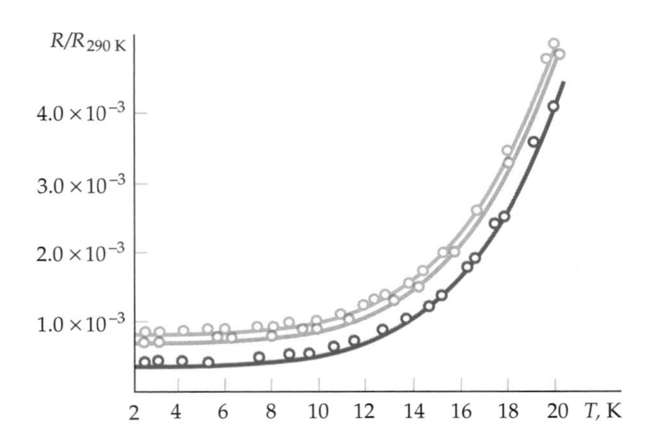

FIGURE 38-13 Relative resistance versus temperature for three samples of sodium. The three curves have the same temperature dependence but different magnitudes because of differing amounts of impurities in the samples.

38-5 BAND THEORY OF SOLIDS

Resistivities vary enormously between insulators and conductors. For a typical insulator, such as quartz, $\rho \sim 10^{16}\ \Omega\cdot\text{m}$, whereas for a typical conductor, $\rho \sim 10^{-8}\ \Omega\cdot\text{m}$. The reason for this enormous variation is the variation in the number density of free electrons n_e. To understand this variation, we consider the effect of the lattice on the electron energy levels.

We begin by considering the energy levels of the individual atoms as they are brought together. The allowed energy levels in an isolated atom are often far apart. For example, in hydrogen, the lowest allowed energy $E_1 = -13.6$ eV is 10.2 eV below the next lowest allowed energy $E_2 = (-13.6\text{ eV})/4 = -3.4$ eV.† Let us consider two identical atoms and focus our attention on one particular energy level. When the atoms are far apart, the energy of a particular level is the same for each atom. As the atoms are brought closer together, the energy level for each atom changes because of the influence of the other atom. As a result, the level splits into two levels of slightly different energies for the two-atom system. If we bring three atoms close together,

* The equipartition theorem does hold for the lattice ions, which obey the Maxwell–Boltzmann energy distribution.

† The energy levels in hydrogen are discussed in Chapter 36.

a particular energy level splits into three separate levels of slightly different energies. Figure 38-14 shows the energy splitting of two energy levels for six atoms as a function of the separation of the atoms.

If we have N identical atoms, a particular energy level in the isolated atom splits into N different, closely spaced energy levels when the atoms are close together. In a macroscopic solid, N is very large—of the order of 10^{23}—so each energy level splits into a very large number of levels called a **band.** The levels are spaced almost continuously within the band. There is a separate band of levels for each particular energy level of the isolated atom. The bands may be widely separated in energy, they may be close together, or they may even overlap, depending on the kind of atom and the type of bonding in the solid.

The lowest energy bands, corresponding to the lowest energy levels of the atoms in the lattice, are filled with electrons that are bound to the individual atoms. The electrons that can take part in conduction occupy the higher energy bands. The highest energy band that contains electrons is called the **valence band.** The valence band may be completely filled with electrons or only partially filled, depending on the kind of atom and the type of bonding in the solid.

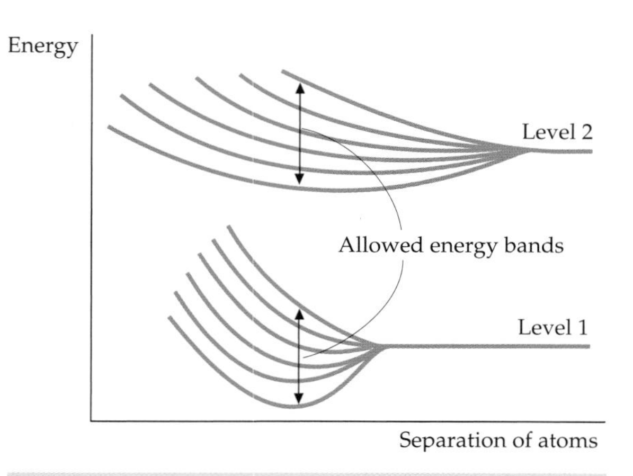

FIGURE 38-14 Energy splitting of two energy levels for six atoms as a function of the separation of the atoms. When there are many atoms, each level splits into a near-continuum of levels called a band.

We can now understand why some solids are conductors and why others are insulators. If the valence band is only partially filled, there are many available empty energy states in the band, and the electrons in the band can easily be raised to a higher energy state by an electric field. Accordingly, this substance is a good conductor. If the valence band is filled and there is a large energy gap between it and the next available band, an applied electric field may be too weak to excite an electron from the upper energy levels of the filled band across the large gap into the energy levels of the empty band, so the substance is an insulator. The lowest band in which there are unoccupied states is called the **conduction band.** In a conductor, the valence band is only partially filled, so the valence band is also the conduction band. An energy gap between allowed bands is called a **forbidden energy band.**

The band structure for a conductor, such as copper, is shown in Figure 38-15*a*. The lower bands (not shown) are filled with the lower energy electrons of the atoms. The valence band is only about half-filled. When an electric field is established in the conductor, the electrons in the conduction band are accelerated, which means that their energies are increased. This is consistent with the exclusion principle because there are many empty energy states just above those occupied by electrons in this band. These electrons are thus the conduction electrons.

Figure 38-15*b* shows the band structure for magnesium, which is also a conductor. In this case, the highest occupied band is completely filled, but there is an empty band above it that overlaps it. The two bands thus form a combined valence–conduction band that is only partially filled.

Figure 38-15*c* shows the band structure for a typical insulator. At $T = 0$ K, the valence band is completely filled. The next energy band having empty energy states, the conduction band, is separated from the valence band by a large energy gap.

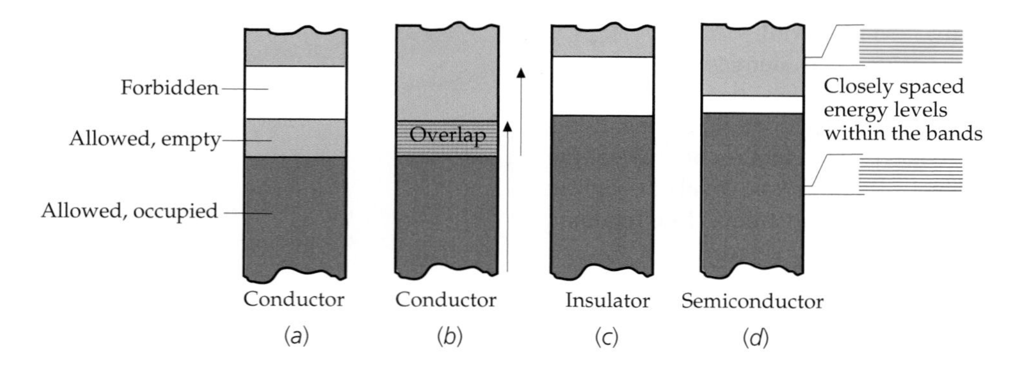

FIGURE 38-15 Four possible band structures for a solid. (*a*) A typical conductor. The valence band is also the conduction band. It is only partially filled, so electrons can be easily excited to nearby energy states. (*b*) A conductor in which the valence band overlaps a conduction band above it. (*c*) A typical insulator. There is a forbidden band that has a large energy gap between the filled valence band and the conduction band. (*d*) A semiconductor. The energy gap between the filled valence band and the conduction band is very small, so some electrons are excited to the conduction band at normal temperatures, leaving holes in the valence band.

At $T = 0$, the conduction band is empty. At ordinary temperatures, a few electrons can be excited to states in that band, but most cannot be excited to states because the energy gap is large compared with the energy an electron might obtain by thermal excitation. Very few electrons can be thermally excited to the nearly empty conduction band, even at fairly high temperatures. When an electric field of ordinary magnitude is established in the solid, electrons cannot be accelerated because there are no empty energy states at nearby energies. We describe this by saying that there are no free electrons. The small conductivity that is observed is due to the very few electrons that are thermally excited into the nearly empty conduction band. When an electric field applied to an insulator is sufficiently strong to cause an electron to be excited across the energy gap to the empty band, dielectric breakdown occurs.

In some substances, the energy gap between the filled valence band and the empty conduction band is very small, as shown in Figure 38-15*d*. At $T = 0$, there are no electrons in the conduction band and the material is an insulator. At ordinary temperatures, however, there are an appreciable number of electrons in the conduction band due to thermal excitation. Such a material is called an **intrinsic semiconductor.** For typical intrinsic semiconductors, such as silicon and germanium, the energy gap is only about 1 eV. In the presence of an electric field, the electrons in the conduction band can be accelerated because there are empty states nearby. Also, for each electron in the conduction band there is a vacancy, or **hole,** in the nearly filled valence band. In the presence of an electric field, electrons in this band can also be excited to a vacant energy level. This contributes to the electric current and is most easily described as the motion of a hole in the direction of the field and opposite to the motion of the electrons. The hole thus acts like a positive charge. To visualize the conduction of holes, think of a two-lane, one-way road that has one lane completely filled with parked cars and the other lane empty. If a car moves out of the completely filled lane into the empty lane, it can move ahead freely. As the other cars move up to occupy the vacated space, the vacated space propagates backward in the direction opposite the motion of the cars. Both the forward motion of the car in the nearly empty lane and the backward propagation of the empty space contribute to a net forward propagation of the cars.

An interesting characteristic of semiconductors is that the resistivity of the substance decreases as the temperature increases, which is contrary to the case for normal conductors. The reason is that as the temperature increases, the number of free electrons increases because there are more electrons in the conduction band. The number of holes in the valence band also increases, of course. In semiconductors, the effect of the increase in the number of charge carriers, both electrons and holes, exceeds the effect of the increase in resistivity due to the increased scattering of the electrons by the lattice ions due to thermal vibrations. Semiconductors therefore have a negative temperature coefficient of resistivity.

38-6 SEMICONDUCTORS

The semiconducting property of intrinsic semiconductors makes them useful as a basis for electronic circuit components whose resistivity can be controlled by application of an external voltage or current. Most such *solid-state devices,* however, such as the semiconductor diode and the transistor, make use of **impurity semiconductors,** which are created through the controlled addition of certain impurities to intrinsic semiconductors. This process is called **doping.** Figure 38-16*a* is a schematic illustration of silicon doped with a small amount of arsenic so that the arsenic atoms replace a few of the silicon atoms in the crystal lattice. The conduction band of pure silicon is virtually empty at ordinary temperatures, so pure silicon is a poor conductor of electricity. However, arsenic has five valence electrons rather than the four valence electrons of silicon. Four of these electrons take part in bonds with the four neighboring silicon atoms, and the fifth electron is very

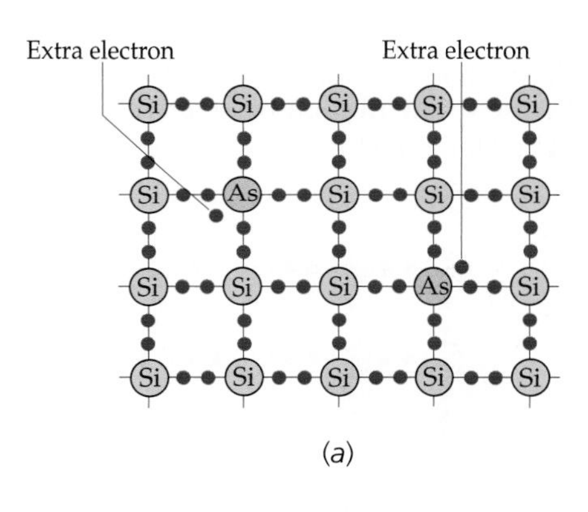

(*a*)

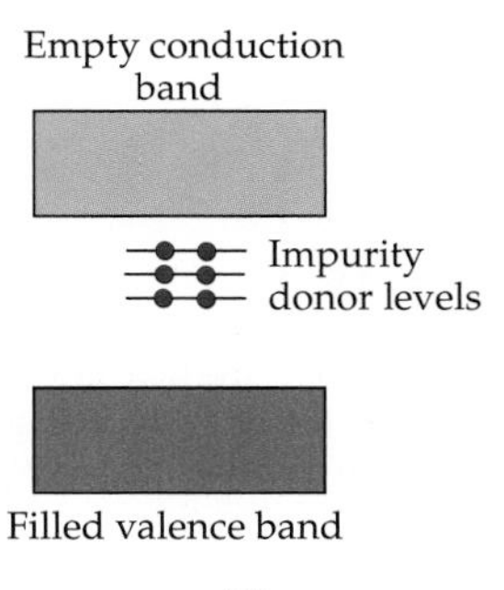

(*b*)

FIGURE 38-16 (*a*) A two-dimensional schematic illustration of silicon doped with arsenic. Because arsenic has five valence electrons, there is an extra, weakly bound electron that is easily excited to the conduction band, where it can contribute to electrical conduction. (*b*) Band structure of an *n*-type semiconductor, such as silicon doped with arsenic. The impurity atoms provide filled energy levels that are just below the conduction band. These levels donate electrons to the conduction band.

loosely bound to the atom. This extra electron occupies an energy level that is just slightly below the conduction band in the solid, and it is easily excited into the conduction band, where it can contribute to electrical conduction.

The effect on the band structure of a silicon crystal achieved by doping it with arsenic is shown in Figure 38-16*b*. The levels shown just below the conduction band are due to the extra electrons of the arsenic atoms. These levels are called **donor levels** because they donate electrons to the conduction band without leaving holes in the valence band. Such a semiconductor is called an ***n*-type semiconductor** because the major charge carriers are negatively charged electrons. The conductivity of a doped semiconductor can be controlled by controlling the amount of impurity added. The addition of just one part per million can increase the conductivity by several orders of magnitude.

Another type of impurity semiconductor can be made by replacing a silicon atom with a gallium atom, which has three valence electrons (Figure 38-17*a*). The gallium atom accepts electrons from the valence band to complete its four covalent bonds, thus creating a hole in the valence band. The effect on the band structure of silicon achieved by doping it with gallium is shown in Figure 38-17*b*. The empty levels shown just above the valence band are due to the holes from the ionized gallium atoms. These levels are called **acceptor levels** because they accept electrons from the filled valence band when those electrons are thermally excited to a higher energy state. This creates holes in the valence band that are free to propagate in the direction of an electric field. Such a semiconductor is called a ***p*-type semiconductor** because the charge carriers are positively charged holes. The fact that conduction is due to the motion of positively charged holes can be verified by the Hall effect. (The Hall effect is discussed in Chapter 26.)

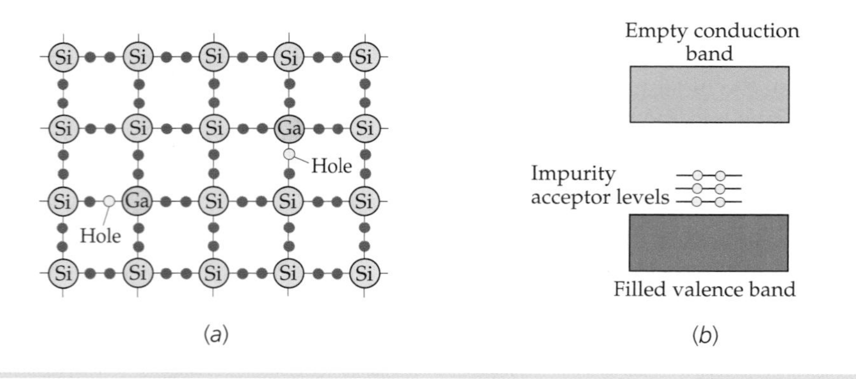

FIGURE 38-17 (*a*) A two-dimensional schematic illustration of silicon doped with gallium. Because gallium has only three valence electrons, there is a hole in one of its bonds. As electrons move into the hole the hole moves about, contributing to the conduction of electrical current. (*b*) Band structure of a *p*-type semiconductor, such as silicon doped with gallium. The impurity atoms provide empty energy levels just above the filled valence band that accept electrons from the valence band.

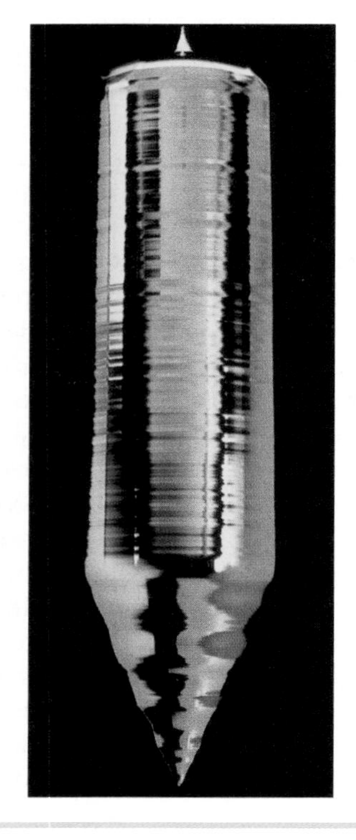

Synthetic crystal silicon is produced beginning with a raw material containing silicon (for instance, common beach sand), separating out the silicon, and melting it. From a seed crystal, the molten silicon grows into a cylindrical crystal, such as the one shown here. The crystals (typically about 1.3 m long) are formed under highly controlled conditions to ensure that they are flawless and the crystals are then sliced into thousands of thin wafers onto which the layers of an integrated circuit are etched. *(Museum of Modern Art.)*

Example 38-7 Number Density of Free Electrons in Arsenic-Doped Silicon — *Try It Yourself*

The number of free electrons in pure silicon is approximately 10^{10} electrons/cm^3 at ordinary temperatures. If one silicon atom out of every 10^6 atoms is replaced by an arsenic atom, how many free electrons per cubic centimeter are there? (The density of silicon is 2.33 g/cm^3 and its molar mass is 28.1 g/mol.)

PICTURE The number of silicon atoms per cubic centimeter, n_{Si}, can be found from $n_{Si} = \rho N_A/M$. Then, because each arsenic atom contributes one free electron, the number of electrons contributed by the arsenic atoms is $10^{-6}\, n_{Si}$.

SOLVE

Cover the column to the right and try these on your own before looking at the answers.

Steps	Answers
1. Calculate the number of silicon atoms per cubic centimeter.	$n_{\mathrm{Si}} = \dfrac{\rho N_{\mathrm{A}}}{M} = \dfrac{(2.33\ \mathrm{g/cm^3})(6.02 \times 10^{23}\ \mathrm{atoms/mol})}{28.1\ \mathrm{g/mol}} = 4.99 \times 10^{22}\ \mathrm{atoms/cm^3}$
2. Multiply by 10^{-6} to obtain the number of arsenic atoms per cubic centimeter, which equals the added number of free electrons per cubic centimeter.	$n_{\mathrm{As}} = 10^{-6} n_{\mathrm{Si}} = 4.99 \times 10^{16}\ \mathrm{atoms/cm^3}$
3. The number of free electrons per cubic centimeter is equal to the number of arsenic atoms per cubic centimeter plus 1×10^{-10} (the number of silicon atoms per cubic centimeter).	$n_{\mathrm{e}} = n_{\mathrm{As}} + 1 \times 10^{-10} n_{\mathrm{Si}} = 4.99 \times 10^{16}\ \mathrm{cm^{-3}} + 1 \times 10^{10}\ \mathrm{cm^{-3}} \approx \boxed{5 \times 10^{16}\ \mathrm{electrons/cm^3}}$

CHECK As expected, the step-3 result is less than the number density of silicon atoms and more than the number density of conduction electrons in pure silicon.

TAKING IT FURTHER Because silicon has so few free electrons per atom, the number density of conduction electrons is increased by a factor of approximately 5 million per cubic centimeter by doping silicon with just one arsenic atom per million silicon atoms.

PRACTICE PROBLEM 38-3 How many free electrons are there per silicon atom in pure silicon?

* 38-7 SEMICONDUCTOR JUNCTIONS AND DEVICES

Semiconductor devices such as diodes and transistors make use of n-type semiconductors and p-type semiconductors joined together, as shown in Figure 38-18. In practice, the two types of semiconductors are often incorporated into a single silicon crystal doped with donor impurities on one side and acceptor impurities on the other side. The region in which the semiconductor changes from a p-type semiconductor to an n-type semiconductor is called a ***pn* junction.**

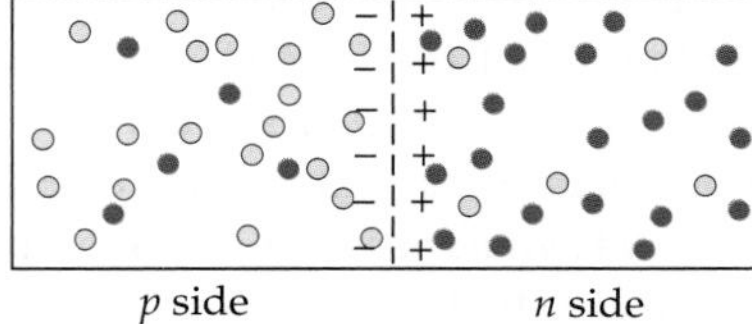

FIGURE 38-18 A pn junction. Because of the difference in their concentrations on either side of the pn junction, holes diffuse from the p side to the n side, and electrons diffuse from the n side to the p side. As a result, there is a double layer of charge at the junction, with the p side being negative and the n side being positive.

When an n-type semiconductor and a p-type semiconductor are placed in contact, the initially unequal concentrations of electrons and holes result in the diffusion of electrons across the junction from the n side to the p side and holes from the p side to the n side until equilibrium is established. The result of this diffusion is a net transport of positive charge from the p side to the n side. Unlike the case when two different metals are in contact, the electrons cannot travel very far from the junction region because the semiconductor is not a particularly good conductor. The diffusion of electrons and holes therefore creates a double layer of charge at the junction similar to that on a parallel-plate capacitor. There is, thus, a potential difference V across the junction, which tends to inhibit further diffusion. In equilibrium, the n side which has a net positive charge will be at a higher potential than the p side which has a net negative charge. In the junction region, between the charge layers, there will be very few charge carriers of either type, so the junction region has a high resistance. Figure 38-19 shows the energy-level diagram for a pn junction. The junction region is also called the **depletion region** because it has been depleted of charge carriers.

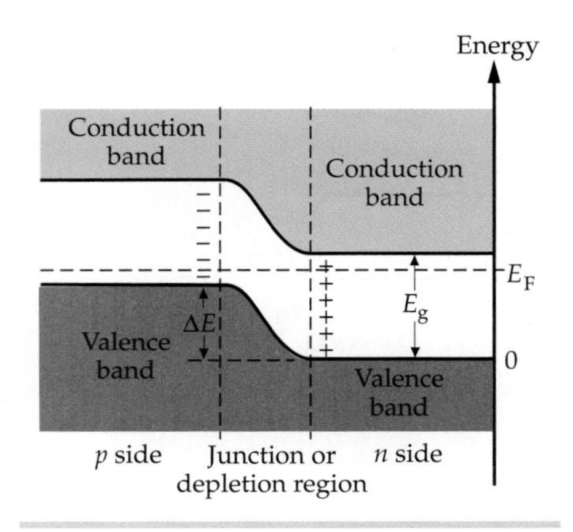

FIGURE 38-19 Electron energy levels for a pn junction.

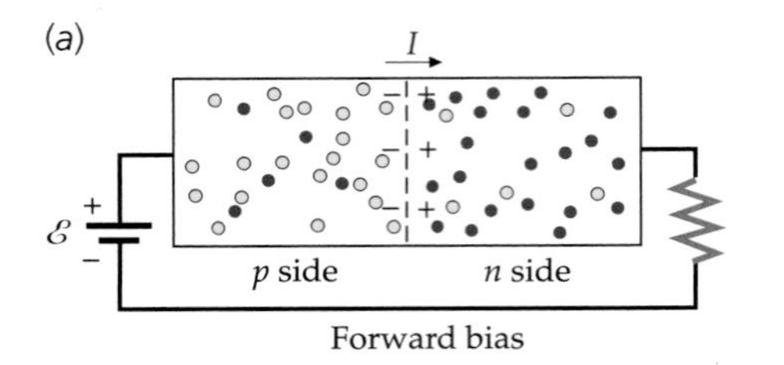

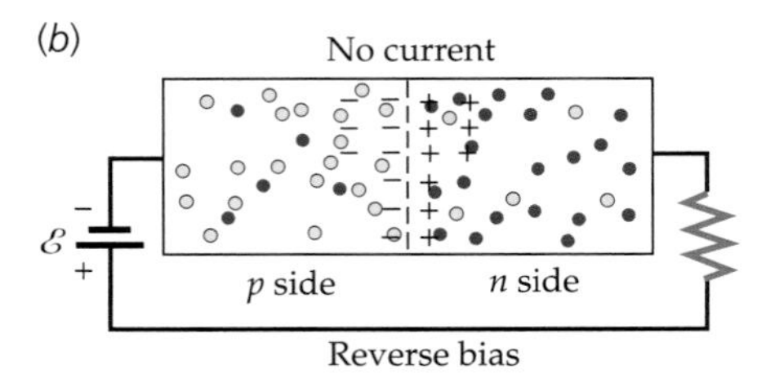

FIGURE 38-20 A *pn*-junction diode. (*a*) Forward-biased *pn* junction. The applied potential difference enhances the diffusion of holes from the *p* side to the *n* side and of electrons from the *n* side to the *p* side, resulting in a current *I*. (*b*) Reverse-biased *pn* junction. The applied potential difference inhibits the further diffusion of holes and electrons across the junction, so there is no current.

*DIODES

In Figure 38-20, an external potential difference has been applied across a *pn* junction by connecting a battery and a resistor to the semiconductor. When the positive terminal of the battery is connected to the *p* side of the junction, as shown in Figure 38-20*a*, the junction is said to be **forward biased.** Forward biasing lowers the potential across the junction. The diffusion of electrons and holes is thereby increased as they attempt to reestablish equilibrium, resulting in a current in the circuit.

If the positive terminal of the battery is connected to the *n* side of the junction, as shown in Figure 38-20*b*, the junction is said to be **reverse biased.** Reverse biasing tends to increase the potential difference across the junction, thereby further inhibiting diffusion. Figure 38-21 shows a plot of current versus voltage for a typical semiconductor junction. Essentially, the junction conducts only in one direction for applied voltages greater than the breakdown voltage. A single-junction semiconductor device is called a **diode.*** Diodes have many uses. One use is to convert alternating current into direct current, a process called *rectification.*

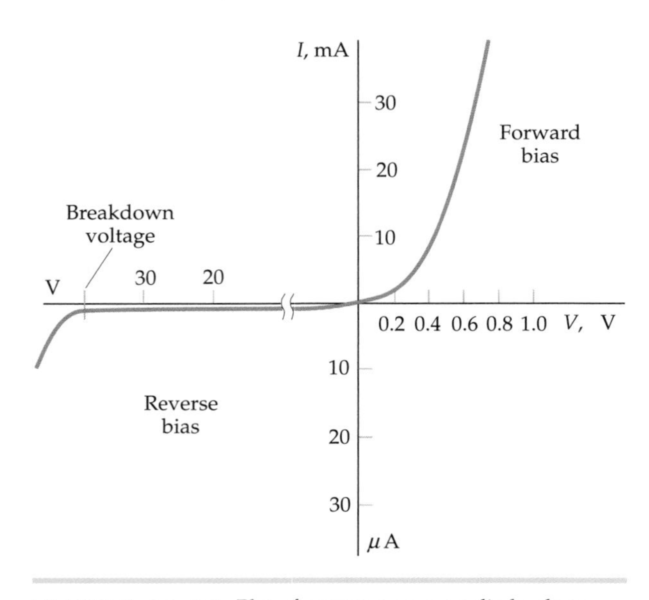

FIGURE 38-21 Plot of current versus applied voltage across a *pn* junction. Note the different scales on both axes for the forward and reverse bias conditions.

Note that the current in Figure 38-21 suddenly increases in magnitude at extreme values of reverse bias. In such large electric fields, electrons are stripped from their atomic bonds and accelerated across the junction. These electrons, in turn, cause others to break loose. This effect is called **avalanche breakdown.** Although such a breakdown can be disastrous in a circuit where it is not intended, the fact that it occurs at a sharply defined voltage makes it of use in a special voltage reference standard known as a **Zener diode.** Zener diodes are also used to protect devices from excessively high voltages.

An interesting effect, one that we discuss only qualitatively, occurs if both the *n* side and the *p* side of a *pn*-junction diode are so heavily doped that the donors on the *n* side provide so many electrons that the lower part of the conduction band is practically filled, and the acceptors on the *p* side accept so many electrons that the upper part of the valence band is nearly empty. Figure 38-22*a* shows the energy-level

FIGURE 38-22 Electron energy levels for a heavily doped *pn*-junction tunnel diode. (*a*) With no bias voltage, some electrons tunnel in each direction. (*b*) With a small bias voltage, the tunneling current is enhanced in one direction, making a sizable contribution to the net current. (*c*) With further increases in the bias voltage, the tunneling current decreases dramatically.

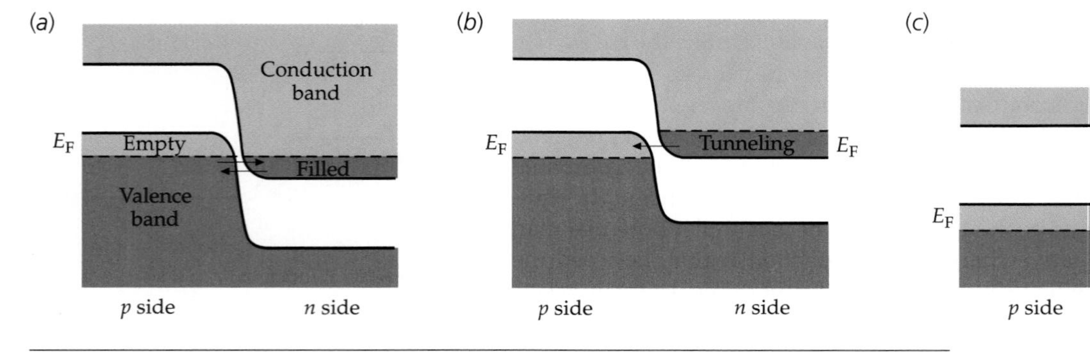

* The name *diode* originates from a vacuum tube device consisting of just two electrodes that also conducts electric current in one direction only.

diagram for this situation. Because the depletion region is now so narrow, electrons can easily penetrate the potential barrier across the junction and tunnel to the other side. The flow of electrons through the barrier is called a **tunneling current,** and such a heavily doped diode is called a **tunnel diode.**

At equilibrium where there is no bias, there is an equal tunneling current in each direction. When a small bias voltage is applied across the junction, the energy-level diagram is as shown in Figure 38-22*b*, and the tunneling of electrons from the *n* side to the *p* side is increased, whereas the tunneling of electrons in the opposite direction is decreased. This tunneling current, in addition to the usual current due to diffusion, results in a considerable net current. When the bias voltage is increased slightly, the energy-level diagram is as shown in Figure 38-22*c*, and the tunneling current is decreased. Although the diffusion current is increased, the net current is decreased. At large bias voltages, the tunneling current is completely negligible, and the total current increases with increasing bias voltage due to diffusion, as in an ordinary *pn*-junction diode. Figure 38-23 shows the current versus voltage curve for a tunnel diode. Such diodes are used in electric circuits because of their very fast response time. When operated near the peak in the current versus voltage curve, a small change in bias voltage results in a large change in the current.

I

V_A V_B V

FIGURE 38-23 Current versus applied (bias) voltage V for a tunnel diode. For $V < V_A$, an increase in the bias voltage V enhances tunneling. For $V_A < V < V_B$, an increase in the bias voltage inhibits tunneling. For $V > V_B$, the tunneling is negligible, and the diode behaves like an ordinary *pn*-junction diode.

Another use for the *pn*-junction semiconductor is the **solar cell,** which is illustrated schematically in Figure 38-24. When a photon of energy greater than the gap energy (1.1 eV in silicon) strikes the *p*-type region, it can excite an electron from the valence band into the conduction band, leaving a hole in the valence band. This region is already rich in holes. Some of the electrons created by the photons will recombine with holes, but some will migrate to the junction. From there, they are accelerated into the *n*-type region by the electric field between the double layer of charge. This creates an excess negative charge in the *n*-type region and an excess positive charge in the *p*-type region. The result is a potential difference between the two regions, which in practice is approximately 0.6 V. If a load resistance is connected across the two regions, a charge flows through the resistance. Some of the incident light energy is thus converted into electrical energy. The current in the resistor is proportional to the rate of arrival of incident photons, which is in turn proportional to the intensity of the incident light.

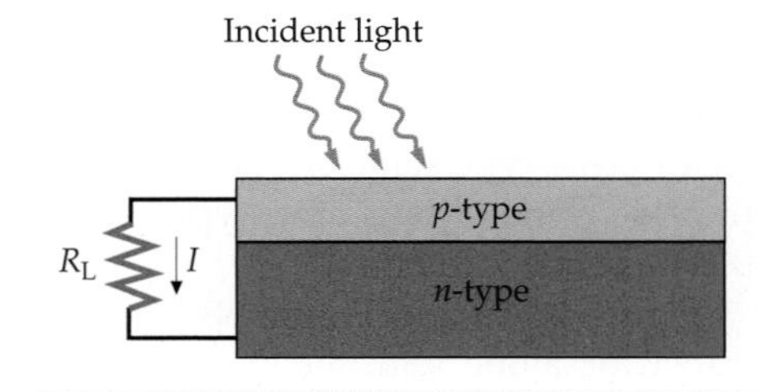

FIGURE 38-24 A *pn*-junction semiconductor as a solar cell. When light strikes the *p*-type region, electron-hole pairs are created, resulting in a current through the load resistance R_L.

There are many other applications of semiconductors with *pn* junctions. Particle detectors, called **surface-barrier detectors,** consist of a *pn*-junction semiconductor that has a large reverse bias so that there is ordinarily no current. When a high-energy particle, such as an electron, passes through the semiconductor, it creates many electron–hole pairs as it loses energy. The resulting current pulse signals the passage of the particle. **Light-emitting diodes** (LEDs) are *pn*-junction semiconductors that have large forward biases that produce large excess concentrations of electrons on the *p* sides and holes on the *n* sides of the junctions. Under these conditions, an LED emits light as the electrons and holes recombine. This is essentially the reverse of the process that occurs in a solar cell, in which electron–hole pairs are created by the absorption of light. LEDs are commonly used as warning indicators and as sources of infrared light beams.

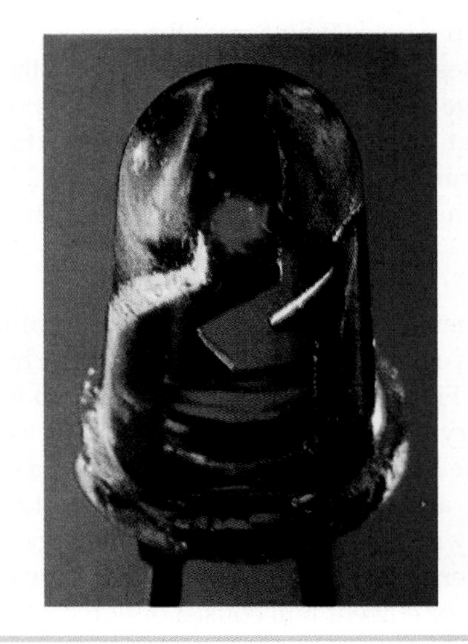

A light-emitting diode (LED). *(© C. Falco/Photo Researchers.)*

*TRANSISTORS

The transistor, a semiconducting device that is used to produce a desired output signal in response to an input signal, was invented in 1948 by William Shockley, John Bardeen, and Walter Brattain and has revolutionized the electronics industry and our everyday world. A *simple bipolar junction transistor** consists of three distinct semiconductor regions called the **emitter,** the **base,** and the **collector.** The base is a very thin region of one type of semiconductor sandwiched between two regions of the opposite type. The emitter semiconductor is much more heavily

* Besides the bipolar junction transistor, there are other categories of transistors, notably, the field–effect transistor.

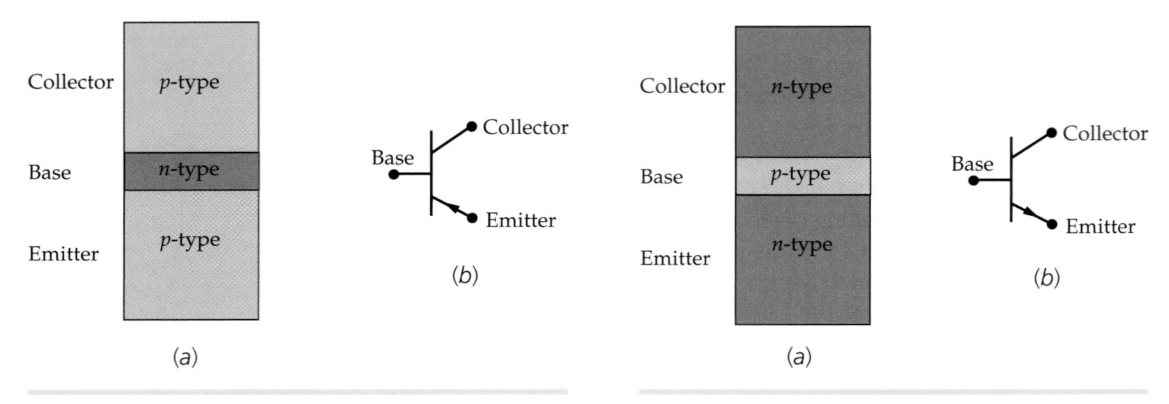

FIGURE 38-25 A *pnp* transistor. (*a*) The heavily doped emitter emits holes that pass through the thin base to the collector. (*b*) Symbol for a *pnp* transistor in a circuit. The arrow points in the direction of the conventional current, which is the same as that of the emitted holes.

FIGURE 38-26 An *npn* transistor. (*a*) The heavily doped emitter emits electrons that pass through the thin base to the collector. (*b*) Symbol for an *npn* transistor. The arrow points in the direction of the conventional current, which is opposite the direction of the emitted electrons.

doped than either the base or the collector. In an *npn* transistor, the emitter and collector are *n*-type semiconductors and the base is a *p*-type semiconductor; in a *pnp* transistor, the base is an *n*-type semiconductor and the emitter and collector are *p*-type semiconductors.

Figure 38-25 and Figure 38-26 show, respectively, a *npn* transistor and an *npn* transistor, along with the symbols used to represent each transistor in circuit diagrams. We see that either transistor consists of two *pn* junctions. We will discuss the operation of a *npn* transistor. The operation of an *npn* transistor is similar.

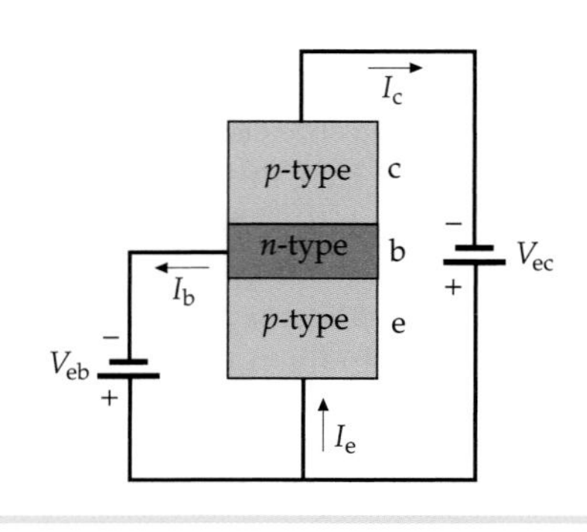

FIGURE 38-27 A *pnp* transistor biased for normal operation. Holes from the emitter can easily diffuse across the base, which is only tens of nanometers thick. Most of the holes flow to the collector, producing the current I_c.

In the normal operation of a *pnp* transistor, the emitter-base junction is forward biased, and the base-collector junction is reverse biased, as shown in Figure 38-27. The heavily doped *p*-type emitter emits holes that flow toward the emitter-base junction. This flow constitutes the emitter current I_e. Because the base is very thin, most of the holes flow across the base into the collector. This flow in the collector constitutes a current I_c. However, some of the holes recombine in the base producing a positive charge that inhibits the further flow of charge. To prevent this, some of the holes that do not reach the collector are drawn off the base as a base current I_b in a wire connected to the base. In Figure 38-27, therefore, I_c is almost but not quite equal to I_e, and I_b is much smaller than either I_c or I_e. It is customary to express I_c as

$$I_c = \beta I_b \qquad 38\text{-}31$$

where β is called the current gain of the transistor. Transistors can be designed to have values of β as low as ten or as high as several hundred.

Figure 38-28 shows a simple *pnp* transistor used as an amplifier. A small, time-varying input voltage v_s is connected in series with a constant bias voltage V_{eb}. The base current is then the sum of a steady current I_b produced by the bias voltage V_{eb} and a time-varying current i_b due to the signal voltage v_s. Because v_s may at any instant be either positive or negative, the bias voltage V_{eb} must be large enough to ensure that there is always a forward bias on the emitter-base junction. The collector current will consist of two parts: a constant direct current $I_c = \beta I_b$ and a time-varyng current $i_c = \beta i_b$. We thus have a current amplifier in which the time-varying output current i_c is β multiplied by the input current i_b. In such an amplifier, the steady currents I_c and I_b, although essential to the operation of the transistor, are usually not of interest. The input signal voltage v_s is related to the base current by Ohm's law:

$$i_b = \frac{v_s}{R_b + r_b} \qquad 38\text{-}32$$

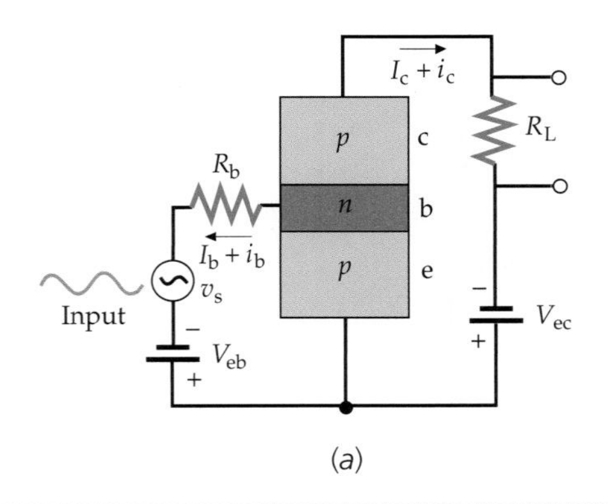

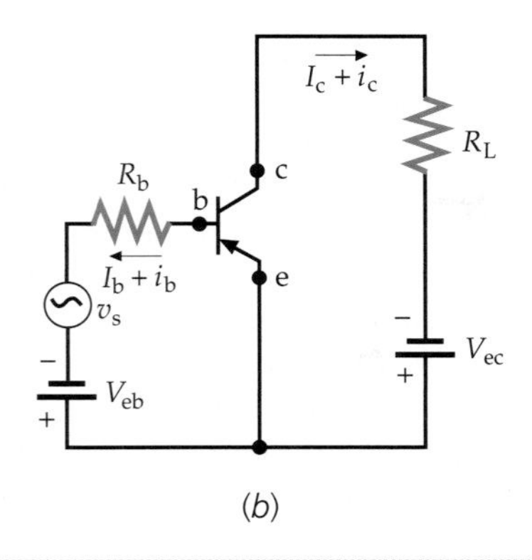

FIGURE 38-28 (*a*) A *pnp* transistor used as an amplifier. A small change i_b in the base current results in a large change i_c in the collector current. Thus, a small signal in the base circuit results in a large signal across the load resistor R_L in the collector circuit. (*b*) The same circuit as in Figure 38-28*a* with the conventional symbol for the transistor.

where r_b is the internal resistance of that part of the transistor between the base and emitter. Similarly, the collector current i_c produces a time-varying voltage v_L across the output or load resistance R_L given by

$$v_L = i_c R_L \tag{38-33}$$

Using Equation 38-31 and Equation 38-32, we have

$$i_c = \beta i_b = \beta \frac{v_s}{R_b + r_b} \tag{38-34}$$

The output voltage is thus related to the input voltage by

$$v_L = \beta \frac{v_s}{R_b + r_b} R_L = \beta \frac{R_L}{R_b + r_b} v_s \tag{38-35}$$

The ratio of the output voltage to the input voltage is the **voltage gain** of the amplifier:

$$\text{Voltage gain} = \frac{v_L}{v_s} = \beta \frac{R_L}{R_b + r_b} \tag{38-36}$$

A typical amplifier (for example, in a tape player) has several transistors, similar to the one shown in Figure 38-28, connected in series so that the output of one transistor serves as the input for the next. Thus, the very small voltage fluctuations produced by the motion of the magnetic tape past the pickup heads controls the large amounts of power required to drive the loudspeakers. The power delivered to the speakers is supplied by the dc sources connected to each transistor.

The technology of semiconductors extends well beyond individual transistors and diodes. Many of the electronic devices we use every day, such as laptop computers and the processors that govern the operation of vehicles and appliances, rely on large-scale integration of many transistors and other circuit components on a single chip. Large-scale integration combined with advanced concepts in semiconductor theory has created remarkable new instruments for scientific research.

38-8 SUPERCONDUCTIVITY

There are some substances for which the resistivity suddenly drops to zero below a certain temperature T_c, which is called the **critical temperature.** This amazing phenomenon, called **superconductivity,** was discovered in 1911 by the Dutch physicist H. Kamerlingh Onnes, who developed a technique for liquefying helium (boiling point equal to 4.2 K) and used his technique to explore the properties of substances at temperatures in that range. Figure 38-29 shows Onnes's plot of the resistance of mercury versus temperature. The critical temperature for mercury is approximately the same as the boiling point of helium, which is 4.2 K. Critical temperatures for other superconducting elements range from less than 0.1 K for hafnium and iridium to 9.2 K for niobium. The temperature range for superconductors is much higher for a number of metallic compounds. For example, the superconducting alloy Nb_3Ge, discovered in 1973, has a critical temperature of 25 K, which was the highest known until 1986, when the discoveries of J. Georg Bednorz and K. Alexander Müller launched the era of high-temperature superconductors, now defined as materials that exhibit superconductivity at temperatures above 77 K (the temperature at which nitrogen boils). The highest temperature at which superconductivity has been demonstrated, using thallium-doped $HgBa_2Ca_2Cu_3O_8$ + delta, is 138 K at atmospheric pressure. At extremely high pressures, some materials exhibit superconductivity at temperatures as high as 164 K.

The resistivity of a superconductor is zero. There can be a current in a superconductor even when there is no emf in the superconducting circuit. Indeed, in superconducting rings in which there was no electric field, steady currents have been observed to persist for years without apparent loss. Despite the cost and inconvenience of refrigeration using expensive liquid helium, many superconducting magnets have been built using superconducting materials, because such magnets require no power expenditure to maintain the large current needed to produce a large magnetic field.

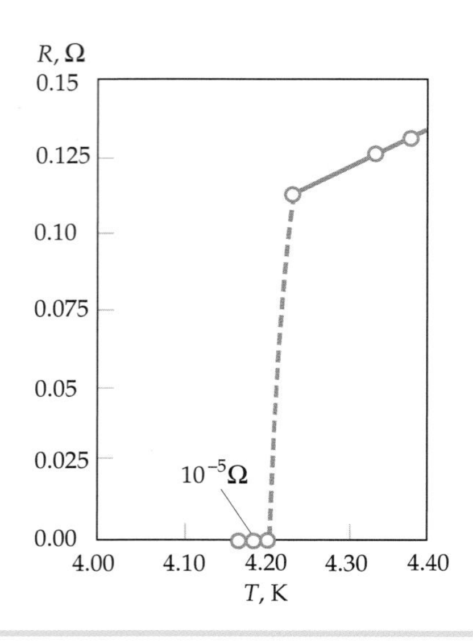

FIGURE 38-29 Plot by H. Kamerlingh Onnes of the resistance of mercury versus temperature, showing the sudden decrease at the critical temperature of $T = 4.2$ K.

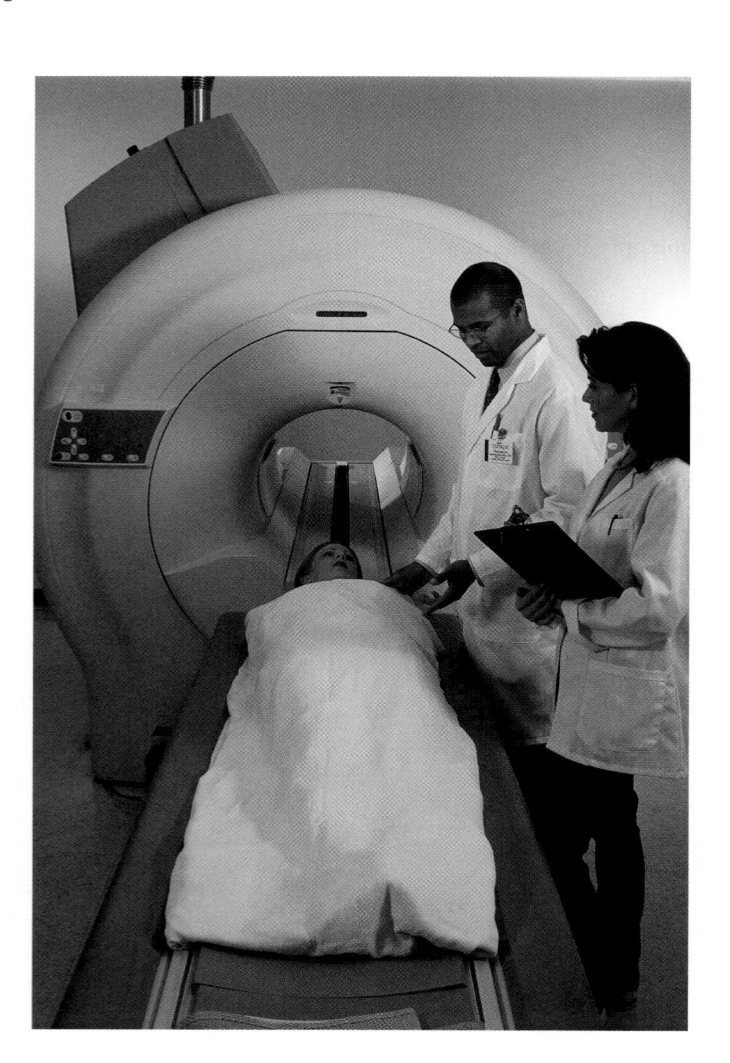

The wires for the magnetic field of a magnetic resonance imaging (MRI) machine carry large currents. To keep the wires from overheating, they are maintained at superconducting temperatures. To accomplish this, they are immersed in liquid helium. *(Corbis.)*

The discovery of high-temperature superconductors has revolutionized the study of superconductivity because relatively inexpensive liquid nitrogen, which boils at 77 K, can be used for a coolant. However, many problems, such as brittleness and the toxicity of the materials, make these new superconductors difficult to use. The search continues for new materials that will be superconductors at even higher temperatures.

THE BCS THEORY

It had been recognized for some time that low temperature superconductivity is due to a collective action of the conducting electrons. In 1957, John Bardeen, Leon Cooper, and Robert Schrieffer published a successful theory of low temperature superconductivity now known by the initials of the inventors as the **BCS theory.** According to this theory, the electrons in a superconductor are coupled in pairs at low temperatures. The coupling comes about because of the interaction between electrons and the crystal lattice. One electron interacts with the lattice and perturbs it. The perturbed lattice interacts with another electron in such a way that there is an attraction between the two electrons that at low temperatures can exceed the Coulomb repulsion between them. The electrons form a bound state called a **Cooper pair.** The electrons in a Cooper pair have equal and opposite spins, so they form a system with zero spin. Each Cooper pair acts as a *single particle* with zero spin, in other words, as a boson. Bosons do not obey the exclusion principle. Any number of Cooper pairs may be in the same quantum state with the same energy. In the ground state of a superconductor (at $T = 0$), all the conduction electrons are in Cooper pairs and all the Cooper pairs are in the same energy state. In the superconducting state, the Cooper pairs are correlated so that they act collectively. An electric current can be produced in a superconductor because all of the electrons in this collective state move together. But energy cannot be dissipated by individual collisions of electron and lattice ions unless the temperature is high enough to break the binding of the Cooper pairs. The required energy is called the *superconducting energy gap* E_g. In the BCS theory, this energy at zero temperature is related to the critical temperature by

$$E_g = \tfrac{7}{2}kT_c \qquad \text{38-37}$$

The energy gap can be determined by measuring the current across a junction between a normal metal and a superconductor as a function of voltage. Consider two metals separated by a layer of insulating material, such as aluminum oxide, that is only a few nanometers thick. The insulating material between the metals forms a barrier that prevents most electrons from traversing the junction. However, waves can tunnel through a barrier if the barrier is not too thick, even if the energy of the wave is less than that of the barrier.

When the materials on either side of the gap are normal nonsuperconducting metals, the current resulting from the tunneling of electrons through the insulating layer obeys Ohm's law for low applied voltages (Figure 38-30*a*). When one of the metals is a normal metal and the other is a superconductor, there is no current (at absolute zero) unless the applied voltage V is greater than a critical voltage $V_c = E_g/(2e)$, where E_g is the superconductor energy gap. Figure 38-30*b* shows the plot of current versus voltage for this situation. The current escalates rapidly when the energy $2eV$ absorbed by a Cooper pair traversing the barrier approaches $E_g = 2eV_c$, the minimum energy needed to break up the pair. (The small current visible in Figure 38-30*b* before the critical voltage is reached is present because at any temperature above absolute zero some of the electrons in the superconductor are thermally excited above the energy gap and are therefore not paired.) At voltages slightly above V_c, the current versus voltage curve becomes that for a normal metal. The superconducting energy gap can thus be measured by measuring the average voltage for the transition region.

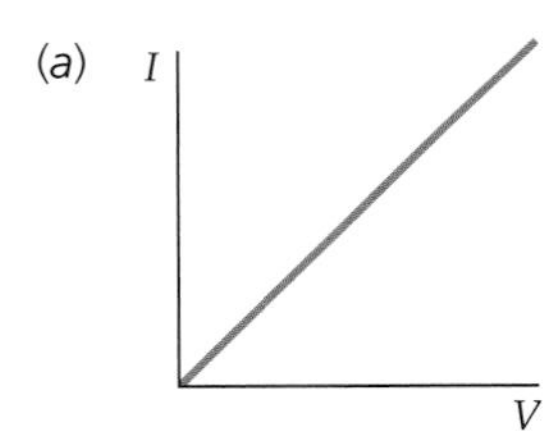

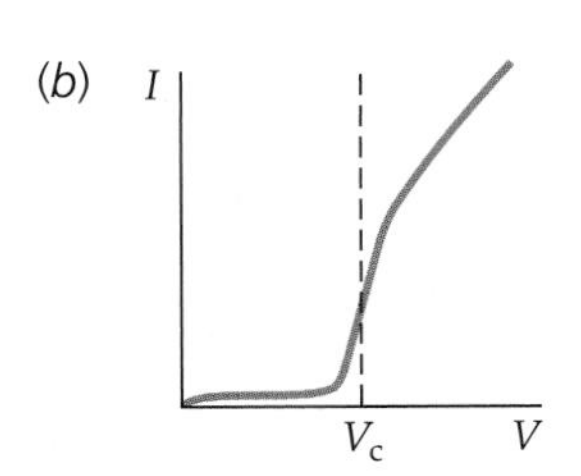

FIGURE 38-30 Tunneling current versus voltage for a junction of two metals separated by a thin oxide layer. (*a*) When both metals are normal metals, the current is proportional to the voltage, as predicted by Ohm's law. (*b*) When one metal is a normal metal and another metal is a superconductor, the current is approximately zero until the applied voltage V approaches the critical voltage $V_c = E_g/(2e)$.

Example 38-8 Superconducting Energy Gap for Mercury

Calculate the superconducting energy gap for mercury ($T_c = 4.2$ K) predicted by the BCS theory.

PICTURE The energy gap is related to the critical temperature by $E_g = 3.5\,kT_c$ (Equation 38-37).

SOLVE

1. The BCS prediction for the energy gap is $\quad E_g = 3.5kT_c$

2. Substitute $T_c = 4.2$ K: $\quad E_g = 3.5kT_c$

$$= 3.5(1.38 \times 10^{-23}\ \text{J/K})(4.2\ \text{K})\left(\frac{1\ \text{ev}}{1.6 \times 10^{-19}\ \text{J}}\right)$$

$$= \boxed{1.3 \times 10^{-3}\ \text{eV}}$$

Note that the energy gap for a typical superconductor is much smaller than the energy gap for a typical semiconductor, which is of the order of 1 eV. As the temperature is increased from $T = 0$, some of the Cooper pairs are broken. Then there are fewer pairs available for each pair to interact with, and the energy gap is reduced until at $T = T_c$ the energy gap is zero (Figure 38-31).

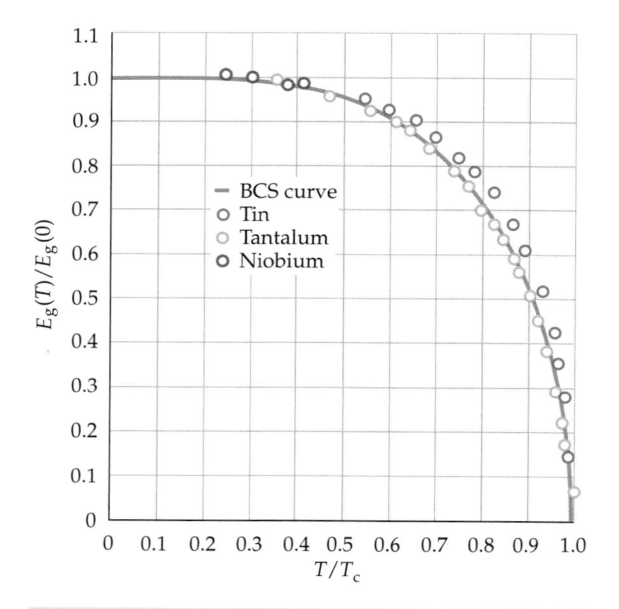

FIGURE 38-31 Ratio of the energy gap at temperature T to that at temperature $T = 0$ as a function of the relative temperature T/T_c. The solid curve is that predicted by the BCS theory.

THE JOSEPHSON EFFECT

When two superconductors are separated by a thin nonsuperconducting barrier (for example, a layer of aluminum oxide a few nanometers thick), the junction is called a **Josephson junction,** based on the prediction in 1962 by Brian Josephson that Cooper pairs could tunnel across such a junction from one superconductor to the other with no resistance. The tunneling of Cooper pairs constitutes a current, which does not require a voltage to be applied across the junction. The current depends on the difference in phase of the wave functions that describe the Cooper pairs. Let ϕ_1 be the phase constant for the wave function of a Cooper pair in one superconductor. All the Cooper pairs in a superconductor act coherently and have the same phase constant. If ϕ_2 is the phase constant for the Cooper pairs in the second superconductor, the current across the junction is given by

$$I = I_{\text{max}} \sin(\phi_2 - \phi_1) \qquad 38\text{-}38$$

where I_{max} is the maximum current, which depends on the thickness of the barrier. This result has been observed experimentally and is known as the **dc Josephson effect.**

Josephson also predicted that if a dc voltage V were applied across a Josephson junction, there would be a current that alternates with frequency f given by

$$f = \frac{2e}{h}V \qquad 38\text{-}39$$

This result, known as the **ac Josephson effect,** has been observed experimentally, and careful measurement of the frequency allows a precise determination of the ratio e/h. Because frequency can be measured very accurately, the ac Josephson effect is also used to establish precise voltage standards. The inverse effect, in which the application of an alternating voltage across a Josephson junction results in a dc current, has also been observed.

Example 38-9 Frequency of Josephson Current

Using $e = 1.602 \times 10^{-19}$ C and $h = 6.626 \times 10^{-34}$ J·s, calculate the frequency of the Josephson current if the applied voltage is 1.000 μV.

PICTURE The frequency f is related to the applied voltage V by $hf = 2eV$ (Equation 38-39).

SOLVE

Substitute the given values into Equation 38-39 to calculate f:

$$f = \frac{2e}{h}V = \frac{2(1.602 \times 10^{-19}\ \text{C})}{6.626 \times 10^{-34}\ \text{J}\cdot\text{s}}(1.000 \times 10^{-6}\ \text{V})$$

$$= 4.835 \times 10^{8}\ \text{Hz} = \boxed{483.5\ \text{MHz}}$$

38-9 THE FERMI–DIRAC DISTRIBUTION

The classical Maxwell–Boltzmann distribution (Equation 17-38) gives the number dN of molecules that have energy E in the range between E and $E + dE$. The number dN is equal to the product of $g(E)\,dE$ where $g(E)$ is the **density of states** (number of energy states in the range dE) and the Boltzmann factor $e^{-E/(kT)}$, which is the probability of a state being occupied. The distribution function for free electrons in a metal is called the **Fermi–Dirac distribution.** The Fermi–Dirac distribution can be written in the same form as the Maxwell–Boltzmann distribution, where the density of states calculated from quantum theory and the Boltzmann factor is replaced by the Fermi factor. Let $n(E)\,dE$ be the number of electrons that have energies between E and $E + dE$. This number is written

$$n(E)dE = f(E)g(E)dE \qquad 38\text{-}40$$

ENERGY DISTRIBUTION FUNCTION

where $g(E)\,dE$ is the number of states that have energies between E and $E + dE$ and $f(E)$ is the probability of a state being occupied, which is the Fermi factor. The density of states in three dimensions is somewhat challenging to calculate, so we just give the result. For electrons in a metal of volume V, the density of states is

$$g(E) = \frac{8\sqrt{2}\pi m_{\text{e}}^{3/2}V}{h^3}E^{1/2} \qquad 38\text{-}41$$

DENSITY OF STATES

As in the classical Maxwell–Boltzmann distribution, the density of states is proportional to $E^{1/2}$.

At $T = 0$, the Fermi factor is given by Equation 38-24:

$$f(E) = \begin{cases} 1 & E < E_{\text{F}} \\ 0 & E > E_{\text{F}} \end{cases}$$

The integral of $n(E)\,dE$ over all energies gives the total number of electrons N. We can derive the equation

$$E_{\text{F}} = \frac{h^2}{8m_{\text{e}}}\left(\frac{3N}{\pi V}\right)^{2/3}$$

(Equation 38-22*a*) for the Fermi energy at $T = 0$ by integrating $n(E)\,dE$ from $E = 0$ to $E = \infty$. We obtain

$$N = \int_0^\infty n(E)dE = \int_0^{E_F} n(E)dE + \int_{E_F}^\infty n(E)dE$$

$$= \int_0^{E_F} \frac{8\sqrt{2}\pi m_e^{3/2}V}{h^3}E^{1/2}\,dE + 0 = \frac{16\sqrt{2}\pi m_e^{3/2}V}{3h^3}E_F^{3/2}$$

Note that at $T = 0$, $n(E)$ is zero for $E > E_F$. Solving for E_F gives the Fermi energy at $T = 0$:

$$E_F = \frac{h^2}{8m_e}\left(\frac{3N}{\pi V}\right)^{2/3} \qquad 38\text{-}42$$

which is Equation 38-22*a*. In terms of the Fermi energy, the density of states (Equation 38-41) is

$$g(E) = \frac{8\sqrt{2}\pi m_e^{3/2}V}{h^3}E^{1/2} = \frac{3}{2}NE_F^{-3/2}E^{1/2} \qquad 38\text{-}43$$

DENSITY OF STATES IN TERMS OF E_F

which is obtained by solving Equation 38-42 for m_e, and then substituting for m_e in Equation 38-41. The average energy at $T = 0$ is calculated from

$$E_{av} = \frac{\int_0^{E_F} Eg(E)dE}{\int_0^{E_F} g(E)dE} = \frac{1}{N}\int_0^{E_F} Eg(E)dE \qquad 38\text{-}44$$

where $N = \int_0^{E_F} g(E)dE$ is the total number of electrons. Substituting for $g(E)$ from Equation 38-43 and then evaluating the integral in Equation 38-44, we obtain Equation 38-23:

$$E_{av} = \tfrac{3}{5}E_F \qquad 38\text{-}45$$

AVERAGE ENERGY AT $T = 0$

At $T > 0$, the Fermi factor is more complicated. It can be shown to be

$$f(E) = \frac{1}{e^{(E-E_F)/(kT)} + 1} \qquad 38\text{-}46$$

FERMI FACTOR

We can see from this equation that for E greater than E_F, $e^{(E-E_F)/(kT)}$ becomes very large as T approaches zero, so at $T = 0$, the Fermi factor is zero for $E > E_F$. On the other hand, for E less than E_F, $e^{(E-E_F)/(kT)}$ approaches 0 as T approaches zero, so at $T = 0$, $f(E) = 1$ for $E < E_F$. Thus, the Fermi factor given by Equation 38-46 holds for all temperatures. Note also that for any nonzero value of T, $f(E) = \frac{1}{2}$ at $E = E_F$.

The complete Fermi-Dirac distribution function is thus

$$n(E)dE = g(E)f(E)dE = \frac{8\sqrt{2}\pi m_e^{3/2}V}{h^3}E^{1/2}\frac{1}{e^{(E-E_F)/(kT)}+1}dE \qquad 38\text{-}47$$

FERMI–DIRAC DISTRIBUTION

We can see that for those few electrons that have energies much greater than the Fermi energy, the Fermi factor approaches $1/e^{(E-E_F)/(kT)} = e^{(E_F-E)/(kT)} = e^{E_F/(kT)}e^{-E/(kT)}$, which is proportional to $e^{-E/(kT)}$. Thus, the high-energy tail of the Fermi–Dirac energy distribution decreases with increasing E as $e^{-E/(kT)}$, just like the classical Maxwell–Boltzmann energy distribution. The reason for this is that in this high-energy region there are many unoccupied energy states and few electrons, so the exclusion principle is not important. Thus, the Fermi–Dirac distribution approaches the classical Maxwell–Boltzmann distribution in the high-energy limit. This result has practical importance because it applies to the conduction electrons in semiconductors.

Example 38-10 Fermi Factor for Copper at 300 K

At what energy is the Fermi factor equal to 0.100 for copper at $T = 300$ K?

PICTURE We set $f(E) = 0.100$ in Equation 38-46, using $T = 300$ K and $E_F = 7.03$ eV from Table 38-1, and solve for E.

SOLVE

1. Solve Equation 38-46 for $e^{(E-E_F)/(kT)}$:

$$f(E) = \frac{1}{e^{(E-E_F)/(kT)}+1}$$

so

$$e^{(E-E_F)/(kT)} = \frac{1}{f(E)} - 1$$

2. Take the logarithm of both sides:

$$\frac{E-E_F}{\mathrm{kT}} = \ln\left[\frac{1}{f(E)} - 1\right]$$

3. Solve for E. For E_F, use the value for E_F at $T = 0$ K listed in Table 38-1:

$$E = E_F + \left[\frac{1}{f(E)} - 1\right]kT$$

$$= 7.03\text{ eV} + \ln\left[\frac{1}{0.100} - 1\right](8.62\times10^{-5}\text{ eV/K})(300\text{ K})$$

$$= \boxed{7.09\text{ eV}}$$

CHECK As expected, the energy is slightly above the Fermi energy when the Fermi factor is equal to 0.100.

Example 38-11 Probability of a Higher Energy State Being Occupied

Find the probability that an energy state in copper 0.100 eV above the Fermi energy is occupied at $T = 300$ K.

PICTURE The probability is the Fermi factor given in Equation 38-46, with $E_F = 7.03$ eV and $E = 7.13$ eV.

SOLVE

1. The probability of an energy state being occupied equals the Fermi factor:

$$P = f(E) = \frac{1}{e^{(E-E_F)/(kT)} + 1}$$

2. Calculate the exponent in the Fermi factor (exponents are always dimensionless):

$$\frac{E - E_F}{kT} = \frac{7.13 \text{ eV} - 7.03 \text{ eV}}{(8.62 \times 10^{-5} \text{ eV/K})(300 \text{ K})} = 3.87$$

3. Use this result to calculate the Fermi factor:

$$f(E) = \frac{1}{e^{(E-E_F)/(kT)} + 1} = \frac{1}{e^{3.87} + 1}$$

$$= \frac{1}{48 + 1} = \boxed{0.020}$$

CHECK The probability that an energy state above the Fermi energy is occupied is less than one-half. As expected, the step 4 result less than one-half.

TAKING IT FURTHER The probability of an electron having an energy 0.100 eV above the Fermi energy at 300 K is only about 2 percent.

Example 38-12 Probability of a Lower Energy State Being Occupied *Try It Yourself*

Find the probability that an energy state in copper 0.10 eV *below* the Fermi energy is occupied at $T = 300$ K.

PICTURE The probability is the Fermi factor given in Equation 38-46, with $E_F = 7.03$ eV and $E = 6.93$ eV.

SOLVE

Cover the column to the right and try these on your own before looking at the answers.

Steps / **Answers**

1. Write the Fermi factor:

$$f(E) = \frac{1}{e^{(E-E_F)/(kT)} + 1}$$

2. Calculate the exponent in the Fermi factor:

$$\frac{E - E_F}{kT} = \frac{6.93 \text{ eV} - 7.03 \text{ eV}}{(8.62 \times 10^{-5} \text{ eV/K})(300 \text{ K})} = -3.87$$

3. Use your result from step 2 to calculate the Fermi factor:

$$f(E) = \frac{1}{e^{(E-E_F)/(kT)} + 1} = \frac{1}{e^{3.87} + 1}$$

$$= \frac{1}{0.021 + 1} = \boxed{0.98}$$

CHECK As expected, the step-3 result is greater than one-half.

TAKING IT FURTHER The probability of an electron having an energy of 0.10 eV *below* the Fermi energy at 300 K is approximately 98 percent.

PRACTICE PROBLEM 38-4 What is the probability of an energy state 0.10 eV below the Fermi energy being unoccupied at 300 K?

Summary

TOPIC	RELEVANT EQUATIONS AND REMARKS	
1. The Structure of Solids	Solids are often found in crystalline form in which a small structure, which is called the unit cell, is repeated over and over. A crystal may have a face-centered-cubic, body-centered-cubic, hexagonal close-packed, or other structure depending on the type of bonding between the atoms, ions, or molecules in the crystal and on the relative sizes of the atoms.	
Potential energy	$U = -\alpha\frac{ke^2}{r} + \frac{A}{r^n}$	38-3
	where r is the center-to-center separation distance between neighboring ions, α is the Madelung constant, which depends on the geometry of the crystal and is of the order of 1.8, and n is approximately 9.	
2. A Microscopic Picture of Conduction		
Resistivity	$\rho = \frac{m_e v_{av}}{n_e e^2 \lambda}$	38-14
	where v_{av} is the average speed of the electrons and λ is their mean free path between collisions with the lattice ions.	
Mean free path	$\lambda = \frac{vt}{n_{ion}\pi r^2 vt} = \frac{1}{n_{ion}\pi r^2} = \frac{1}{n_{ion}A}$	38-16
	where n_{ion} is the number of lattice ions per unit volume, r is their effective radius, and A is their effective cross-sectional area.	
3. Classical Interpretation of v_{av} and λ	v_{av} is determined from the Maxwell–Boltzmann distribution, and r is the actual radius of a lattice ion. (This interpretation is not consistent with measured results.)	
4. Quantum Interpretation of v_{av} and λ	v_{av} is determined from the Fermi–Dirac distribution and is approximately constant independent of temperature. The mean free path is determined from the scattering of electron waves, which occurs only because of deviations from a perfectly ordered array. The radius r is the amplitude of vibration of the lattice ion, which is proportional to $\sqrt{T}$, so A is proportional to T.	
5. Free Electrons		
Fermi energy E_F at $T = 0$	E_F is the energy of the last filled (or half-filled) energy state.	
E_F at $T > 0$	E_F is the energy at which the probability of being occupied is $\frac{1}{2}$.	
Approximate magnitude of E_F	E_F is between 5 eV and 10 eV for most metals.	
Dependence of E_F on the number density (N/V) of free electrons	$E_F = \frac{h^2}{8m_e}\left(\frac{3N}{\pi V}\right)^{2/3}$	38-22*a*
Average energy at $T = 0$	$E_{av} = \frac{3}{5}E_F$	38-23
Fermi factor at $T = 0$	The Fermi factor $f(E)$ is the probability of a state being occupied. $f(E) = \begin{cases} 1 & E < E_F \\ 0 & E > E_F \end{cases}$	38-24
Fermi temperature	$T_F = \frac{E_F}{k}$	38-25
Fermi speed	$u_F = \sqrt{\frac{2E_F}{m_e}}$	38-26

TOPIC	RELEVANT EQUATIONS AND REMARKS
Contact potential	When two different metals are placed in contact, electrons flow from the metal with the higher Fermi energy to the metal with the lower Fermi energy until the Fermi energies of the two metals are equal. In equilibrium, there is a potential difference between the metals that is equal to the difference in the work function of the two metals divided by the electronic charge e: $$V_{\text{contact}} = \frac{\phi_1 - \phi_2}{e} \qquad 38\text{-}27$$
Specific heat due to conduction electrons	$$c'_V = \tfrac{1}{2}\pi^2 R \frac{T}{T_F} \qquad 38\text{-}29$$
6. Band Theory of Solids	When many atoms are brought together to form a solid, the individual energy levels are split into bands of allowed energies. The splitting depends on the type of bonding and the lattice separation. The highest energy band that contains electrons is called the valence band. (The lowest energy band that is not filled with electrons is called the conduction band.) In a conductor, the valence band is only partially filled, so there are many available empty energy states for excited electrons. In an insulator, the valence band is completely filled and there is a large energy gap between it and the next allowed band, the conduction band. In a semiconductor, the energy gap between the filled valence band and the empty conduction band is small; so, at ordinary temperatures, an appreciable number of electrons are thermally excited into the conduction band.
7. Semiconductors	The conductivity of a semiconductor can be greatly increased by doping. In an n-type semiconductor, the doping adds electrons at energies just below that of the conduction band. In a p-type semiconductor, holes are added at energies just above that of the valence band.
8. *Semiconductor Junctions and Devices	
**pn* Junctions	Semiconductor devices such as diodes and transistors make use of n-type semiconductors and p-type semiconductors. The two types of semiconductors are typically a single silicon crystal doped with donor impurities on one side and acceptor impurities on the other side. The region in which the semiconductor changes from a p-type semiconductor to an n-type semiconductor is called a junction. Junctions are used in diodes, solar cells, surface barrier detectors, LEDs, and transistors.
*Diodes	A diode is a single-junction device that carries current in one direction only.
*Zener diodes	A Zener diode is a diode with a very high reverse bias. It breaks down suddenly at a distinct voltage and can therefore be used as a voltage reference standard.
*Tunnel diodes	A tunnel diode is a diode that is heavily doped so that electrons tunnel through the depletion barrier. At normal operation, a small change in bias voltage results in a large change in current.
*Transistors	A transistor consists of a very thin semiconductor of one type sandwiched between two semiconductors of the opposite type. Transistors are used in amplifiers because a small variation in the base current results in a large variation in the collector current.
9. Superconductivity	In a superconductor, the resistance drops suddenly to zero below a critical temperature T_c. Superconductors with critical temperatures as high as 138 K have been discovered.
The BCS theory	Superconductivity is described by a theory of quantum mechanics called the BCS theory in which the free electrons form Cooper pairs. The energy needed to break up a Cooper pair is called the superconducting energy gap E_g. When all the electrons are paired, individual electrons cannot be scattered by a lattice ion, so the resistance is zero.
Tunneling	When a normal conductor is separated from a superconductor by a thin layer of oxide, electrons can tunnel through the energy barrier if the applied voltage across the layer is $E_g/(2e)$, where E_g is the energy needed to break up a Cooper pair. The energy gap E_g can be determined by a measurement of the tunneling current versus the applied voltage.

TOPIC	RELEVANT EQUATIONS AND REMARKS
Josephson junction	A system of two superconductors separated by a thin layer of nonconducting material is called a Josephson junction.
dc Josephson effect	A dc current is observed to tunnel through a Josephson junction even in the absence of voltage across the junction.
ac Josephson effect	When a dc voltage V is applied across a Josephson junction, an ac current is observed with a frequency $$f = \frac{2e}{h}V \quad \text{38-39}$$ Measurement of the frequency of this current allows a precise determination of the ratio e/h.
10. The Fermi–Dirac Distribution	The number of electrons with energies between E and $E + dE$ is given by $$n(E)\,dE = f(E)\,g(E)\,dE \quad \text{38-40}$$ where $g(E)$ is the density of states and $f(E)$ is the Fermi factor.
Density of states	$$g(E) = \frac{8\sqrt{2}\pi m_e^{3/2}V}{h^3}E^{1/2} \quad \text{38-41}$$
Fermi factor at $T > 0$	$$f(E) = \frac{1}{e^{(E-E_F)/(kT)} + 1} \quad \text{38-46}$$

Answers to Practice Problems

38-1	$E_F = 9.40$ eV
38-2	5.53 eV
38-3	2×10^{-13} electrons/atom
38-4	One minus the probability of the energy state being occupied. That is, $1 - 0.98 = 0.02$ or 2 percent.

Problems

In a few problems, you are given more data than you actually need; in a few other problems, you are required to supply data from your general knowledge, outside sources, or informed estimate.

Interpret as significant all digits in numerical values that have trailing zeros and no decimal points.

- • Single-concept, single-step, relatively easy
- •• Intermediate-level, may require synthesis of concepts
- ••• Challenging
- SSM Solution is in the *Student Solutions Manual*
- Consecutive problems that are shaded are paired problems.

CONCEPTUAL PROBLEMS

1 • In the classical model of conduction, the electron loses energy on average during a collision because it loses the drift velocity it had acquired since the last collision. Where does this energy appear?

2 • A metal is a good conductor because the valence energy band for electrons is (*a*) empty, (*b*) partly filled, (*c*) filled, but there is only a small gap to a higher empty band, (*d*) completely filled, (*e*) none of the above.

3 • Thomas refuses to believe that a potential difference can be created simply by bringing two different metals into contact with each other. John talks him into making a small wager and is about to win the bet. (*a*) Which two metals from Table 38-2 would demonstrate his point most effectively? (*b*) What is the value of that contact potential?

4 • (*a*) In Problem 3, which choices of different metals would make the least impressive demonstration? (*b*) What is the value of that contact potential?

5 • When a sample of pure copper is cooled from 300 K to 4 K, its resistivity decreases more than the resistivity of a sample of brass when it is cooled through the same temperature difference. Why? SSM

6 • Insulators are poor conductors of electricity because (*a*) there is a small energy gap between the filled valence band and the next higher band where electrons can exist, (*b*) there is a large energy gap between the completely filled valence band and the next higher band where electrons can exist, (*c*) the valence band has a few vacancies for electrons, (*d*) the valence band is only partly filled, (*e*) none of the above.

7 • How does the sign of the change in the resistivity of a sample of copper compare with the sign of the change in the resistivity of a sample of silicon when the temperatures of both samples increase? SSM

8 • True or false:

(*a*) Solids that are good electrical conductors are usually good heat conductors.

(*b*) At $T = 0$, an intrinsic semiconductor is an insulator.

(*c*) The Fermi energy is the average energy of an electron in a solid.
(*d*) At $T = 0$, the value of the Fermi factor can be either 1 or 0.
(*e*) Semiconductors conduct current in one direction only.
(*f*) The classical free-electron theory adequately explains the heat capacity of metals.
(*g*) The contact potential between two metals is proportional to the difference in the work functions of the two metals.

9 • Which of the following elements are most likely to act as acceptor impurities in germanium? (*a*) bromine, (*b*) gallium, (*c*) silicon, (*d*) phosphorus, (*e*) magnesium

10 • Which of the following elements are most likely to serve as donor impurities in germanium? (*a*) bromine, (*b*) gallium, (*c*) silicon, (*d*) phosphorus, (*e*) magnesium

11 • An electron hole is created when a photon is absorbed by a semiconductor. How does this hole enable the semiconductor to conduct electricity?

12 • Examine the positions of phosphorus, boron, thallium, and antimony in Table 36-1. (*a*) Which of these elements can be used to dope silicon to create an *n*-type semiconductor? (*b*) Which of these elements can be used to dope silicon to create a *p*-type semiconductor?

13 • When photons of visible light strike the *p*-type semiconductor in a *pn* junction solar cell, (*a*) only free electrons are created, (*b*) only positive holes are created, (*c*) both electrons and holes are created, (*d*) protons are created, (*e*) none of the above.

ESTIMATION AND APPROXIMATION

14 • The ratio of the resistivity of the most resistive (least conductive) material to that of the least resistive material (excluding superconductors) is approximately 10^{24}. You can develop a feeling for how remarkable this range is by considering what the ratio is of the largest values to smallest values of other material properties. Choose any three properties of materials, and using tables in this book or some other resource, calculate the ratio of the largest instance of the property to the smallest instance of that property (other than zero) and rank these in decreasing order. Can you find any other property that shows a range as large as that of electrical resistivity?

15 • A device is said to be "ohmic" if a graph of current versus applied voltage is a straight line through the origin. The resistance R of the device is the reciprocal of the slope of this line. A *pn* junction is an example of a nonohmic device, as may be seen from Figure 38-21. For nonohmic devices, it is sometimes convenient to define the *differential resistance* as the reciprocal of the slope of the I versus V curve. Using the curve in Figure 38-21, estimate the differential resistance of the *pn* junction at applied voltages of -20 V, $+0.2$ V, $+0.4$ V, $+0.6$ V, and $+0.8$ V.

THE STRUCTURE OF SOLIDS

16 • Calculate the center-to-center separation distance r_0 between the K^+ and the Cl^- ions in KCl. Do this by assuming that each ion occupies a cubic volume of side r_0. The molar mass of KCl is 74.55 g/mol and its density is 1.984 g/cm³.

17 • The center-to-center separation distance between the Li^+ and Cl^- ions in LiCl is 0.257 nm. Use that value and the molar mass of LiCl (42.4 g/mol) to compute the density of LiCl.

18 • Find the value of n in Equation 38-6 that gives the measured dissociation energy of 741 kJ/mol for LiCl, which has the same structure as NaCl and for which $r_0 = 0.257$ nm.

19 •• (*a*) Use Equation 38-6 and calculate $U(r_0)$ for calcium oxide, CaO, where $r_0 = 0.208$ nm. Assume $n = 8$. (*b*) If n increases from 8 to 10, what is the fractional change in $U(r_0)$?

A MICROSCOPIC PICTURE OF CONDUCTION

20 • A measure of the density of the free electrons in a metal is the distance r_s, which is defined as the radius of the sphere whose volume equals the volume per conduction electron. (*a*) Show that $r_s = [3/(4\pi n)]^{1/3}$, where n is the free-electron number density. (*b*) Calculate r_s for copper in nanometers.

21 • (*a*) Given a mean free path $\lambda = 0.400$ nm and a mean speed $v_{av} = 1.17 \times 10^5$ m/s for the charge flow in copper at a temperature of 300 K, calculate the classical value for the resistivity ρ of copper. (*b*) The classical model suggests that the mean free path is temperature independent and that v_{av} depends on temperature. According to this model, what would ρ be at 100 K?

FREE ELECTRONS IN A SOLID

22 •• Silicon has a molar mass of 28.09 g/mol and a density of 2.41×10^3 kg/m³. Each atom of silicon has four valence electrons and the Fermi energy of the material is 4.88 eV. (*a*) Given that the electron mean free path at room temperature is $\lambda = 27.0$ nm, estimate the resistivity. (*b*) The accepted value for the resistivity of silicon is 640 Ω·m (at room temperature). How does this accepted value compare to the value calculated in Part (*a*) ?

23 • Calculate the number density of free electrons in (*a*) Ag ($\rho = 10.5$ g/cm³) and (*b*) Au ($\rho = 19.3$ g/cm³), assuming one free electron per atom, and compare your results with the values listed in Table 38-1.

24 • The density of aluminum is 2.7 g/cm³. How many free electrons are present per aluminum atom?

25 • The density of tin is 7.3 g/cm³. How many free electrons are present per tin atom?

26 • Calculate the Fermi temperature for (*a*) Mg, (*b*) Mn, and (*c*) Zn.

27 • What is the speed of a conduction electron whose energy is equal to the Fermi energy E_F for (*a*) Na, (*b*) Au, and (*c*) Sn?

28 • Calculate the Fermi energy for (*a*) Al, (*b*) K, and (*c*) Sn using the number densities given in Table 38-1.

29 • Find the average energy of the conduction electrons at $T = 0$ in (*a*) copper and (*b*) lithium.

30 • Calculate (*a*) the Fermi temperature and (*b*) the Fermi energy at $T = 0$ for iron.

31 •• (*a*) Assuming that each gold atom in a sample of gold metal contributes one free electron, calculate the free-electron density in gold knowing that its atomic mass is 196.97 g/mol and its density is 19.3×10^3 kg/m³. (*b*) If the Fermi speed for gold is 1.39×10^6 m/s, what is the Fermi energy in electron volts? (*c*) By what factor is the Fermi energy higher than the kT energy at room temperature? (*d*) Explain the difference between the Fermi energy and the kT energy. SSM

32 •• The bulk modulus B of a material can be defined by $B = -V\partial P/\partial V$ (*a*) Use the monatomic ideal-gas relation $PV = \frac{2}{3}NE_{av}$, where E_{av} is the average kinetic energy, Equation 38-22 and Equation 38-23 to show that $P = \frac{2}{5}NE_F/V = CV^{-5/3}$, where C is a constant independent of V. (*b*) Show that the bulk modulus of the free electrons in a solid metal is therefore $B = \frac{5}{3}P = \frac{2}{3}NE_F/V$. (*c*) Compute the bulk modulus in newtons per square meter for the free electrons in a sample of copper and compare your result with the measured value of 140×10^9 N/m².

33 •• The pressure of a monatomic ideal gas is related to the average kinetic energy of the gas particles by $PV = \frac{2}{3}NE_{av}$, where n is the number of particles and E_{av} is the average kinetic energy. Use this information to calculate the pressure of the free electrons in a sample of copper in newtons per square meter, and compare your result with atmospheric pressure, which is about 10^5 N/m^2. (*Note:* The units are most easily handled by using the conversion factors 1 N/m^2 = 1 J/m^3 and 1 eV = 1.602 × 10^{-19} J.) SSM

34 • Calculate the contact potential between (*a*) Ag and Cu, (*b*) Ag and Ni, and (*c*) Ca and Cu.

HEAT CAPACITY DUE TO ELECTRONS IN A METAL

35 •• Gold has a Fermi energy of 5.53 eV. Determine the molar specific heat at constant volume for gold at room temperature. SSM

QUANTUM THEORY OF ELECTRICAL CONDUCTION

36 • The resistivities and Fermi speeds of Na, Au, and Sn at T = 273 K are 4.2 μΩ·cm, 2.04 μΩ·cm, and 10.6 μΩ·cm, and 1.07 × 10^6 m/s, 1.39 × 10^6 m/s, and 1.89 × 10^6 m/s, respectively. Use those values to find the mean free paths for the conduction electrons in these elements.

37 •• The resistivity of pure copper increases by approximately 1.0 × 10^{-8} Ω·m with the addition of 1.0 percent (by number of atoms) of an impurity distributed throughout the metal. The mean free path λ depends on both the impurity and the oscillations of the lattice ions according to the equation $1/\lambda = 1/\lambda_t + 1/\lambda_i$, where λ_t is the mean free path associated with the thermal vibrations of the ions and λ_i is the mean free path associated with the impurities. (*a*) Estimate λ_i using Equation 38-14 and the data given in Table 38-1. (*b*) If r is the effective radius of an impurity lattice ion seen by an electron, the scattering cross section is πr^2. Estimate this area, using the fact that r is related to λ_i by Equation 38-16. SSM

BAND THEORY OF SOLIDS

38 • Electromagnetic radiation is incident on the surface of a semiconductor. The maximum wavelength of this light that is required if electrons are to cross the energy gap between the valence band and the conduction band is 380.0 nm. What is the energy gap, in electron volts, for the semiconductor?

39 • An electron occupies the highest energy level of the valence band in a silicon sample. What is the maximum photon wavelength that will excite the electron across the energy gap if the gap is 1.14 eV? SSM

40 • An electron occupies the highest energy level of the valence band in a germanium sample. What is the maximum photon wavelength that will excite the electron into the conduction band? In germanium, the energy gap between the valence and conduction bands is 0.74 eV.

41 • An electron occupies the highest energy level of the valence band in a diamond sample. What is the maximum photon wavelength that will excite the electron into the conduction band? In diamond, the energy gap between the valence and conduction bands is 7.0 eV.

42 •• A photon of wavelength 3.35 μm has just enough energy to raise an electron from the valence band to the conduction band in a lead sulfide sample. (*a*) Find the energy gap between the bands in lead sulfide. (*b*) Find the temperature T for which kT equals that energy gap.

SEMICONDUCTORS

43 • The donor energy levels in an n-type semiconductor are 0.0100 eV below the conduction band. Find the temperature for which kT = 0.0100 eV.

44 •• When a thin slab of semiconducting material is illuminated with monochromatic electromagnetic radiation, most of the radiation is transmitted through the slab if the wavelength is greater than 1.85 mm. For wavelengths less than 1.85 mm, most of the incident radiation is absorbed. Determine the energy gap of the semiconductor.

45 •• The relative binding of the extra electron in the arsenic atom that replaces an atom in silicon or germanium can be understood from a calculation of the first Bohr radius of the electron in these materials. Four of arsenic's valence electrons form covalent bonds, so the fifth electron sees a center of attraction with a charge of $+e$. This model is a modified hydrogen atom. In the Bohr model of the hydrogen atom, the electron moves in free space at a radius a_0 given by $a_0 = 4\pi\epsilon_0\hbar^2/(m_e e^2)$ (Equation 36-12). When an electron moves in a crystal, we can approximate the effect of the other atoms by replacing ϵ_0 with $\kappa\epsilon_0$ and m_e with an effective mass for the electron. For silicon, κ is 12 and the effective mass is approximately $0.2m_e$. For germanium, κ is 16 and the effective mass is approximately $0.1m_e$. Estimate the Bohr radii for the valence electron as it orbits the impurity arsenic atom in silicon and in germanium.

46 •• The ground-state energy of the hydrogen atom is given by $E_1 = -m_e e^4/(8\epsilon_0^2 h^2)$ (Equations 36-15 and 36-16 where $4\pi\epsilon_0$ is substituted for k^{-1}). Modify this equation using information in Problem 45 by replacing ϵ_0 with $\kappa\epsilon_0$ and m_e with an effective mass for the electron to estimate the binding energy of the extra electron of an impurity arsenic atom in (*a*) silicon and (*b*) germanium.

47 •• A doped n-type silicon sample has 1.00 × 10^{16} electrons per cubic centimeter in the conduction band and has a resistivity of 5.00 × 10^{23} Ω·m at 300 K. Find the mean free path of the electrons. Use the effective mass of $0.2m_e$ for the mass of the electrons. (See Problem 45.) Compare this mean free path with that of conduction electrons in copper at 300 K.

48 ••• In the Hall effect, the Hall coefficient R_H is the proportionality constant between the transverse electric field and the product of the applied magnetic field and the current density. That is, $E_y = R_H B_z J_x$, where the current density, the transverse electric field, and the applied magnetic field are in the $+x$, $-y$, and $+z$ directions, respectively. (The Hall effect is presented in Chapter 26.) The measured Hall coefficient of a doped silicon sample is 0.0400 V·m/(A·T) at room temperature. If all the doping impurities have contributed to the total number of charge carriers of the sample, find (*a*) the type of impurity (donor or acceptor) used to dope the sample and (*b*) the concentration of the impurities.

*SEMICONDUCTOR JUNCTIONS AND DEVICES

49 •• Simple theory for the current versus the bias voltage across a pn junction yields the equation $I = I_0(e^{eV_b/kT} - 1)$. Sketch I versus V_b for both positive and negative values of V_b using that equation.

50 • The base current in an npn transistor circuit is 25.0 mA. If 88.0 percent of the electrons entering the base from the emitter reach the collector, what is the base current?

51 •• In Figure 38-28 for the npn-transistor amplifier, suppose R_b = 2.00 kΩ and R_L = 10.0 kΩ. Suppose further that a 10.0-μA ac base current i_b generates a 0.500-mA ac collector current i_c. What is the voltage gain of the amplifier? SSM

52 •• Germanium can be used to measure the energy of incident photons. Consider a 660-keV gamma ray emitted from ^{137}Cs. (*a*) Given that the band gap in germanium is 0.72 eV, how many electron-hole pairs can be generated as this gamma ray travels through germanium? (*b*) The number N of pairs in Part (*a*) will have statistical fluctuations given by $\pm\sqrt{N}$. What then is the energy resolution of the detector in that photon energy region?

53 •• Make a sketch showing the valence and conduction band edges and Fermi energy of a *pn*-junction diode when biased (*a*) in the forward direction and (*b*) in the reverse direction.

54 •• A good silicon diode has the current-voltage characteristic given by $I = I_0(e^{eV_b/kT} - 1)$. Let $kT = 0.025$ eV (room temperature) and the saturation current $I_0 = 1.0$ nA. (*a*) Show that for small reverse-bias voltages, the resistance is 25 MΩ. *Hint: Do a Taylor-series expansion of the exponential function about* $V_b = 0$, *or use the expansion for* e^x *found in Table M-4 of the Math Tutorial.* (*b*) Find the dc resistance for a reverse bias of 0.50 V. (*c*) Find the resistance V/I for a 0.50-V forward bias. What is the current in this case? (*d*) Calculate the differential resistance dV/dI for a 0.50-V forward bias.

55 •• A long slab of silicon of thickness $T = 1.0$ mm and width $w = 1.0$ cm is placed in a magnetic field $B = 0.40$ T. The slab is in the xy plane, where the length of the slab is parallel with the x axis and the magnetic field points in the $+z$ direction. When a current of 0.20 A exists in the sample in the $+x$ direction, a potential difference of 5.0 mV develops across the width of the sample and the electric field in the sample points in the $+y$ direction. Determine the semiconductor type (n or p) and the concentration of charge carriers. (The Hall effect is presented in Chapter 26.)

THE BCS THEORY

56 • (*a*) Use Equation 38-37 to calculate the superconducting energy gap for tin, and compare your result with the measured value of 6.00×10^{-4} eV. (*b*) Use the measured value to calculate the minimum value of the wavelength of a photon that has sufficient energy to break up Cooper pairs in lead ($T_c = 3.72$ K) at $T = 0$.

57 • (*a*) Use Equation 38-37 to calculate the superconducting energy gap for lead, and compare your result with the measured value of 2.73×10^{-3} eV. (*b*) Use the measured value to calculate the minimum value of the wavelength of a photon that has sufficient energy to break up Cooper pairs in tin ($T_c = 7.19$ K) at $T = 0$. SSM

THE FERMI–DIRAC DISTRIBUTION

58 •• The number of electrons in the conduction band of an insulator or intrinsic semiconductor is governed chiefly by the Fermi factor. Because the valence band in these materials is nearly filled and the conduction band is nearly empty, the Fermi energy E_F is generally midway between the top of the valence band and the bottom of the conduction band; that is, it is at $E_g/2$, where E_g is the band gap between the two bands and the energy is measured from the top of the valence band. (*a*) In silicon, $E_g \approx 1.0$ eV. Show that in this case the Fermi factor for electrons at the bottom of the conduction band is given by $\exp(-E_g/2kT)$ and evaluate this factor. Discuss the significance of the result if there are 10^{22} valence electrons per cubic centimeter and the probability of finding an electron in the conduction band is given by the Fermi factor. (*b*) Repeat the calculation in Part (*a*) for an insulator with a band gap of 6.0 eV.

59 •• Approximately how many energy states that have energies between 2.00 eV and 2.20 eV are available to electrons in a cube of silver measuring 1.00 mm on a side?

60 •• Show that at $E = E_F$, the expression for the Fermi factor (Equation 38-24) is equal to 0.5.

61 •• (*a*) Using the equation $E_F = [h^2/(8m_e)][3N/(\pi V)]^{2/3}$ (Equation 38-22*a*), calculate the Fermi energy for silver. (*b*) Determine the average kinetic energy of a free electron and (*c*) find the Fermi speed for silver. SSM

62 •• What is the difference between the energies at which the Fermi factor is 0.9 and 0.1 at 300 K in (*a*) copper, (*b*) potassium, and (*c*) aluminum.

63 •• What is the probability that a conduction electron in silver will have a kinetic energy of 4.90 eV at $T = 300$ K? SSM

64 •• Show that $g(E) = \frac{3}{2}NE_F^{-3/2}E^{1/2}$ (Equation 38-43) follows from Equation 38-41 for $g(E)$, and from Equation 38-22*a* for E_F.

65 •• Carry out the integration $E_{av} = (1/N)\int_0^{E_F} Eg(E)dE$ to show that the average energy at $T = 0$ is $\frac{3}{5}E_F$.

66 •• The density of the electron states in a metal can be written $g(E) = AE^{1/2}$, where A is a constant and E is measured from the bottom of the conduction band. (*a*) Show that the total number of states is $\frac{2}{3}AE_F^{3/2}$. (*b*) Approximately what fraction of the conduction electrons are within kT of the Fermi energy? (*c*) Evaluate that fraction for copper at $T = 300$ K.

67 •• What is the probability that a conduction electron in silver will have a kinetic energy of 5.49 eV at $T = 300$ K?

68 •• Using the expression $g(E) = (8\sqrt{2}\pi m_e^{3/2}V/h^3)E^{1/2}$ (Equation 38-41) for the density of states, estimate the fraction of the conduction electrons in copper that can absorb energy from collisions with the vibrating lattice ions at (*a*) 77 K and (*b*) 300 K.

69 •• In an intrinsic semiconductor, the Fermi energy is about midway between the top of the valence band and the bottom of the conduction band. In germanium, the forbidden energy band has a width of 0.70 eV. Show that at room temperature the distribution function of electrons in the conduction band is given by the Maxwell–Boltzmann distribution function.

70 ••• The root-mean-square (rms) value of a variable is obtained by calculating the average value of the square of that variable and then taking the square root of the result. Use that procedure to determine the rms energy of a Fermi distribution. Express your result in terms of E_F and compare it to the average energy. Why do E_{av} and E_{rms} differ?

GENERAL PROBLEMS

71 • The density of potassium is 0.851 g/cm³. How many free electrons are there per potassium atom in a crystal of potassium?

72 • Calculate the number density of free electrons for (*a*) magnesium, which has a density of 1.74 g/cm³, and (*b*) zinc, which has a density of 7.14 g/cm³. For the calculations assume there are two free electrons per atom, and compare your results with the values listed in Table 38-1.

73 •• Estimate the fraction of free electrons in copper that are in energy states above the Fermi energy at (*a*) 300 K (about room temperature) and (*b*) 1000 K.

74 •• A certain free-electron energy state of manganese has a 10.0 percent chance of being occupied when the temperature of the manganese is $T = 1300$ K. What is the energy of the state?

75 •• The semiconducting compound CdSe is widely used for light-emitting diodes (LEDs). The energy gap in CdSe is 1.80 eV. What is the frequency of the light emitted by a CdSe LED?

76 ••• A 2.00-cm² wafer of pure silicon is irradiated with electromagnetic radiation having a wavelength of 775 nm. The intensity of the radiation is 4.00 W/m² and every photon that strikes the sample is absorbed and creates an electron-hole pair. (*a*) How many electron-hole pairs are produced in one second? (*b*) If the number of electron-hole pairs in the sample is 6.25×10^{11} in the steady state, at what rate do the electron-hole pairs recombine? (*c*) If every recombination event results in the radiation of one photon, at what rate is energy radiated by the sample?

THE ANDROMEDA GALAXY BY MEASURING THE FREQUENCY OF THE LIGHT COMING TO US FROM DISTANT OBJECTS, WE ARE ABLE TO DETERMINE HOW FAST THE OBJECTS ARE APPROACHING TOWARD US OR RECEDING FROM US. *(NASA.)*

Relativity

? Have you wondered how the frequency of the light enables us to determine the speed of recession of a distant galaxy? (See Example 39-5.)

The theory of relativity consists of two rather different theories, the special theory and the general theory. The special theory, developed by Albert Einstein and others in 1905, describes measurements made in different inertial reference frames moving with constant velocity relative to one another. Its consequences, which can be derived with a minimum of mathematics, are applicable in a wide variety of situations encountered in physics and in engineering. On the other hand, the general theory, also developed by Einstein and others around 1916, describes accelerated reference frames and gravity. A thorough understanding of the general theory requires sophisticated mathematics, and the applications of the theory are mainly in the area of gravitation. The general theory is of great importance in cosmology, but it is rarely encountered in other areas of physics or in engineering. The general theory is applied, however, in the engineering of the Global Positioning System (GPS).*

* The satellites used in GPS contain atomic clocks.

In this chapter, we concentrate on the special theory (often referred to as special relativity*). General relativity theory (general relativity) will be discussed briefly near the end of the chapter. Special relativity is first presented in Chapter R (which precedes Chapter 11). You should consider reviewing the material in Chapter R before proceeding in this chapter.*

39-1 NEWTONIAN RELATIVITY

Newton's first law does not distinguish between a particle at rest and a particle moving with constant velocity. If there is no net external force acting, the particle will remain in its initial state, either at rest or moving with its initial velocity. A particle at rest relative to you is moving with constant velocity relative to an observer who is moving with constant velocity relative to you. How might we distinguish whether you and the particle are at rest and the second observer is moving with constant velocity, or the second observer is at rest and you and the particle are moving?

Let us consider some simple experiments. Suppose we have a railway boxcar moving along a straight, flat track with a constant velocity v. (The velocity v is a signed quantity, and the sign indicates the direction of the motion of the boxcar along the track.) We note that a ball at rest in the boxcar remains at rest relative to the car. If we drop the ball, it falls straight down, relative to the boxcar, with an acceleration g due to gravity. Of course, when viewed from the reference frame of the track, the ball moves along a parabolic path because it has an initial velocity v to the right. No mechanics experiment that we can do—measuring the period of a pendulum, observing the collisions between two objects, or whatever—will tell us whether the boxcar is moving and the track is at rest or the track is moving and the boxcar is at rest. If we have a coordinate system attached to the track and another attached to the boxcar, Newton's laws hold in either system.

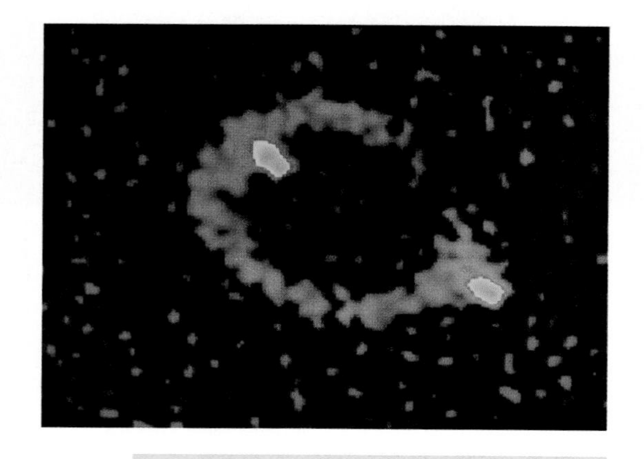

The ringlike structure of the radio source MG1131 + 0456 is thought to be due to *gravitational lensing*, first proposed by Albert Einstein in 1936, in which a source is imaged into a ring by a large, massive object in the foreground. *(NRAO/AUI.)*

A set of coordinate systems at rest relative to each other is called a *reference frame*. A reference frame in which Newton's laws hold is called an *inertial reference frame*.* All reference frames moving at constant velocity relative to an inertial reference frame are also inertial reference frames. If we have two inertial reference frames moving with constant velocity relative to each other, there are no mechanics experiments that can tell us which is at rest and which is moving or if they are both moving. This result is known as the principle of **Newtonian relativity:**

> Absolute motion cannot be detected.
>
> PRINCIPLE OF NEWTONIAN RELATIVITY

This principle was well known by Galileo, Newton, and others in the seventeenth century. By the late nineteenth century, however, this view had changed. It was then generally thought that Newtonian relativity was not valid and that absolute motion could be detected in principle by a measurement of the speed of light.

ETHER AND THE SPEED OF LIGHT

We saw in Chapter 15 that the velocity of a wave depends on the properties of the medium in which the wave travels and not on the velocity of the source of the waves. For example, the velocity of sound relative to still air depends on the temperature of the air. Light and other electromagnetic waves (for example, radio

* Reference frames were first discussed in Section 3-1. Inertial reference frames were also discussed in Section 4-1.

waves and X rays) travel through a vacuum with a speed $c = 3.00 \times 10^8$ m/s that is predicted by James Clerk Maxwell's equations for electricity and magnetism. But what is this speed relative to? What is the equivalent of still air for a vacuum? A proposed medium for the propagation of light was called the *ether;* it was thought to pervade all space. The velocity of light relative to the ether was assumed to be c, as predicted by Maxwell's equations. The velocity of any object relative to the ether was considered to be the absolute velocity of the object.

Albert Michelson, first in 1881 and then again with Edward Morley in 1887, set out to measure the velocity of Earth relative to the ether by an ingenious experiment in which the velocity of light relative to Earth was compared for two light beams, one parallel to the direction of Earth's motion relative to the Sun and the other perpendicular to the direction of Earth's motion. Despite painstakingly careful measurements, they could detect no difference. The experiment has since been repeated under various conditions by a number of people, and no difference has ever been found. The absolute motion of Earth relative to the ether cannot be detected.

39-2 EINSTEIN'S POSTULATES

In 1905, at the age of 26, Albert Einstein published a paper on the electrodynamics of moving bodies.* In this paper, he postulated that absolute motion cannot be detected by any experiment. That is, there is no ether. Earth can be considered to be at rest and the velocity of light will be the same in any direction.† His theory of special relativity can be derived from two postulates. Simply stated, these postulates are as follows:

> Postulate 1: Absolute uniform motion cannot be detected.
>
> Postulate 2: The speed of light is independent of the motion of the source.
>
> EINSTEIN'S POSTULATES

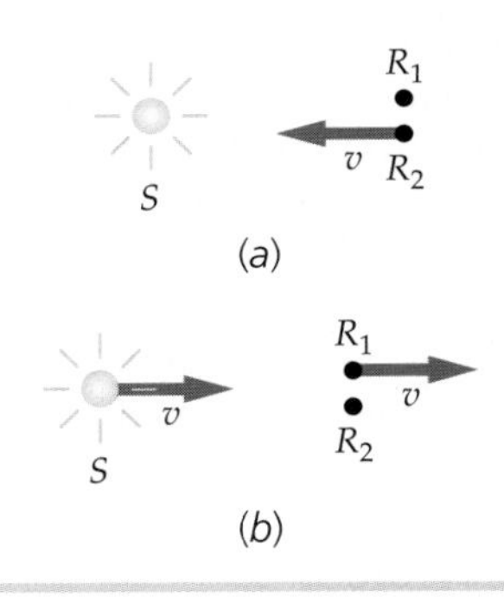

FIGURE 39-1 (*a*) A stationary light source S, a stationary observer R_1, and a second observer R_2 moving toward the source with speed v. (*b*) In the reference frame in which the observer R_2 is at rest, the light source S and observer R_1 move to the right with speed v. If absolute motion cannot be detected, the two views are equivalent. Because the speed of light does not depend on the motion of the source, observer R_2 measures the same value for that speed as observer R_1.

Postulate 1 is merely an extension of the Newtonian principle of relativity to include all types of physical measurements (not just those measurements that are mechanical). Postulate 2 describes a common property of many waves. For example, the speed of sound waves does not depend on the motion of the sound source. The sound waves from a car horn travel through the air with the same speed, relative to the air, independent of whether the car is moving relative to the air or not. The speed of the waves depends only on the properties of the air, such as its temperature.

Although each postulate seems quite reasonable, many of the implications of the two postulates together are quite surprising and contradict what is often called common sense. For example, one important implication of these postulates is that every observer measures the same value for the speed of light independent of the relative motion of the source and the observer. Consider a light source S and two observers, R_1 at rest relative to S and R_2 moving toward S with speed v, as shown in Figure 39-1*a*. The speed of light measured by R_1 is $c = 3.00 \times 10^8$ m/s. What is the speed measured by R_2? The answer is *not* $c + v$. By postulate 1, Figure 39-1*a* is equivalent to Figure 39-1*b*, in which R_2 is at rest and the source S and R_1 are

* *Annalen der Physik*, vol. 17, 1905, p. 841. For a translation from the original German, see W. Perrett and G. B. Jeffery (trans.), *The Principle of Relativity: A Collection of Original Memoirs on the Special and General Theory of Relativity* by H. A. Lorentz, A. Einstein, H. Minkowski, and W. Weyl, Dover, New York, 1923.

† Einstein did not set out to explain the results of the Michelson–Morley experiment. His theory arose from his considerations of the theory of electricity and magnetism and the unusual property of electromagnetic waves that they propagate in a vacuum. In his first paper, which contains the complete theory of special relativity, he made only a passing reference to the Michelson–Morley experiment, and in later years he could not recall whether he was aware of the details of the experiment before he published his theory.

moving with speed v. That is, because absolute motion cannot be detected, it is not possible to say which is really moving and which is at rest. By postulate 2, the speed of light from a moving source is independent of the motion of the source. Thus, looking at Figure 39-1*b*, we see that R_2 measures the speed of light to be c, just as R_1 does. This result is often considered as an alternative to Einstein's second postulate:

> Postulate 2 (alternate): Every observer measures the same value c for the speed of light.

This result contradicts our intuitive ideas about relative velocities. If a car moves at 50 km/h away from an observer and another car moves at 80 km/h in the same direction, the velocity of the second car relative to the first car is 30 km/h. This result is easily measured and conforms to our intuition. However, according to Einstein's postulates, if a light beam is moving in the direction of the cars, observers in both cars will measure the same speed for the light beam. Our intuitive ideas about the combination of velocities are approximations that hold only when the speeds are very small compared with the speed of light. Even in an airplane moving with the speed of sound, to measure the speed of light accurately enough to distinguish the difference between the results c and $c + v$, where v is the speed of the plane, would require a measurement with six-digit accuracy.

39-3 THE LORENTZ TRANSFORMATION

Einstein's postulates have important consequences for measuring time intervals and space intervals, as well as relative velocities. Throughout this chapter, we will be comparing measurements of the positions and times of events (such as lightning flashes) made by observers who are moving relative to each other. We will use a rectangular coordinate system xyz that has an origin O and is called the S reference frame; we will use another system $x'y'z'$ that has an origin O', is called the S' frame, and is moving with a constant velocity $\vec{v}$ relative to the S frame. Relative to the S' frame, the S frame is moving with a constant velocity $-\vec{v}$. For simplicity, we will consider the S' frame to be moving along the x axis in the $+x$ direction relative to S, where the $+x'$ direction is the same as the $+x$ direction. In each frame, we will assume that there are as many observers as are needed who have measuring devices, such as clocks and metersticks, that are identical when compared at rest (see Figure 39-2).

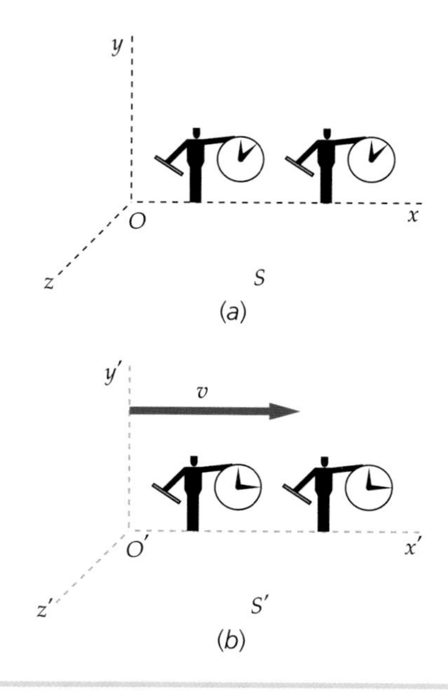

FIGURE 39-2 Coordinate reference frames S and S' moving with relative speed v. In each frame, there are observers who have metersticks and clocks that are identical when compared at rest.

We will use Einstein's postulates to find the general relation between the coordinates x, y, and z and the time t of an event as seen in reference frame S and the coordinates x', y', and z' and the time t' of the same event as seen in reference frame S', which is moving with uniform velocity relative to S. For convenience, we assume that the origins are coincident at time $t = t' = 0$. The classical relation, called the **Galilean transformation,** is

$$x = x' + vt', \quad y = y', \quad z = z', \quad t = t' \qquad \text{39-1a}$$

GALILEAN TRANSFORMATION

The inverse transformation is

$$x' = x - vt, \quad y' = y, \quad z' = z, \quad t' = t \qquad \text{39-1b}$$

These equations are consistent with experimental observations as long as v is much less than c. They lead to the familiar classical rules for velocities. If a particle has velocity $u_x = dx/dt$ in frame S, its velocity in frame S' is

$$u'_x = \frac{dx'}{dt'} = \frac{dx'}{dt} = \frac{d}{dt}(x - vt) = u_x - v \qquad \text{39-2}$$

If we differentiate this equation again, we find that the acceleration of the particle is the same in both frames:

$$a_x = \frac{du_x}{dt} = \frac{du'_x}{dt'} = a'_x$$

It should be clear that the Galilean transformation is not consistent with Einstein's postulates of special relativity. If light moves along the x axis with speed $u'_x = c$ in S', these equations imply that the speed in S' is $u_x = c + v$ rather than $u_x = c$, which is consistent with Einstein's postulates and with experiment. The classical transformation equations must therefore be modified to make them consistent with Einstein's postulates. We will give a brief outline of one method of obtaining the relativistic transformation.

We assume that the relativistic transformation equation for x is the same as the classical equation (Equation 39-1*a*) except for a constant multiplier on the right side. That is, we assume the equation is of the form

$$x = \gamma(x' + vt') \qquad \text{39-3}$$

where γ is a constant that can depend on v and c but not on the coordinates. The inverse transformation must look the same except for the plus sign:

$$x' = \gamma(x - vt) \qquad \text{39-4}$$

Let us consider a light pulse that starts at the origin of S at $t = 0$. Because we have assumed that the two origins are coincident at $t = t' = 0$, the pulse also starts at the origin of S' at $t' = 0$. Einstein's postulates require that the equation for the x component of the wave front of the light pulse is $x = ct$ in frame S and $x' = ct'$ in frame S'. Substituting ct for x and ct' for x' in Equation 39-3 and Equation 39-4, we obtain

$$ct = \gamma(c + v)t' \qquad \text{39-5}$$

and

$$ct' = \gamma(c - v)t \qquad \text{39-6}$$

We divide both sides of Equations 39-5 and 39-6 by t, and then eliminate the ratio t'/t from the two equations and determine γ. Thus,

$$\gamma = \frac{1}{\sqrt{1 - (v^2/c^2)}} \qquad \text{39-7}$$

(Note that γ is always greater than 1, and when v is much less than c, $\gamma \approx 1$.) The relativistic transformation for x and x' is therefore given by Equation 39-3 and Equation 39-4, where γ is given by Equation 39-7. We can obtain equations for t and t' by combining Equation 39-3 with the inverse transformation given by Equation 39-4. Substituting $x = \gamma(x' + vt')$ for x in Equation 39-4, we obtain

$$x' = \gamma[\gamma(x' + vt') - vt] \qquad \text{39-8}$$

which can be solved for t in terms of x' and t'. The complete relativistic transformation is

$$x = \gamma(x' + vt'), \quad y = y', \quad z = z' \qquad \text{39-9}$$

$$t = \gamma\left(t' + \frac{vx'}{c^2}\right) \qquad \text{39-10}$$

LORENTZ TRANSFORMATION

The inverse transformation is

$$x' = \gamma(x - vt), \quad y' = y, \quad z' = z \qquad \text{39-11}$$

$$t' = \gamma\left(t - \frac{vx}{c^2}\right) \qquad \text{39-12}$$

The transformation described by Equations 39-9 through 39-12 is called the **Lorentz transformation.** It relates the space and time coordinates x, y, z, and t of an event in frame S to the coordinates x', y', z', and t' of the same event as seen in frame S', which is moving along the x axis with speed v relative to frame S.

We will now look at some applications of the Lorentz transformation.

TIME DILATION

Consider two events, one that occurs on the x' axis at point x'_0 at time t'_1 in frame S' and another that occurs on the x' axis at point x'_0 at time t'_2 in frame S'. (Both events occur at point x'_0 in S'.) We can find the times t_1 and t_2 for the events in S from Equation 39-10. We have

$$t_1 = \gamma\left(t'_1 + \frac{vx'_0}{c^2}\right)$$

and

$$t_2 = \gamma\left(t'_2 + \frac{vx'_0}{c^2}\right)$$

so

$$t_2 - t_1 = \gamma(t'_2 - t'_1)$$

The time between two events that happen at the *same place* in a reference frame is called **proper time Δt_p** between the events. In this case, the time interval $t'_2 - t'_1$ measured in frame S' is proper time. The time interval Δt measured in any other reference frame is always longer than the proper time. This expansion is called **time dilation:**

$$\Delta t = \gamma\, \Delta t_p \qquad 39\text{-}13$$

TIME DILATION

Example 39-1 Spatial Separation and Temporal Separation of Two Events

Two events occur at the same point x'_0 at times t'_1 and t'_2 in frame S', which is traveling in the $+x$ direction at speed v relative to frame S. (*a*) What is the spatial separation of the events in frame S? (*b*) What is the temporal separation of the events in frame S?

PICTURE The spatial separation in S is $x_2 - x_1$, where x_2 and x_1 are the coordinates of the events in S, which are found using Equation 39-9.

SOLVE

(*a*) 1. The position x_1 in S is given by Equation 39-9 with $x'_1 = x'_0$:

$$x_1 = \gamma(x'_0 + vt'_1)$$

2. Similarly, the position x_2 in S is given by:

$$x_2 = \gamma(x'_0 + vt'_2)$$

3. Subtract to find the spatial separation:

$$\Delta x = x_2 - x_1 = \gamma v(t'_2 - t'_1) = \boxed{\frac{v(t'_2 - t'_1)}{\sqrt{1 - (v^2/c^2)}}}$$

(*b*) Using the time dilation formula, relate the two time intervals. The two events occur at the same place in S', so the proper time between the two events is $\Delta t_p = t'_2 - t'_1$:

$$\Delta t = t_2 - t_1 = \gamma(t'_2 - t'_1) = \boxed{\frac{(t'_2 - t'_1)}{\sqrt{1 - (v^2/c^2)}}}$$

CHECK Taking the limits of the Part-(*a*) and Part-(*b*) results as c approaches infinity gives $\Delta x = v(t_2' - t_1')$ and $\Delta t = t_2' - t_1'$, respectively. Combining these expressions gives $\Delta x = v\Delta t$. This is just the classical (nonrelativistic) equation that displacement equals velocity multiplied by time that is developed in Chapter 2 for one-dimensional motion. In addition, the equation $\Delta t = t_2' - t_1'$ is just the classical result that the length of the time between events is the same in both reference frames.

TAKING IT FURTHER Dividing the Part-(*a*) result by the Part-(*b*) result gives $\Delta x/\Delta t = v$. The spatial separation Δx of the two events in S is the distance a fixed point, such as x_0' in S', moves in S during the time interval between the events in S.

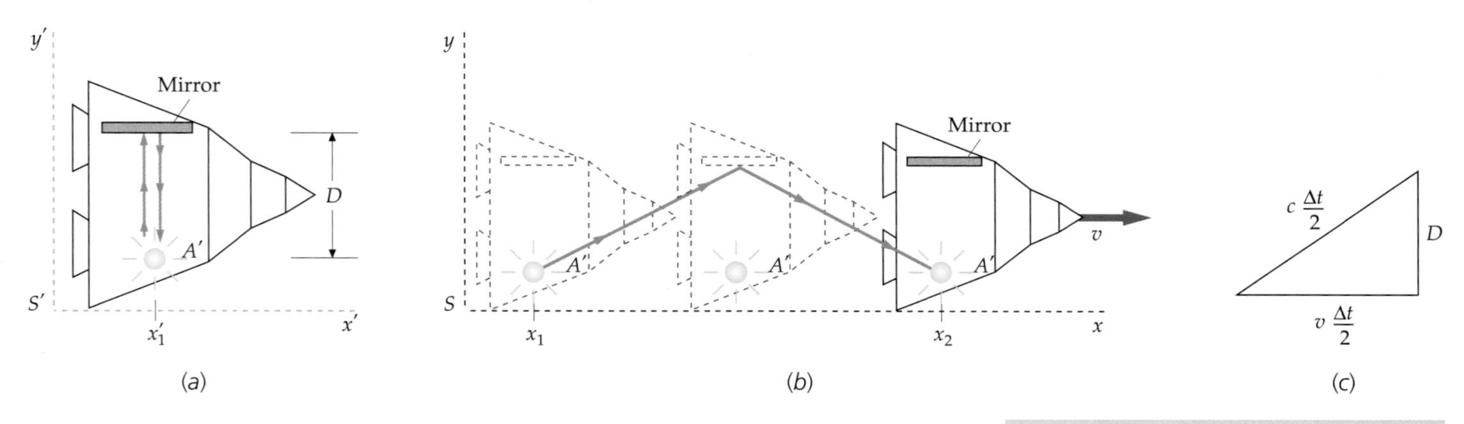

FIGURE 39-3 (*a*) Observer A' and the mirror are in a spaceship at rest in frame S'. The time it takes for the light pulse to reach the mirror and return is measured by A' to be $2D/c$. (*b*) In frame S, the spaceship is moving to the right with speed v. If the speed of light is the same in both frames, the time it takes for the light to reach the mirror and return is longer than $2D/c$ in S because the distance traveled is greater than $2D$. (*c*) A right triangle for computing the time Δt in frame S.

We can understand time dilation directly from Einstein's postulates without using the Lorentz transformation. Figure 39-3*a* shows an observer A' a distance D from a mirror. The observer and the mirror are in a spaceship that is at rest in frame S'. The observer explodes a flash gun and measures the time interval $\Delta t'$ between the original flash (Event 1) and his seeing the return flash from the mirror (Event 2). Because light travels with speed c, this time is

$$\Delta t' = \frac{2D}{c}$$

We now consider the same two events, the original flash of light and the receiving of the return flash, as observed in reference frame S, in which observer A' and the mirror are moving to the right with speed v, as shown in Figure 39-3*b*. Events 1 and 2 happen at positions x_1 and x_2, respectively, in frame S. During the time interval Δt (as measured in S) between the original flash and the return flash, observer A' and his spaceship have moved to the right a distance $v\,\Delta t$. In Figure 39-3, we can see that the path traveled by the light is longer in S than in S'. However, by Einstein's postulates, light travels with the same speed c in frame S as it does in frame S'. Because light travels farther in S at the same speed, it takes longer in S to reach the mirror and return. The time interval in S is thus longer than it is in S'. From the triangle in Figure 39-3*c*, we have

$$\left(\frac{c\,\Delta t}{2}\right)^2 = D^2 + \left(\frac{v\,\Delta t}{2}\right)^2$$

or

$$\Delta t = \frac{2D}{\sqrt{c^2 - v^2}} = \frac{2D}{c}\frac{1}{\sqrt{1 - (v^2/c^2)}}$$

Using $\Delta t' = 2D/c$, we obtain

$$\Delta t = \frac{\Delta t'}{\sqrt{1 - (v^2/c^2)}} = \gamma\,\Delta t'$$

Example 39-2 How Long Is a One-Hour Nap?

Try It Yourself

Astronauts in a spaceship traveling at $v = 0.600c$ relative to Earth sign off from space control, saying that they are going to nap for 1.00 h and then call back. How long does their nap last as measured on Earth?

PICTURE Because the astronauts go to sleep (Event 1) and wake up (Event 2) at the same place in the reference frame of the ship, the time interval for their nap of 1.00 h, as measured by a clock on the ship, is the proper time between the two events. In the reference frame of Earth, they move a considerable distance during the time between the two events. The time interval measured in Earth's frame is measured using two clocks that are stationary relative to Earth. Clock 1 is located at the position of Event 1 and measures the time of occurrence of Event 1. Clock 2 is located at the position of Event 2 and measures the time of occurrence of Event 2. The difference between the two times is longer than the proper time between the two events by the factor γ.

SOLVE

Cover the column to the right and try these on your own before looking at the answers.

Steps	Answers
1. Relate the time interval measured on Earth Δt to the proper time Δt_p (Equation 39-13).	$\Delta t = \gamma\, \Delta t_p$
2. Calculate γ for $v = 0.6c$ (Equation 39-7).	$\gamma = 1.25$
3. Substitute the value for γ to calculate the time of the nap in Earth's frame.	$\Delta t = \gamma\, \Delta t_p = \boxed{1.25\text{ h}}$

CHECK The time interval is longer in the reference frame in which the two events occur at different locations as expected.

PRACTICE PROBLEM 39-1 If the spaceship is moving at $v = 0.800c$, how long would a 1.00 h nap last as measured on Earth?

LENGTH CONTRACTION

A phenomenon closely related to time dilation is **length contraction.** The length of an object measured in the reference frame in which the object is at rest is called its **proper length** L_p. In a reference frame in which the object is moving parallel to its length, the measured length is shorter than its proper length. Consider a rod at rest in frame S' with one end at x'_2 and the other end at x'_1. The length of the rod in this frame is its proper length $L_p = x'_2 - x'_1$. Some care must be taken to find the length of the rod in frame S. In that frame, the rod is moving to the right with speed v, the speed of frame S'. The length of the rod in frame S is defined as $L = x_2 - x_1$, where x_2 is the position of one end at some time t_2, and x_1 is the position of the other end *at the same time* $t_1 = t_2$ as measured in frame S. To calculate $x_2 - x_1$ at some time t we use Equation 39-11:

$$x'_2 = \gamma(x_2 - vt_2)$$

and

$$x'_1 = \gamma(x_1 - vt_1)$$

Because $t_2 = t_1$, by subtracting the second equation from the first we obtain

$$x'_2 - x'_1 = \gamma(x_2 - x_1)$$

Solving for $x_2 - x_1$ gives

$$x_2 - x_1 = \frac{1}{\gamma}(x_2' - x_1') = (x_2' - x_1')\sqrt{1 - \frac{v^2}{c^2}}$$

or

$$L = \frac{1}{\gamma}L_P = L_P\sqrt{1 - \frac{v^2}{c^2}} \qquad \text{39-14}$$

LENGTH CONTRACTION

Because $1/\gamma$ is less than one, it follows that the length of the rod is smaller when it is measured in a frame in which it is moving parallel to its length. Before Einstein's paper was published, Hendrik A. Lorentz and George F. FitzGerald tried to explain the null result of the Michelson–Morley experiment by assuming that distances in the direction of motion contracted by the amount given in Equation 39-14. This length contraction is now known as the **Lorentz–FitzGerald contraction.**

Example 39-3 The Length of a Moving Meterstick

A stick that has a proper length of 1.00 m moves in a direction parallel to its length with speed v relative to you. The length of the stick as measured by you is 0.914 m. What is the speed v?

PICTURE We can find v directly from Equation 39-14.

SOLVE

1. Equation 39-14 relates the lengths L and L_p and the speed v: $L = L_p\sqrt{1 - \frac{v^2}{c^2}}$

2. Solve for v: $v = c\sqrt{1 - \frac{L^2}{L_p^2}} = c\sqrt{1 - \frac{(0.914\text{ m})^2}{(1.00\text{ m})^2}} = \boxed{0.406c}$

CHECK As expected, the speed is a significant fraction of c.

An interesting example of time dilation or length contraction is the generation of muons as secondary radiation from cosmic rays. Muons decay according to the statistical law of radioactivity:

$$N(t) = N_0 e^{-t/\tau} \qquad \text{39-15}$$

where N_0 is the number of muons at time $t = 0$, $N(t)$ is the number remaining at time t, and τ is the mean lifetime, which is approximately 2.2 μs for muons at rest. Because muons are generated (from the decay of pions) high in the atmosphere, usually several thousand meters above sea level, few muons should reach sea level. A typical muon moving with speed 0.9978c would travel only about 660 m in 2.2 μs. However, the lifetime of the muon measured in Earth's reference frame is increased by the factor $1/\sqrt{1 - (v^2/c^2)}$, which is 15 for this particular speed. The mean lifetime measured in Earth's reference frame is therefore 33 μs, and a muon with speed 0.9978c travels approximately 10 000 m during this time. From the muon's point of view, it exists for only 2.2 μs, but the atmosphere is rushing past it with a speed of 0.9978c. The distance of 10 000 m in Earth's frame is thus contracted to only 660 m in the muon's frame, as indicated in Figure 39-4.

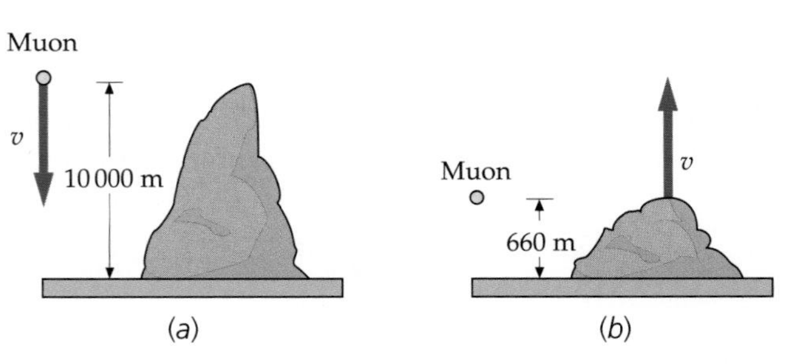

FIGURE 39-4 Although muons are created high above Earth and their mean lifetime is only about 2.2 μs when at rest, many appear at Earth's surface. (*a*) In Earth's reference frame, a typical muon moving at 0.9978c has a mean lifetime of 33 μs and travels 10 000 m during that time. (*b*) In the reference frame of the muon, the distance traveled by Earth is only 660 m in the muon's lifetime of 2.2 μs.

It is easy to distinguish experimentally between the classical and relativistic predictions of the observation of muons at sea level. Suppose that we observe 10^8 muons at an altitude of 10 000 m during some time interval with a muon detector. How many would we expect to observe at sea level during the same time interval? According to the nonrelativistic prediction, the time it takes for the muons to travel 10 000 m is $(10\,000\text{ m})/(0.998c) \approx 33\ \mu\text{s}$, which is 15 lifetimes. Substituting $N_0 = 1.0 \times 10^8$ and $t = 15\tau$ into Equation 39-15, we obtain

$$N = N_0 e^{-t/\tau} = 1.0 \times 10^8 e^{-15} = 31$$

We would thus expect all but about 31 of the original 100 million muons to decay before reaching sea level.

According to the relativistic prediction, Earth must travel only the contracted distance of 660 m in the rest frame of the muon. This trip takes only $2.2\ \mu\text{s} = 1\tau$. Therefore, the number of muons expected at sea level is

$$N = N_0 e^{-t/\tau} = 1.0 \times 10^8 e^{-1} = 37 \times 10^6$$

Thus, relativity predicts that we would observe 37 million muons during the same time interval. Experiments have confirmed the relativistic prediction of 37 million muons.

See* Math Tutorial *for more information on the

The Exponential Function

THE RELATIVISTIC DOPPLER EFFECT

For light or other electromagnetic waves in a vacuum, a distinction between the motion of the source and the motion of the receiver cannot be made. Therefore, the expressions we derived in Chapter 15 for the Doppler effect cannot be correct for light. The reason it is not correct is that in Chapter 15 we assumed the time intervals in the reference frames of the source and receiver to be the same.

Consider a source moving toward a receiver with speed v, relative to the receiver. If the source emits N electromagnetic-wave crests during a time Δt_R (measured in the frame of the receiver), the first crest will travel a distance $c\,\Delta t_R$ and the source will travel a distance $v\,\Delta t_R$ measured in the frame of the receiver. The wavelength in this reference frame will be

$$\lambda' = \frac{c\,\Delta t_R - v\,\Delta t_R}{N}$$

The frequency f' observed by the receiver will therefore be

$$f' = \frac{c}{\lambda'} = \frac{c}{(c - v)}\frac{N}{\Delta t_R} = \frac{1}{1 - (v/c)}\frac{N}{\Delta t_R}$$

If the frequency of the source in the reference frame of the source is f_0, it will emit $N = f_0\,\Delta t_S$ waves in the time Δt_S measured by the source. Then

$$f' = \frac{1}{1 - (v/c)}\frac{N}{\Delta t_R} = \frac{1}{1 - (v/c)}\frac{f_0\,\Delta t_S}{\Delta t_R} = \frac{f_0}{1 - (v/c)}\frac{\Delta t_S}{\Delta t_R}$$

Here Δt_S is the proper time interval (the first wave and the Nth wave are emitted at the same place in the reference frame of the source). Times Δt_S and Δt_R are related by Equation 39-13 for time dilation:

$$\Delta t_R = \gamma\,\Delta t_S = \frac{\Delta t_S}{\sqrt{1 - (v^2/c^2)}}$$

Thus, when the source and the receiver are moving toward one another we obtain

$$f' = \frac{f_0}{1 - (v/c)}\sqrt{1 - (v^2/c^2)} = \sqrt{\frac{1 + (v/c)}{1 - (v/c)}}\,f_0 \quad \text{approaching} \qquad \text{39-16}a$$

This differs from our classical equation only in the time-dilation factor. It is left as a problem (Problem 25) for you to show that the same results are obtained if the calculations are done in the reference frame of the receiver.

When the source and the receiver are moving away from one another, the same analysis shows that the observed frequency is given by

$$f' = \frac{f_0}{1 + (v/c)}\sqrt{1 - (v/c^2)} = \sqrt{\frac{1 - (v/c)}{1 + (v/c)}}f_0 \quad \text{receding} \qquad \text{39-16}b$$

An application of the relativistic Doppler effect is the **redshift** observed in the light from distant galaxies. Because the galaxies are moving away from us, the light they emit is shifted toward the longer wavelengths. (Because light that has the longest visible wavelengths appears red, this is referred to as a redshift.) The speed of the galaxies relative to us can be determined by measuring this shift.

Example 39-4 Red Light/Green Light

You are spending the day shadowing two police officers. You have just witnessed the officers pulling over a car that went through a red light. The driver claims that the red light looked green because the car was moving toward the stoplight, which shifted the wavelength of the observed light. You quickly do some calculations to see if the driver has a reasonable case.

PICTURE We can use the Doppler shift formula for approaching objects in Equation 39-16*a*. This will tell us the velocity, but we need to know the frequencies of the light. We can make good guesses for the wavelengths of red light and green light and use $c = f\lambda$ to determine the frequencies.

SOLVE

1. The observer is approaching the light source, so we use the Doppler formula (Equation 39-16*a*) for approaching sources:

$$f' = \sqrt{\frac{1 + (v/c)}{1 - (v/c)}}f_0$$

2. Substitute c/λ for f, then simplify:

$$\frac{c}{\lambda'} = \sqrt{\frac{1 + (v/c)}{1 - (v/c)}}\frac{c}{\lambda_0}$$

$$\left(\frac{\lambda_0}{\lambda'}\right)^2 = \frac{1 + (v/c)}{1 - (v/c)}$$

3. Cross multiply and solve for v/c:

$$(\lambda_0)^2\left(1 - \frac{v}{c}\right) = (\lambda')^2\left(1 + \frac{v}{c}\right)$$

$$(\lambda_0)^2 - (\lambda')^2 = \left[(\lambda_0)^2 + (\lambda')^2\right]\left(\frac{v}{c}\right)$$

$$\frac{v}{c} = \frac{(\lambda_0)^2 - (\lambda')^2}{(\lambda_0)^2 + (\lambda')^2} = \frac{1 - (\lambda'/\lambda_0)^2}{1 + (\lambda'/\lambda_0)^2}$$

4. The values for the wavelengths for the colors of the visible spectrum can be found in Table 30-1. The wavelengths for red are 625 nm or longer, and the wavelengths for green are 530 nm or shorter. Solve for the speed needed to shift the wavelength from 625 nm to 530 nm:

$$\frac{\lambda'}{\lambda_0} = \frac{530\text{ nm}}{625\text{ nm}} = 0.848$$

$$\frac{v}{c} = \frac{1 - 0.848^2}{1 + 0.848^2} = 0.163$$

$$v = 0.163c = 4.90 \times 10^7\text{ m/s} = 1.10 \times 10^8\text{ mi/h}$$

5. This speed is beyond any possible speed for a car:

The driver does not have a plausible case.

CHECK A car cannot travel at relativistic speeds, so the answer to this problem was obvious.

Example 39-5 Finding Speed from the Doppler Shift — *Try It Yourself*

The emission spectrum of hydrogen includes a line that has the wavelength $\lambda_0 = 656$ nm. In light reaching us from a distant galaxy, the wavelength of that spectral line is measured to be $\lambda' = 1458$ nm. Find the speed at which the distant galaxy is receding from Earth.

PICTURE Wavelength is related to frequency by $c = f\lambda$ and the received frequency is related to the unshifted frequency by the Doppler shift equation for a receding source (Equation 39-16*b*).

SOLVE

Cover the column to the right and try these on your own before looking at the answers.

Steps	Answers
1. Use Equation 39-16*b* to relate the speed v to the received frequency f' and the unshifted frequency f_0.	$f' = \sqrt{\frac{1-(v/c)}{1+(v/c)}}f_0$
2. Substitute $f' = c/\lambda'$ and $f_0 = c/\lambda_0$ and solve for v/c.	$\frac{v}{c} = \frac{1-(\lambda_0/\lambda')^2}{1+(\lambda_0/\lambda')^2} = 0.664$
	$v = \boxed{0.664c}$

CHECK As expected, the result is a significant fraction of c. This result is expected because the wavelength of the received light is large in comparison to the wavelength of the same spectral line in the reference frame of the source.

39-4 CLOCK SYNCHRONIZATION AND SIMULTANEITY

We saw in Section 39-3 that proper time is the time interval between two events that occur at the same point in some reference frame. It can therefore be measured on a single clock. (Remember, in each frame there is, in principle, a stationary clock at each point in space, and the time of an event in a given frame is measured by the clock at that point.) However, in another reference frame moving relative to the first, the same two events occur at different places, so two stationary clocks are needed in this reference frame to record the times. The time of each event is measured on a different clock, and the interval is found by subtraction of the measured times. This procedure requires that the clocks be **synchronized.** We will show in this section that

> Two clocks that are synchronized in one reference frame are typically not synchronized in any other frame moving relative to the first frame.
>
> SYNCHRONIZED CLOCKS

Here is a corollary to this result:

> Two events that are simultaneous in one reference frame typically are not simultaneous in another frame that is moving relative to the first.*
>
> SIMULTANEOUS EVENTS

* This is true unless the x coordinates of the two events are equal, where the x axis is parallel to the relative velocity of the two frames.

Comprehension of these facts usually resolves all relativity paradoxes. Unfortunately, the intuitive (and incorrect) belief that simultaneity is an absolute relation is difficult to overcome.

Suppose we have two clocks at rest, one at point A and the other at point B, where points A and B are a distance L apart in frame S. How can we synchronize the two clocks? If an observer at A looks at the clock at B and sets her clock to read the same time, the clocks will not be synchronized because of the time L/c it takes light to travel from one clock to another. To synchronize the clocks, the observer at A must set her clock ahead by the time L/c. Then she will see that the clock at B reads a time that is L/c behind the time on her clock, but she will calculate that the clocks are synchronized when she allows for the time L/c for the light to reach her. Any other observers in S (except those equidistant from the two clocks) will see the clocks reading different times, but they will also calculate that the clocks are synchronized when they correct for the time it takes the light to reach them. An equivalent method for synchronizing two clocks would be for an observer at point C, a point midway between the clocks, to send a light signal and for the observers at A and B to set their clocks to some prearranged time when they receive the signal.

We now examine the question of **simultaneity.** Suppose observers at A and B agree to explode flashguns at t_0 (having previously synchronized their clocks). The observer at C will see the light from the two flashes at the same time, and because he is equidistant from A and B, he will conclude that the flashes were simultaneous. Other observers in frame S will see the light from A or B first, depending on their location, but after correcting for the time the light takes to reach them, they also will conclude that the flashes were simultaneous. We can thus define simultaneity as follows:

> Two events in a reference frame are simultaneous if light signals from the events reach an observer halfway between the events at the same time.
>
> DEFINITION—SIMULTANEITY

To show that two events that are simultaneous in frame S are not simultaneous in another frame S' moving relative to S, we will use an example introduced by Einstein. A train is moving with speed v past a station platform. We will consider the train to be at rest in S' and the platform to be at rest in S. We have observers A', B', and C' at the front, back, and middle of the train (Figure 39-5). We now suppose that the train and platform are struck by lightning at the front and back of the train and that the lightning bolts are simultaneous in the frame of the platform S. That is, an observer C on the platform halfway between the positions A and B, where the lightning strikes, sees the light from the two strikes at the same time.

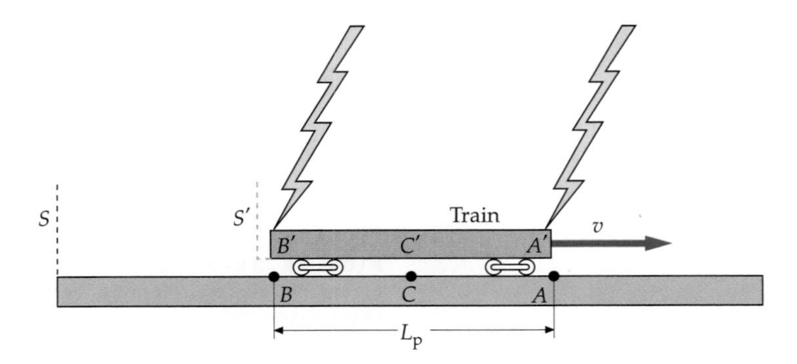

FIGURE 39-5 In frame S attached to the platform, simultaneous lightning bolts strike the ends of a train traveling with speed v. The light from the simultaneous events reaches observer C, standing on the platform midway between the events, at the same time. The distance between the bolts is $L_{\text{p platform}}$.

It is convenient to suppose that the lightning scorches both the train and platform so that the events can be easily located. Because C' is in the middle of the train, halfway between the places on the train that are scorched, the events are simultaneous in S' only if C' sees the flashes at the same time. However, the flash from the front of the train is seen by C' before the flash from the back of the train. We can understand this by considering the motion of C' as seen in frame S (Figure 39-6). By the time the light from the front flash reaches C', C' has moved some distance toward the front flash and some distance away from the back flash. Thus, the light from the back flash has not yet reached C', as indicated in the figure. Observer C' must therefore conclude that the events are not simultaneous and that the front of the train was struck before the back. Furthermore, all observers in S' on the train will agree with C' when they have corrected for the time it takes the light to reach them.

Figure 39-7 shows the events of the lightning bolts as seen in the reference frame of the train (S'). In this frame the platform is moving, so the distance between the scorch marks on the platform is contracted. The platform is shorter than it is in S, and, because the train is at rest, the train is longer than its contracted length in S. When the lightning bolt strikes the front of the train at A', the front of the train is at point A, and the back of the train has not yet reached point B. Later, when the lightning bolt strikes the back of the train at B', the back has reached point B on the platform.

The time discrepancy of two clocks that are synchronized in frame S as seen in frame S' can be found from the Lorentz transformation equations. Suppose we have clocks at points x_1 and x_2 that are synchronized in S. What are the times t_1 and t_2 on the clocks as observed from frame S' at a time t'_0?

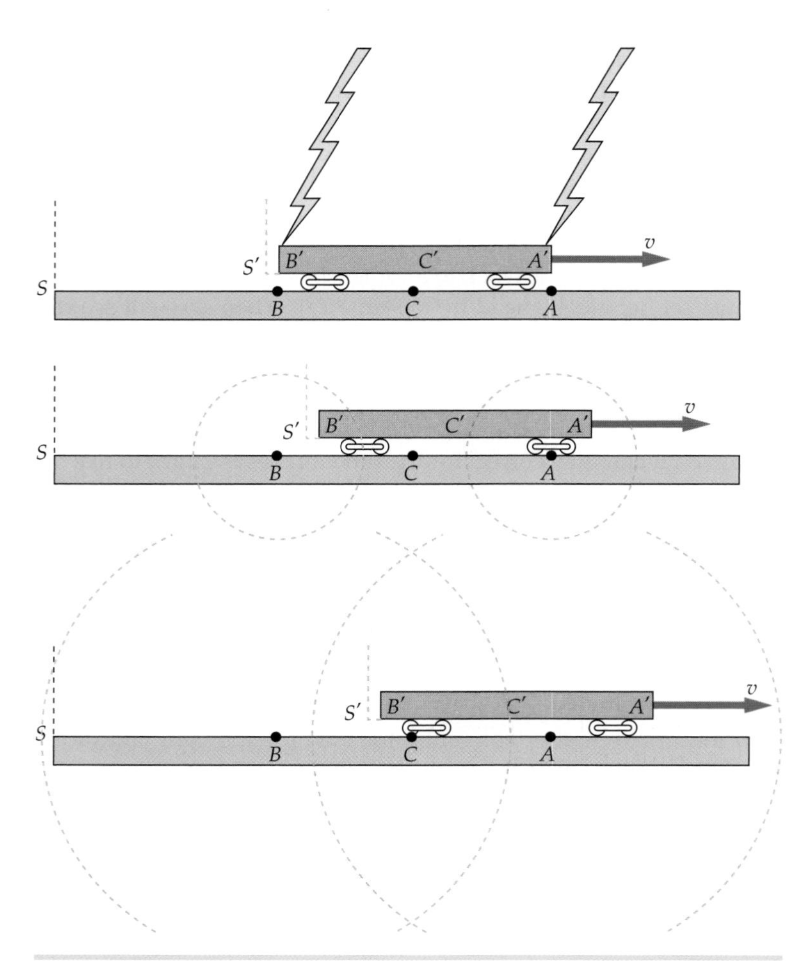

FIGURE 39-6 In frame S attached to the platform, the light from the lightning bolt at the front of the train reaches observer C', standing on the train at its midpoint, before the light from the bolt at the back of the train. Because C' is midway between the events (which occur at the front and rear of the train), the events are not simultaneous for him.

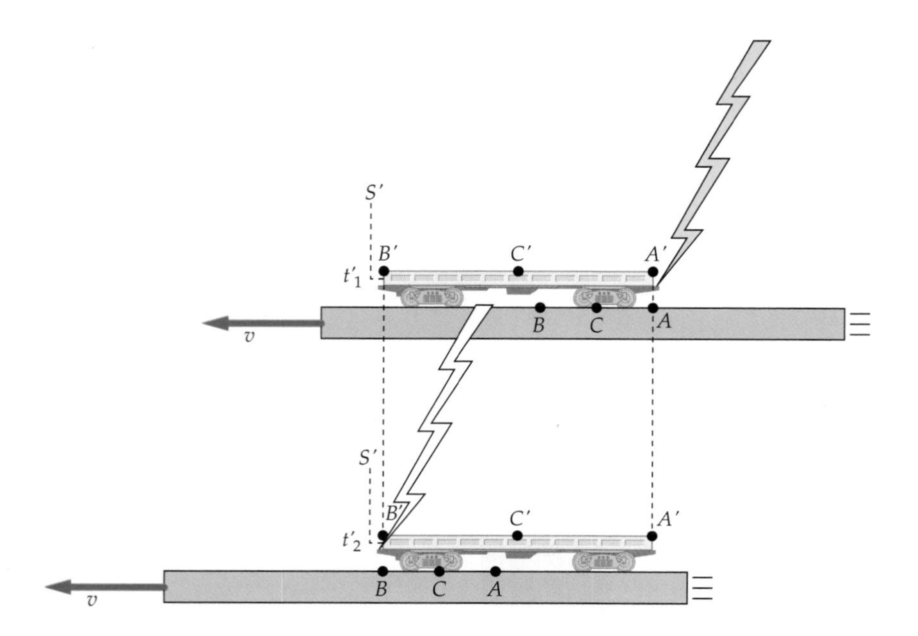

FIGURE 39-7 The lightning bolts of Figure 39-5 as seen in frame S' of the train. In this frame, the distance between A and B on the platform is less than $L_{\text{p platform}}$, and the proper length of the train $L_{\text{p train}}$ is longer than $L_{\text{p platform}}$. The first lightning bolt strikes the front of the train when A' and A are coincident. The second bolt strikes the rear of the train when B' and B are coincident.

From Equation 39-12, when $t'_1 = t'_2 = t'_0$, we have

$$t'_0 = \gamma\left(t_1 - \frac{vx_1}{c^2}\right)$$

and

$$t'_0 = \gamma\left(t_2 - \frac{vx_2}{c^2}\right)$$

Subtracting the first equation from the second, and then rearranging, gives

$$t_2 - t_1 = \frac{v}{c^2}(x_2 - x_1)$$

Note that the chasing clock (at x_2) leads the other (at x_1) by an amount that is proportional to their proper separation $L_p = x_2 - x_1$.

If two clocks are synchronized in the frame in which they are both at rest, in a frame in which they are moving along the line through both clocks, the chasing clock leads (shows a later time) by an amount

$$\Delta t_S = L_p \frac{v}{c^2} \qquad \text{39-17}$$

where L_p is the proper distance between the clocks.

CHASING CLOCK SHOWS LATER TIME

A numerical example should help clarify time dilation, clock synchronization, and the internal consistency of these results.

Example 39-6 Synchronizing Clocks

Observer A' in a spaceship that has a flashgun and a mirror is shown in Figure 39-3 (see page 1325). Observer A' is standing next to the flashgun. The distance from the gun to the mirror is 15 light-minutes (written 15 $c \cdot$ min) and the spaceship, at rest in frame S', travels with speed $v = 0.80c$ relative to a very long space platform that is at rest in frame S. The platform has two synchronized clocks, one clock at position x_1, the position of the spaceship when the observer explodes the flashgun, and the other clock at position x_2, the position of the spaceship when the light returns to the gun from the mirror. Find the time intervals between the events (exploding the flashgun and receiving the return flash from the mirror) (*a*) in the frame of the spaceship and (*b*) in the frame of the platform. Find (*c*) the distance traveled by the spaceship and (*d*) the amount by which the clocks on the platform are out of synchronization according to observers on the spaceship.

PICTURE The events occur at the same place on the spaceship, so the time between the events in frame S' is the proper time between the events.

SOLVE

(*a*) 1. In the frame of the spaceship, the light travels from the gun to the mirror and back, a total distance $D = 30\ c \cdot$ min. The time required is D/c:

$$\Delta t' = \frac{D}{c} = \frac{30\ c \cdot \text{min}}{c} = \boxed{30\ \text{min}}$$

2. Because the two events happen at the same place in the spaceship, the time interval is proper time:

$$\Delta t_p = \boxed{30\ \text{min}}$$

(*b*) 1. In frame S, the time between the events is longer by the factor γ:

$$\Delta t = \gamma\, \Delta t_p = \gamma(30\ \text{min})$$

2. Calculate γ:

$$\gamma = \frac{1}{\sqrt{1 - (v^2/c^2)}} = \frac{1}{\sqrt{1 - (0.80)^2}} = \frac{1}{\sqrt{0.36}} = \frac{5}{3}$$

3. Use the value of γ to calculate the time between the events as observed in frame S:

$$\Delta t = \gamma\, \Delta t_p = \tfrac{5}{3}(30\ \text{min}) = \boxed{50\ \text{min}}$$

(*c*) In frame S, the distance traveled by the spaceship is $v\,\Delta t$:

$$x_2 - x_1 = v\,\Delta t = (0.80c)(50\text{ min}) = \boxed{40\,c\cdot\text{min}}$$

(*d*) 1. The amount that the clocks on the platform are out of synchronization is related to the proper distance between the clocks L_p:

$$\Delta t_s = L_p\frac{v}{c^2}$$

2. The Part-(*c*) result is the proper distance between the clocks on the platform:

$$L_p = x_2 - x_1 = 40\,c\cdot\text{min}$$

so

$$\Delta t_s = L_p\frac{v}{c^2} = (40\,c\cdot\text{min})\frac{(0.80c)}{c^2} = \boxed{32\text{ min}}$$

TAKING IT FURTHER Observers on the platform would say that the spaceship's clock is running slow because it records a time of only 30 min between the events, whereas the time measured by observers on the platform is 50 min.

Figure 39-8 shows the situation in Example 39-6 viewed from the spaceship in S'. The platform is traveling past the ship with speed $0.8c$. There is a clock at point x_1, which coincides with the ship when the flashgun is exploded, and another at point x_2, which coincides with the ship when the return flash is received from the mirror. We assume that the clock at x_1 reads 12:00 noon at the time of the light flash. The clocks at x_1 and x_2 are synchronized in S but not in S'. In S', the clock at x_2, which is chasing the one at x_1, leads by 32 min; it would thus read 12:32 to an observer in S'. When the spaceship coincides with x_2, the clock there reads 12:50. The time between the events is therefore 50 min in S. Note that according to observers in S', this clock ticks off 50 min − 32 min = 18 min for a trip that takes 30 min in S'. Thus, observers in S' see this clock run slow by the factor 30/18 = 5/3.

Every observer in one frame sees the clocks in the other frame run slow. According to observers in S, who measure 50 min for the time interval, the time interval in S' (30 min) is too small, so they see the single clock in S' run too slow by the factor 5/3. According to the observers in S', the observers in S measure a time that is too *long* despite the fact that their clocks run too slow because the clocks in S are out of synchronization. The clocks tick off only 18 min, but the second clock leads the first clock by 32 min, so the time interval is 50 min.

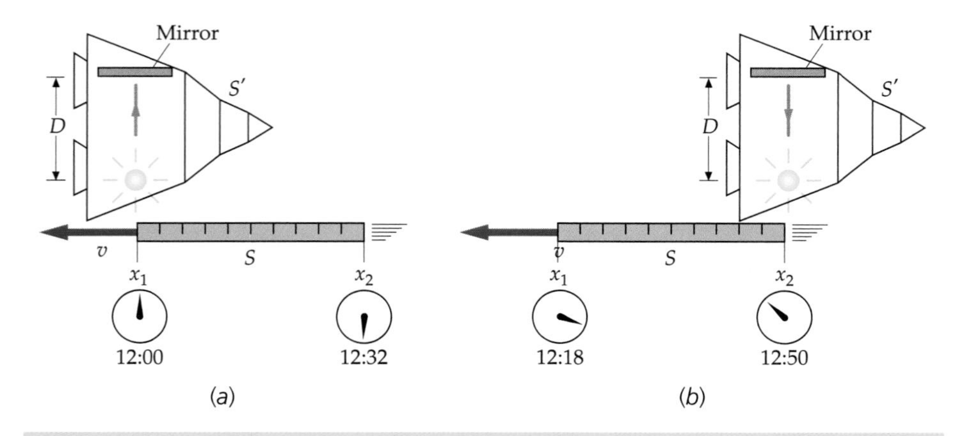

FIGURE 39-8 Clocks on a platform as observed from the spaceship's frame of reference S'. During the time $\Delta t' = 30$ min it takes for the platform to pass the spaceship, the clocks on the platform run slow and tick off (30 min)/γ = 18 min. But the clocks are unsynchronized, with the chasing clock leading by $L_p v/c^2$, which for this case is 32 min. The time it takes for the spaceship to go from x_1 to x_2, as measured on the platform, is therefore 32 min + 18 min = 50 min.

THE TWIN PARADOX

Homer and Ulysses are identical twins. Ulysses travels at high speed to a planet beyond the solar system and returns while Homer remains at home. When they are together again, which twin is older or are they the same age? The correct answer is that Homer, the twin who stays at home, is older. This problem, with variations, has been the subject of spirited debate for decades, though there are very few who disagree with the answer. The problem appears to be a paradox because of the seemingly symmetric roles played by the twins and the asymmetric result in their aging. The paradox is resolved when the asymmetry of the twins' roles is noted. The relativistic result conflicts with common sense based on our strong but incorrect belief in absolute simultaneity. We will consider a particular case with some numerical magnitudes that, though impractical, make the calculations easy.

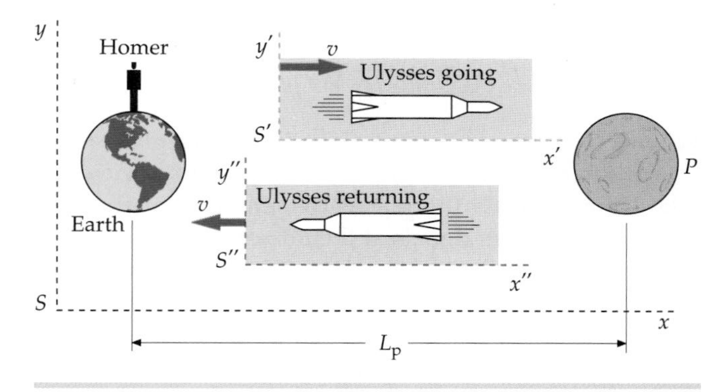

FIGURE 39-9 The twin paradox. Earth and a distant planet are fixed in frame S. Ulysses coasts in frame S' to the planet and then coasts back in frame S''. His twin Homer stays on Earth. When Ulysses returns, he is younger than his twin. The roles played by the twins are not symmetric. Homer remains at rest in one inertial reference frame, but Ulysses must go from being at rest in one inertial reference frame to another if he is to return home.

In reference frame S, Earth, planet P and Homer are at rest and Earth and planet P are a distance L_p apart (Figure 39-9). Homer is on Earth. Reference frames S' and S'' are moving with speed v toward and away from planet P, respectively. Ulysses quickly accelerates to speed v, then coasts, at rest in S', until he reaches the planet, where he quickly decelerates to a stop and is momentarily at rest in S. To return, Ulysses quickly accelerates to speed v toward Earth and then coasts, at rest in S'', until he reaches Earth, where he quickly decelerates to a stop. We can assume that the acceleration (and deceleration) times are negligible compared with the coasting times. We use the following values for illustration: $L_p = 8$ light-years ($8\ c \cdot y$) and $v = 0.8c$. Then $\sqrt{1 - (v^2/c^2)} = 3/5$ and $\gamma = 5/3$.

It is easy to analyze the problem from Homer's point of view on Earth. According to Homer's clock, Ulysses coasts in S' for a time $L_p/v = 10$ y and in S'' for an equal time. Thus, Homer is 20 y older when Ulysses returns. The time interval in S' between Ulysses's leaving Earth and his arriving at the planet is shorter because it is proper time. The time it takes to reach the planet by Ulysses's clock is

$$\Delta t' = \frac{\Delta t}{\gamma} = \frac{10\text{ y}}{5/3} = 6\text{ y}$$

Because the same time is required for the return trip, Ulysses will have recorded 12 y for the round trip and will be 8 years younger than Homer upon his return.

From Ulysses's point of view, the distance between Earth and the planet is contracted and is only

$$L' = \frac{L_p}{\gamma} = \frac{8\ c \cdot \text{y}}{5/3} = 4.8\ c \cdot \text{y}$$

At $v = 0.8c$, it takes only $L'/v = 4.8\ c \cdot y/0.8\ c = 6$ y each way.

The real challenge in this problem is for Ulysses to understand why his twin aged 20 y during his absence. If we consider Ulysses as being at rest and Homer as moving away, Homer's clock should run slow and measure only $\frac{3}{5}(6\text{ y}) = 3.6$ y. Then why shouldn't Homer age only 7.2 y during the round trip? This, of course, is the paradox. The difficulty with the analysis from the point of view of Ulysses is that he does not remain in a single inertial reference frame. What happens while Ulysses is stopping and starting? To investigate this problem in detail, we would need to treat accelerated reference frames, a subject dealt with in the study of general relativity and beyond the scope of this book. However, we can get some insight into the problem by having the twins send regular signals to each other so that they can record the other's age continuously. If they arrange to send a signal once a year, each can determine the age of the other merely by counting the signals received. The arrival frequency of the signals will not be 1 per year because of the Doppler shift. The frequency observed will be given by Equations 39-16*a* and 39-16*b*.

Using $v/c = 0.8$ (so $v^2/c^2 = 0.64$), we have for the case in which the twins are receding from each other

$$f' = \frac{f_0}{1 + (v/c)}\sqrt{1 - (v^2/c^2)} = \frac{\sqrt{1 - 0.64}}{1 + 0.8}f_0 = \frac{1}{3}f_0$$

When they are approaching, Equation 39-16*a* gives $f' = 3f_0$.

Consider the situation first from the point of view of Ulysses. During the 6 y it takes him to reach the planet (remember that the distance is contracted in his frame), he receives signals at the rate of $\frac{1}{3}$ signal per year, and so he receives 2 signals. As soon as Ulysses turns around and starts back to Earth, he begins to receive 3 signals per year. In the 6 y it takes him to return he receives 18 signals, giving a total of 20 for the trip. He accordingly expects his twin to have aged 20 years.

We now consider the situation from Homer's point of view. He receives signals at the rate of $\frac{1}{3}$ signal per year not only for the 10 y it takes Ulysses to reach the planet but also for the time it takes for the last signal sent by Ulysses before he turns around to get back to Earth. (He cannot know that Ulysses has turned around until the signals begin reaching him with increased frequency.) Because the planet is 8 light-years away, there is an additional 8 y of receiving signals at the rate of $\frac{1}{3}$ signal per year. During the first 18 y, Homer receives 6 signals. In the final 2 y before Ulysses arrives, Homer receives 6 signals, or 3 per year. (The first signal sent after Ulysses turns around takes 8 y to reach Earth, whereas Ulysses, traveling at $0.8c$, takes 10 y to return and therefore arrives just 2 y after Homer begins to receive signals at the faster rate.) Thus, Homer expects Ulysses to have aged 12 y. In this analysis, the asymmetry of the twins' roles is apparent. When they are together again, both twins agree that the one who has been accelerated will be younger than the one who stayed home.

The predictions of the special theory of relativity concerning the twin paradox have been tested using small particles that can be accelerated to such large speeds that γ is appreciably greater than 1. Unstable particles can be accelerated and trapped in circular orbits in a magnetic field, for example, and their lifetimes can then be compared with those of identical particles at rest. In all such experiments, the accelerated particles live longer on the average than the particles at rest, as predicted. These predictions have also been confirmed by the results of an experiment in which high-precision atomic clocks were flown around the world in commercial airplanes, but the analysis of that experiment is complicated due to the necessity of including gravitational effects treated in the general theory of relativity.

39-5 THE VELOCITY TRANSFORMATION

We can find how velocities transform from one reference frame to another by differentiating the Lorentz transformation equations. Suppose a particle has velocity $u'_x = dx'/dt'$ in frame S', which is moving to the right with speed v relative to frame S. The particle's velocity in frame S is

$$u_x = \frac{dx}{dt}$$

From the Lorentz transformation equations (Equations 39-9 and 39-10), we have

$$dx = \gamma(dx' + v\,dt')$$

and

$$dt = \gamma\left(dt' + \frac{v\,dx'}{c^2}\right)$$

The velocity relative to frame S is thus

$$u_x = \frac{dx}{dt} = \frac{\gamma(dx' + v\,dt')}{\gamma\left(dt' + \frac{v\,dx'}{c^2}\right)} = \frac{\frac{dx'}{dt'} + v}{1 + \frac{v}{c^2}\frac{dx'}{dt'}} = \frac{u'_x + v}{1 + \frac{vu'_x}{c^2}}$$

If a particle has components of velocity along the y or z axes, we can use the same relation between dt and dt', with $dy = dy'$ and $dz = dz'$, to obtain

$$u_y = \frac{dy}{dt} = \frac{dy'}{\gamma\left(dt' + \frac{v\,dx'}{c^2}\right)} = \frac{\frac{dy'}{dt'}}{\gamma\left(1 + \frac{v}{c^2}\frac{dx'}{dt'}\right)} = \frac{u'_y}{\gamma\left(1 + \frac{vu'_x}{c^2}\right)}$$

and

$$u_z = \frac{u'_z}{\gamma\left(1 + \frac{vu'_x}{c^2}\right)}$$

The complete relativistic velocity transformation is

$$u_x = \frac{u'_x + v}{1 + \frac{vu'_x}{c^2}} \qquad \text{39-18}a$$

$$u_y = \frac{u'_y}{\gamma\left(1 + \frac{vu'_x}{c^2}\right)} \qquad \text{39-18}b$$

$$u_z = \frac{u'_z}{\gamma\left(1 + \frac{vu'_x}{c^2}\right)} \qquad \text{39-18}c$$

RELATIVISTIC VELOCITY TRANSFORMATION

The inverse velocity transformation equations are

$$u'_x = \frac{u_x - v}{1 - \frac{vu_x}{c^2}} \qquad \text{39-19}a$$

$$u'_y = \frac{u_y}{\gamma\left(1 - \frac{vu_x}{c^2}\right)} \qquad \text{39-19}b$$

$$u'_z = \frac{u_z}{\gamma\left(1 - \frac{vu_x}{c^2}\right)} \qquad \text{39-19}c$$

These equations differ from the classical and intuitive result $u_x = u'_x + v$, $u_y = u'_y$, and $u_z = u'_z$ because the denominators in the equations are not equal to 1. When v and u'_x are small compared with the speed of light c, $\gamma \approx 1$ and $vu'_x/c^2 \ll 1$. Then the relativistic and classical velocity transformation equations are the same.

Example 39-7 Relative Velocity at Nonrelativistic Speeds

A supersonic plane moves away from you, and in the $+x$ direction, at a speed of 1000 m/s (about 3 times the speed of sound) relative to you. A second plane, traveling in the same direction and ahead of the first plane, moves away from you, and away from the first plane, at a speed of 500 m/s relative to the first plane. How fast is the second plane moving relative to you?

PICTURE The speeds are so small compared with c that we expect the classical equations for combining velocities to be accurate. We show this by calculating the correction term in the denominator of Equation 39-18*a*. Let frame S be your rest frame and frame S' be the rest frame of the first plane. Then v, the velocity of S' relative to S, is 1000 m/s. The second plane has velocity $u'_x = 500$ m/s relative to S'.

SOLVE

1. Let S and S' be the reference frames of you and the first plane, respectively. Also, let u_x and u'_x be the velocities of the second plane relative to S and S', respectively. Equation 39-18*a* can be used to find u_x. The velocity of the second plane relative to you is v:

$$u_x = \frac{u'_x + v}{1 + \dfrac{vu'_x}{c^2}}$$

2. If the correction term in the denominator is negligible (compared to 1), Equation 39-18*a* gives the classical formula for combining velocities. Calculate the value of the correction term:

$$\frac{vu'_x}{c^2} = \frac{(1000)(500)}{(3.00 \times 10^8)^2} = 5.56 \times 10^{-12}$$

3. The correction term is so small that the classical and relativistic results are virtually the same:

$$u_x \approx u'_x + v$$
$$= 500\text{ m/s} + 1000\text{ m/s} = \boxed{1500\text{ m/s}}$$

Example 39-8 Relative Velocity at Relativistic Speeds

Work Example 39-7 if the first plane moves with speed $v = 0.80c$ relative to you and the second plane moves with the same speed $0.80c$ relative to the first plane.

PICTURE These speeds are not small compared with c, so we need to use the relativistic expression (Equation 39-18*a*). We again assume that you are at rest in frame S and the first plane is at rest in frame S' that is moving at $v = 0.80c$ relative to you. The velocity of the second plane relative to S' is $u'_x = 0.80c$.

SOLVE

Use Equation 39-18*a* to calculate the speed of the second plane relative to you:

$$u_x = \frac{u'_x + v}{1 + \dfrac{vu'_x}{c^2}} = \frac{0.80c + 0.80c}{1 + \dfrac{(0.80c)(0.80c)}{c^2}} = \frac{1.60c}{1.64} = \boxed{0.98c}$$

CHECK As expected, the result is less than c.

The result in Example 39-8 is quite different from the classically expected result of $0.80c + 0.80c = 1.60c$. In fact, it can be shown from Equations 39-18*a*–*c* that if the speed of an object is less than c in one frame, it is less than c in all other frames moving relative to that frame with a speed less than c. (See Problem 59.) We will see in Section 39-7 that it takes an infinite amount of energy to accelerate a particle to the speed of light. The speed of light c is thus an upper, unattainable limit for the speed of a particle with mass. (There are massless particles, such as photons, that always move at the speed of light.)

Example 39-9 Relative Speed of a Photon

A photon moves along the x' axis in frame S', with speed $u'_x = c$. What is its speed in frame S?

PICTURE Use Equation 39-18*a* to calculate the speed of the photon in S.

SOLVE

The speed in S is given by Equation 39-18*a*:

$$u_x = \frac{u'_x + v}{1 + \frac{vu'_x}{c^2}} = \frac{c + v}{1 + \frac{vc}{c^2}} = \frac{c + v}{1 + \frac{v}{c}} = \frac{c + v}{\frac{1}{c}(c + v)} = \boxed{c}$$

CHECK The speed in both frames is c, independent of v. This is in accord with Einstein's postulates.

Example 39-10 Rockets Passing in Opposite Directions

Two spaceships, each 100 m long when measured at rest, travel toward each other, each with a speed of $0.85c$ relative to Earth. (*a*) What is the length of each spaceship as measured by someone at rest relative to Earth? (*b*) How fast is each spaceship traveling as measured by an observer at rest relative to the other spaceship? (*c*) What is the length of one spaceship when measured by an observer at rest relative to the other spaceship? (*d*) At time $t = 0$ on Earth, the front ends of the ships are next to each other as they just begin to pass each other. At what time on Earth are their back ends next to each other?

PICTURE (*a*) The length of each spaceship as measured on Earth is the contracted length $\sqrt{1 - (u^2/c^2)}L_p$ (Equation 39-14), where u is the speed of either spaceship relative to Earth. To solve Part (*b*), let Earth be at rest in frame S, and let the spaceship on the left (spaceship 1) be at rest in frame S', which is moving at speed $v = 0.85c$ relative to S. Then the spaceship on the right (spaceship 2) moves with velocity $u_{1x} = -0.85c$ (Figure 39-10). (*c*) The length of spaceship 2 as seen by an observer at rest relative to spaceship 1 is $\sqrt{1 - (u_{2x}^2/c^2)}L_p$.

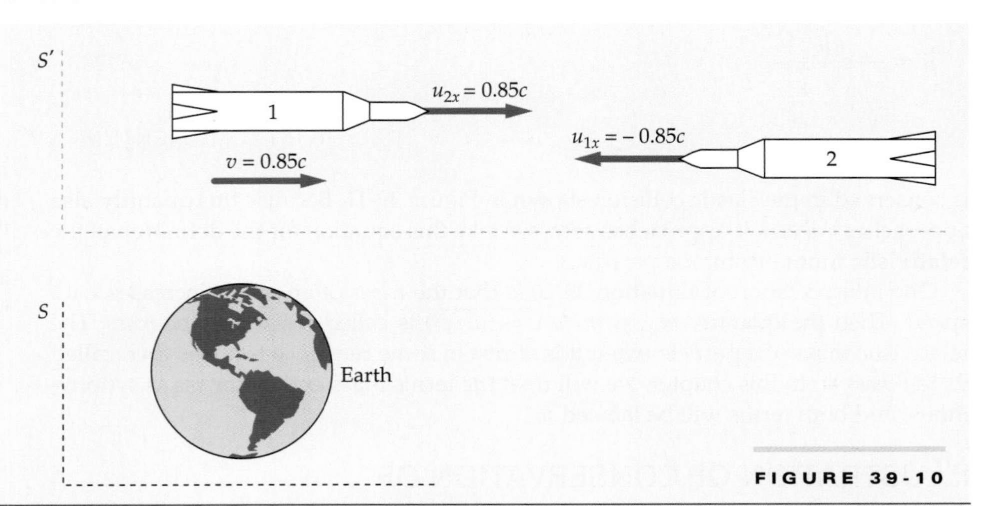

FIGURE 39-10

SOLVE

(*a*) The length of each spaceship in S, the reference frame of Earth, is the proper length divided by γ.

$$L = \frac{1}{\gamma}L_p = \sqrt{1 - \frac{|u_{2x}|^2}{c^2}}L_p = \sqrt{1 - \frac{(0.85c)^2}{c^2}}(100\text{ m}) = \boxed{53\text{ m}}$$

(*b*) Use the velocity transformation formula (Equation 39-19*a*) to find the velocity u'_{2x} of spaceship 2 as seen in frame S':

$$u'_{2x} = \frac{u_{2x} - v}{1 - \frac{vu_{2x}}{c^2}} = \frac{-0.85c - 0.85c}{1 - \frac{(0.85c)(-0.85c)}{c^2}} = \frac{-1.70c}{1.7225} = -0.987c$$

so

$$|u'_{2x}| = \boxed{0.99c}$$

(*c*) In the frame of spaceship 1, spaceship 2 is moving with speed $|u'| = 0.987c$. Use this to calculate the length of spaceship 2 as seen by an observer at rest relative to spaceship 1:

$$L = \frac{1}{\gamma}L_p = \sqrt{1 - \frac{|u'_{2x}|^2}{c^2}}L_p = \sqrt{1 - \frac{(0.987c)^2}{c^2}}(100\text{ m}) = \boxed{16\text{ m}}$$

(*d*) If the front ends of the spaceships are together at $t = 0$ on Earth, their back ends will be together after the time it takes either spaceship to move the length of the spaceship in Earth's frame:

$$t = \frac{L}{u} = \frac{53\text{ m}}{0.85c} = \frac{53\text{ m}}{(0.85)(3.00 \times 10^8\text{ m/s})} = \boxed{2.1 \times 10^{-7}\text{ s}}$$

CHECK As expected, the Part (*c*) result is less than the Part (*a*) result, and both results are less than the proper length of 100 m.

39-6 RELATIVISTIC MOMENTUM

We have seen in previous sections that Einstein's postulates require important modifications in our ideas of simultaneity and in our measurements of time and length. Einstein's postulates also require modifications in our concepts of mass, momentum, and energy. In classical mechanics, the momentum of a particle is defined as the product of its mass and its velocity, $m\vec{u}$, where $\vec{u}$ is the velocity. In an isolated system of particles, with no net force acting on the system, the total momentum of the system remains constant.

The reason that the total momentum of a system is important in classical mechanics is that it is conserved when there are no external forces acting on the system, as is the case in collisions. But we have just seen that $\Sigma m_i\vec{u}_i$ is conserved only in the approximation that $u \ll c$. We will define the relativistic momentum $\vec{p}$ of a particle to have the following properties:

1. In collisions, $\vec{p}$ is conserved.
2. As u/c approaches zero, $\vec{p}$ approaches $m\vec{u}$.

We will show that the quantity

$$\vec{p} = \frac{m\vec{u}}{\sqrt{1 - \frac{u^2}{c^2}}} \qquad 39\text{-}20$$

RELATIVISTIC MOMENTUM

is conserved in the elastic collision shown in Figure 39-11. Because this quantity also approaches $m\vec{u}$ as u/c approaches zero, we take this equation for the definition of the **relativistic momentum** of a particle.

One interpretation of Equation 39-20 is that the mass of an object increases with speed. Then the quantity $m_{\text{rel}} = m/\sqrt{1 - (u^2/c^2)}$ is called the *relativistic mass*. The relativistic mass of a particle when it is at rest in some reference frame is then called its *rest mass* m. In this chapter, we will treat the terms mass and rest mass as synonymous, and both terms will be labeled m.

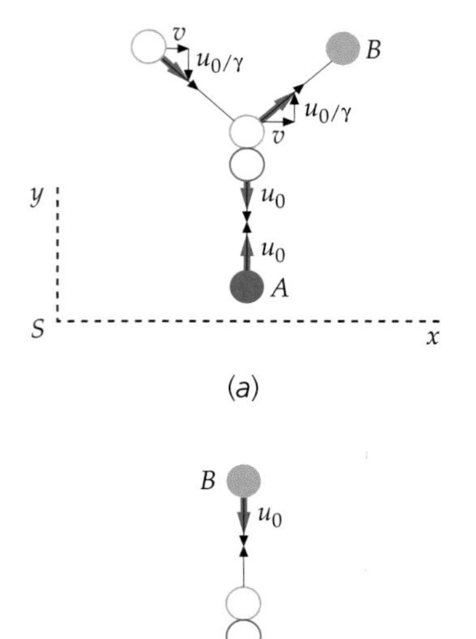

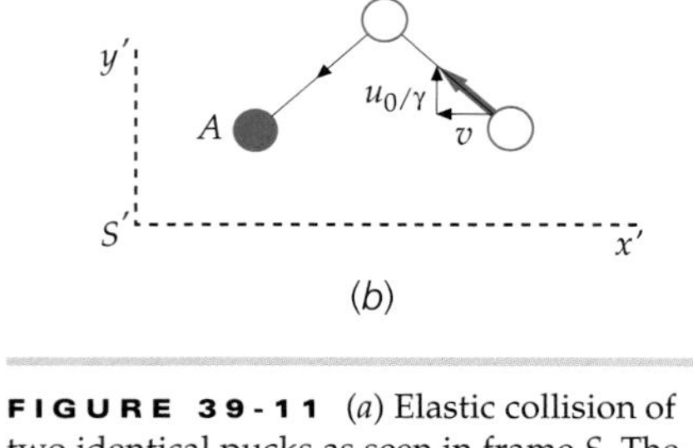

FIGURE 39-11 (*a*) Elastic collision of two identical pucks as seen in frame S. The vertical component of the velocity of puck B is u_0/γ in S if it is u_0 in S'. (*b*) The same collision as seen in S'. In this frame, puck A has a vertical component of velocity equal to u_0/γ.

ILLUSTRATION OF CONSERVATION OF THE RELATIVISTIC MOMENTUM

We consider two observers: one observer at rest in reference frame S, and the other observer at rest in frame S', which is moving to the right in the $+x$ direction with speed v relative to frame S. Each has a puck of mass m that can slide freely across a flat horizontal surface. The two pucks are identical when compared at rest. One observer launches puck A in the $+y$ direction with a speed u_0 relative to himself and the other launches puck B in the $-y$ direction with a speed u_0 relative to himself, so that each puck makes an elastic collision with the other puck, and returns to the person that launched it. Figure 39-11 shows how the collision looks in each reference frame.

We will compute the y component of the relativistic momentum of each puck in the reference frame S for the collision and show that the y component of the total relativistic momentum is zero. The speed of puck A in S is u_0, so the y component of its relativistic momentum is

$$p_{Ay} = \frac{mu_0}{\sqrt{1 - (u_0^2/c^2)}}$$

The speed of puck B in S is more complicated. Its x component is v and its y component is $-u_0/\gamma$ (Equation 39-18*b*). Thus,

$$u_B^2 = u_{Bx}^2 + u_{By}^2 = v^2 + \left[-u_0\sqrt{1 - (v^2/c^2)}\right]^2 = v^2 + u_0^2 - \frac{u_0^2 v^2}{c^2}$$

Using this result to compute $\sqrt{1 - (u_B^2/c^2)}$ we obtain

$$1 - \frac{u_B^2}{c^2} = 1 - \frac{v^2}{c^2} - \frac{u_0^2}{c^2} + \frac{u_0^2 v^2}{c^4} = \left(1 - \frac{v^2}{c^2}\right)\left(1 - \frac{u_0^2}{c^2}\right)$$

and

$$\sqrt{1 - (u_B^2/c^2)} = \sqrt{1 - (v^2/c^2)}\,\sqrt{1 - (u_0^2/c^2)} = \left(\frac{1}{\gamma}\right)\sqrt{1 - (u_0^2/c^2)}$$

The y component of the relativistic momentum of puck B as seen in S is therefore

$$p_{By} = \frac{mu_{By}}{\sqrt{1 - (u_B^2/c^2)}} = \frac{-mu_0/\gamma}{(1/\gamma)\sqrt{1 - (u_0^2/c^2)}} = \frac{-mu_0}{\sqrt{1 - (u_0^2/c^2)}}$$

Because $p_{By} = -p_{Ay}$, the y component of the total momentum of the two pucks is zero. If the y component of the momentum of each puck is reversed by the collision, the total momentum will remain zero and momentum will be conserved.

39-7 RELATIVISTIC ENERGY

In classical mechanics, the work done by the net force acting on a particle equals the change in the kinetic energy of the particle. In relativistic mechanics, we equate the net force to the rate of change of the relativistic momentum. The work done by the net force can then be calculated and set equal to the change in kinetic energy.

As in classical mechanics, we will define kinetic energy as the work done by the net force in accelerating a particle from rest to some final velocity u_f. Considering one dimension only, we have

$$K = \int_{u=0}^{u=u_f} F_{\text{net}}\, ds = \int_{u=0}^{u-u_f} \frac{dp}{dt}\, ds = \int_{u=0}^{u-u_f} u\, dp = \int_{u=0}^{u-u_f} u\, d\left(\frac{mu}{\sqrt{1 - (u^2/c^2)}}\right) \qquad 39\text{-}21$$

where we have used $u = ds/dt$. It is left as a problem (Problem 35) for you to show that

$$d\left(\frac{mu}{\sqrt{1 - (u^2/c^2)}}\right) = m\left(1 - \frac{u^2}{c^2}\right)^{-3/2} u\, du$$

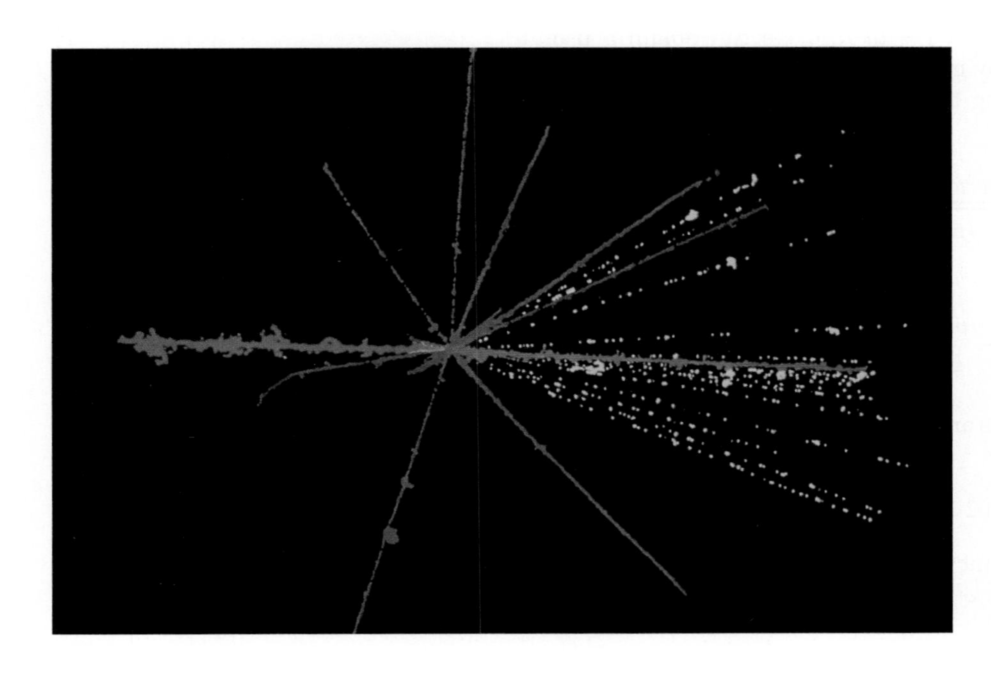

The creation of elementary particles demonstrates the conversion of kinetic energy to rest energy. In this 1950 photograph of a cosmic ray shower, a high-energy sulfur nucleus (red) collides with a nucleus in a photographic emulsion and produces a spray of particles, including a fluorine nucleus (green), other nuclear fragments (blue), and approximately 16 pions (yellow). *(© C. Powell, P. Fowler, and D. Perkins. Science Photo Library/Photo Researchers.)*

If we substitute that expression into the integrand in Equation 39-21, we obtain

$$K = \int_{u=0}^{u=u_f} u\, d\left(\frac{mu}{\sqrt{1-(u^2/c^2)}}\right) = \int_0^{u_f} m\left(1-\frac{u^2}{c^2}\right)^{-3/2} u\,du$$

$$= mc^2\left(\frac{1}{\sqrt{1-(u_f^2/c^2)}} - 1\right)$$

or

$$K = \frac{mc^2}{\sqrt{1-(u^2/c^2)}} - mc^2 \qquad 39\text{-}22$$

RELATIVISTIC KINETIC ENERGY

(In this expression the final speed u_f is arbitrary, so the subscript f is not needed.)

The expression for kinetic energy consists of two terms. The first term depends on the speed of the particle. The second term, mc^2, is independent of the speed. The quantity mc^2 is called the **rest energy** E_0 of the particle. The rest energy is the product of the mass and c^2:

$$E_0 = mc^2 \qquad 39\text{-}23$$

REST ENERGY

The total **relativistic energy** E is then defined to be the sum of the kinetic energy and the rest energy:

$$E = K + mc^2 = \frac{mc^2}{\sqrt{1-(u^2/c^2)}} \qquad 39\text{-}24$$

RELATIVISTIC ENERGY

Thus, the work done by an unbalanced force increases the energy from the rest energy mc^2 to the final energy $mc^2/\sqrt{1-(u^2/c^2)}$. We can obtain a useful expression for the velocity of a particle by multiplying Equation 39-20 for the relativistic momentum by c^2 and comparing the result with Equation 39-24 for the relativistic energy. We have

$$pc^2 = \frac{mc^2u}{\sqrt{1-(u^2/c^2)}} = Eu$$

Dividing both sides by cE gives

$$\frac{u}{c} = \frac{pc}{E} \qquad 39\text{-}25$$

Energies in atomic and nuclear physics are usually expressed in units of electron volts (eV) or mega-electron volts (MeV):

$$1\text{ eV} = 1.602 \times 10^{-19}\text{ J}$$

A convenient unit for the masses of atomic particles is eV/c^2 or MeV/c^2, which is the rest energy of the particle divided by c^2. The rest energies of some elementary particles and light nuclei are given in Table 39-1.

Table 39-1 Rest Energies of Some Elementary Particles and Light Nuclei

Particle	Symbol	Rest energy, MeV
Photon	γ	0
Electron (positron)	e or $e^-(e^+)$	0.5110
Muon	$\mu^\pm$	105.7
Pion	π^0	135.0
	$\pi^\pm$	139.6
Proton	^{1}H or p	938.272
Neutron	n	939.565
Deuteron	^{2}H or d	1875.613
Triton	^{3}H or t	2808.920
Helion	^{3}He or h	2808.391
Alpha particle	^{4}He or α	3727.379

Example 39-11 Total Energy, Kinetic Energy, and Momentum

An electron (rest energy 0.511 MeV) moves with speed $u = 0.800c$. Find (*a*) its total energy, (*b*) its kinetic energy, and (*c*) the magnitude of its momentum.

PICTURE This problem involves substituting into Equations 39-20 to 39-25.

SOLVE

(*a*) The total energy is given by Equation 39-24:

$$E = \frac{mc^2}{\sqrt{1-(u^2/c^2)}} = \frac{0.511\text{ MeV}}{\sqrt{1-0.64}} = \frac{0.511\text{ MeV}}{0.6} = \boxed{0.852\text{ MeV}}$$

(*b*) The kinetic energy is the total energy minus the rest energy:

$$K = E - mc^2 = 0.852\text{ MeV} - 0.511\text{ MeV} = \boxed{0.341\text{ MeV}}$$

(*c*) The magnitude of the momentum is found from Equation 39-20. We can simplify the momentum expression by multiplying both numerator and denominator by c^2 and using the Part-(*a*) result:

$$p = \frac{mu}{\sqrt{1-(u^2/c^2)}}$$

$$= \frac{mc^2}{\sqrt{1-(u^2/c^2)}}\frac{u}{c^2} = (0.852\text{ MeV})\frac{0.8c}{c^2} = \boxed{0.682\text{ MeV}/c}$$

CHECK The kinetic energy is less than the total energy as expected.

TAKING IT FURTHER The technique used to solve Part (*c*) (multiplying numerator and denominator by c^2) is equivalent to using Equation 39-25.

The expression for kinetic energy given by Equation 39-22 does not look much like the classical expression $\frac{1}{2}mu^2$. However, when u is much less than c, we can approximate $1/\sqrt{1-(u^2/c^2)}$ using the binomial expansion

$$(1+x)^n = 1 + nx + \frac{n(n-1)}{2}x^2 + \cdots \approx 1 + nx \qquad x \ll 1 \qquad 39\text{-}26$$

Then

$$\frac{1}{\sqrt{1-(u^2/c^2)}} = \left(1 - \frac{u^2}{c^2}\right)^{-1/2} \approx 1 + \frac{1}{2}\frac{u^2}{c^2} \qquad u \ll c$$

See **Math Tutorial** *for more information on the* **Binomial Expansion**

From this result, when u is much less than c, the expression for relativistic kinetic energy becomes

$$K = mc^2\left[\frac{1}{\sqrt{1-(u^2/c^2)}} - 1\right] \approx mc^2\left[1 + \frac{1}{2}\frac{u^2}{c^2} - 1\right] = \frac{1}{2}mu^2 \qquad u \ll c$$

Thus, at low speeds, the relativistic expression is the same as the classical expression.

We note from Equation 39-24 that as the speed u approaches the speed of light c, the energy of the particle becomes very large (because $1/\sqrt{1-(u^2/c^2)}$ becomes very large). At $u = c$, the energy becomes infinite. A simple interpretation of the result is that it takes an infinite amount of energy to accelerate a particle (that has mass) to the speed of light.

In practical applications, the momentum or energy of a particle is often known rather than the speed. Equation 39-20 for the relativistic momentum and Equation 39-24 for the relativistic energy can be combined to eliminate the speed u. The result is

$$E^2 = p^2c^2 + (mc^2)^2 \qquad 39\text{-}27$$

RELATION FOR TOTAL ENERGY, MOMENTUM, AND REST ENERGY

This useful equation can be conveniently remembered from the right triangle shown in Figure 39-12.

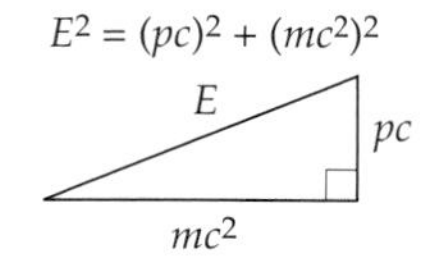

FIGURE 39-12 Right triangle to remember Equation 39-27.

PRACTICE PROBLEM 39-2

A proton (mass equal to 938 MeV/c^2) moving at speed u has a total energy of 1400 MeV. Find (*a*) $1/\sqrt{1-(u^2/c^2)}$, (*b*) the momentum of the proton, and (*c*) the speed u of the proton.

If the energy of a particle is much greater than its rest energy mc^2, the second term on the right side of Equation 39-27 can be neglected, giving the useful approximation

$$E \approx pc \qquad E \gg mc^2 \qquad 39\text{-}28$$

Equation 39-28 is an exact relation between energy and momentum for particles that do not have mass, such as photons.

MASS AND ENERGY

Einstein considered the relation $E_0 = mc^2$ (Equation 39-23) relating the energy of a particle to its mass to be the most significant result of the theory of relativity. Energy and inertia, which were formerly two distinct concepts, are related through this famous equation. As discussed in Chapter 7, the conversion of rest energy to kinetic energy with a corresponding decrease in mass is a common occurrence in radioactive decay and nuclear reactions, including nuclear fission and nuclear fusion. We illustrated this in Section 7-4 with the deuteron, whose mass is 2.22 MeV/c^2 less than the mass of its parts—a proton and a neutron. When a neutron and a proton combine to form a deuteron, 2.22 MeV of energy is released. The breaking up of a deuteron into a neutron and a proton requires 2.22 MeV of energy input. The proton and the neutron are thus bound together in a deuteron by a binding energy of 2.22 MeV. Any stable composite particle, such as a deuteron or an alpha particle (2 neutrons plus 2 protons), that is made up of other particles has a mass and rest energy that are less than the sum of the masses and rest energies of its parts. The difference in these rest energies is the binding energy of the composite particle. The binding energies of atoms and molecules are of the order of a few electron volts, which explains why there is only a negligible difference in mass between the composite particle and its parts. The binding energies of nuclei are of the order of several MeV, which explains why there is a noticeable difference in mass between the composite particle and its parts. Some very heavy nuclei, such as radium, are radioactive and decay into a less massive nucleus plus an alpha particle. In this case, the original nucleus has a rest energy greater than that of the decay particles. The excess energy appears as the kinetic energy of the decay products.

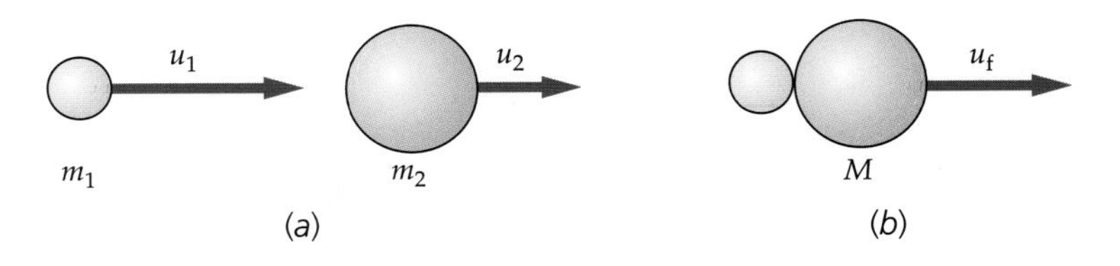

FIGURE 39-13 A perfectly inelastic collision between two particles. One particle of mass m_1 collides with another particle of mass m_2. After the collision, the particles stick together, forming a composite particle of mass M that moves with speed u_f so that relativistic momentum is conserved. Kinetic energy is lost in the process. If we assume that the total energy is conserved, the loss in kinetic energy must equal c^2 multiplied by the increase in the mass of the system.

To further illustrate the connection between mass and energy, we consider a perfectly inelastic collision of two particles. Classically, kinetic energy is lost during such a collision. Relativistically, this loss in kinetic energy shows up as an increase in rest energy of the system; that is, the total energy of the system is conserved. Consider a particle of mass m_1 moving with initial speed u_1 that collides with a particle of mass m_2 moving with initial speed u_2. The particles collide and stick together, forming a particle of mass M that moves with speed u_f, as shown in Figure 39-13. The initial total energy of particle 1 is

$$E_1 = K_1 + m_1c^2$$

where K_1 is its initial kinetic energy. Similarly the initial total energy of particle 2 is

$$E_2 = K_2 + m_2c^2$$

The total initial energy of the system is

$$E_i = E_1 + E_2 = K_1 + m_1c^2 + K_2 + m_2c^2 = K_i + M_ic^2$$

where $K_i = K_1 + K_2$ and $M_i = m_1 + m_2$ are the initial kinetic energy and initial mass of the system. The final total energy of the system is

$$E_f = K_f + M_fc^2$$

If we set the final total energy equal to the initial total energy, we obtain

$$K_f + M_fc^2 = K_i + M_ic^2$$

Rearranging gives $K_f - K_i = -(M_f - M_i)c^2$, which can be expressed

$$\Delta K + (\Delta M)c^2 = 0 \qquad \text{39-29}$$

where $\Delta M = M_f - M_i$ is the change in mass of the system.

Example 39-12 Totally Inelastic Collision

A particle of mass 2.00 MeV/c^2 and kinetic energy 3.00 MeV collides with a stationary particle of mass 4.00 MeV/c^2. After the collision, the two particles stick together. Find (*a*) the magnitude of the initial momentum of the system, (*b*) the final velocity of the two-particle system, and (*c*) the mass of the two-particle system.

PICTURE (*a*) The initial momentum of the system is the initial momentum of the incoming particle, which can be found from the total energy of the particle. (*b*) The final velocity of the system can be found from its total energy and momentum using $u/c = pc/E$ (Equation 39-25). The energy is found from conservation of energy, and the momentum from conservation of momentum. (*c*) Because the final energy and momentum are known, the final mass can be found using $E^2 = p^2c^2 + (mc^2)^2$.

SOLVE

(*a*) 1. The initial momentum of the system is the initial momentum of the incoming particle. The momentum of a particle is related to its energy and mass (Equation 39-27):

$$E_1^2 = p_1^2c^2 + (m_1c^2)^2$$
$$p_1c = \sqrt{E_1^2 - (m_1c^2)^2}$$

2. The total energy of the moving particle is the sum of its kinetic energy and its rest energy:

$$E_1 = 3.00\text{ MeV} + 2.00\text{ MeV} = 5.00\text{ MeV}$$

3. Use the total energy to calculate the magnitude of the momentum:

$$p_1c = \sqrt{E_1^2 - (m_1c^2)^2} = \sqrt{(5.00\text{ MeV})^2 - (2.00\text{ MeV})^2} = \sqrt{21.0}\text{ MeV}$$
$$p_1 = \boxed{4.58\text{ MeV}/c}$$

(*b*) 1. We can find the final velocity of the system from its total energy E_f and its momentum p_f using Equation 39-25:

$$\frac{u_f}{c} = \frac{p_fc}{E_f}$$

2. By the conservation of total energy, the final energy of the system equals the initial total energy of the two particles:

$$E_f = E_i = E_1 + E_2 = 5.00\text{ MeV} + 4.00\text{ MeV} = 9.00\text{ MeV}$$

3. By the conservation of momentum, the final momentum of the two-particle system equals the initial momentum:

$$p_f = 4.58\text{ MeV}/c$$

4. Calculate the velocity of the two-particle system from its total energy and momentum using $u/c = pc/E$:

$$\frac{u_f}{c} = \frac{p_fc}{E_f} = \frac{4.58\text{ MeV}}{9.00\text{ MeV}} = 0.509$$
$$u_f = \boxed{0.509c}$$

(*c*) We can find the mass M_f of the final two-particle system from Equation 39-27 using $pc = 4.58$ MeV and $E = 9.00$ MeV:

$$E_f^2 = (p_fc)^2 + (M_fc^2)^2$$
$$(9.00\text{ MeV})^2 = (4.58\text{ MeV})^2 + (M_fc^2)^2$$
$$M_f = \boxed{7.75\text{ MeV}/c^2}$$

TAKING IT FURTHER Note that the mass of the system increased from 6.00 MeV/c^2 to 7.75 MeV/c^2. This mass increase, multiplied by c^2, equals the loss in kinetic energy of the system, as you will show in the following exercise.

PRACTICE PROBLEM 39-3 (*a*) Find the final kinetic energy of the two-particle system in Example 39-12. (*b*) Find the loss in kinetic energy, K_{loss}, in the collision. (*c*) Show that $K_{loss} = (\Delta M)c^2$, where ΔM is the change in mass of the system.

Example 39-13 Momentum and Total-Energy Conservation

A 1.00×10^6-kg rocket has 1.00×10^3 kg of fuel on board. The rocket is parked in space when it suddenly becomes necessary to accelerate. The rocket engines ignite, and the 1.00×10^3 kg of fuel are consumed. The exhaust (spent fuel) is ejected during a very short time interval at a speed of 0.500*c* relative to *S*—the inertial reference frame in which the rocket is initially at rest. (*a*) Calculate the change in the mass of the rocket–fuel system. (*b*) Calculate the final speed of the rocket u_R relative to *S*. (*c*) Again, calculate the final speed of the rocket relative to *S*, this time using classical (Newtonian) mechanics.

PICTURE The speed of the rocket and the change in the mass of the system can be calculated using conservation of momentum and conservation of energy. In reference frame *S*, the total momentum of the rocket plus fuel remains zero. After the burn, the magnitude of the momentum of the rocket equals that of the ejected fuel. Let $m_R = 1.00 \times 10^6$ kg be the mass of the rocket, not including the mass of the fuel, let $m_{Fi} = 1.00 \times 10^3$ kg be the mass of the fuel *before* the burn, and let m_{Ff} be the mass of the fuel *after* the burn. The mass of the rocket, m_R, remains fixed, but during the burn the mass of the fuel decreases. (The fuel has less chemical energy after the burn, and so has less mass as well.)

SOLVE

(*a*) 1. The magnitudes of the momentum of the rocket and the momentum of the ejected fuel are equal. For the reasons stated above, the mass of the rocket, not including the 1.00×10^3 kg of fuel, does not change during the burn:

$$p_R = p_F = p$$

$$\frac{m_R u_R}{\sqrt{1 - (u_R^2/c^2)}} = \frac{m_{Ff} u_F}{\sqrt{1 - (u_F^2/c^2)}} = p$$

$m_R = 1.00 \times 10^6$ kg, $u_F = 0.500c$, and u_R is the final speed of the rocket.

2. The total energy of the system does not change:

$$E_f = E_i$$

3. The initial energy is the rest energy of the rocket and fuel before the burn. The final energy is the energy of the rocket plus energy of the fuel. The energy of each is related to its momentum by Equation 39-27:

$$E_i = m_R c^2 + m_{Fi} c^2 = (m_R + m_{Fi})c^2$$

$$E_{Rf}^2 = p^2c^2 + (m_R c^2)^2$$

$$E_{Ff}^2 = p^2c^2 + (m_{Ff} c^2)^2$$

so

$$E_f = E_{Rf} + E_{Ff}$$

$$E_f = \sqrt{p^2c^2 + (m_R c^2)} + \sqrt{p^2c^2 + (m_{Ff}c^2)^2}$$

4. Equate the initial and final energies:

$$\sqrt{p^2c^2 + (m_R c^2)^2} + \sqrt{p^2c^2 + (m_{Ff}c^2)^2} = (m_R + m_{Fi})c^2$$

5. The step-4 result and $p = \frac{m_{Ff} u_F}{1 - (u_F^2/c^2)}$ (the step-1 result) constitute two simultaneous equations with unknowns p and m_{Ff}. Solving for m_{Ff} gives:

$$m_{Ff} = 866 \text{ kg}$$

so

$$m_{\text{loss}} = m_{Fi} = 1000 \text{ kg} - 866 \text{ kg} = \boxed{134 \text{ kg}}$$

(*b*) 1. To solve for u_R, we use Equation 39-25:

$$\frac{u_R}{c} = \frac{pc}{E_{Rf}}$$

2. To solve for p, we substitute the value for m_{Ff} into the Part (*a*), step-1 result:

$$p = \frac{m_{Ff} u_F}{\sqrt{1 - (u_F^2/c^2)}} = \frac{(866 \text{ kg})0.500c}{\sqrt{1 - 0.250}}$$

$$= (5.00 \times 10^2 \text{ kg})c$$

3. We use the value for p to solve for E_{Rf}:

$$E_{Rf}^2 = p^2c^2 + (m_R c^2)^2$$

$$= (5.00 \times 10^2 \text{ kg})^2c^4 + (1.00 \times 10^6 \text{ kg})^2c^4$$

$$= (1.00 \times 10^{12} \text{ kg}^2)c^4$$

so

$$E_{Rf} = (1.00 \times 10^6 \text{ kg})c^2$$

4. Using our Part (*b*), step-1 result, we solve for u_R:

$$u_R = \frac{pc^2}{E_{Rf}} = \frac{(5.00 \times 10^2 \text{ kg})c^3}{(1.00 \times 10^6 \text{ kg})c^2}$$

$$= \boxed{5.00 \times 10^{-4}c = 1.50 \times 10^{-5} \text{ m/s}}$$

(*c*) Equate the magnitude of the classical expressions for the momentum of the rocket and burned fuel and solve for u_R:

$$m_R m_R = m_F u_F$$

$$u_R = \frac{m_F}{m_R} u_F = \frac{1.00 \times 10^3 \text{ kg}}{1.00 \times 10^6 \text{ kg}} 0.500c$$

$$= 5.00 \times 10^{-4}c$$

$$= \boxed{1.50 \times 10^5 \text{ m/s}}$$

CHECK We find the result of the relativistic calculation of the final rocket speed to differ from the classical result. If carried out to five figures, the relativistic calculation gives $u_R = 4.9994 \times 10^{-4}\, c$ for the final speed of the rocket. However, the classical calculation gives $u_R = 5.0000 \times 10^{-4}\, c$. These two values differ by less than one part in 8000.

CONCEPT CHECK 39-1

If the matter being ejected were a 1.00×10^3-kg rigid block launched by a spring with one end attached to the rocket, would the mass of the block change or would the mass of the spring change?

39-8 GENERAL RELATIVITY

The generalization of the theory of relativity to noninertial reference frames by Einstein in 1916 is known as the general theory of relativity. It is much more difficult mathematically than the special theory of relativity, and there are fewer situations in which it can be tested. Nevertheless, its importance calls for a brief qualitative discussion.

The basis of the general theory of relativity is the **principle of equivalence:**

> A homogeneous gravitational field is completely equivalent to a uniformly accelerated reference frame.
>
> PRINCIPLE OF EQUIVALENCE

This principle arises in Newtonian mechanics because of the apparent identity of gravitational mass and inertial mass. In a uniform gravitational field, all objects fall with the same acceleration $\vec{g}$ independent of their masses because the gravitational force is proportional to the (gravitational) mass, whereas the acceleration varies inversely with the (inertial) mass. Consider a compartment in space undergoing a uniform acceleration $\vec{a}$, as shown in Figure 39-14*a*. No mechanics experiment can be performed inside the compartment that will distinguish whether the compartment is actually accelerating in space or is at rest (or is moving with uniform velocity) in the presence of a uniform gravitational field $\vec{g} = -\vec{a}$, as shown in Figure 39-14*b*. If objects are dropped in the compartment, they will fall to the floor with an acceleration $\vec{g} = -\vec{a}$. If people stand on a spring scale, it will read their weight of magnitude $ma = mg$.

Einstein assumed that the principle of equivalence applies to all physics and not just to mechanics. In effect, he assumed that there is no experiment of any kind that can distinguish uniformly accelerated motion from the presence of a gravitational field.

One consequence of the principle of equivalence—the deflection of a light beam in a gravitational field—was one of the first to be tested experimentally. In a region that has no gravitational field, a light beam will travel in a straight line at speed c. The principle of equivalence tells us that a region that has no gravitational field exists only in a compartment that is in free fall. Figure 39-15 shows a beam of light entering a compartment that is accelerating relative to a nearby reference frame in free fall. Successive positions of the compartment at equal time intervals are shown in Figure 39-15*a*. Because the compartment is accelerating, the distance it moves in each time interval increases with time. The path of the beam of light as observed from inside the compartment is therefore a parabola, as shown in Figure 39-15*b*.

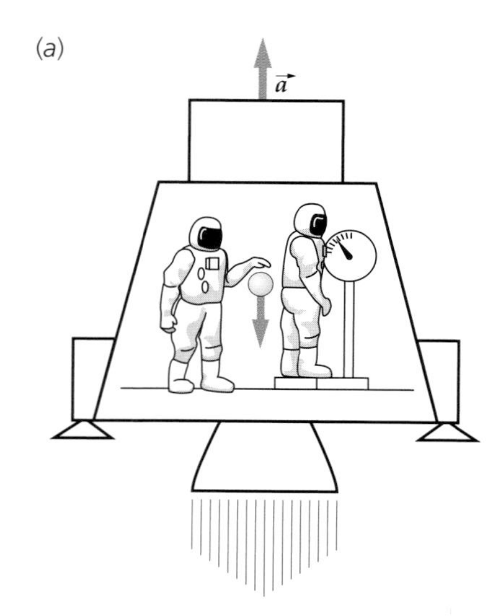

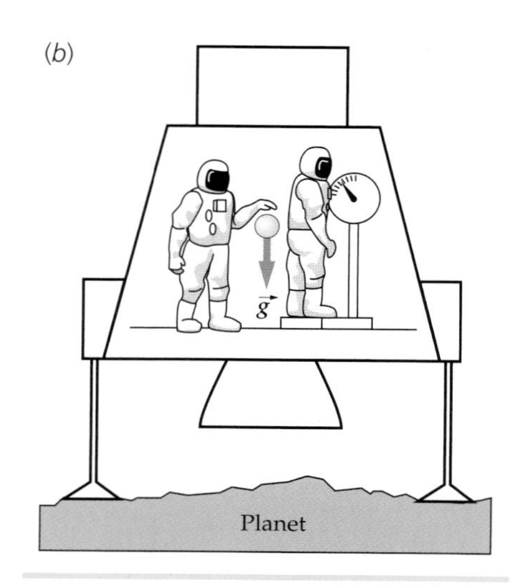

FIGURE 39-14 The results of experiments in a uniformly accelerated reference frame (*a*) cannot be distinguished from those in a uniform gravitational field (*b*) if the acceleration $\vec{a}$ and the gravitational field $\vec{g}$ have the same magnitude.

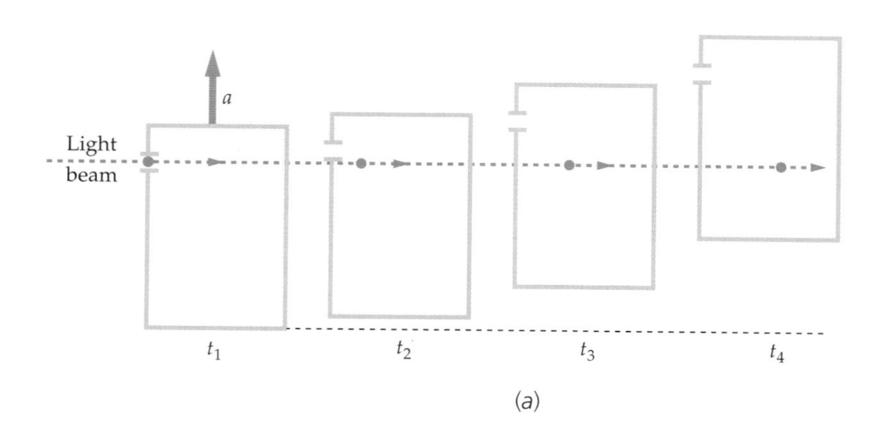

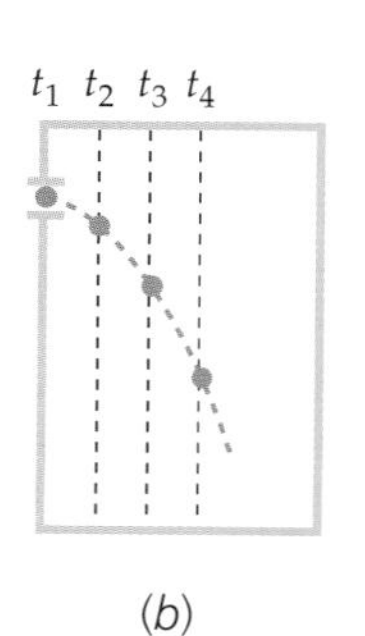

FIGURE 39-15 (*a*) A light beam moving in a straight line through a compartment that is undergoing uniform acceleration relative to a nearby reference frame in free fall. The position of the beam is shown at equally spaced times t_1, t_2, t_3, and t_4. (*b*) In the reference frame of the compartment, the light travels in a parabolic path as a ball would if it were projected horizontally. The vertical displacements are greatly exaggerated for emphasis.

But according to the principle of equivalence, there is no way to distinguish between an accelerating compartment and one moving with uniform velocity in a uniform gravitational field. We conclude, therefore, that a beam of light will accelerate in a gravitational field, just like objects that have mass. For example, near the surface of Earth, light will fall with an acceleration of 9.81 m/s^2. This is difficult to observe because of the enormous speed of light. In a distance of 3000 km, which takes light about 0.01 s to traverse, a beam of light should fall approximately 0.5 mm. Einstein pointed out that the deflection of a light beam in a gravitational field might be observed when light from a distant star passes close to the Sun, as illustrated in Figure 39-16. Because of the brightness of the Sun, this cannot ordinarily be seen. Such a deflection was first observed in 1919 during an eclipse of the Sun. This well-publicized observation brought instant worldwide fame to Einstein.

A second prediction from Einstein's theory of general relativity, which we will not discuss in detail, is the excess precession of the perihelion of the orbit of Mercury of about 0.01° per century. This effect had been known and unexplained for some time, so, in a sense, explaining it constituted an immediate success of the theory.

A third prediction of general relativity concerns the change in time intervals and frequencies of light in a gravitational field. In Chapter 11, we found that the gravitational potential energy between two masses M and m a distance r apart is

$$U = -\frac{GMm}{r}$$

where G is the universal gravitational constant, and the point of zero potential energy has been chosen to be when the separation of the masses is infinite. The potential energy per unit mass near a mass M is called the *gravitational potential* ϕ:

$$\phi = -\frac{GM}{r} \qquad 39\text{-}30$$

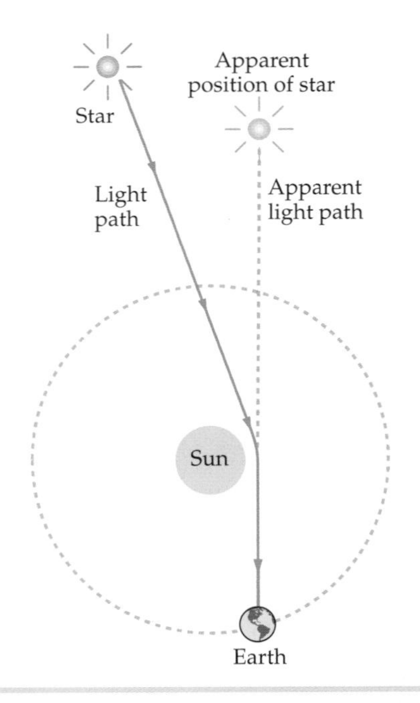

FIGURE 39-16 The deflection (greatly exaggerated) of a beam of light due to the gravitational attraction of the Sun.

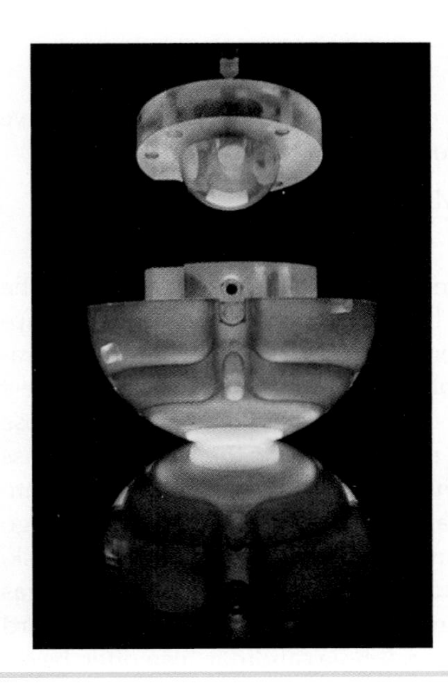

The quartz sphere in the top part of the container is probably the world's most perfectly round object. It is designed to spin as a gyroscope in a satellite orbiting Earth. General relativity predicts that the rotation of Earth will cause the axis of rotation of the gyroscope to precess in a circle at a rate of approximately 1 revolution in 100 000 years. *(Michael Freeman.)*

According to the general theory of relativity, clocks run more slowly in regions of lower gravitational potential. (Because the gravitational potential is negative, as can be seen from Equation 39-30, the nearer the mass the more negative, and therefore, the lower the gravitational potential.) If Δt_1 is a time interval between two events measured by a clock where the gravitational potential is ϕ_1 and Δt_2 is the interval between the same two events as measured by a clock where the gravitational potential is ϕ_2, general relativity predicts that the fractional difference between the times will be approximately*

$$\frac{\Delta t_2 - \Delta t_1}{\Delta t} = \frac{1}{c^2}(\phi_2 - \phi_1) \qquad 39\text{-}31$$

A clock in a region of low gravitational potential will therefore run more slowly than a clock in a region of higher gravitional potential. Because a vibrating atom can be considered to be a clock, the frequency of vibration of an atom in a region of low potential, such as near the Sun, will be lower than the frequency of vibration of the same atom on Earth. This shift toward a lower frequency, and therefore a longer wavelength, is called the **gravitational redshift.**

This extremely accurate hydrogen maser clock was launched in a satellite in 1976, and its time was compared to that of an identical clock on Earth. In accordance with the prediction of general relativity, the clock on Earth, where the gravitational potential was lower, lost approximately 4.3×10^{-10} s each second compared with the clock orbiting Earth at an altitude of approximately 10000 km. *(NASA.)*

As our final example of the predictions of general relativity, we mention **black holes,** which were first predicted by J. Robert Oppenheimer and Hartland Snyder in 1939. According to the general theory of relativity, if the density of an object such as a star is great enough, its gravitational attraction will be so great that once inside a critical radius, nothing can escape, not even light or other electromagnetic radiation. (The effect of a black hole on objects outside the critical radius is the same as that of any other mass.) A remarkable property of such an object is that nothing that happens inside it can be communicated to the outside. As sometimes occurs in physics, a simple but incorrect calculation gives the correct results for the relation between the mass and the critical radius of a black hole. In Newtonian mechanics, the speed needed for a particle to escape from the surface of a planet or a star of mass M and radius R is given by Equation 11-21:

$$v_e = \sqrt{\frac{2GM}{R}}$$

If we set the escape speed equal to the speed of light and solve for the radius, we obtain the critical radius R_S, called the **Schwarzschild radius:**

$$R_S = \frac{2GM}{c^2} \qquad 39\text{-}32$$

For an object that has a mass equal to five times that of our Sun (theoretically the minimum mass for a black hole) to be a black hole, its radius would have to be approximately 15 km. Because no radiation is emitted from a black hole and its radius is expected to be small, the detection of a black hole is not easy. The best chance of detection occurs in a binary-star system in which a black hole is a close companion to a normal star. Then both stars revolve around their center of mass and the gravitational field of the black hole will pull gas from the normal star into the black hole. However, to conserve angular momentum, the gas does not go straight into the black hole. Instead, the gas orbits around the black hole in a disk, called an accretion disk, while slowly being pulled closer to the black hole. The gas in this disk emits X rays because the temperature of the gas being pulled inward reaches several million kelvins. The mass of a black-hole candidate can often be estimated. An estimated mass of at least five solar masses, along with the emission of X rays, establishes a strong inference that the candidate is, in fact, a black hole. In addition to the black holes just described, there are supermassive black holes that exist at the centers of galaxies. At the center of the Milky Way is a supermassive black hole that has a mass of about two million solar masses.

* Because this shift is usually very small, it does not matter by which interval we divide on the left side of the equation.

Summary

TOPIC	RELEVANT EQUATIONS AND REMARKS	
1. Einstein's Postulates	The special theory of relativity is based on two postulates of Albert Einstein. All of the results of special relativity can be derived from these postulates. Postulate 1: Absolute uniform motion cannot be detected. Postulate 2: The speed of light is independent of the motion of the source. An important implication of these postulates is Postulate 2 (alternate): Every observer measures the same value c for the speed of light.	
2. The Lorentz Transformation	$x = \gamma(x' + vt'), \qquad y = y', \qquad z = z'$	39-9
	$t = \gamma\left(t' + \frac{vx'}{c^2}\right)$	39-10
	$\gamma = \frac{1}{\sqrt{1 - (v^2/c^2)}}$	39-7
Inverse transformation	$x' = \gamma(x - vt), \qquad y' = y, \qquad z' = z$	39-11
	$t' = \gamma\left(t - \frac{vx}{c^2}\right)$	39-12
3. Time Dilation	The time interval measured between two events that occur at the same point in space in some reference frame is called the proper time interval Δt_{p} between those two events. In another reference frame in which the same two events occur at different places, the time interval Δt between the events is longer by the factor γ. $\Delta t = \gamma\, \Delta t_{\mathrm{p}}$	39-13
4. Length Contraction	The length of an object measured in the reference frame in which the object is at rest is called its proper length L_{p}. When measured in another reference frame, the length of the object along the direction parallel to the velocity of the object is $L = \frac{L_{\mathrm{p}}}{\gamma}$	39-14
5. The Relativistic Doppler Effect	$f' = \frac{\sqrt{1 - (v^2/c^2)}}{1 - (v/c)} f_0$ approaching	39-16*a*
	$f' = \frac{\sqrt{1 - (v^2/c^2)}}{1 + (v/c)} f_0$ receding	39-16*b*
6. Clock Synchronization and Simultaneity	Two events that are simultaneous in one reference frame typically are not simultaneous in another frame that is moving relative to the first. If two clocks are synchronized in the frame in which they are at rest, they will be out of synchronization in another frame. In the frame in which they are moving, the chasing clock leads by an amount $\Delta t_{\mathrm{S}} = L_{\mathrm{p}} \frac{v}{c^2}$ where L_{p} is the proper distance between the clocks.	39-17
7. The Velocity Transformation	$u_x = \frac{u'_x + v}{1 + (vu'_x/c^2)}$	39-18*a*
	$u_y = \frac{u'_y}{\gamma[1 + (vu'_x/c^2)]}$	39-18*b*
	$u_z = \frac{u'_z}{\gamma[1 + (vu'_x/c^2)]}$	39-18*c*

TOPIC	RELEVANT EQUATIONS AND REMARKS	
Inverse velocity transformation	$u'_x = \dfrac{u_x - v}{1 - (vu_x/c^2)}$	39-19*a*
	$u'_y = \dfrac{u_y}{\gamma[1 - (vu_x/c^2)]}$	39-19*b*
	$u'_z = \dfrac{u_z}{\gamma[1 - (vu_x/c^2)]}$	39-19*c*
8. Relativistic Momentum	$\vec{p} = \dfrac{m\vec{u}}{\sqrt{1 - (u^2/c^2)}}$	39-20
	where m is the mass of the particle.	
9. Relativistic Energy		
Kinetic energy	$K = \dfrac{mc^2}{\sqrt{1 - (u^2/c^2)}} - mc^2$	39-22
Rest energy	$E_0 = mc^2$	39-23
Total Relativistic energy	$E = K + E_0 = \dfrac{mc^2}{\sqrt{1 - (u^2/c^2)}}$	39-24
10. Useful Formulas for Speed, Energy, and Momentum	$\dfrac{u}{p} = \dfrac{pc}{E}$	39-25
	$E^2 = p^2c^2 + (mc^2)^2$	39-27
	$E \approx pc \qquad E \gg mc^2$	39-28

Answer to Concept Check

39-1 Only the rest mass of the spring would change.

Answers for Practice Problems

39-1 1.67 h

39-2 (*a*) 1.49, (*b*) $p = 1.04 \times 10^3$ MeV/c, and (*c*) $u = 0.74c$

39-3 (*a*) $K_f = E_f - M_fc^2 = 9.00\text{ MeV} - 7.75\text{ MeV} = 1.25\text{ MeV}$,

(*b*) $K_{loss} = K_i - K_f = 3.00\text{ MeV} - 1.25\text{ MeV} = 1.75\text{ MeV}$, and

(*c*) $(\Delta M)c^2 = (M_f - M_i)c^2 = 7.75\text{ MeV} - (2.00\text{ MeV} + 4.00\text{ MeV}) = 1.75\text{ MeV} = K_{loss}$

Problems

In a few problems, you are given more data than you actually need; in a few other problems, you are required to supply data from your general knowledge, outside sources, or informed estimate.

Interpret as significant all digits in numerical values that have trailing zeros and no decimal points.

- • Single-concept, single-step, relatively easy
- •• Intermediate-level, may require synthesis of concepts
- ••• Challenging
- SSM Solution is in the *Student Solutions Manual*
- Consecutive problems that are shaded are paired problems.

CONCEPTUAL PROBLEMS

1 • The approximate total energy of a particle of mass m moving at speed $u \ll c$ is (a) $mc^2 + \frac{1}{2}mu^2$, (b) $\frac{1}{2}mu^2$, (c) cmu, (d) mc^2, (e) $\frac{1}{2}cmu$. SSM

2 • A set of twins work in an office building. One twin works on the top floor and the other twin works in the basement. Considering general relativity, which twin will age more quickly? (a) They will age at the same rate. (b) The twin who works on the top floor will age more quickly. (c) The twin who works in the basement will age more quickly. (d) It depends on the speed of the office building. (e) None of the above.

3 • True or false:

(a) The speed of light is the same in all reference frames.
(b) The time interval between two events is never shorter than the proper time interval between the two events.
(c) Absolute motion can be determined by means of length contraction.
(d) The light-year is a unit of distance.
(e) Simultaneous events must occur at the same place.
(f) If two events are not simultaneous in one frame, they cannot be simultaneous in any other frame.
(g) The mass of a system that consists of two particles tightly bound together by attractive forces is less than the sum of the masses of the individual particles when separated.

4 • An observer sees a system moving past her that consists of a mass oscillating on the end of a spring and measures the period T of the oscillations. A second observer, who is moving with the mass–spring system, also measures its period. The second observer will find a period that is (a) equal to T, (b) less than T, (c) greater than T, (d) either (a) or (b) depending on whether the system was approaching or receding from the first observer, (e) Not enough information is given to answer the question.

5 • The Lorentz transformation for y and z is the same as the classical result: $y = y'$ and $z = z'$. Yet the relativistic velocity transformation does not give the classical result $u_y = u'_y$ and $u_z = u'_z$. Explain why this result occurs.

ESTIMATION AND APPROXIMATION

6 •• The Sun radiates energy at the rate of approximately 4×10^{26} W. Assume that this energy is produced by a reaction whose net result is the fusion of four protons to form a single ^{4}He nucleus and the release of 25 MeV of energy that is radiated into space. Calculate the Sun's loss of mass per day.

7 •• The most distant galaxies that can be seen by the Hubble telescope are moving away from us and have a redshift parameter of about $z = 5$. [The redshift parameter z is defined as $(f - f')/f'$, where f is the frequency measured in the rest frame of the emitter and f' is the frequency measured in the rest frame of the receiver.] (a) What is the speed of the galaxies relative to us (expressed as a fraction of the speed of light)? (b) *Hubble's law* states that the recession speed is given by the expression $v = Hx$, where v is the speed of recession, x is the distance, and H, the Hubble constant, is equal to 75 km/s/Mpc, where 1 pc = 3.26 $c \cdot$y. (The abbreviation for parsec is pc.) Estimate the distance of such a galaxy from us using the information given. SSM

TIME DILATION AND LENGTH CONTRACTION

8 • The proper mean lifetime of a muon is 2.2 μs. Muons in a beam are traveling through a laboratory at $0.95c$. (a) What is their mean lifetime as measured in the laboratory? (b) How far do they travel, on average, before they decay?

9 •• In the Stanford linear collider, small bundles of electrons and positrons are fired at each other. In the laboratory's frame of reference, each bundle is approximately 1.0 cm long and 10 μm in diameter. In the collision region, each particle has an energy of 50 GeV, and the electrons and the positrons are moving in opposite directions. (a) How long and how wide is each bundle in its own reference frame? (b) What must be the minimum proper length of the accelerator for a bundle to have both its ends simultaneously in the accelerator in its own reference frame? (The actual proper length of the accelerator is less than 1000 m.) (c) What is the length of a positron bundle in a reference frame that moves with the electron bundle?

10 • Use the binomial expansion equation

$$(1 + x)^n = 1 + nx + \frac{n(n-1)}{2}x^2 + \ldots \approx 1 + nx \qquad x \ll 1$$

to derive the following results for the case when v is much less than c.

$$(a)\ \gamma \approx 1 + \frac{1}{2}\frac{v^2}{c^2}$$

$$(b)\ \frac{1}{\gamma} \approx 1 - \frac{1}{2}\frac{v^2}{c^2}$$

$$(c)\ \gamma - 1 \approx 1 - \frac{1}{\gamma} \approx \frac{1}{2}\frac{v^2}{c^2}$$

11 •• Star A and Star B are at rest relative to Earth. Star A is 27 $c \cdot$y from Earth, and as viewed from Earth, Star B is located beyond (behind) Star A. (a) A spaceship is making a trip from Earth to Star A at a speed such that the trip from Earth to Star A takes 12 y according to clocks on the spaceship. At what speed, relative to Earth, must the spaceship travel? (Assume that the times for the accelerations are very short compared to the overall trip time.) (b) Upon reaching Star A, the spaceship speeds up and departs for Star B at a speed such that the gamma factor, γ, is twice that of Part (a). The trip from Star A to Star B takes 5.0 y (spaceship's time). How far, in $c \cdot$y, is Star B from Star A in the rest frame of Earth and the two stars? (c) Upon reaching Star B, the spaceship departs for Earth at the same speed as in Part (b). It takes it 10 y (spaceship's time) to return to Earth. If you were born on Earth the day the ship left Earth and you remain on Earth, how old are you on the day the ship returns to Earth?

12 • A spaceship travels to a star 35 $c \cdot$y away at a speed of 2.7×10^8 m/s. How long does the spaceship take to get to the star (a) as measured on Earth and (b) as measured by a passenger on the spaceship?

13 •• Unobtainium (Un) is an unstable particle that decays into normalium (Nr) and standardium (St) particles. (a) An accelerator produces a beam of Un that travels to a detector located 100 m away from the accelerator. The particles travel with a velocity of $v = 0.866c$. How long do the particles take (in the laboratory frame) to get to the detector? (b) By the time the particles get to the detector, half of the particles have decayed. What is the half-life of Un? (*Note:* half-life as it would be measured in a frame moving with the particles) (c) A new detector is going to be used, which is located 1000 m away from the accelerator. How fast should the particles be moving if half of the particles are to make it to the new detector? SSM

14 •• A clock on Spaceship A measures the time interval between two events, both of which occur at the location of the clock. You are on Spaceship B. According to your careful measurements, the time interval between the two events is 1.00 percent longer than that measured by the two clocks on Spaceship A. How fast is Spaceship A moving relative to Spaceship B. (*Hint: Use one or more of the results of Problem 10.*)

15 •• If a plane flies at a speed of 2000 km/h, how long must the plane fly before its clock loses 1.00 s because of time dilation? (*Hint: Use one or more of the results of Problem 10.*)

THE LORENTZ TRANSFORMATION, CLOCK SYNCHRONIZATION, AND SIMULTANEITY

16 •• Show that when $v \ll c$ the relativistic transformation equations for x, t, and u_x reduce to the classical transformation equations.

17 •• A spaceship of proper length $L_p = 400$ m moves past a transmitting station at a speed of $0.760c$. (The transmitting station broadcasts signals that travel at the speed of light.) A clock is attached to the nose of the spaceship and a second clock is attached to the transmitting station. The instant that the nose of the spaceship passes the transmitter, the clock attached to the transmitter and the clock attached to the nose of the spaceship are set equal to zero. The instant that the tail of the spaceship passes the transmitter a signal is sent by the transmitter that is subsequently detected by a receiver in the nose of the spaceship. (*a*) When, according to the clock attached to the nose of spaceship, is the signal sent? (*b*) When, according to the clocks attached to the nose of spaceship, is the signal received? (*c*) When, according to the clock attached to the transmitter, is the signal received by the spaceship? (*d*) According to an observer that works at the transmitting station, how far from the transmitter is the nose of the spaceship when the signal is received? SSM

18 •• In frame S, event B occurs 2.0 μs after event A, and event A occurs at the origin whereas event B occurs on the x axis at $x = 1.5$ km. How fast and in what direction must an observer be traveling along the x axis so that events A and B occur simultaneously? Is it possible for event B to precede event A for some observer?

19 •• Observers in reference frame S see an explosion located on the x axis at $x_1 = 480$ m. A second explosion occurs, 5.0 μs later, at $x_2 = 1200$ m. In reference frame S', which is moving along the x axis in the $+x$ direction at speed v, the two explosions occur at the same point in space. What is the separation in time between the two explosions as measured in S'?

20 ••• In reference frame S, events 1 and 2 are separated by a distance $D = x_2 - x_1$ and a time $T = t_2 - t_1$. (*a*) Use the Lorentz transformation to show that in frame S', which is moving along the x axis with speed v relative to S, the time separation is $t'_2 - t'_1 = \gamma(T - vD/c^2)$. (*b*) Show that the events can be simultaneous in frame S' only if D is greater than cT. (*c*) If one of the events is the *cause* of the other, the separation D must be less than cT, because D/c is the smallest time that a signal can take to travel from x_1 to x_2 in frame S. Show that if D is less than cT, t'_2 is greater than t'_1 in all reference frames. This shows that if the cause precedes the effect in one frame, it must precede it in all reference frames. (*d*) Suppose that a signal could be sent with speed $c' > c$ so that in frame S the cause precedes the effect by the time $T = D/c'$. Show that there is then a reference frame moving with speed v less than c in which the effect precedes the cause.

21 ••• A rocket that has a proper length of 700 m is moving to the right at a speed of $0.900c$. It has two clocks—one in the nose and one in the tail—that have been synchronized in the frame of the rocket. A clock on the ground and the clock in the nose of the rocket both read zero as they pass by each other. (*a*) At the instant the clock on the ground reads zero, what does the clock in the tail of the rocket read according to observers on the ground? When the clock in the tail of the rocket passes the clock on the ground, (*b*) what does the clock in the tail read according to observers on the ground, and (*c*) what does the clock in the nose read according to observers on the ground, and (*d*) what does the clock in the nose read according to observers on the rocket? (*e*) At the instant the clock in the nose of the rocket reads 1.00 h, a light signal is sent from the nose of the rocket to an observer standing by the clock on the ground. What does the clock on the ground read when the observer on the ground receives the signal? (*f*) When the observer on the ground receives the signal, he immediately sends a return signal to the nose of the rocket. What is the reading of the clock in the nose of the rocket when that signal is received at the nose of the rocket?

THE VELOCITY TRANSFORMATION AND THE RELATIVISTIC DOPPLER EFFECT

22 •• **SPREADSHEET** A spaceship, at rest in a certain reference frame S, is given a speed increase of $0.50c$ (call this increase boost 1). Relative to its new rest frame, the spaceship is given a further $0.50c$ increase 10 seconds later (as measured in its new rest frame; call this increase boost 2). This process is continued indefinitely, at 10-s intervals, as measured in the rest frame of the spaceship. (Assume that the boosts take a very short time compared to 10 s.) (*a*) Using a spreadsheet program, calculate and graph the speed of the spaceship in reference frame S as a function of the boost number for boost 1 to boost 10. (*b*) Graph the gamma factor in the same manner. (*c*) How many boosts does it take until the speed of the ship in S is greater than $0.999c$? (*d*) How far does the spaceship move between boost 1 and boost 6, as measured in reference frame S? What is the average speed of the spaceship between boost 1 and boost 6, as measured in S?

23 • Light is emitted by a sodium sample that is moving toward Earth with speed v. The wavelength of the light is 589 nm in the rest frame of the sample. The wavelength measured in the frame of Earth is 547 nm. Find v.

24 • A distant galaxy is moving away from us at a speed of 1.85×10^7 m/s. Calculate the fractional redshift $(\lambda' - \lambda_0)/\lambda_0$ that we observe the light from the galaxy to have.

25 •• Derive $f' = f_0\sqrt{1 - (v^2/c^2)}/[1 - (v/c)]$ (Equation 39-16*a*) for the frequency received by an observer moving with speed v toward a stationary source of electromagnetic waves.

26 • Show that if v is much less than c, the Doppler shift is given approximately by

$$\Delta f/f \approx \pm v/c$$

27 •• A clock is placed in a satellite that orbits Earth with an orbital period of 90 min. By what time interval will this clock differ from an identical clock on Earth after 1.0 y? (Assume that special relativity applies and neglect general relativity.) SSM

28 •• For light that is Doppler-shifted with respect to an observer, we define the redshift parameter $z = (f - f')/f'$, where f is the frequency of the light measured in the rest frame of the emitter and f' is the frequency measured in the rest frame of the receiver. If the emitter is moving directly away from the receiver, show that the relative velocity between the emitter and the receiver is $v = c(u^2 - 1)/(u^2 + 1)$, where $u = z + 1$.

29 • A light beam moves along the y' axis with speed c in frame S', which is moving in the $+x$ direction with speed v relative to frame S. (*a*) Find the x and y components of the velocity of the light beam in frame S. (*b*) Show that, according to the velocity transformation equations, the magnitude of the velocity of the light beam in S is c.

30 •• A spaceship is moving east at speed $0.90c$ relative to Earth. A second spaceship is moving west at speed $0.90c$ relative to Earth. What is the speed of one spaceship relative to the other spaceship?

31 •• A particle moves with speed $0.800c$ in the $+x''$ direction along the x'' axis of frame S'', which moves with the same speed and in the same direction along the x' axis relative to frame S'. Frame S' moves with the same speed and in the same direction along the x axis relative to frame S. (*a*) Find the speed of the particle relative to frame S'. (*b*) Find the speed of the particle relative to frame S. SSM

RELATIVISTIC MOMENTUM AND RELATIVISTIC ENERGY

32 • A proton that has a rest energy equal to 938 MeV has a total energy of 2200 MeV. (*a*) What is its speed? (*b*) What is its momentum?

33 • If the kinetic energy of a particle equals twice its rest energy, what percentage error is made by using $p = mu$ for the magnitude of its momentum?

34 •• In a certain reference frame, a particle has momentum of 6.00 MeV/c and total energy of 8.00 MeV. (*a*) Determine the mass of the particle. (*b*) What is the total energy of the particle in a reference frame in which its momentum is 4.00 MeV/c? (*c*) What is the relative speed of the two reference frames?

35 •• Show that

$$d\left(\frac{mu}{\sqrt{1-(u^2/c^2)}}\right) = m\left(1-\frac{u^2}{c^2}\right)^{-3/2} du$$

Note: This relation was used to derive the relativistically correct expression for kinetic energy (Equation 39-22).

36 •• The K^0 particle has a mass of 497.7 MeV/c^2. It decays into a π^- and π^+, each having mass 139.6 MeV/c^2. Following the decay of a K^0, one of the pions is at rest in the laboratory. Determine the kinetic energy of the other pion after the decay and of the K^0 prior to the decay.

37 •• In reference frame S', two protons, each moving at $0.500c$, approach each other head-on. (*a*) Calculate the total kinetic energy of the two protons in frame S'. (*b*) Calculate the total kinetic energy of the protons as seen in reference frame S, which is moving with one of the protons. SSM

38 •• An antiproton $\bar{p}$ has the same mass m as a proton p. The antiproton is created during the reaction $p + p \rightarrow p + p + p + \bar{p}$. During an experiment, protons at rest in the laboratory are bombarded with protons of kinetic energy K_L, which must be great enough so that an amount of kinetic energy equal to $2mc^2$ can be converted into the rest energy of the two particles. In the frame of the laboratory, the total kinetic energy cannot be converted into rest energy because of conservation of momentum. However, in the zero-momentum reference frame in which the two initial protons are moving toward each other with equal speed u, the total kinetic energy can be converted into rest energy. (*a*) Find the speed of each proton u so that the total kinetic energy in the zero-momentum frame is $2mc^2$. (*b*) Transform to the laboratory's frame in which one proton is at rest, and find the speed u' of the other proton. (*c*) Show that the kinetic energy of the moving proton in the laboratory's frame is $K_L = 6mc^2$.

39 ••• A particle of mass 1.00 MeV/c^2 and kinetic energy 2.00 MeV collides with a stationary particle of mass 2.00 MeV/c^2. After the collision, the particles stick together. Find (*a*) the speed of the first particle before the collision, (*b*) the total energy of the first particle before the collision, (*c*) the initial total momentum of the system, (*d*) the total kinetic energy after the collision, and (*e*) the mass of the system after the collision.

GENERAL RELATIVITY

40 •• Light traveling in the direction of increasing gravitational potential undergoes a frequency redshift. Calculate the shift in wavelength if a beam of light of wavelength $\lambda = 632.8$ nm is sent up a vertical shaft of height $L = 100$ m.

41 •• Let us revisit a problem from Chapter 3: Two cannons are pointed directly toward each other, as shown in Figure 39-17. When fired, the cannonballs will follow the trajectories shown. Point P is the point where the trajectories cross each other. Ignore any effects due to air resistance. Using the principle of equivalence, show that if the cannons are fired simultaneously (in the rest frame of the cannons), the cannonballs will hit each other at point P.

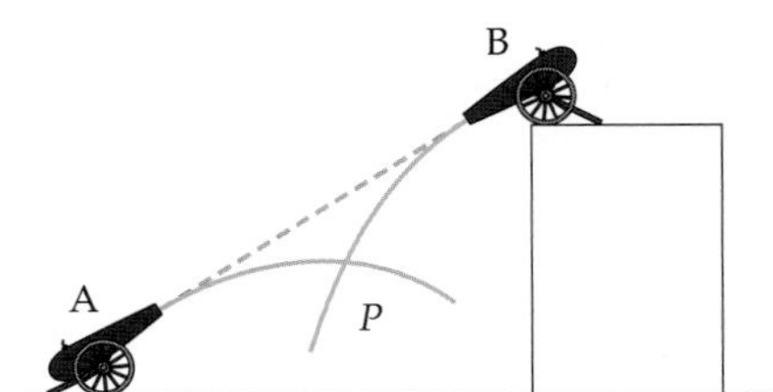

FIGURE 39-17 Problem 41

42 ••• A horizontal turntable rotates with angular speed ω. There is a clock at the center of the turntable and an identical clock mounted on the turntable a distance r from the center. In an inertial reference frame, in which the clock at the center is at rest, the clock at distance r is moving with speed $u = r\omega$. (*a*) Show that from time dilation according to special relativity, the time between ticks, Δt_0 for the clock at rest and Δt_R for the moving clock, are related by

$$\frac{\Delta t_R - \Delta t_0}{\Delta t_0} = -\frac{r^2\omega^2}{2c^2} \qquad r\omega \ll c$$

(*b*) In a reference frame rotating with the table, both clocks are at rest. Show that the clock at distance r experiences a pseudoforce $F_r = mr\omega^2$ in the rotating frame and that this is equivalent to a difference in gravitational potential between r and the origin of $\phi_r - \phi_0 = -\frac{1}{2}r^2\omega^2$. (*c*) Use the difference in gravitational potential given in Part (*b*) to show that in this frame the difference in time intervals is the same as in the inertial frame.

GENERAL PROBLEMS

43 • How fast must a muon travel so that its mean lifetime is 46 μs if its mean lifetime at rest is 2.2 μs?

44 • A distant galaxy is moving away from Earth with a speed that results in each wavelength received on Earth being shifted so that $\lambda' = 2\lambda_0$. Find the speed of the galaxy relative to Earth.

45 •• Frames S and S' are moving relative to each other along the x and x' axes (which superpose). Observers at rest in the two frames set their clocks to $t = 0$ when the two origins coincide. In frame S, event 1 occurs at $x_1 = 1.0\ c \cdot$y and $t_1 = 1.00$ y and event 2 occurs at $x_2 = 2.0\ c \cdot$y and $t_2 = 0.50$ y. The events occur simultaneously in frame S'. (*a*) Find the magnitude and direction of the velocity of S' relative to S. (*b*) At what time do both events occur as measured in S'? SSM

46 •• An interstellar spaceship travels from Earth to a star system 12 light-years away (as measured in Earth's frame). The trip takes 15 y as measured by clocks on the spaceship. (*a*) What is the speed of the spaceship relative to Earth? (*b*) When the spaceship arrives, it sends an electromagnetic signal to Earth. How long after the spaceship leaves Earth will observers on Earth receive the signal?

47 •• The neutral pion π^0 has a mass of 135.0 MeV/c^2. This particle can be created in a proton–proton collision:

$$\text{p} + \text{p} \rightarrow \text{p} + \text{p} + \pi^0$$

Determine the threshold kinetic energy for the creation of a π^0 in a collision of a moving proton and a stationary proton. (See Problem 38.)

48 •• A rocket that has a proper length of 1000 m moves away from a space station and in the $+x$ direction at $0.60c$ relative to an observer on the station. An astronaut stands at the rear of the rocket and fires a dart toward the front of the rocket at $0.80c$ relative to the rocket. How long does it take the dart to reach the front of the rocket (*a*) as measured in the frame of the rocket, (*b*) as measured in the frame of the space station, and (*c*) as measured in the frame of the dart?

49 ••• Using a simple thought experiment, Einstein showed that there is mass associated with electromagnetic radiation. Consider a box of length L and mass M resting on a frictionless surface. Attached to the left wall of the box is a light source that emits a directed pulse of radiation of energy E, which is completely absorbed at the right wall of the box. According to classical electromagnetic theory, the radiation carries momentum of magnitude $p = E/c$ (Equation 30-24). The box recoils when the pulse is emitted by the light source. (*a*) Find the recoil velocity of the box so that momentum is conserved when the light is emitted. (Because p is small and M is large, you may use classical mechanics.) (*b*) When the light is absorbed at the right wall of the box the box stops, so the total momentum of the system remains zero. If we neglect the very small velocity of the box, the time it takes for the radiation to travel across the box is $\Delta t = L/c$. Find the distance moved by the box in that time. (*c*) Show that if the center of mass of the system is to remain at the same place, the radiation must carry mass $m = E/c^2$. SSM

50 ••• Using the relativistic conservation of momentum and energy and the relation between energy and momentum for a photon $E = pc$, prove that a free electron (an electron not bound to an atomic nucleus) cannot absorb or emit a photon.

51 ••• When a moving particle that has a kinetic energy greater than the threshold kinetic energy K_{th} strikes a stationary target particle, one or more particles may be created in the inelastic collision. Show that the threshold kinetic energy of the moving particle is given by

$$K_{\text{th}} = \frac{(\Sigma m_{\text{in}} + \Sigma m_{\text{fin}})(\Sigma m_{\text{fin}} + \Sigma m_{\text{in}})c^2}{2m_{\text{target}}}$$

Here Σm_{in} is the sum of the masses of the particles prior to the collision, Σm_{fin} is the sum of the masses of the particles following the collision, and m_{target} is the mass of the target particle. Use this expression to determine the threshold kinetic energy of protons incident on a stationary proton target for the production of a proton–antiproton pair; compare your result with the result of Problem 38. SSM

52 ••• A particle of mass M decays into two identical particles, each of mass m, where $m = 0.30M$. Prior to the decay, the particle of mass M has a total energy of $4.0mc^2$ in the laboratory reference frame. The velocities of the decay products are along the direction of motion of M. Find the velocities of the decay products in the laboratory reference frame.

53 ••• A rod of proper length L_{p} makes an angle θ with the x axis in frame S. Show that the angle θ' made with the x' axis in frame S', which is moving in the $+x$ direction with speed v, is given by $\tan\theta' = \gamma \tan\theta$ and that the length of the stick in S' is $L' = L_{\text{p}}(\gamma^{-2}\cos^2\theta + \sin^2\theta)^{1/2}$.

54 ••• Show that if a particle moves at an angle θ with the x axis with speed u in frame S, it moves at an angle θ' with the x' axis in S' given by

$$\tan\theta' = \gamma^{-1}\sin\theta/[\cos\theta - (v/u)].$$

55 ••• For the special case of a particle moving with speed u along the y axis in frame S, show that its momentum and energy in frame S', a frame that is moving along the x axis with velocity v, are related to its momentum and energy in S by the transformation equations

$$p'_x = \gamma\left(p_x - \frac{vE}{c^2}\right) \qquad p'_y = p_y \qquad p'_z = p_z \qquad \frac{E'}{c} = \gamma\left(\frac{E}{c} - \frac{vp_x}{c}\right)$$

Compare these equations with the Lorentz transformation equations for x', y', z', and t'. Notice that the quantities p_x, p_y, p_z and E/c transform in the same way as do x, y, z, and ct. SSM

56 ••• The equation for the spherical wavefront of a light pulse that begins at the origin at time $t = 0$ is $x^2 + y^2 + z^2 - (ct)^2 = 0$. Frame S' moves with velocity v along the x axis. Using the Lorentz transformation, show that such a light pulse also has a spherical wavefront in frame S' by showing that $x'^2 + y'^2 + z'^2 - (ct')^2 = 0$.

57 ••• In Problem 56, you showed that the quantity $x^2 + y^2 + z^2 - (ct)^2$ has the same value (zero) in both S and S'. A quantity that has the same value in all inertial frames is called a *Lorentz invariant*. From the results of Problem 55, the quantity $p_x^2 + p_y^2 + p_z^2 - E^2/c^2$ must also be a Lorentz invariant. Show that this quantity has the value $-m^2c^2$ in both the S and S' reference frames.

58 ••• A long rod that is parallel to the x axis is released from rest. Subsequently, it is in free fall with an acceleration of magnitude g in the $-y$ direction. An observer in a rocket ship moving with speed v parallel to the x axis passes by. Using the Lorentz transformations, show that the observer on the rocket ship will measure the rod to be bent into a parabolic shape. Is the parabola concave upward or concave downward?

59 •• Show that if u'_x and v in $u_x = (u'_x + v)/[1 + (vu'_x/c^2)]$ (Equation 39-18*a*) are both positive and less than c, then u_x is positive and less than c. (*Hint: Let* $u'_x = (1 - \varepsilon_1)c$ *and* $v = (1 - \varepsilon_2)c$, *where* ε_1 *and* ε_2 *are positive numbers that are less than 1.*)

60 ••• In reference frame S, the acceleration of a particle is $\vec{a} = a_x\hat{i} + a_y\hat{j} + a_z\hat{k}$. Derive expressions for the acceleration components a'_x, a'_y, and a'_z of the particle in reference frame S' that is moving relative to S in the x direction with velocity v.

THE DIABLO CANYON NUCLEAR POWER PLANT NEAR SAN LUIS OBISPO, CALIFORNIA. *(Tony Hertz/Alamy.)*

? How much energy is released during the fission of one gram of ^{235}U? (See Example 40-6).

Nuclear Physics

To many chemists, the atomic nucleus is modeled as a point charge that has most of the mass of the atom. In this chapter, we will look at the nucleus from the physicist's perspective and see how the protons and neutrons that make up the nucleus have played important roles in our everyday life as well as in the history and structure of the universe.

In this chapter, we study the properties of atomic nuclei, examine radioactivity, and explore nuclear reactions. We also discuss fission and fusion. The fission of very heavy nuclei, such as uranium, is a major source of power today, while the fusion of very light nuclei is the energy source that powers the stars, including our Sun, and may hold the key to our energy needs of the future.

40-1 PROPERTIES OF NUCLEI

The nucleus of an atom has just two kinds of particles, protons and neutrons,* which have approximately the same mass (the neutron is approximately 0.2 percent more massive). The proton has a charge of $+e$, and the neutron is uncharged.

* The most prevalent hydrogen nucleus has a single proton.

The number of protons, Z, is the atomic number of the atom, which also equals the number of electrons in the atom. The number of neutrons that a nucleus has, N, is approximately equal to Z for light nuclei. For heavier nuclei, the number of neutrons is increasingly greater than Z. The total number of nucleons* $A = N + Z$ is called the **nucleon number** or **mass number** of the nucleus. A particular nuclear species is called a **nuclide.** Two or more nuclides that have the same atomic number Z but have different values for N and A are called **isotopes.** A particular nuclide is designated by its atomic symbol (for example, H for hydrogen and He for helium) and its mass number A as a superscript. The lightest element, hydrogen, has three isotopes: protium, ^{1}H, whose nucleus is just a single proton; deuterium, ^{2}H, whose nucleus is composed of one proton and one neutron; and tritium, ^{3}H, whose nucleus is composed of one proton and two neutrons. Although the mass of the deuterium atom is about twice the mass of the protium atom and the mass of the tritium atom is about three times the mass of protium, these three atoms have nearly identical chemical properties because they each have one electron. On the average, there are about three stable isotopes for each element, although some atoms have only one stable isotope while others have five or six. The most common isotope of the second lightest element, helium, is ^{4}He. The ^{4}He nucleus is also known as an α particle. Another isotope of helium is ^{3}He, and the ^{3}He nucleus is also known as helion.

Nucleons exert a strong attractive force on other nucleons. This force, called the **strong nuclear force** or the **hadronic force,** is much stronger than the electrostatic force of repulsion between the protons and is very much stronger than the gravitational forces between the nucleons. (Gravity is so comparably weak that it can always be neglected in nuclear physics.) The strong nuclear force is roughly the same between two neutrons, two protons, or a neutron and a proton. Two protons, of course, also exert a repulsive electrostatic force on each other due to their charges, which tends to weaken the attraction between them somewhat. The strong nuclear force decreases rapidly with distance, and it is negligible when two nucleons are more than a few femtometers apart.

SIZE, SHAPE, AND DENSITY

The size and shape of the nucleus can be determined by bombarding it with high-energy particles and observing the scattering. The results depend somewhat on the kind of experiment. For example, a scattering experiment using electrons measures the charge distribution of the nucleus, whereas a scattering experiment using neutrons determines the region of influence of the strong nuclear force. A wide variety of experiments suggest that most nuclei are approximately spherical, with radii given approximately by

$$R = R_0 A^{1/3} \qquad 40\text{-}1$$

NUCLEAR RADIUS

where R_0 is approximately 1.2 fm. The fact that the radius of a spherical nucleus is proportional to $A^{1/3}$ implies that the volume of the nucleus is proportional to A. Because the mass of the nucleus is also approximately proportional to A, the densities of all nuclei are approximately the same. This is analogous to a drop of liquid, which also has constant density independent of its size. The **liquid-drop model** of the nucleus has proved quite successful in explaining nuclear behavior, especially the fission of heavy nuclei.

---

* The word *nucleon* refers to either a neutron or a proton that is part of a nucleus.

N AND Z Numbers

For light nuclei, the greatest stability is achieved when the numbers of protons and neutrons are approximately equal, $N \approx Z$. For heavier nuclei, instability caused by the electrostatic repulsion between the protons is minimized when there are more neutrons than protons. We can see this by looking at the N and Z numbers for the most abundant isotopes of some representative elements: for ${}^{16}_{8}\text{O}$, $N = 8$ and $Z = 8$; for ${}^{40}_{20}\text{Ca}$, $N = 20$ and $Z = 20$; for ${}^{56}_{26}\text{Fe}$, $N = 30$ and $Z = 26$; for ${}^{207}_{82}\text{Pb}$, $N = 125$ and $Z = 82$; and for ${}^{238}_{92}\text{U}$, $N = 146$ and $Z = 92$. (The atomic number Z has been included here as a subscript of the atomic symbol for emphasis. It is not actually needed because the atomic number is implied by the atomic symbol.)

Figure 40-1 shows a plot of N versus Z for the known stable nuclei. The curve follows the straight line $N = Z$ for small values of N and Z. We can understand this tendency for N and Z to be equal by considering the total energy of A particles in a one-dimensional box. For $A = 8$, Figure 40-2 shows the energy levels for eight neutrons and for four neutrons and four protons. Because of the exclusion principle, only two identical particles (that have opposite spins) can be in the same space state. Because protons and neutrons are not identical, we can put two each in a state, as shown in Figure 40-2b. Thus, the total energy for four protons and four neutrons is less than the total energy for eight neutrons (or eight protons), as shown in Figure 40-2a. When the Coulomb energy of repulsion, which is proportional to Z^2, is included, this result changes somewhat. For large values of A and Z, the total energy may be increased less by adding two neutrons than by adding one neutron and one proton because of the electrostatic repulsion involved in the latter case. This explains why $N > Z$ for the larger values of A (for the heavier nuclei).

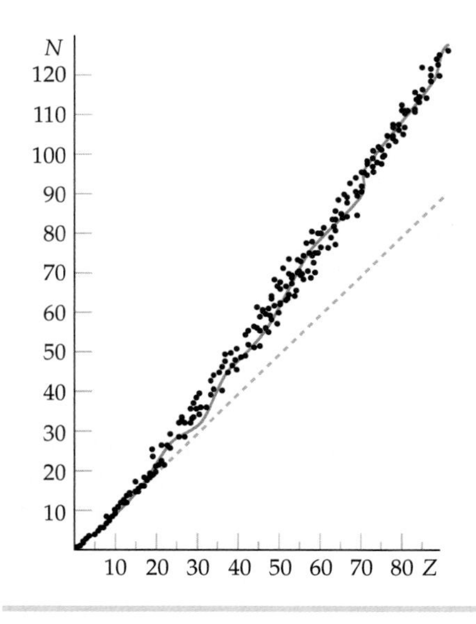

FIGURE 40-1 Plot of number of neutrons N versus number of protons Z for the stable nuclides. The dashed line is $N = Z$.

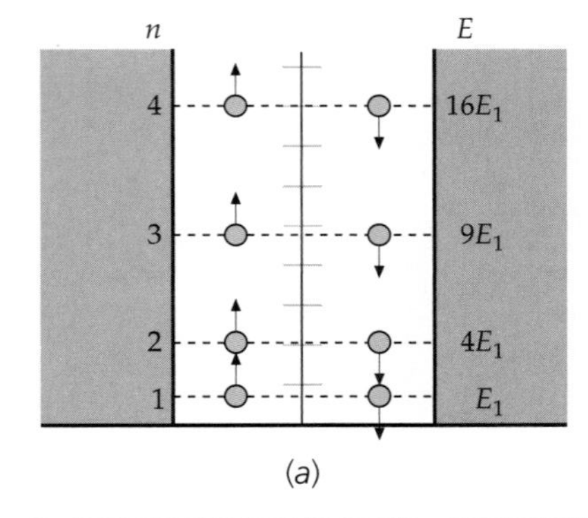

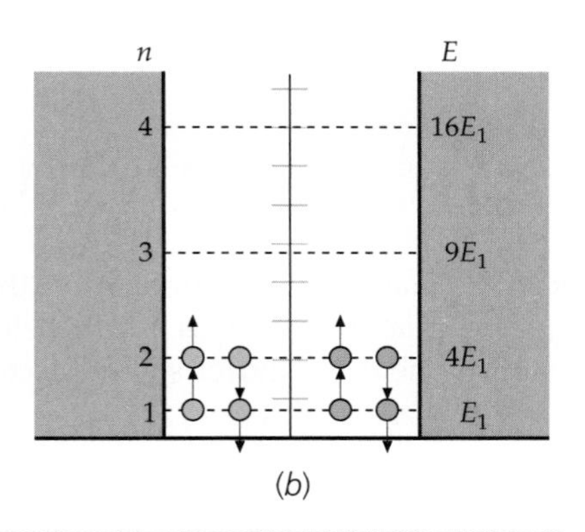

FIGURE 40-2 (a) Eight neutrons in a one-dimensional box. In accordance with the exclusion principle, only two neutrons (that have opposite spins) can be in a given energy level. (b) Four neutrons and four protons in a one-dimensional box. Because protons and neutrons are not identical particles, two of each can be in the same energy level. The total energy is much less for this case than for the case shown in Figure 40-2a.

PRACTICE PROBLEM 40-1

(a) Calculate the total energy of the eight neutrons in the one-dimensional box shown in Figure 40-2a. (b) Calculate the total energy of the four neutrons and four protons in the one-dimensional box shown in Figure 40-2b.

MASS AND BINDING ENERGY

The mass of a nucleus is less than the sum of the masses of its parts by E_b/c^2, where E_b is the binding energy and c is the speed of light. When two or more nucleons fuse together to form a nucleus, the total mass decreases and energy is released. Conversely, to break up a nucleus into its parts, energy is absorbed by the system and the mass of the system increases.

Atomic masses and nuclear masses are often given in unified atomic mass units (u), defined as one-twelfth the mass of a ^{12}C atom. The rest energy of one such mass unit is

$$(1\text{ u})c^2 = 931.5\text{ MeV} \qquad 40\text{-}2$$

Consider ^{4}He, for example, which consists of two protons and two neutrons. The mass of an atom can be accurately measured in a mass spectrometer. The mass of the ^{4}He atom is 4.002 603 u and the mass of the ^{1}H atom is 1.007 825 u. These values include the masses of the electrons in the atom. The mass of the neutron is 1.008 665 u. The sum of the masses of two ^{1}H atoms and two neutrons is 2(1.007 825 u) + 2(1.008 665 u) = 4.032 980 u, which is greater than the mass of the ^{4}He atom by 0.030 377 u.* We can find the binding energy of the ^{4}He nucleus from this mass difference of 0.030 377 u by using the mass conversion factor $(1\text{ u})c^2 = 931.5$ MeV from Equation 40-2. Then

$$(0.030\,377\text{ u})c^2 = (0.030\,377\text{ u})c^2 \times \frac{931.5\text{ MeV}/c^2}{1\text{ u}} = 28.30\text{ MeV}$$

The total binding energy of ^{4}He is thus 28.30 MeV. In general, the binding energy of a nucleus of an atom of atomic mass M_A having Z protons and N neutrons is found by calculating the difference between the sum of the masses of the nucleons and the mass of the nucleus and then multiplying by c^2:

$$E_b = (ZM_H + Nm_n - M_A)c^2 \qquad 40\text{-}3$$

TOTAL NUCLEAR BINDING ENERGY

where M_H is the mass of the ^{1}H atom and m_n is the mass of the neutron. (Note that the mass of the Z electrons in the term ZM_H is canceled by the mass of the Z electrons in the term M_A.†) The atomic masses of the neutron and of some selected isotopes are listed in Table 40-1.

Example 40-1 Binding Energy of the Last Neutron

Find the binding energy of the last neutron in a ^{4}He nucleus.

PICTURE The binding energy is energy equivalent of the mass of a ^{3}He atom plus the mass of a neutron minus the mass of a ^{4}He atom. We find the masses from Table 40-1 and multiply by c^2 to obtain the energy equivalents.

---

* Note that by using the masses of two ^{1}H atoms rather than two protons, the masses of the electrons in the atom are accounted for. We do this because it is atomic masses, not nuclear masses, that are measured directly and listed in mass tables.

† The mass associated with the binding energies of the electrons are not accounted for in this calculation.

SOLVE

1. Add the mass of the neutron to that of ^{3}He:

$$m_{^3\text{He}} + m_\text{n} = 3.016\,030\text{ u} + 1.008\,665\text{ u}$$
$$= 4.024\,695\text{ u}$$

2. Subtract the mass of ^{4}He from the result:

$$\Delta m = (m_{^3\text{He}} + m_\text{n}) - m_{^4\text{He}}$$
$$= 4.024\,695\text{ u} - 4.002\,603\text{ u} = 0.022\,092\text{ u}$$

3. Multiply this mass difference by c^2 and convert to MeV:

$$E_\text{b} = (\Delta m)c^2$$
$$= (0.022\,092\text{ u})c^2 \times \frac{931.5\text{ MeV}/c^2}{1\text{ u}}$$
$$= \boxed{20.58\text{ MeV}}$$

CHECK As expected, the step-3 result of 20.58 MeV is less than the total binding energy of a ^{4}He nucleus. (The total binding energy of a ^{4}He nucleus is 28.30 MeV, a value that is calculated preceding Equation 40-3.)

Table 40-1 Atomic Masses of the Neutron and Selected Isotopes*

Element	Symbol	Z	Atomic mass, u
Neutron	n	0	1.008 665
Hydrogen			
Protium	^{1}H	1	1.007 825
Deuterium	^{2}H or D	1	2.014 102
Tritium	^{3}H or T	1	3.016 050
Helium	^{3}He	2	3.016 030
	^{4}He	2	4.002 603
Lithium	^{6}Li	3	6.015 125
	^{7}Li	3	7.016 004
Boron	^{10}B	5	10.012 939
Carbon	^{12}C	6	12.000 000
	^{13}C	6	13.003 354
	^{14}C	6	14.003 242
Nitrogen	^{13}N	7	13.005 738
	^{14}N	7	14.003 074
Oxygen	^{16}O	8	15.994 915
Sodium	^{23}Na	11	22.989 771
Potassium	^{39}K	19	38.963 710
Iron	^{56}Fe	26	55.939 395
Copper	^{63}Cu	29	62.929 592
Silver	^{107}Ag	47	106.905 094
Gold	^{197}Au	79	196.966 541
Lead	^{208}Pb	82	207.976 650
Polonium	^{212}Po	84	211.989 629
Radon	^{222}Rn	86	222.017 531
Radium	^{226}Ra	88	226.025 360
Uranium	^{238}U	92	238.048 608
Plutonium	^{242}Pu	94	242.058 725

*Mass values obtained at <http://physics.nist.gov/PhysRefData/Compositions/index.html>.

Figure 40-3 shows the binding energy per nucleon E_b/A versus A. The mean value is approximately 8.3 MeV. The flatness of the curve for $A > 50$ shows that E_b is approximately proportional to A. This indicates that there is saturation of nuclear forces in the nucleus as would be the case if each nucleon were attracted only to its nearest neighbors. Such a situation also leads to a constant nuclear density consistent with the measurements of the radius. If, for example, there were no saturation and each nucleon bonded to each other nucleon, there would be $A - 1$ bonds for each nucleon and a total of $A(A - 1)$ bonds altogether. The total binding energy, which is a measure of the energy needed to break all these bonds, would then be proportional to $A(A - 1)$, and E_b/A would not be approximately constant. The steep rise in the curve for low A is due to the increase in the number of nearest neighbors and therefore to the increased number of bonds per nucleon. The gradual decrease at high A is due to the Coulomb repulsion of the protons, which increases as Z^2 and decreases the binding energy. For very large A, this Coulomb repulsion is so great that a nucleus that has an A greater than approximately 300 is unstable and undergoes spontaneous fission.

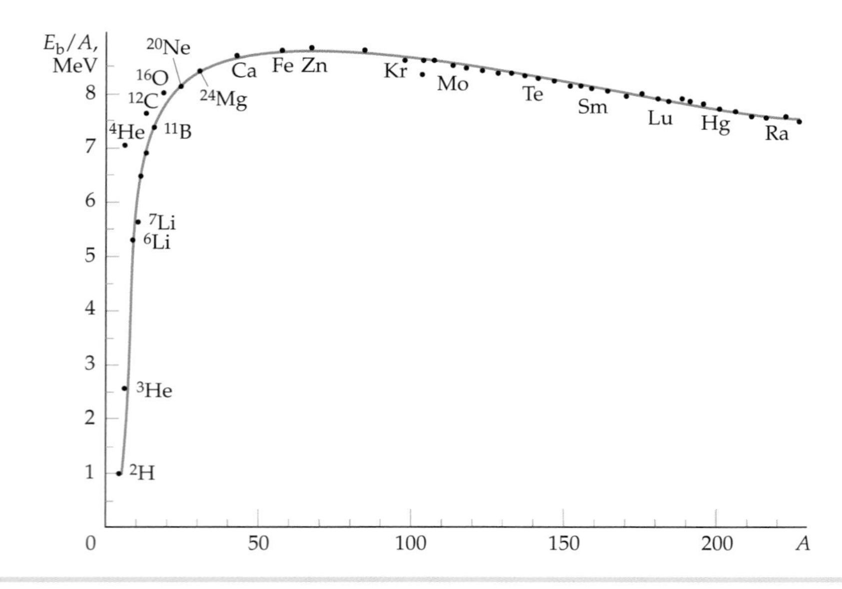

FIGURE 40-3 The binding energy per nucleon versus the nucleon number A. For nuclei that have values of A greater than 50, the curve is approximately constant, indicating that the total binding energy is approximately proportional to A.

40-2 RADIOACTIVITY

Many nuclei are radioactive; that is, they decay into other nuclei by the emission of particles such as photons, electrons, neutrons, or α particles. The terms α decay, β decay, and γ decay were used before it was discovered that α particles are ^{4}He nuclei, β particles are either electrons (β^-) or positrons* (β^+), and γ rays are photons. The rate of decay of a radioactive sample decreases exponentially with increasing time. *This exponential time dependence is characteristic of all radioactivity and indicates that radioactive decay is a statistical process.* Because each nucleus is well shielded from others by the atomic electrons, pressure and temperature changes have little or no effect on the rate of radioactive decay or other nuclear properties.

* The positron has the same mass as an electron and it has a charge of $+e$.

Let N be the number of radioactive nuclei at some time t. If the decay of an individual nucleus is a random event, we expect the number of nuclei that decay in some time interval dt to be proportional both to N and to dt. Because of these decays, the number N will decrease. The change in N between time t and time $t + dt$ is given by

$$dN = -\lambda N\,dt \qquad 40\text{-}4$$

where λ is a constant of proportionality called the **decay constant.** The rate of change of N, dN/dt, is proportional to N. This is characteristic of exponential decay. To solve Equation 40-4 for N, we first divide each side by N, thus separating the variables N and t:

$$\frac{dN}{N} = -\lambda\,dt$$

Integrating, we obtain

$$\int_{N_0}^{N'} \frac{dN}{N} = -\lambda \int_0^{t'} dt$$

or

$$\ln\frac{N'}{N_0} = -\lambda t' \qquad 40\text{-}5$$

where N' is the number of nuclei that remain at time t'. For convenience, we drop the primes from N' and t'. This introduces no ambiguity because the parameters N and t have been integrated out of the equation. Taking the exponential of each side, we obtain

$$\frac{N}{N_0} = e^{-\lambda t}$$

See
Math Tutorial _for more information on_
Exponential Functions

or

$$N = N_0 e^{-\lambda t} \qquad 40\text{-}6$$

The number of radioactive decays per second is called the **decay rate** R:

$$R = -\frac{dN}{dt} = \lambda N = \lambda N_0 e^{-\lambda t} = R_0 e^{-\lambda t} \qquad 40\text{-}7$$

DECAY RATE

where

$$R_0 = \lambda N_0 \qquad 40\text{-}8$$

is the decay rate at time $t = 0$. The decay rate R is the quantity that is determined experimentally. The decay rate is also called the **activity** of the sample.

The average or **mean lifetime** τ is equal to the reciprocal of the decay constant (see Problem 40):

$$\tau = \frac{1}{\lambda} \qquad 40\text{-}9$$

The mean lifetime is analogous to the time constant in the exponential decrease in the charge on a capacitor in an RC circuit that we discussed in Section 25-6. After a time equal to the mean lifetime, the number of radioactive nuclei and the decay rate are each equal to $e^{-1} = 37$ percent of their original values. The **half-life** $t_{1/2}$ is defined as the time it takes for the number of nuclei and the decay rate to decrease by half. Setting $t = t_{1/2}$ and $N = N_0/2$ in Equation 40-6 gives

$$\frac{N_0}{2} = N_0 e^{-\lambda t_{1/2}} \qquad \text{40-10}$$

or

$$e^{+\lambda t_{1/2}} = 2$$

Solving for $t_{1/2}$ gives

$$t_{1/2} = \frac{\ln 2}{\lambda} = (\ln 2)\tau = 0.693\tau \qquad \text{40-11}$$

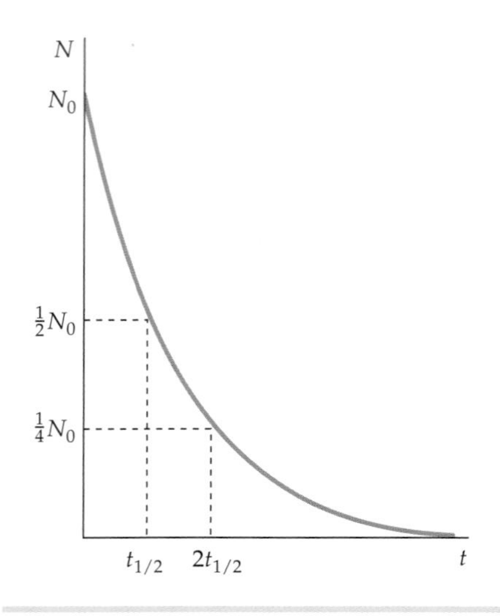

FIGURE 40-4 Exponential radioactive decay. After each half-life $t_{1/2}$, the number of nuclei remaining has decreased by one-half. The decay rate $R = \lambda N$ has the same time dependence as does N.

Figure 40-4 shows a plot of N versus t. If we multiply the numbers on the N axis by λ, this graph becomes a plot of R versus t. After each time interval of one half-life, both the number of nuclei left and the decay rate have decreased to half of their previous values. For example, if the decay rate is R_0 initially, it will be $\frac{1}{2}R_0$ after one half-life, $(\frac{1}{2})(\frac{1}{2})R_0$ after two half-lives, and so forth. After n half-lives, the decay rate will be

$$R = \left(\tfrac{1}{2}\right)^n R_0 \qquad \text{40-12}$$

The half-lives of radioactive nuclei vary from very small times (less than 1 μs) to very large times (greater than 10^{10} y).

Example 40-2 Counting Rate for Radioactive Decay

A radioactive source has a half-life of 1.0 min. At time $t = 0$, the radioactive source is placed near a detector, and the counting rate (the number of decay particles detected per unit time) is observed to be 2000 counts/s. Find the counting rate at times $t = 1.0$ min, $t = 2.0$ min, $t = 3.0$ min, and $t = 10$ min.

PICTURE The counting rate r is proportional to the decay rate R, and the decay rate is given by $R = (\frac{1}{2})^n R_0$ (Equation 40-12), where n is the time divided by 1.0 min.

SOLVE

1. Because the half-life is 1.0 min, the counting rate will be half as great at $t = 1.0$ min as at $t = 0$:

$$r_1 = \tfrac{1}{2}r_0 = \tfrac{1}{2}(2000 \text{ counts/s}) = \boxed{1.0 \times 10^3 \text{ counts/s at 1.0 min}}$$

2. At $t = 2.0$ min, the rate is half that at 1 min. It decreases by one-half each minute:

$$r_2 = \left(\tfrac{1}{2}\right)^2 r_0 = \tfrac{1}{4}(2000 \text{ counts/s}) = \boxed{5.0 \times 10^2 \text{ counts/s at 2.0 min}}$$

$$r_3 = \left(\tfrac{1}{2}\right)^3 r_0 = \tfrac{1}{8}(2000 \text{ counts/s}) = \boxed{2.5 \times 10^2 \text{ counts/s at 3.0 min}}$$

3. At $t = 10$ min, the rate will be $(\frac{1}{2})^{10}$ multiplied by the initial rate:

$$r_{10} = \left(\tfrac{1}{2}\right)^{10} r_0 = \tfrac{1}{1024}(2000 \text{ counts/s}) = 1.95 \text{ counts/s} \approx \boxed{2.0 \text{ counts/s at 10 min}}$$

CHECK As expected, the counting rate decreases as the number of minutes increases.

CONCEPT CHECK 40-1

A radioactive isotope has a half-life of 10 s. You are observing a sample of this isotope. After approximately one minute of observation, there is only one atom of this isotope left in your sample. How many atoms of this isotope will be left in your sample 15 s later?

Example 40-3 Detection-Efficiency Considerations

If the detection efficiency in Example 40-2 is 20 percent, (*a*) how many radioactive nuclei are there at time $t = 0$ and (*b*) at time $t = 1.0$ min? (c) How many nuclei decay in the first minute?

PICTURE The detection efficiency depends on the probability that a radioactive decay particle will enter the detector and the probability that upon entering the detector it will produce a count. If the efficiency is 20 percent, the decay rate must be five times the counting rate.

SOLVE

(*a*) 1. The number of radioactive nuclei is related to the decay rate R and the decay constant λ:

$$R = \lambda N$$

2. The decay constant is related to the half-life:

$$\lambda = \frac{\ln 2.0}{t_{1/2}} = \frac{0.693}{1.0\text{ min}} = 0.693\text{ min}^{-1}$$

3. Because the detection efficiency is 20 percent, the decay rate is five times the counting rate. Calculate the initial decay rate:

$$R_0 = (5\text{ decays/count}) \times (2000\text{ counts/s}) = 1.0 \times 10^4\text{ decays/s}$$

4. Substitute to calculate the initial number of radioactive nuclei N_0 at $t = 0$:

$$N_0 = \frac{R_0}{\lambda} = \frac{1.0 \times 10^4\text{ s}^{-1}}{0.693\text{ min}^{-1}} \times \frac{60\text{ s}}{1\text{ min}}$$

$$= 8.66 \times 10^5 = \boxed{8.7 \times 10^5}$$

(*b*) At time $t = 1\text{ min} = t_{1/2}$, there are half as many radioactive nuclei as at $t = 0$:

$$N_1 = \tfrac{1}{2}(8.66 \times 10^5) = 4.33 \times 10^5 = \boxed{4.3 \times 10^5}$$

(*c*) The number of nuclei that decay in the first minute is $N_0 - N_1$:

$$\Delta N = N_0 - N_1 = 8.66 \times 10^5 - 4.33 \times 10^5 = \boxed{4.3 \times 10^5}$$

CHECK The results for Parts (*b*) and (*c*) are equal, as expected. At the end of one half-life, half of the nuclei have decayed and the other half remain.

The SI unit of radioactive decay is the **becquerel** (Bq), which is defined as one decay per second:

$$1\text{ Bq} = 1\text{ decay/s} \qquad \text{40-13}$$

A historical unit that applies to all types of radioactivity is the **curie** (Ci), which is defined as

$$1\text{ Ci} = 3.7 \times 10^{10}\text{ decays/s} = 3.7 \times 10^{10}\text{ Bq} \qquad \text{40-14}$$

The curie is the rate at which radiation is emitted by 1 g of radium. Because this is a very large unit, the millicurie (mCi) or microcurie (μCi) are often used.

BETA DECAY

Beta decay occurs in nuclei that have too many neutrons or too few neutrons for stability. During β decay, A remains the same while Z either increases by 1 (β^- decay) or decreases by 1 (β^+ decay).

An example of β decay is the decay of a free neutron into a proton and an electron. (The half-life of a free neutron is about 10.8 min.) The energy of β decay is 0.782 MeV, which is the difference between the rest energy of the neutron and the rest energy of the proton and an electron. More generally, during β^- decay, a nucleus of mass number A and atomic number Z decays into a nucleus, referred to as the **daughter nucleus,** of mass number A and atomic number $Z' = Z + 1$ and an electron is emitted. (The original nucleus is called the **parent.**) If the decay energy were shared by only the daughter nucleus and the emitted electron, the energy of the electron would be uniquely determined by the conservation of energy and momentum. Experiments show, however, the energies of the electrons emitted during the β^- decay of a nucleus are observed to vary from zero to the maximum energy available. A typical energy spectrum for the electrons is shown in Figure 40-5.

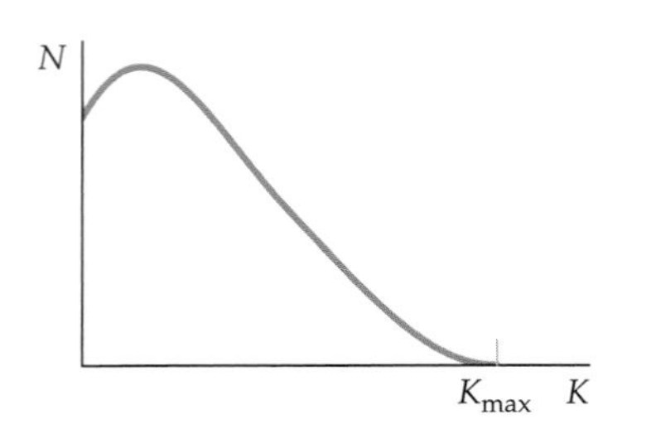

FIGURE 40-5 Number of electrons emitted during β^- decay versus kinetic energy. The fact that all the electrons do not have the same energy K_{max} suggests that another particle, one that shares the energy available for decay, is emitted.

To explain the fact that energy seemed not to be conserved during β decay, Wolfgang Pauli in 1930 suggested that a third particle, which he called the **neutrino,** is also emitted. Because the measured maximum energy of the emitted electrons is equal to the total available for the decay, the rest energy and therefore the mass of the neutrino was assumed to be zero. (It is now known that the mass of the neutrino is very small but not zero.) In 1948, measurements of the momenta of the emitted electron and the recoiling nucleus showed that the neutrino was also needed for the conservation of linear momentum during β decay. The neutrino was first observed experimentally in 1957. It is now known that there are at least three kinds of neutrinos, one (ν_e) associated with electrons, one (ν_μ) associated with muons, and one (ν_τ) associated with the tau particle, τ. Moreover, each neutrino has an antiparticle, written $\bar{\nu}_e$, $\bar{\nu}_\mu$, and $\bar{\nu}_\tau$. It is the electron antineutrino that is emitted during the decay of a neutron, which is written*

$$\mathrm{n} \rightarrow \mathrm{p} + \mathrm{e}^- + \bar{\nu}_\mathrm{e} \qquad \text{40-15}$$

During β^+ decay, a proton changes into a neutron, and a positron (and a neutrino) is emitted. A free proton cannot decay by positron emission because of conservation of energy (the mass of the neutron and the positron is greater than the mass of the proton); however, because of binding-energy effects, a proton inside a nucleus can decay. A typical β^+ decay is

$$^{13}_{7}\mathrm{N} \rightarrow {}^{13}_{6}\mathrm{C} + \mathrm{e}^+ + \nu_\mathrm{e} \qquad \text{40-16}$$

! Do not confuse the symbols e^- and e^+ with the symbol e. The symbols e^- and e^+ denote particles (the electron and the positron), whereas the symbol e denotes an amount of charge.

The electrons or the positrons emitted during β decay do not exist inside the nucleus. They are created during the process of decay, just as photons are created when an atom makes a transition from a higher energy state to a lower energy state.

An important example of β decay is that of ^{14}C, which is used in radioactive carbon dating:

$$^{14}\mathrm{C} \rightarrow {}^{14}\mathrm{N} + \mathrm{e}^- + \bar{\nu}_\mathrm{e} \qquad \text{40-17}$$

The half-life for this decay is 5730 y. The radioactive isotope ^{14}C is produced in the upper atmosphere during nuclear reactions caused by cosmic rays. The chemical reactivity of a carbon atom that has a ^{14}C nucleus is the same as the chemical reactivity of a carbon atom that has a ^{12}C nucleus. For example, atoms that have these nuclei combine with oxygen to form CO_2 molecules. Because living organisms continually exchange CO_2 with the atmosphere, the ratio of ^{14}C to ^{12}C in a living organism is the same as the equilibrium ratio in the atmosphere, which is about 1.3×10^{-12}. After an organism dies, it no longer absorbs ^{14}C from the atmosphere, so the ratio of ^{14}C to ^{12}C continually decreases due to the radioactive decay of ^{14}C. The number of ^{14}C decays per minute per gram of carbon in a living organism can be calculated from the known half-life of ^{14}C and the number of ^{14}C nuclei in a gram of carbon. The result is that there are approximately 15.0 decays per minute per gram of carbon in a living organism. Using this result and the measured number of decays per minute per gram of carbon in a nonliving sample of bone, wood, or other object having carbon, we can determine the age of the sample. For example, if the measured rate were 7.5 decays per minute per gram, the age of the sample would be one half-life = 5730 years.

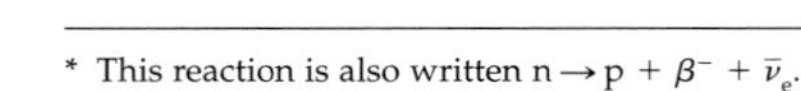

* This reaction is also written $\mathrm{n} \rightarrow \mathrm{p} + \beta^- + \bar{\nu}_\mathrm{e}$.

Example 40-4 How Old Is the Artifact? *Context-Rich*

You have a summer job working in an archeological research lab. Your supervisor calls to tell you that they found a new bone at their current site and asks you to determine the age of the bone from a sample that she will send you. When the bone sample arrives, you take a section that contains 200 grams of carbon and you find a beta decay rate of 400 decays/min.

PICTURE There are approximately 15.0 decays per minute per gram of carbon in a living organism, and the half-life of carbon-14 is 5730 y. We need to determine the number of half-lives that have occurred since the death of the organism. We do this by using the equality $R_n = (1/2)^n R_0$ (Equation 40-12), where R_n is the current decay rate, R_0 is the initial decay rate, and n is the number of half-lives. We can determine the initial decay rate by multiplying the decay rate per gram by the mass of the carbon of the sample.

SOLVE

1. Write the decay rate after n half-lives in terms of the initial decay rate:

$$R_n = \left(\tfrac{1}{2}\right)^n R_0$$

2. Calculate the initial decay rate (the decay for 200 g of carbon when the organism died):

$$R_0 = [(15 \text{ decays/min})/\text{g}](200 \text{ g}) = 3000 \text{ decays/min}$$

3. Substitute the values for R_0 and R_n into the step-1 equation and solve for n:

$$R_n = \left(\tfrac{1}{2}\right)^n R_0$$

$$400 \text{ decays/min} = \left(\tfrac{1}{2}\right)^n 3000 \text{ decays/min}$$

$$\left(\tfrac{1}{2}\right)^n = \frac{400}{3000}$$

$$2^n = \frac{3000}{400} = 7.5$$

4. We solve for n by taking the logarithm of each side:

$$n \ln 2 = \ln 7.5 \quad \Rightarrow \quad n = \frac{\ln 7.5}{\ln 2} = 2.91$$

5. The age of the bone is $nt_{1/2}$:

$$t = nt_{1/2} = 2.91(5730 \text{ y}) = \boxed{1.67 \times 10^4 \text{ y}}$$

CHECK If the bone were from a recently living organism, we would expect the decay rate to be a steady [(15 decays/min)/g](200 g) = 3000 decays/min. The current decay rate is given as 400 decays/min. Because 400/3000 is roughly 1/8 (actually 1/7.5), the sample must be approximately three half-lives old, which is about 3(5730 y). This is in agreement with the step-5 result of 2.91(5730 y).

PRACTICE PROBLEM 40-2 The Check of Example 40-4 states, "Because 400/3000 is roughly 1/8 (actually 1/7.5), the sample must be approximately three half-lives old . . . ." Explain why this ratio of 1/8 implies an age equal to three half-lives.

GAMMA DECAY

During γ decay, a nucleus in an excited state decays to a lower-energy state by the emission of a photon. This process is the nuclear counterpart of spontaneous emission of photons by atoms and molecules. Unlike β decay or α decay, neither the mass number A nor the atomic number Z change during γ decay. Because the spacing of the nuclear energy levels is of the order of 1 MeV (as compared with spacing of the order of 1 eV in atoms), the wavelengths of the emitted photons are of the order of 1 pm (1 pm = 10^{-12} m):

$$\lambda = \frac{hc}{E} \approx \frac{1240 \text{ eV}\cdot\text{nm}}{1 \text{ MeV}} = 0.00124 \text{ nm} = 1.24 \text{ pm}$$

The mean lifetime for γ decay is often very short. It is usually observed only because it follows either α decay or β decay. For example, if a radioactive parent nucleus decays by β decay to an excited state of the daughter nucleus, the daughter nucleus then decays to its ground state by γ emission. Direct measurements of mean lifetimes as short as approximately 10^{-11} s are possible. Measurements of mean lifetimes shorter than 10^{-11} s are difficult, but they can sometimes be made by indirect methods.

A few γ emitters have very long lifetimes, of the order of hours. The energy states that do have such long lifetimes are called **metastable states.**

ALPHA DECAY

All very heavy nuclei ($Z > 83$) are potentially unstable via α decay because the mass of the original radioactive nucleus is greater than the sum of the masses of the decay products—an α particle and the daughter nucleus. Consider the decay of ^{232}Th ($Z = 90$) into ^{228}Ra ($Z = 88$) and an α particle. This process is written as

$$^{232}\text{Th} \rightarrow {}^{228}\text{Ra} + \alpha = {}^{228}\text{Ra} + {}^{4}\text{He} \qquad 40\text{-}18$$

The mass of the ^{232}Th atom is 232.038 050 u. The mass of the daughter atom ^{228}Ra is 228.031 064 u. Adding 4.002 603 u (the mass of ^{4}He) to the mass of ^{228}Ra, we get 232.033 667 u for the total mass of the decay products. This value is less than the mass of ^{232}Th by 0.004 383 u, which multiplied by 931.5 MeV/c^2 gives 4.08 MeV/c^2 for the excess mass of ^{232}Th when compared to the total mass of the decay products. The isotope ^{232}Th is therefore potentially unstable to α decay. This decay does in fact occur in nature with the emission of an α particle of kinetic energy 4.08 MeV. (The kinetic energy of the α particle is actually somewhat less than 4.08 MeV because some of the released energy is taken up by the recoiling ^{228}Ra nucleus.)

When a nucleus emits an α particle, both N and Z decrease by 2 and A decreases by 4. The daughter of a radioactive nucleus is often itself radioactive and decays by either α decay or β decay or both. If the original nucleus has a mass number A that is 4 times an integer, the daughter nucleus and all those in the decay chain will also have mass numbers equal to 4 multiplied by an integer. Similarly, if the mass number of the original nucleus is $4n + 1$, where n is an integer, all the nuclei in the decay chain will have mass numbers given by $4n + 1$, where n decreases by one at each α decay. We can see, therefore, that there are four possible α-decay chains, depending on whether A equals $4n$, $4n + 1$, $4n + 2$, or $4n + 3$, where n is an integer. All but one of these decay chains are found on Earth. The $4n + 1$ series is not found because its longest-lived member (other than the stable end product ^{209}Bi) is ^{237}Np, which has a half-life of only 2×10^6 y. Because this period is much less than the age of Earth, this series has disappeared.

Figure 40-6 shows the thorium series, for which $A = 4n$. It begins with an α decay from ^{232}Th to ^{228}Ra. The daughter nuclide of an α decay is on the left or neutron-rich side of the stability curve (the dashed line in the figure), so it often decays by β^- decay. In the thorium series, ^{228}Ra decays by β^- decay to ^{228}Ac, which in turn decays by β^- decay to ^{228}Th. There are then four α decays to ^{212}Pb, which decays by β^- decay to ^{212}Bi. The series branches at ^{212}Bi, which decays either by α decay to ^{208}Tl or by β^- decay to ^{212}Po. The branches meet at the stable lead isotope ^{208}Pb.

The energies of α particles from natural radioactive sources range from approximately 4 MeV to 7 MeV, and the half-lives of the sources range from approximately 10^{-5} s to 10^{10} y. In general, the smaller the energy of the emitted α particle, the longer the half-life. As we discussed in Section 35-4, the enormous variation

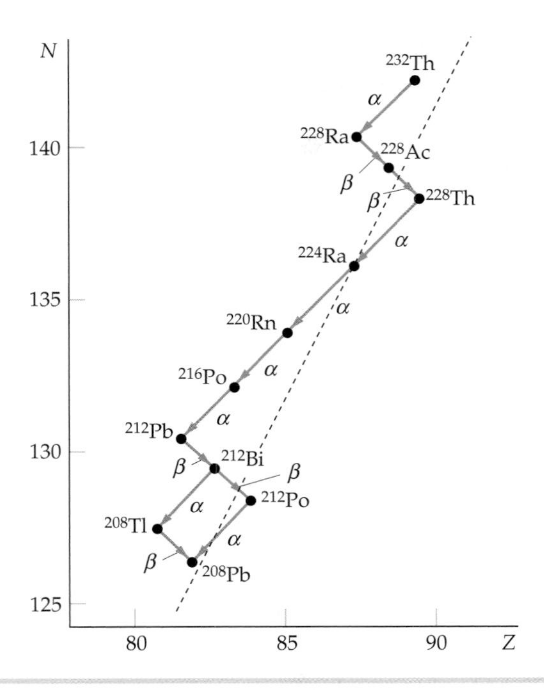

FIGURE 40-6 The thorium (4n) α decay series. The dashed line is the curve of stability.

in half-lives was explained by George Gamow in 1928. He considered α decay to be a process in which an α particle is first formed inside a nucleus and then tunnels through the Coulomb barrier (Figure 40-7). A slight increase in the energy of the α particle reduces the relative height $U_{max} - E$ of the barrier and also the thickness $r_1 - R$. Because the probability of penetration is so sensitive to the relative height and thickness of the barrier, a small increase in E leads to a large increase in the probability of barrier penetration and therefore to a significantly shorter lifetime. Gamow was able to derive an expression for the half-life as a function of E that is in excellent agreement with experimental results.

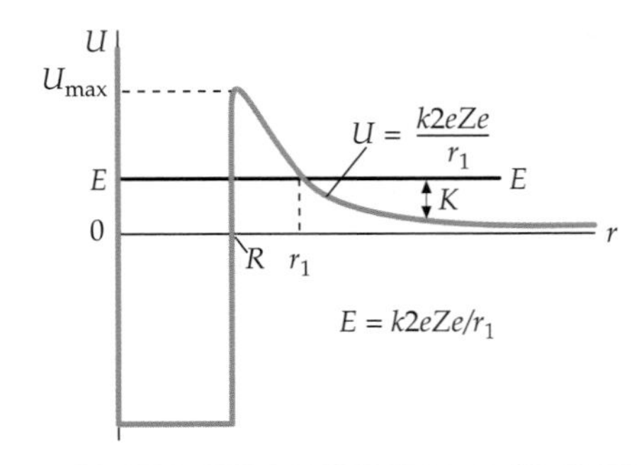

FIGURE 40-7 A model of the potential energy for an α particle and a nucleus. The strong attractive nuclear force that exists for values of r less than the nuclear radius R is indicated by the potential well. Outside the nucleus, the nuclear force is negligible, and the potential energy is given by the Coulomb potential energy function $U = +k2eZe/r$, where Ze is the nuclear charge and $2e$ is the charge of the α particle. The kinetic energy K of the α particle is equal to the energy E when the α particle is far away from the nucleus. A small increase in E reduces the relative height $U_{max} - E$ of the barrier and also reduces its thickness $r_1 - R$, leading to a much greater probability of penetration. An increase in the energy of the emitted α particles by a factor of 2 results in a reduction of the half-life by a factor of more than 10^{20}.

40-3 NUCLEAR REACTIONS

Information about nuclei is typically obtained by bombarding the nuclei with various particles and observing the results. Although the first experiments of this type were limited by the need to use naturally occurring radiation, they produced many important discoveries. In 1932, J. D. Cockcroft and E. T. S. Walton succeeded in producing the reaction

$$\mathrm{p} + {}^{7}\mathrm{Li} \rightarrow {}^{8}\mathrm{Be} \rightarrow {}^{4}\mathrm{He} + {}^{4}\mathrm{He}$$

using artificially accelerated protons. At about the same time, the Van de Graaff electrostatic generator (by R. Van de Graaff in 1931) and the first cyclotron (by E. O. Lawrence and M. S. Livingston in 1932) were built. Since then, enormous advances in the technology for accelerating and detecting particles have been made, and many nuclear reactions have been studied.

When a particle is incident on a nucleus, several different things can happen. The incident particle may be scattered, either elastically or inelastically, or the incident particle may be absorbed by the nucleus, and another particle or particles may be emitted. In inelastic scattering, the nucleus is left in an excited state and subsequently decays by emitting photons (or other particles).

The amount of energy released or absorbed during a reaction (in the center of mass reference frame) is called the ***Q* value** of the reaction. The Q value equals c^2 multiplied by the mass difference. When energy is released during a reaction, the reaction is said to be an **exothermic reaction.** During an exothermic reaction, the total mass of the incoming particles is greater than the total mass of the outgoing particles, and the Q value is positive. If the total mass of the incoming particles is less than that of the outgoing particles, energy is required for the reaction to take place, and the reaction is said to be an **endothermic reaction.** The Q value of an endothermic reaction is negative. In general, if Δm is the change in mass, the Q value is

$$Q = -(\Delta m)c^2 \qquad \text{40-19}$$

Q VALUE

An endothermic reaction cannot take place below a specific threshold energy. In the laboratory reference frame in which stationary particles are bombarded by incoming particles, the threshold energy is somewhat greater than $|Q|$ because the outgoing particles must have some kinetic energy to conserve momentum.

A measure of the effective size of a nucleus for a particular nuclear reaction is the **cross section** σ. If I is the number of the incident particles per unit time per unit area (the incident intensity) and R is the number of reactions per unit time per nucleus, the cross section is

$$\sigma = \frac{R}{I} \qquad \text{40-20}$$

The cross section σ has the dimensions of area. Because nuclear cross sections are of the order of the square of the nuclear radius, a convenient unit for them is the **barn,** which is defined as

$$1 \text{ barn} = 10^{-28}\ \text{m}^2 \qquad \text{40-21}$$

The cross section for a particular reaction is a function of energy. For an endothermic reaction, it is zero for energies below the threshold energy.

Example 40-5 Exothermic or Endothermic?

Find the Q value of the reaction $\mathrm{p} + {}^{7}\mathrm{Li} \rightarrow {}^{4}\mathrm{He} + {}^{4}\mathrm{He}$ and state whether the reaction is exothermic or endothermic.

PICTURE We find the masses of the atoms from Table 40-1 and calculate the difference in the total mass of the outgoing particles and the incoming particles. The Q value is related to the change in mass Δm by $Q = -(\Delta m)c^2$. If we use the mass of protium rather than the mass of the proton, there will be four electrons on each side of the reaction, so the electron masses will cancel.

SOLVE

1. Find the mass of each atom from Table 40-1:
$$^1\text{H} \quad 1.007\,825\text{ u}$$
$$^7\text{Li} \quad 7.016\,004\text{ u}$$
$$^4\text{He} \quad 4.002\,603\text{ u}$$

2. Calculate the initial mass m_i of the incoming particles:
$$m_i = 1.007\,825\text{ u} + 7.016\,004\text{ u} = 8.023\,829\text{ u}$$

3. Calculate the final mass m_f:
$$m_f = 2(4.002\,603\text{ u}) = 8.005\,206\text{ u}$$

4. Calculate the change in mass:
$$\Delta m = m_f - m_i = 8.005\,206\text{ u} - 8.023\,829\text{ u}$$
$$= -0.018\,623\text{ u}$$

5. Calculate the Q value:
$$Q = -(\Delta m)c^2 = (+0.018\,623\text{ u})c^2 \times \frac{931.5\text{ MeV}}{1\text{ u}}$$
$$= \boxed{17.35\text{ MeV}}$$

$\boxed{Q \text{ is positive, so the reaction is exothermic.}}$

CHECK Because the initial mass is greater than the final mass, the initial energy is greater than the final energy and the reaction is exothermic, yielding 17.35 MeV.

REACTIONS WITH NEUTRONS

Nuclear reactions that involve neutrons are important for understanding nuclear reactors. The most likely reaction between a nucleus and a neutron that has an energy of more than about 1 MeV is scattering. However, even if the scattering is elastic, the neutron loses some energy to the nucleus because the nucleus recoils. If a neutron is scattered many times in a material, its energy decreases until the neutron is of the order of the energy of thermal motion kT, where k is Boltzmann's constant and T is the absolute temperature. (At ordinary room temperatures, kT is approximately 0.025 eV.) The neutron is then equally likely to gain or lose energy from a nucleus when it is elastically scattered. A neutron that has an energy of the order of kT is called a **thermal neutron.**

At low energies, a neutron is likely to be captured, producing an excited nucleus. A γ ray is then emitted from the excited nucleus. Figure 40-8 shows the neutron-capture cross section for silver as a function of the energy of the neutron. The large

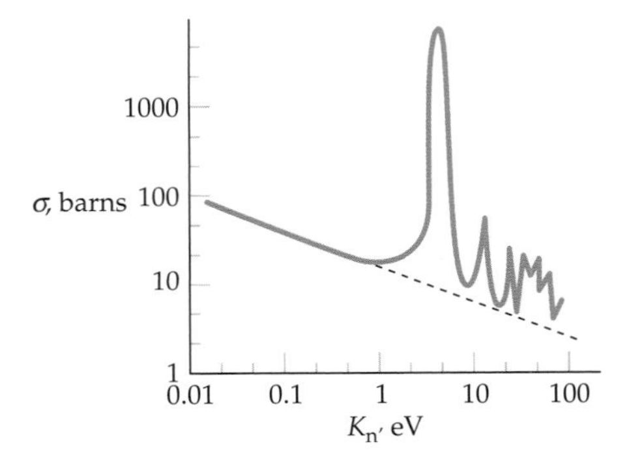

FIGURE 40-8 Neutron-capture cross section for silver as a function of the energy of the incident neutron. The straight line indicates the $1/v$ dependence of the cross section, which is proportional to the time spent by the neutron in the vicinity of the silver nucleus. Superimposed on this dependence are a large resonance and several smaller resonances.

peak in this curve is called a **resonance.** Except for the resonance, the cross section varies fairly smoothly with energy, decreasing with increasing energy roughly as $1/v$, where v is the speed of the neutron. We can understand this energy dependence as follows: Consider a neutron moving with speed v near a nucleus of diameter $2R$. The time it takes the neutron to pass the nucleus is $2R/v$. Thus, the neutron-capture cross section is proportional to the time spent by the neutron in the vicinity of the silver nucleus. The dashed line in Figure 40-8 indicates this $1/v$ dependence. At the maximum of the resonance, the value of the cross section is very large ($\sigma > 5000$ barns) compared with a value of only about 10 barns just past the resonance. Many elements show similar resonances in their neutron-capture cross sections. For example, the maximum cross section for ^{113}Cd is approximately 57 000 barns. This material is thus very useful for shielding against low-energy neutrons.

An important nuclear reaction that involves neutrons is fission, which is discussed in the next section.

40-4 FISSION AND FUSION

Figure 40-9 shows a plot of the nuclear mass difference per nucleon $(M - Zm_p - Nm_n)/A$ in units of MeV/c^2 versus A. This curve is just the negative of the binding-energy curve shown in Figure 40-3. From Figure 40-9, we can see that the values for the mass difference per nucleon for both very heavy ($A \approx 200$) and very light ($A \leq 20$) nuclides are greater than the values for nuclides of intermediate mass. Thus, energy is released when a very heavy nucleus, such as ^{235}U, breaks up into two lighter nuclei—during a process called **fission**—or when two very light nuclei, such as ^{2}H and ^{3}H, fuse together to form a nucleus of greater mass—during a process called **fusion.**

The applications of both fission and fusion to the generation of electrical power and the development of nuclear weapons have had a profound effect on our lives since the early twentieth century. The application of these reactions to the development of energy resources may have an even greater effect in the future. We will look at some of the features of fission and fusion that are important for their application in reactors to generate power.

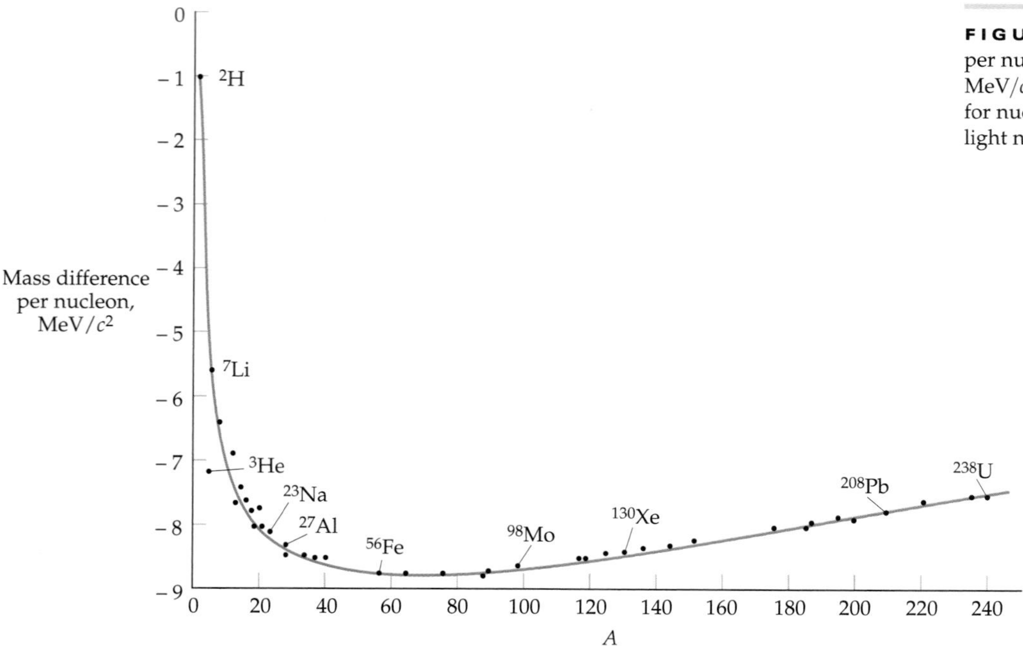

FIGURE 40-9 Plot of mass difference per nucleon $(M - Zm_p - Nm_n)/A$ in units of MeV/c^2 versus A. The mass per nucleon is less for nuclei of intermediate mass than for very light nuclei or very heavy nuclei.

(a)

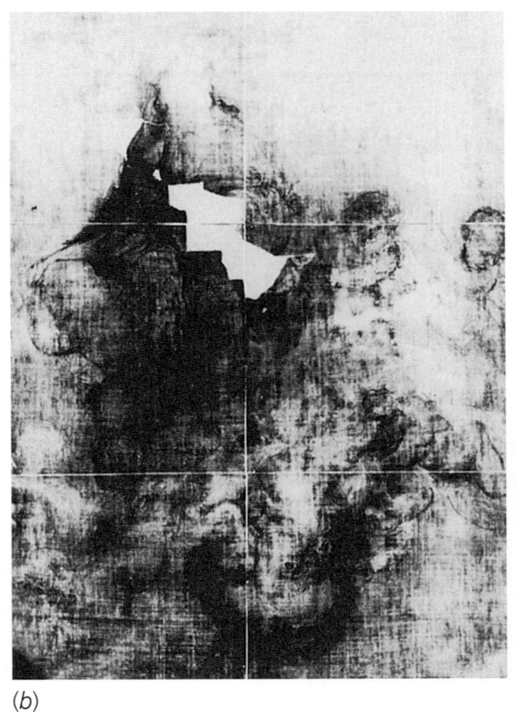

(b)

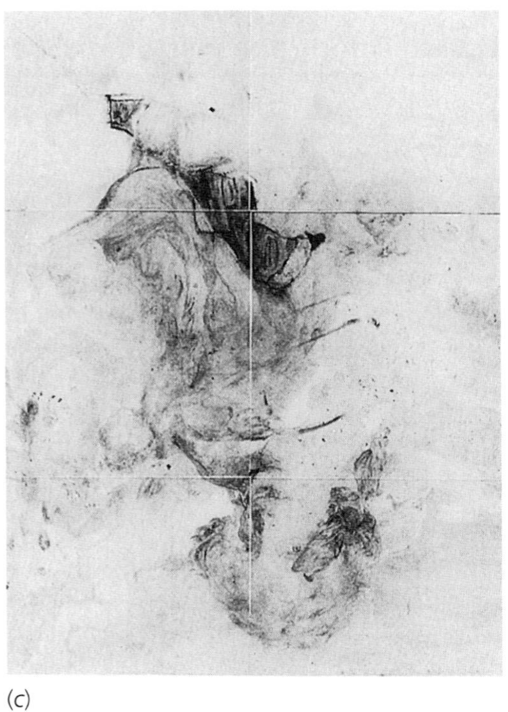

(c)

Hidden layers in paintings are analyzed by bombarding the painting with neutrons and observing the radiative emissions from nuclei that have captured a neutron. Different elements used in the painting have different half-lives. (*a*) Van Dyck's painting *Saint Rosalie Interceding for the Plague-Stricken of Palermo.* The black-and-white images in (*b*) and (*c*) were formed using a special film sensitive to electrons emitted by the radioactively decaying elements. Image (*b*), taken a few hours after the neutron irradiation, reveals the presence of manganese, found in umber, which is a dark earth pigment used for the painting's base layer. (Blank areas show where modern repairs, free of manganese, have been made.) The image in (*c*) was taken 4 days later, after the umber emissions had died away and when phosphorus, found in charcoal and boneblack, was the main radiating element. Upside down is revealed a sketch of Van Dyck himself. The self-portrait, executed in charcoal, had been overpainted by the artist. *((a) © 1991 by the Metropolitan Museum of Art. (b) and (c) Courtesy of Paintings Conservation Department, Metropolitan Museum of Art.)*

FISSION

Very heavy nuclei ($Z > 92$) are subject to spontaneous fission. They break apart into two nuclei even if the nuclei are not disturbed. We can understand this by considering the analogy of a charged liquid drop. If the drop is not too large, surface tension can overcome the repulsive forces of the charges and hold the drop together. There is, however, a certain maximum size beyond which the drop will be unstable and will spontaneously break apart. Because of spontaneous fission, an upper limit exists on the size of a nucleus and therefore on the number of elements that are possible.

Some heavy nuclei—uranium and plutonium, in particular—can be induced to fission by the capture of neutrons. During the fission of ^{235}U, for example, the uranium nucleus is excited by the capture of a neutron, causing it to split into two nuclei and emit several neutrons. The Coulomb force of repulsion drives the fission fragments apart, with the released energy eventually appearing as thermal energy. Consider, for example, the fission of a nucleus of mass number $A = 200$ into two nuclei of mass number $A = 100$. Because the rest energy for $A = 200$ is about 1 MeV per nucleon greater than that for $A = 100$, approximately 200 MeV per nucleus is released during such a fission. This is a large amount of energy. By contrast, during the chemical reaction of combustion, only about 4 eV of energy is released per molecule of oxygen consumed.

Example 40-6 Energy Released During the Fission of ^{235}U

Calculate the total energy (in kilowatt-hours) released during the fission of 1.00 g of ^{235}U, assuming that 200 MeV is released per fission.

PICTURE We need to find the number of uranium nuclei in one gram of ^{235}U, which we find using the fact that there are Avogadro's number ($N_A = 6.02 \times 10^{23}$) of nuclei in 235 grams.

SOLVE

1. The total energy is the number of nuclei multiplied by the energy per nucleus:

$$E = NE_{\text{nucleus}} = N(200\text{MeV/nucleus})$$

2. Calculate N:

$$N = \frac{6.02 \times 10^{23}\ \text{nuclei/mol}}{235\ \text{g/mol}} \times 1.00\ \text{g} = 2.56 \times 10^{21}\ \text{nuclei}$$

3. Calculate the energy per gram in eV and convert to kW · h:

$$E = \frac{200 \times 10^{6}\ \text{eV}}{1\ \text{nucleus}} \times 2.56 \times 10^{21}\ \text{nuclei}$$
$$= 5.12 \times 10^{29}\ \text{eV} = 8.19 \times 10^{10}\ \text{J}$$
$$= 8.19 \times 10^{7}\ \text{kW}\cdot\text{s} = \boxed{2.28 \times 10^{4}\ \text{kW}\cdot\text{h}}$$

The fission of uranium was discovered in 1938 by Otto Hahn and Fritz Strassmann, who found that medium-mass elements (for example, barium and lanthanum) were produced in the bombardment of uranium with neutrons. The discovery that several neutrons were emitted during the fission process led to speculation concerning the possibility of using those neutrons to cause further

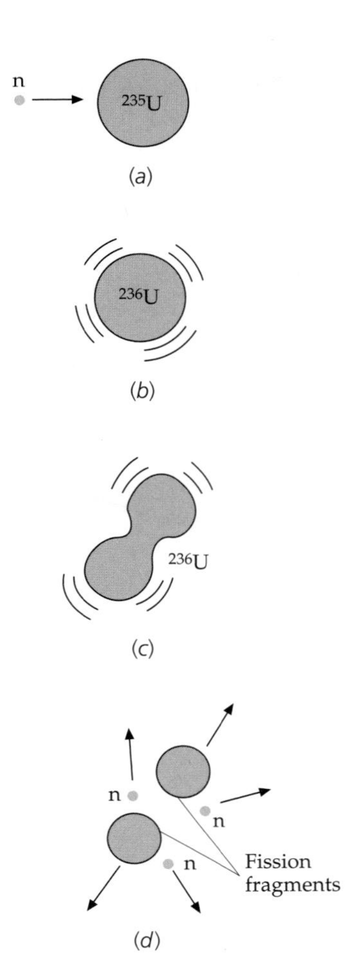

FIGURE 40-10 Schematic illustration of nuclear fission. (*a*) The absorption of a neutron by ^{235}U leads to (*b*) ^{236}U in an excited state. (*c*) The oscillation of ^{236}U has become unstable. (*d*) The nucleus splits apart into two nuclei that are less massive than the original nucleus and emits several neutrons that can produce fission in other nuclei.

fissions, thereby producing a chain reaction. When ^{235}U captures a neutron, the resulting ^{236}U nucleus emits γ rays as it de-excites to the ground state approximately 15 percent of the time and undergoes fission approximately 85 percent of the time. The fission process is somewhat analogous to the oscillation of a liquid drop, as shown in Figure 40-10. If the oscillations are violent enough, the drop splits in two. Using the liquid-drop model, Niels Bohr and John Wheeler calculated the critical energy E_c needed by the ^{236}U nucleus to undergo fission. (^{236}U is the nucleus formed momentarily by the capture of a neutron by ^{235}U.) For this nucleus, the critical energy is 5.3 MeV, which is less than the 6.4 MeV of excitation energy produced when ^{235}U captures a neutron. The capture of a neutron by ^{235}U therefore produces an excited state of the ^{236}U nucleus that has more than enough energy to break apart. On the other hand, the critical energy for fission of the ^{239}U nucleus is 5.9 MeV. The capture of a neutron by a ^{238}U nucleus produces an excitation energy of only 5.2 MeV. Therefore, when a neutron is captured by ^{238}U to form ^{239}U, the excitation energy is not great enough for fission to occur. In this case, the excited ^{239}U nucleus de-excites by γ emission and then decays to ^{239}Np by β decay, and then again to ^{239}Pu by β decay.

A fissioning nucleus can split into a pair of medium-mass nuclei, as shown in Figure 40-11. Depending on the particular reaction, 1, 2, or 3 neutrons may be emitted. The average number of neutrons emitted in the fission of ^{235}U is approximately 2.5. A typical fission reaction is

$$n + {}^{235}\mathrm{U} \rightarrow {}^{141}\mathrm{Ba} + {}^{92}\mathrm{Kr} + 3n$$

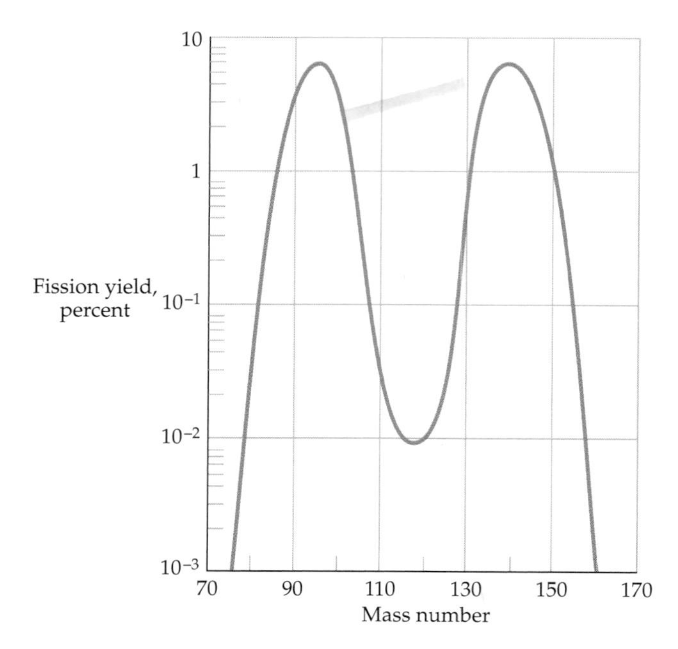

FIGURE 40-11 Distribution of the possible fission fragments of ^{235}U. The splitting of ^{235}U into two fragments of unequal mass is more likely than its splitting into fragments of equal mass.

NUCLEAR FISSION REACTORS

To sustain a chain reaction in a fission reactor, one of the neutrons (on average) that is emitted during and following* the fission of ^{235}U must be captured by another ^{235}U nucleus and cause it to fission. The **reproduction constant** k of a reactor is defined as the average number of neutrons from each fission that cause a subsequent fission. The maximum possible value of k for a uranium reactor is 2.5, but it is normally less than this for two important reasons: (1) Some of the neutrons may escape from the region containing fissionable nuclei and (2) some of the neutrons may be captured by nonfissioning nuclei in the reactor. If k is exactly 1, the reaction will be self-sustaining. If k is less than 1, the reaction will die out. If k is significantly greater than 1, the reaction rate will increase rapidly and become uncontrollable. In the design of nuclear bombs, such a runaway reaction is desired. In power reactors, the value of k must be kept very nearly equal to 1.

Because the neutrons emitted during and following fission have energies of the order of 1 MeV, whereas the chance for neutron capture leading to fission in ^{235}U is largest at small energies, the chain reaction can be sustained only if the neutrons are slowed down before they escape from the reactor. At high energies (1 MeV to 2 MeV), neutrons lose energy rapidly by inelastic scattering from ^{238}U, the principal constituent of natural uranium. (Natural uranium contains 99.3 percent ^{238}U and only 0.7 percent fissionable ^{235}U.) Once the neutron energy is below the excitation energies of the nuclei in the reactor (about 1 MeV), the main process of energy loss is by elastic scattering, in which a fast neutron collides with a nucleus at rest and transfers some of its kinetic energy to that nucleus. Such energy transfers are efficient only if the masses of the two bodies are comparable. A neutron will not transfer much energy in an elastic collision with a heavy uranium nucleus. Such a collision is like one between a marble and a billiard ball. The marble will be deflected by the much more massive billiard ball, and very little of its kinetic energy will be transferred to the billiard ball. A **moderator** consisting of material, such as water or carbon, that has light nuclei is therefore placed around the fissionable material in the core of the reactor to slow down the neutrons.

* Neutrons are sometimes emitted by the fission products. These neutrons are typically emitted a few seconds following the fission.

The inside of a nuclear power plant in Kent, England. A technician is standing on the reactor charge transfer plate, into which uranium fuel rods fit. *(© Jerry Mason/ Photo Researchers.)*

The neutrons are slowed down by elastic collisions with the nuclei of the moderator until they are in thermal equilibrium with the moderator. Because of the relatively large neutron-capture cross section of the hydrogen nucleus in water, reactors that use ordinary water as a moderator cannot easily achieve $k \approx 1$ unless they use enriched uranium, in which the ^{235}U content has been increased from 0.7 percent to between 1 percent and 4 percent. Natural uranium can be used if heavy water (D_2O) is used instead of ordinary (light) water (H_2O) as the moderator. Although heavy water is expensive, most Canadian reactors use heavy water for a moderator to avoid the cost of constructing uranium-enrichment facilities.

Figure 40-12 shows some of the features of a pressurized-water reactor commonly used in the United States to generate electricity. Fission in the core heats the water to a high temperature in the primary loop, which is closed. This water, which also serves as the moderator, is under high pressure to prevent the water from boiling.

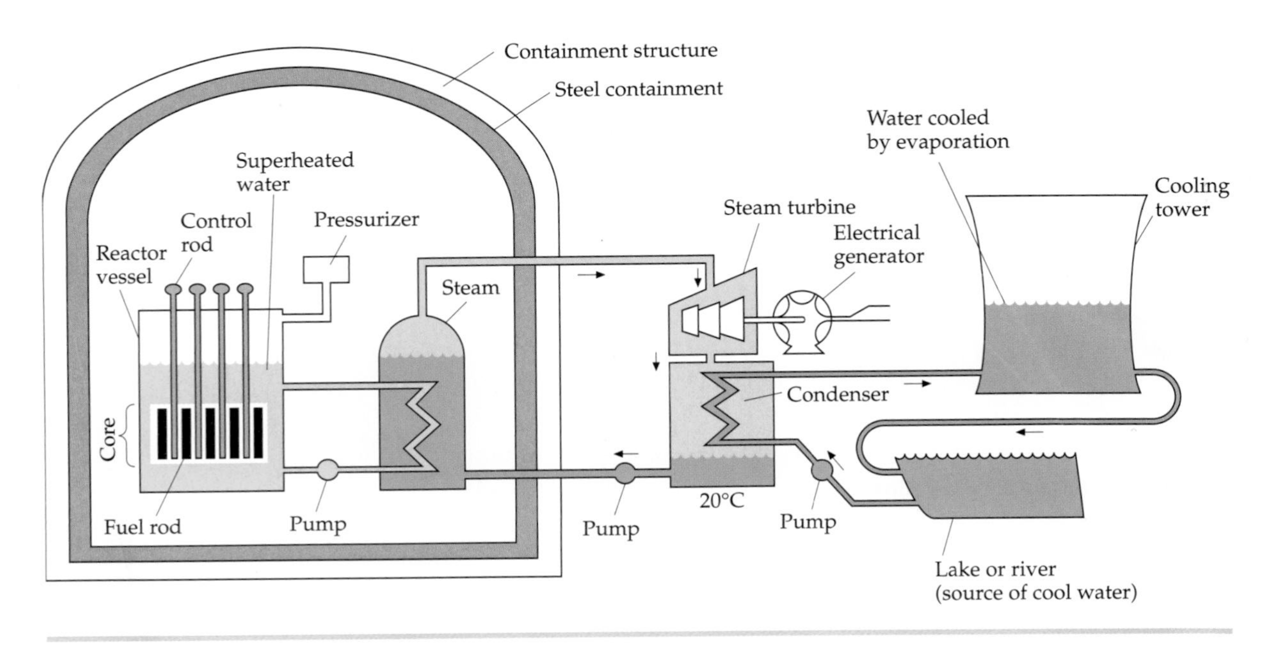

FIGURE 40-12 Simplified drawing of a pressurized-water reactor. The water in contact with the reactor core serves as both the moderator and the heat-transfer material. It is isolated from the water used to produce the steam that drives the turbines. Many features, such as the backup cooling mechanisms, are not shown here.

The hot water is pumped to a heat exchanger, where it heats the water in the secondary loop and converts the water to steam, which is then used to drive the turbines that produce electrical power. Note that the water in the secondary loop is isolated from the water in the primary loop to prevent its contamination by the radioactive nuclei in the reactor core.

The ability to control the reproduction factor k precisely is important if a power reactor is to be operated safely. Both natural negative-feedback mechanisms and mechanical methods of control are used. If k is greater than 1, the reaction rate increases and the temperature of the reactor increases. If water is used as a moderator, its density decreases with increasing temperature and the water becomes a less effective moderator. A second important control method is the use of control rods made of a material, such as cadmium, that has a very large neutron-capture cross section. To decrease the reaction rate, the control rods are inserted so that more neutrons are captured by the rods and k becomes less than 1. To increase the reaction rate, the rods are gradually withdrawn from the reactor; fewer neutrons are captured by the control rods and k becomes greater than 1.

Mechanical control of the reaction rate of a nuclear reactor using control rods is possible only because some of the neutrons emitted during the fission process are **delayed neutrons.** The time needed for a neutron to slow down from 1 MeV or 2 MeV to the thermal-energy level and then be captured is only of the order of a millisecond. If all the neutrons emitted during fission were prompt neutrons, that is, emitted immediately during the fission process, mechanical control would not be possible because the reactor would run away before the rods could be inserted farther. However, approximately 0.65 percent of the neutrons emitted are delayed by an average time of about 14 s. Those neutrons are emitted not during the fission process itself but during the decay of the fission fragments. The effect of the delayed neutrons can be seen in the following examples.

Example 40-7 Doubling Time

If the average time between fission generations (the time it takes for a neutron emitted during one fission to cause another) is $t_1 = 1\text{ ms} = 0.001\text{ s}$ and if the average number of neutrons from each fission that cause a subsequent fission is 1.001, how long will it take for the reaction rate to double?

PICTURE The reaction rate is the number of nuclei that fission per unit time. The time to double the reaction rate is the product of the number of generations N needed to double the reaction rate and the generation time. If $k = 1.001$, the reaction rate after N generations is 1.001^N. We find the number of generations by setting 1.001^N equal to 2 and solving for N.

SOLVE

1. Set 1.001^N equal to 2 and solve for N:

$$(1.001)^N = 2$$

$$N \ln 1.001 = \ln 2$$

$$N = \frac{\ln 2}{\ln 1.001} = 693$$

2. Multiply the number of generations by the generation time:

$$t = Nt_1 = 693(0.001\text{ s}) = \boxed{0.7\text{ s}}$$

CHECK The step-2 result of 0.7 s is approximately 700 times the average time between generations. This many generations is plausible because the reproduction factor k is so close to 1.

TAKING IT FURTHER The doubling time of about 0.7 s is not enough time for insertion of control rods.

Example 40-8 Delayed Neutrons and Control-Rod Insertion

Try-It-Yourself

Assuming that 0.65 percent of the neutrons emitted are delayed by 14 s, find the average generation time and the doubling time if $k = 1.001$.

PICTURE The doubling time is Nt_{av}, where t_{av} is the average time between generations. Since 99.35 percent of the generation times are 0.001 s and 0.65 percent are 14 s, the average generation time is 0.9935(0.001 s) + 0.0065(14 s).

SOLVE

Cover the column to the right and try these on your own before looking at the answers.

Steps	Answers
1. Compute the average generation time.	$t_{av} = 0.9935(0.001\text{ s}) + 0.0065(14\text{ s}) = 0.092\text{ s}$
2. Use your result to find the time for 693 generations.	$t = 63.8\text{ s} \approx \boxed{60\text{ s}}$

CHECK The number of delayed neutrons is approximately 0.7 percent of the total number of neutrons, but the generation time of 1 ms is approximately 0.007 percent of 14 s. Thus, an increase in the doubling time by a factor of about 100 is plausible.

TAKING IT FURTHER A doubling time of 60 s is plenty of time for mechanical insertion of control rods.

Because of the limited supply of natural uranium, the small fraction of ^{235}U in natural uranium, and the limited capacity of enrichment facilities, reactors based on the fission of ^{235}U cannot meet our energy needs for very long. A promising alternative is the **breeder reactor.** When the relatively plentiful but nonfissionable ^{238}U nucleus captures a neutron, it decays by β decay (with a half-life of 20 min) to ^{239}Np, which in turn decays by β decay (with a half-life of 2.35 days) to the fissionable nuclide ^{239}Pu. Because ^{239}Pu fissions with fast neutrons, no moderator is needed. A reactor initially fueled with a mixture of ^{238}U and ^{239}Pu will breed as much fuel as it uses or more if one or more of the neutrons emitted in the fission of ^{239}Pu is captured by ^{238}U. Practical studies indicate that a typical breeder reactor can be expected to double its fuel supply in 7 to 10 years.

There are two major safety problems inherent with breeder reactors. The fraction of delayed neutrons is only 0.3 percent for the fission of ^{239}Pu, so the time between generations is much less than that for ordinary reactors. Mechanical control is therefore much more difficult. Also, because the operating temperature of a breeder reactor is relatively high and a moderator is not desired, a heat-transfer material, such as liquid sodium metal, is used rather than water (which is the moderator as well as the heat-transfer material in an ordinary reactor). If the temperature of the reactor increases, the resulting decrease in the density of the heat-transfer material leads to positive feedback, because it will absorb fewer neutrons than before. Because of these safety considerations, breeder reactors are not yet in commercial use in the United States. There are, however, several in operation in France, Great Britain, and the former Soviet Union.

FUSION

During fusion, two light nuclei, such as deuterium (^{2}H) and tritium (^{3}H), fuse together to form a heavier nucleus. A typical fusion reaction is

$$^2\text{H} + {}^3\text{H} \rightarrow {}^4\text{He} + \text{n} + 17.6\text{ MeV}$$

The energy released in fusion depends on the particular reaction. For the $^2H + {}^3H$ reaction, the energy released is 17.6 MeV. Although this energy is less than the energy released during a fission reaction, it is a greater amount of energy per unit mass. The energy released during this fusion reaction is (17.6 MeV)/(5 nucleons) = 3.52 MeV per nucleon. This is approximately 3.5 times as great as the 1 MeV per nucleon released in fission.

The production of power from the fusion of light nuclei holds great promise because of the relative abundance of the fuel and the absence of some of the dangers inherent in fission reactors. Unfortunately, the technology necessary to make fusion a practical source of energy has not yet been developed. We will consider the $^2H + {}^3H$ reaction; other reactions present similar problems.

Because of the Coulomb repulsion between the ^2H and ^3H nuclei, very large kinetic energies, of the order of 1 MeV, are needed to get the nuclei close enough together for the attractive nuclear forces to become effective and to cause fusion. Such energies can be obtained in an accelerator, but because the scattering of one nucleus by the other is much more probable than fusion, the bombardment of one nucleus by another in an accelerator requires the input of more energy than is recovered. To obtain energy from fusion, the particles must be heated to a temperature great enough for the fusion reaction to occur as the result of random thermal collisions. Because a significant number of particles have kinetic energies greater than the mean kinetic energy, $\frac{3}{2}kT$, and because some particles can tunnel through the Coulomb barrier, a temperature T corresponding to $kT \approx 10$ keV is adequate to ensure that a reasonable number of fusion reactions will occur if the density of the particles is sufficiently high. The temperature corresponding to $kT = 10$ keV is of the order of 10^8 K. These temperatures occur in the interiors of stars, where such reactions are common. At these temperatures, a gas consists of positive ions and electrons and is called a **plasma.** One of the problems arising in attempts to produce controlled fusion reactions is the problem of confining the plasma long enough for the reactions to take place. In the interior of the Sun, the plasma is confined by the enormous gravitational field of the Sun. In a laboratory on Earth, confinement is a difficult problem.

The energy required to heat a plasma is proportional to the number density of its ions, n, whereas the collision rate is proportional to n^2 (the square of the number density). If τ is the confinement time, the output energy is proportional to $n^2\tau$. If the output energy is to exceed the input energy, we must have

$$C_1 n^2 \tau > C_2 n$$

where C_1 and C_2 are constants. In 1957, the British physicist J. D. Lawson evaluated these constants from estimates of the efficiencies of various hypothetical fusion reactors and derived the following relation between density and confinement time, known as **Lawson's criterion:**

$$n\tau > 10^{20}\ \mathrm{s \cdot particles/m^3} \qquad \text{40-22}$$

LAWSON'S CRITERION

If Lawson's criterion is met and the thermal energy of the ions is great enough ($kT \approx 10$ keV), the energy released by a fusion reactor will just equal the energy input; that is, the reactor will just break even. For the reactor to be practical, much more energy must be released.

Two schemes for achieving Lawson's criterion are currently under investigation. In one scheme, **magnetic confinement,** a magnetic field is used to confine the plasma (see Section 26-2). In the most common arrangement, first developed in the former Soviet Union and called a *tokamak*, the plasma is confined in a large toroid.

The magnetic field is a combination of the doughnut-shaped magnetic field due to the windings of the toroid and the self-field due to the current of the circulating plasma. The break-even point has almost been achieved using magnetic confinement, but we are still a long way from building a practical fusion reactor.

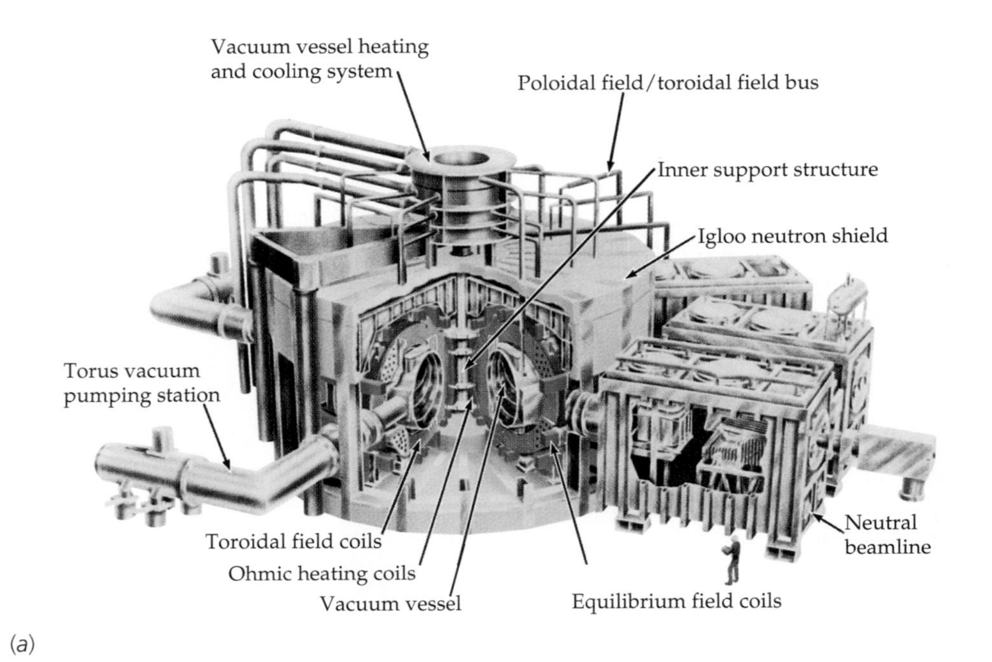

(*a*)

(*b*)

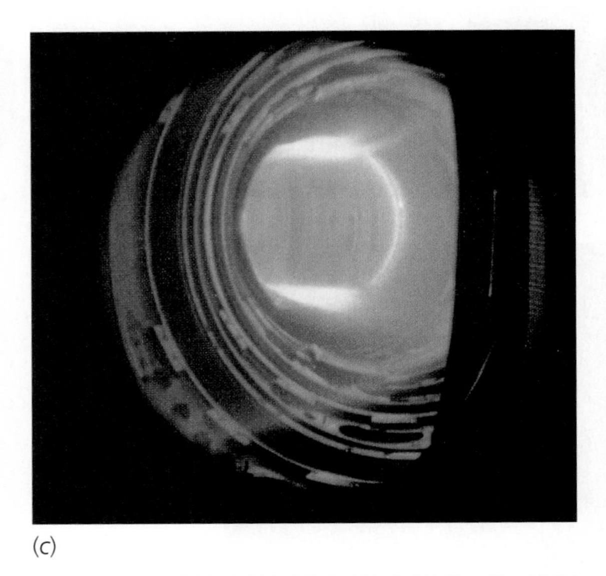

(*c*)

(*a*) Schematic of the Tokamak Fusion Test Reactor (TFTR). The toroidal coils, surrounding the doughnut-shaped vacuum vessel, are designed to conduct current for 3-s pulses, separated by waiting times of 5 min. Pulses peak at 73 000 A, producing a magnetic field of 5.2 T. This magnetic field is the principal means of confining the deuterium–tritium plasma that circulates within the vacuum vessel. Current for the pulses is delivered by converting the rotational energy of two 600-ton flywheels. Sets of poloidal coils, perpendicular to the toroidal coils, carry an oscillating current that generates a current through the confined plasma itself, heating it ohmically. Additional poloidal fields help stabilize the confined plasma. Between four and six neutral-beam injection systems (only one of which is shown in the schematic) are used to inject high-energy deuterium atoms into the deuterium–tritium plasma, heating beyond what could be obtained ohmically, ultimately to the point of fusion. (*b*) The TFTR itself. The diameter of the vacuum vessel is 7.7 m. (*c*) An 800-kA plasma, lasting 1.6 s, as it discharges within the vacuum vessel. *((All) Courtesy of the Princeton Plasma Physics Laboratory.)*

In a second scheme, called **inertial confinement,** a pellet of solid deuterium and tritium is bombarded from all sides by intense pulsed laser beams of energies of the order of 10^4 J lasting about 10^{-8} s. (Intense beams of ions are also used.) Computer simulation studies indicate that the pellet should be compressed to approximately 10^4 times its normal density and heated to a temperature greater than 10^8 K. This should produce approximately 10^6 J of fusion energy in 10^{-11} s, which is so brief that confinement is achieved by inertia alone.

Because the break-even point is just barely being achieved in magnetic-confinement fusion, and because the building of a fusion reactor involves many practical problems that have not yet been solved, the availability of fusion to meet our energy needs is not expected for at least several decades. However, fusion holds great promise as an energy source for the future.

(*a*)

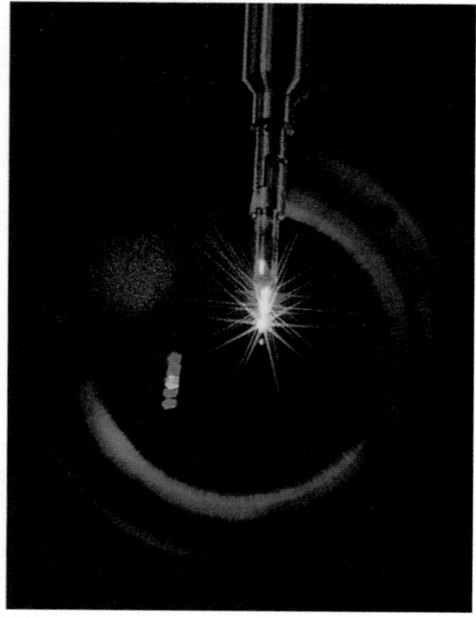

(b)

(*a*) The Nova target chamber, an aluminum sphere approximately 5 m in diameter, inside which 10 beams from the world's most powerful laser converge onto a hydrogen-containing pellet 0.5 mm in diameter. (*b*) The resulting fusion reaction is visible as a tiny star, lasting 10^{-10} s, releasing 10^{13} neutrons. *((All) Courtesy of the Lawrence Livermore National Laboratory/U.S. Department of Energy.)*

Summary

TOPIC	RELEVANT EQUATIONS AND REMARKS	
1. **Properties of Nuclei**	A nucleus has N neutrons, Z protons, and a mass number $A = N + Z$. For light nuclei, N and Z are approximately equal, whereas for heavy nuclei, N is greater than Z.	
Isotopes	Isotopes are two or more nuclei that have the same atomic number Z but have different values of N and A.	
Size and shape	Most nuclei are approximately spherical in shape and have a volume that is proportional to A. Because the mass is proportional to A, nuclear density is independent of A.	
Radius	$R = R_0 A^{1/3} \approx (1.2\text{ fm})A^{1/3}$	40-1
Mass and binding energy	The mass of a stable nucleus is less than the sum of the masses of its nucleons. The mass difference Δm multiplied by c^2 equals the binding energy E_b of the nucleus. The binding energy is approximately proportional to the mass number A.	
2. **Radioactivity**	Unstable nuclei are radioactive and decay by emitting α particles (^{4}He nuclei), β particles (electrons or positrons), or γ rays (photons). All radioactivity is statistical in nature and follows an exponential decay law: $N = N_0 e^{-\lambda t}$	40-6
Decay rate	$R = \lambda N = R_0 e^{-\lambda t}$	40-7
Mean lifetime	$\tau = \frac{1}{\lambda}$	40-9
Half-life	$t_{1/2} = \tau \ln 2 = 0.693\tau$ The half-lives of α decay range from a fraction of a second to millions of years. For β decay, the half-lives range up to hours or days. For γ decay, the half-lives are usually less than a microsecond.	40-11
Decay-rate units	The number of decays per second of 1 g of radium is the curie (Ci). $1\text{ Ci} = 3.7 \times 10^{10}\text{ decays/s} = 3.7 \times 10^{10}\text{ Bq}$ $(1\text{ Bq} = 1\text{ decay/s})$	
3. **Nuclear Reactions**		
Q value	The Q value equals c^2 multiplied by the total mass of the incoming particles less the total mass of the outgoing particles in the center of mass reference frame. If the net mass change is Δm, the Q value is $Q = -(\Delta m)c^2$	40-19
Exothermic reaction	If total mass decreases during a reaction, Q is positive and measures the energy released.	
Endothermic reaction	If total mass increases during a reaction, Q is negative. Then $\lvert Q \rvert$ is the threshold energy for the reaction in the center of mass reference frame.	
4. **Fission**	Fission occurs when some heavy elements, such as ^{235}U or ^{239}Pu, capture a neutron and split apart into two nuclei. The two nuclei then fly apart because of electrostatic repulsion. A chain reaction is possible because several neutrons are emitted by a nucleus when it undergoes fission. A chain reaction can be sustained in a reactor if, on the average, one of the emitted neutrons is slowed down by scattering in the reactor and is then captured by another fissionable nucleus. Very heavy nuclei ($Z > 92$) are subject to spontaneous fission.	

TOPIC	RELEVANT EQUATIONS AND REMARKS
5. Fusion	A large amount of energy is released when two light nuclei, such as ^{2}H and ^{3}H, fuse together. Fusion takes place spontaneously inside the Sun and other stars, where the temperature is great enough (about 10^8 K) for thermal motion to bring the charged hydrogen ions close enough together to fuse. Although controlled fusion holds great promise as a future energy source, practical difficulties have thus far hindered its development.
Lawson criterion	The minimum product of particle density n and confinement time τ to get more energy out of a fusion reactor than is put in is $n\tau > 10^{20}$ s·particles/m^3.

Answer to Concept Check

41-1 The number left can be either one or zero, where zero left is more probable than one left.

Answers to Practice Problems

40-1 (*a*) $60E_1$ (*b*) $20E_1$

40-2 It is because $\frac{1}{8} = \left(\frac{1}{2}\right)^3$, so $n = 3$.

Problems

In a few problems, you are given more data than you actually need; in a few other problems, you are required to supply data from your general knowledge, outside sources, or informed estimate.

Interpret as significant all digits in numerical values that have trailing zeros and no decimal points.

- • Single-concept, single-step, relatively easy
- •• Intermediate-level, may require synthesis of concepts
- ••• Challenging
- SSM Solution is in the *Student Solutions Manual*
- Consecutive problems that are shaded are paired problems.

CONCEPTUAL PROBLEMS

1 • Isotopes of nitrogen, iron and tin have stable isotopes ^{14}N, ^{56}Fe and ^{118}Sn. Give the symbols for two other isotopes of (*a*) nitrogen, (*b*) iron, and (*c*) tin.

2 • Why is the decay chain $A = 4n + 1$ not found in nature?

3 • A decay by α emission is often followed by β decay. When this occurs, it is by β^- and not β^+ decay. Why?

4 • The half-life of ^{14}C is much less than the age of the universe, yet ^{14}C is found in nature. Why?

5 • What effect would a long-term variation in cosmic-ray activity have on the accuracy of ^{14}C dating?

6 • Why does an element that has $Z = 130$ not exist?

7 • Why is a moderator needed in an ordinary nuclear fission reactor?

8 • Explain why water is more effective than lead in slowing down fast neutrons.

9 • The stable isotope of sodium is ^{23}Na. What kind of beta decay would you expect of (*a*) ^{22}Na and (*b*) ^{24}Na?

10 • What is the advantage of a breeder reactor over an ordinary reactor? What are the disadvantages of a breeder reactor?

11 • True or false:

(*a*) In a breeder reactor, fuel can be produced as fast as it is consumed.
(*b*) The atomic nucleus is composed of protons, neutrons, and electrons.
(*c*) The mass of a ^{2}H nucleus is less than the mass of a ^{1}H nucleus plus the mass of a neutron.
(*d*) After two half-lives, all the radioactive nuclei in a given sample have decayed.

12 • Why is it that extreme changes in the temperature or the pressure of a radioactive sample have little or no effect on the radioactivity?

ESTIMATION AND APPROXIMATION

13 • We found in Chapter 25 that the ratio of the resistivity of the most insulating material to the resistivity of the least resistive material (excluding superconductors) is approximately 10^{22}. Few properties of materials show such a wide range of values. Using information in the textbook or other resources, find the ratio of largest to smallest for some nuclear properties of matter. Some examples might be the range of mass densities found in an atom, the half-life of radioactive nuclei, or the range of nuclear masses.

14 •• According to the United States Department of Energy, the U.S. population consumes approximately 10^{20} joules of energy each year. Estimate the mass (in kilograms) of (*a*) uranium that would be needed to produce this much energy using nuclear fission and (*b*) deuterium and tritium that would be needed to produce this much energy using nuclear fusion.

PROPERTIES OF NUCLEI

15 • Calculate the binding energy and the binding energy per nucleon from the masses given in Table 40-1 for (*a*) ^{12}C, (*b*) ^{56}Fe, and (*c*) ^{238}U. SSM

16 • Calculate the binding energy and the binding energy per nucleon from the masses given in Table 40-1 for (*a*) ^{6}Li, (*b*) ^{39}K, and (*c*) ^{208}Pb.

17 • Use the radius formula $R = R_0A^{1/3}$ (Equation 40-1), where $R_0 = 1.2$ fm, to compute the radii of the following nuclei: (*a*) ^{16}O, (*b*) ^{56}Fe, and (*c*) ^{197}Au.

18 • During a fission process, a ^{239}Pu nucleus splits into two nuclei whose mass number ratio is 3 to 1. Calculate the radii of the nuclei formed during the process.

19 •• The neutron, when isolated from an atomic nucleus, decays into a proton, an electron, and an antineutrino as follows: $n \rightarrow {}^{1}H + e^{-} + \bar{\nu}$. The thermal energy of a neutron is of the order of kT, where k is the Boltzmann constant. (*a*) In both joules and electron volts, calculate the energy of a thermal neutron at 25°C. (*b*) What is the speed of that thermal neutron? (*c*) A beam of monoenergetic thermal neutrons is produced at 25°C and has an intensity I. After traveling 1350 km, the beam has an intensity of $\frac{1}{2}I$. Using this information, estimate the half-life of the neutron. Express your answer in minutes. SSM

20 • Use $R = R_0A^{1/3}$ (Equation 40-1), where $R_0 = 1.2$ fm, for the radius of a spherical nucleus to calculate the density of nuclear matter. Express your answer in grams per cubic centimeter.

21 •• In 1920, 12 years before the discovery of the neutron, Ernest Rutherford argued that proton–electron pairs might exist in the confines of the nucleus in order to explain the mass number, A, being greater than the nuclear charge, Z. He also used this argument to account for the source of beta particles in radioactive decay. Rutherford's scattering experiments in 1910 showed that the nucleus had a diameter of approximately 10 fm. Using this nuclear diameter, the uncertainty principle, and that beta particles have an energy range of 0.02 MeV to 3.40 MeV, show why the hypothetical electrons cannot be confined to a region occupied by the nucleus. SSM

22 •• Consider the following fission process: ${}^{235}_{92}U + n \rightarrow {}^{95}_{37}Rb + {}^{137}_{55}Cs + 4n$. Determine the electrostatic potential energy, in MeV, of the reaction products when the surfaces of the ^{95}Rb nucleus and the ^{137}Cs nucleus are just touching immediately after being formed during the fission process.

RADIOACTIVITY

23 • Homer enters the visitors' chambers, and his Geiger beeper sounds. He shuts off the beeper, removes the device from his shoulder patch, and holds it near the only new object in the room—an orb that is to be presented as a gift from the visiting Cartesians. Pushing a button marked "monitor," Homer reads that the device is reading a counting rate of 4000 counts/s above the background counting rate. After 10 min, the counting rate has dropped to 1000 counts/s above the background rate. (*a*) What is the half-life of the source? (*b*) How high will the counting rate be (above the background counting rate) 20 min after the monitoring device was switched on?

24 • A certain source gives 2000 counts/s at time $t = 0$. Its half-life is 2.0 min. How many counts per second will it give after (*a*) 4.0 min, (*b*) 6.0 min, and (*c*) 8.0 min?

25 • The counting rate from a radioactive source is 8000 counts/s at time $t = 0$, and 10 min later the rate is 1000 counts/s. (*a*) What is the half-life? (*b*) What is the decay constant? (*c*) What is the counting rate after 20 min?

26 • The half-life of radium is 1620 y. Calculate the number of disintegrations per second of 1.00 g of radium and show that the disintegration rate is approximately 1.0 Ci.

27 • A radioactive piece of silver foil ($t_{1/2} = 2.4$ min) is placed near a Geiger counter and 1000 counts/s are observed at time $t = 0$. (*a*) What is the counting rate at $t = 2.4$ min and at $t = 4.8$ min? (*b*) If the counting efficiency is 20 percent, how many radioactive silver nuclei are there at time $t = 0$? At time $t = 2.4$ min? (*c*) At what time will the counting rate be about 30 counts/s?

28 • Use Table 40-1 to calculate the energy release, in MeV, for the α decay of (*a*) ^{226}Ra and (*b*) ^{242}Pu.

29 •• Plutonium is very toxic to the human body. Once it enters the body it collects primarily in the bones, although it also can be found in other organs. Red blood cells are synthesized within the marrow of the bones. The isotope ^{239}Pu is an alpha emitter that has a half-life of 24 360 years. Because alpha particles are an ionizing radiation, the blood-making ability of the marrow is, in time, destroyed by the presence of ^{239}Pu. In addition, many kinds of cancers will also develop in the surrounding tissues because of the ionizing effects of the alpha particles. (*a*) If a person accidentally ingested 2.0 μg of ^{239}Pu and all of it is absorbed by the bones of the person, how many alpha particles are produced per second within the body of the person? (*b*) When, in years, will the activity be 1000 alpha particles per second? SSM

30 •• Consider an alpha-emitting parent nucleus ${}^{A}_{Z}X$ initially at rest. The nucleus decays into a daughter nucleus Y and an alpha particle as follows: ${}^{A}_{Z}X \rightarrow {}^{A-4}_{Z-2}Y + {}^{4}_{2}\alpha + Q$. (*a*) Show that the alpha particle has a kinetic energy of $(A - 4)Q/A$. (*b*) Show that the kinetic energy of the recoiling daughter nucleus is given by $K_Y = 4Q/A$.

31 • The fissile material ^{239}Pu is an alpha emitter. Write the reaction that describes ^{239}Pu undergoing alpha decay. Given that ^{239}Pu, ^{235}U, and an alpha particle have respective masses of 239.052 156 u, 235.043 923 u, and 4.002 603 u, use the relations appearing in Problem 30 to calculate the kinetic energies of the alpha particle and the recoiling daughter nucleus. SSM

32 • Through a friend in the security department at the museum, Angela obtains a sample of a wooden tool handle that contains 175 g of carbon. The decay rate of the ^{14}C in the sample is 8.1 Bq. How long ago was the wood in the handle last alive?

33 • A sample of a radioactive isotope is found to have an activity of 115.0 Bq immediately after it is pulled from the reactor that formed the isotope. Its activity 2 h 15 min later is measured to be 85.2 Bq. (*a*) Calculate the decay constant and the half-life of the sample. (*b*) How many radioactive nuclei were there in the sample initially?

34 •• A 1.00-mg sample of substance that has an atomic mass of 59.934 u and emits β particles has an activity of 1.131 Ci. Find the decay constant for the substance in reciprocal seconds and find the half-life in years.

35 •• Radiation has been used for a long time in medical therapy to control the development and growth of cancer cells. Cobalt-60, a gamma emitter that emits photons that have energies of 1.17 MeV and 1.33 MeV, is used to irradiate and destroy deep-rooted cancers. Small needles made of ^{60}Co of a specified activity are encased in gold and used as body implants in tumors for time periods that are related to tumor size, tumor cell reproductive rate, and the activity of the needle. (*a*) A 1.00-μg sample of ^{60}Co, that has a half-life of

5.27 y and that is used to irradiate a small internal tumor with gamma rays, is prepared in the cyclotron of a medical center. Determine the activity of the sample in curies. (*b*) What will the activity of the sample be 1.75 y from now? SSM

36 •• (*a*) Show that if the decay rate is R_0 at time $t = 0$ and R_1 at some later time $t = t_1$, the decay constant is given by $\lambda = t_1^{-1}\ln(R_0/R_1)$ and the half-life is given by $t_{1/2} = t_1 \ln(2)/\ln(R_0/R_1)$. (*b*) Use these results to find the decay constant and the half-life if the decay rate is 1200 Bq at $t = 0$ and 800 Bq at $t_1 = 60.0$ s.

37 •• A wooden casket is thought to be 18 000 years old. How much carbon would have to be recovered from the object to yield a ^{14}C counting rate of no less than 5 counts/min with a detection efficiency of 20 percent?

38 •• A sample of radioactive material is initially found to have an activity of 115.0 decays/min. After 4 d 5 h, its activity is measured to be 73.5 decays/min. (*a*) Calculate the half-life of the material. (*b*) How long (from the initial time) will it take for the sample to reach an activity of 10.0 decays/min?

39 •• The rubidium isotope ^{87}Rb is a β^- emitter that has a half-life of 4.9×10^{10} y. It decays into ^{87}Sr. This nuclear decay is used to determine the age of rocks and fossils. Rocks containing the fossils of early animals have a ratio of ^{87}Sr to ^{87}Rb of 0.0100. Assuming that there was no ^{87}Sr present when the rocks were formed, calculate the age of the fossils.

40 ••• Consider a single nucleus of a radioactive isotope that has a decay rate equal to λ. The nucleus has not decayed at $t = 0$. The probability that the nucleus will decay between time t and time $t + dt$ is equal to $\lambda e^{-\lambda t}dt$. (*a*) Show that this statement is consistent with the fact that the probability is 1 that the nucleus will decay between $t = 0$ and $t = \infty$. (*b*) Show that the expected lifetime of the nucleus is equal to $1/\lambda$. *Hint: The expected lifetime is equal to* $\int_0^\infty t\lambda e^{-\lambda t}dt$ *divided by* $\int_0^\infty \lambda e^{-\lambda t}dt$. (*c*) A sample of material contains a number of these radioactive nuclei at time $t = 0$. What is the mean lifetime of the radioactive nuclei in the sample?

NUCLEAR REACTIONS

41 • Using Table 40-1, find the Q values for the following reactions: (*a*) $^1H + {}^3H \rightarrow {}^3He + n + Q$ and (*b*) $^2H + {}^2H \rightarrow {}^3He + n + Q$.

42 • Using Table 40-1, find the Q values for the following reactions: (*a*) $^2H + {}^2H \rightarrow {}^3H + {}^1H + Q$, (*b*) $^2H + {}^3He \rightarrow {}^4He + {}^1H + Q$, and (*c*) $^6Li + n \rightarrow {}^3H + {}^4He + Q$.

43 •• (*a*) Use the values 14.003 242 u and 14.003 074 u for the atomic masses of ^{14}C and ^{14}N, respectively, to calculate the Q value (in MeV) for the β-decay reaction $^{14}_{6}C \rightarrow {}^{14}_{7}N + e^- + \bar{v}_e$. (*b*) Explain why you should not add the mass of the electron to that of atomic ^{14}N for the calculation in Part (*a*). SSM

44 •• (*a*) Use the values 13.005 738 u and 13.003 354 u for the atomic masses of $^{13}_{7}N$ and $^{13}_{6}C$, respectively, to calculate the Q value (in MeV) for the β-decay reaction

$$^{13}_{7}N \rightarrow {}^{13}_{6}C + e^+ + v_e$$

(*b*) Explain why you need to add twice the mass of an electron to the mass of $^{13}_{6}C$ during the calculation of the Q value for the reaction in Part (*a*).

FISSION AND FUSION

45 • Assuming an average energy of 200 MeV per fission, calculate the number of fissions per second needed for a 500-MW reactor. SSM

46 • If the reproduction factor in a reactor is 1.1, find the number of generations needed for the power level to (*a*) double, (*b*) increase by a factor of 10, and (*c*) increase by a factor of 100. Find the time needed in each case if (*d*) there are no delayed neutrons, so that the time between generations is 1.0 ms, and (*e*) there are delayed neutrons that make the average time between generations 100 ms.

47 •• Consider the following fission reaction: $^{235}_{92}U + n \rightarrow {}^{95}_{42}Mo + {}^{139}_{57}La + 2n + Q$. The masses of the neutron, ^{235}U, ^{95}Mo, and ^{139}La are 1.008 665 u, 235.043 923 u, 94.905 842 u, and 138.906 348 u, respectively. Calculate the Q value, in MeV, for the fission reaction. SSM

48 •• In 1989, researchers claimed to have achieved fusion in an electrochemical cell at room temperature. Their now thoroughly discredited claim was that a power output of 4.00 W was produced by deuterium fusion reactions in the palladium electrode of their apparatus. The two most likely reactions are

$$^2H + {}^2H \rightarrow {}^3He + n + 3.27\text{ MeV}$$

and

$$^2H + {}^2H \rightarrow {}^3H + {}^1H + 4.03\text{ MeV}$$

Of the deuterium nuclei that participated in these reactions, assume half of the deuterium nuclei participated in the first reaction and the other half participated in the second reaction. How many neutrons per second would we expect to be emitted in the generation of 4.00 W of power?

49 •• A fusion reactor that uses only deuterium for fuel would have the two reactions in Problem 48 taking place in the reactor. The ^3H produced in the second reaction reacts immediately with another ^2H in the reaction

$$^3H + {}^2H \rightarrow {}^4He + n + 17.6\text{ MeV}$$

The ratio of ^2H to ^1H atoms in naturally occurring hydrogen is 1.5×10^{-4}. How much energy would be produced from 4.0 L of water if all of the ^2H nuclei undergo fusion?

50 ••• The fusion reaction between ^2H and ^3H is

$$^3H + {}^2H \rightarrow {}^4He + n + 17.6\text{ MeV}$$

Using the conservation of momentum and the given Q value, find the final energies of both the ^4He nucleus and the neutron, assuming the initial kinetic energy of the system is 1.00 MeV and the initial momentum of the system is zero.

51 ••• Energy is generated in the Sun and other stars by fusion. One of the fusion cycles, the proton–proton cycle, consists of the following reactions:

$$^1H + {}^1H \rightarrow {}^2H + e^+ + \nu_e$$

$$^1H + {}^2H \rightarrow {}^3He + \gamma$$

followed by

$$^1H + {}^3He \rightarrow {}^4He + e^+ + \nu_e$$

(*a*) Show that the net effect of these reactions is

$$4{}^1H \rightarrow {}^4He + 2e^+ + 2\nu_e + \gamma$$

(*b*) Show that 24.7 MeV is released during this cycle (not counting the additional energy of 1.02 MeV that is released when each positron meets an electron and the two annihilate). (*c*) The Sun radiates energy at the rate of approximately 4.0×10^{26} W. Assuming that this is due to the conversion of four protons into helium, γ rays, and neutrinos, which releases 26.7 MeV, what is the rate of proton consumption in the Sun? How long will the Sun last if it continues to radiate at its present level? (Assume that protons constitute about half of the total mass, 2.0×10^{30} kg, of the Sun.)

GENERAL PROBLEMS

52 • (*a*) Show that $ke^2 = 1.44\ \text{MeV}\cdot\text{fm}$, where k is the Coulomb constant and e is the magnitude of the electron charge. (*b*) Show that $hc = 1240\ \text{MeV}\cdot\text{fm}$.

53 • The counting rate from a radioactive source is 6400 counts/s. The half-life of the source is 10 s. Make a plot of the counting rate as a function of time for times up to 1 min. What is the decay constant for the source? SSM

54 • Find the energy needed to remove a neutron from (*a*) ^{4}He and (*b*) ^{7}Li.

55 • The isotope ^{14}C decays according to $^{14}\text{C} \rightarrow {}^{14}\text{N} + \text{e}^- + \bar{v}_\text{e}$. The atomic mass of ^{14}N is 14.003 074 u. Determine the maximum kinetic energy of the electron. (Neglect recoil of the nitrogen atom.)

56 • The density of a neutron star is the same as the density of a nucleus. If our Sun were to collapse to a neutron star, what would be the radius of that object?

57 •• Show that the ^{109}Ag nucleus is stable and does not undergo alpha decay, $^{109}_{47}\text{Ag} \rightarrow {}^4_2\text{He} + {}^{105}_{45}\text{Rh} + Q$. The mass of the ^{109}Ag nucleus is 108.904 756 u, and the products of the decay are 4.002 603 u and 104.905 250 u, respectively. SSM

58 •• Gamma rays can be used to induce photofission (fission triggered by the absorption of a photon) in nuclei. Calculate the threshold photon wavelength for the following nuclear reaction: $^2\text{H} + \gamma \rightarrow {}^1\text{H} + \text{n}$. Use Table 40-1 for the masses of the interacting particles.

59 • The relative abundance of ^{40}K (potassium 40) is 1.2×10^{-4}. The isotope ^{40}K has a molar mass of 40.0 g/mol, is radioactive, and has a half-life of 1.3×10^9 y. Potassium is an essential element of every living cell. In the human body the mass of potassium constitutes approximately 0.36 percent of the total mass. Determine the activity of this radioactive source in a student whose mass is 60 kg.

60 •• When a positron makes contact with an electron, the electron–positron pair annihilate by way of the reaction $\beta^+ + \beta^- \rightarrow 2\gamma$. Calculate the minimum total energy, in MeV, of the two photons created when a positron–electron pair annihilate.

61 •• The isotope ^{24}Na is a β emitter and has a half-life of 15 h. A saline solution containing the radioactive isotope has an activity of 600 kBq and is injected into the bloodstream of a patient. Ten hours later, the activity of 1 mL of blood from the individual yields a counting rate of 12 counts/s at a counting efficiency of 20 percent. Determine the volume of blood in the patient.

62 •• (*a*) Determine the distance of closest approach of an 8.0-MeV α particle in a head-on collision with a stationary nucleus of ^{197}Au and with a stationary nucleus of ^{10}B, neglecting the recoil of the struck nuclei. (*b*) Repeat the calculation taking into account the recoil of the struck nuclei.

63 •• Twelve nucleons are in a one-dimensional infinite square well of length $L = 3.0$ fm. (*a*) Using the approximation that the mass of a nucleon is 1.0 u, find the lowest energy of a nucleon in the well. Express your answer in MeV. What is the ground-state energy of the system of 12 nucleons in the well if (*b*) all the nucleons are neutrons so that there can be no more than 2 in each spatial state and (*c*) 6 of the nucleons are neutrons and 6 are protons so that there can be as many as 4 nucleons in each spatial state? (Neglect the energy of Coulomb repulsion of the protons.)

64 •• The helium nucleus or α particle is a very tightly bound system. Nuclei with $N = Z = 2n$, where n is an integer (for example, ^{12}C, ^{16}O, ^{20}Ne, and ^{24}Mg), can be modeled as agglomerates of α particles. (*a*) Use this model to estimate the binding energy of a pair of α particles from the atomic masses of ^{4}He and ^{16}O. Assume that the four α particles in ^{16}O form a regular tetrahedron that has one α particle at each vertex. (*b*) From the result obtained in Part (*a*) determine, on the basis of the model, the binding energy of ^{12}C and compare your result with the result obtained from the atomic mass of ^{12}C.

65 •• Nuclei of a radioactive isotope that has a decay constant of λ are produced in an accelerator at a constant rate R_p. The number of radioactive nuclei N then obeys the equation $dN/dt = R_\text{p} - \lambda N$. (*a*) If N is zero at $t = 0$, sketch N versus t for the situation. (*b*) The isotope ^{62}Cu is produced at a rate of 100 per second by placing ordinary copper (^{63}Cu) in a beam of high-energy photons. The reaction is

$$\gamma + {}^{63}\text{Cu} \rightarrow {}^{62}\text{Cu} + \text{n}$$

The isotope ^{62}Cu decays by β decay and has a half-life of 10 min. After a time long enough so that $dN/dt \approx 0$, how many ^{62}Cu nuclei are there?

66 •• The total energy consumed in the United States in 1 y is approximately 7.0×10^{19} J. How many kilograms of ^{235}U would be needed to provide this amount of energy if we assume that 200 MeV of energy is released by each fissioning uranium nucleus, that all of the uranium atoms undergo fission, and that all of the energy-conversion mechanisms used are 100 percent efficient?

67 •• (*a*) Find the wavelength of a particle in the ground state of a one-dimensional infinite square well of length $L = 2.00$ fm. (*b*) Find the momentum in units of MeV/c for a particle that has this wavelength. (*c*) Show that the total energy of an electron that has this wavelength is approximately $E \approx pc$. (*d*) What is the kinetic energy of an electron in the ground state of the well? This calculation shows that if an electron were confined in a region of space as small as a nucleus, it would have a very large kinetic energy.

68 •• If ^{12}C, ^{11}B, and ^{1}H have respective masses of 12.000 000 u, 11.009 306 u, and 1.007 825 u, determine the minimum energy, Q, in MeV, required to remove a proton from a ^{12}C nucleus.

69 ••• Assume that a neutron decays into a proton and an electron without the emission of a neutrino. The kinetic energy shared by the proton and the electron is then 0.782 MeV. In the rest frame of the neutron, the total momentum is zero, so the momentum of the proton must be equal and opposite the momentum of the electron. This determines the ratio of the kinetic energies of the two particles, but because the electron is relativistic, the exact calculation of these relative kinetic energies is somewhat challenging. (*a*) Assume that the kinetic energy of the electron is 0.782 MeV and calculate the momentum p of the electron in units of MeV/c. *Hint: Use* $E^2 = p^2c^2 + (mc^2)^2$ *(Equation 39-27)*. (*b*) Using your result from Part (*a*), calculate the kinetic energy $p^2/2m_\text{p}$ of the proton. (*c*) Because the total kinetic energy of the electron and the proton is 0.782 MeV, the calculation in Part (*b*) gives a correction to the assumption that the kinetic energy of the electron is 0.782 MeV. What percentage of 0.782 MeV is this correction? SSM

70 ••• In the laboratory reference frame, a neutron of mass m moving with speed v_L makes an elastic head-on collision with a nucleus of mass M that is at rest. (*a*) Show that the speed of the center of mass in the lab frame is $V = mv_\text{L}/(m + M)$. (*b*) What is the speed of the nucleus in the center-of-mass frame before the collision and after the collision? (*c*) What is the speed of the nucleus in the laboratory frame after the collision? (*d*) Show that the energy of the nucleus after the collision in the laboratory frame is

$$\frac{1}{2}M(2V)^2 = \frac{4mM}{(m + M)^2}\left(\frac{1}{2}mv_\text{L}^2\right)$$

(*e*) Show that the fraction of the energy lost by the neutron in the elastic collision is

$$\frac{-\Delta E}{E} = \frac{4mM}{(m + M)^2} = \frac{4(m/M)}{[1 + (m/M)]^2}$$

71 ••• (*a*) Use the result from Part (*e*) of Problem 70 to show that after N head-on collisions of a neutron with carbon nuclei at rest, the energy of the neutron is approximately $(0.714)^N E_0$, where E_0 is its original energy. (*b*) How many head-on collisions are required to reduce the energy of the neutron from 2.0 MeV to 0.020 eV, assuming stationary carbon nuclei?

72 ••• On the average, a neutron loses 63 percent of its energy in a collision with a hydrogen atom and 11 percent of its energy in a collision with a carbon atom. Calculate the number of collisions needed to reduce the energy of a neutron from 2.0 MeV to 0.020 eV if the neutron collides with (*a*) hydrogen atoms and (*b*) carbon atoms.

73 ••• Frequently, the daughter nucleus of a radioactive parent nucleus is itself radioactive. Suppose the parent nucleus, designated by P, has a decay constant λ_P, while the daughter nucleus, designated by D, has a decay constant λ_D. The number of daughter nuclei N_D are then given by the solution to the differential equation

$$dN_D/dt = \lambda_P N_P - \lambda_D N_D$$

where N_P is the number of parent nuclei. (*a*) Justify this differential equation. (*b*) Show that the solution for the equation is

$$N_D(t) = \frac{N_{P0}\lambda_P}{\lambda_D - \lambda_P}(e^{-\lambda_P t} - e^{-\lambda_D t})$$

where N_{P0} is the number of parent nuclei present at $t = 0$ when there are no daughter nuclei. (*c*) Show that the expression for N_D in Part (*b*) gives $N_D(t) > 0$ whether $\lambda_P > \lambda_D$ or $\lambda_D > \lambda_P$. (*d*) Make a plot of $N_P(t)$ and $N_D(t)$ as a function of time when $\tau_D = 3\tau_P$, where τ_D and τ_P are the mean lifetimes of the daughter and parent nuclei, respectively. **SSM**

74 ••• Suppose isotope A decays to isotope B and has a decay constant λ_A, and isotope B in turn decays and has a decay constant λ_B. Suppose a sample contains, at $t = 0$, only isotope A nuclei. Derive an expression for the time at which the number of isotope B nuclei will be a maximum. (See Problem 73.)

75 ••• An example of the situation discussed in Problem 73 is the radioactive isotope ^{229}Th, an α emitter that has a half-life of 7300 y. Its daughter, ^{225}Ra, is a β emitter that has a half-life of 14.8 d. In this instance, as in many instances, the half-life of the parent is much longer than the half-life of the daughter. Using the expression given in Problem 73, Part (*b*), starting with a sample of pure ^{229}Th containing N_{P0} nuclei, show that the number, N_D, of ^{225}Ra nuclei will, after several years, be given by

$$N_D = \frac{\lambda_P}{\lambda_D} N_P$$

where N_P is the number of ^{229}Th nuclei. The number of daughter nuclei are said to be in *secular equilibrium*.

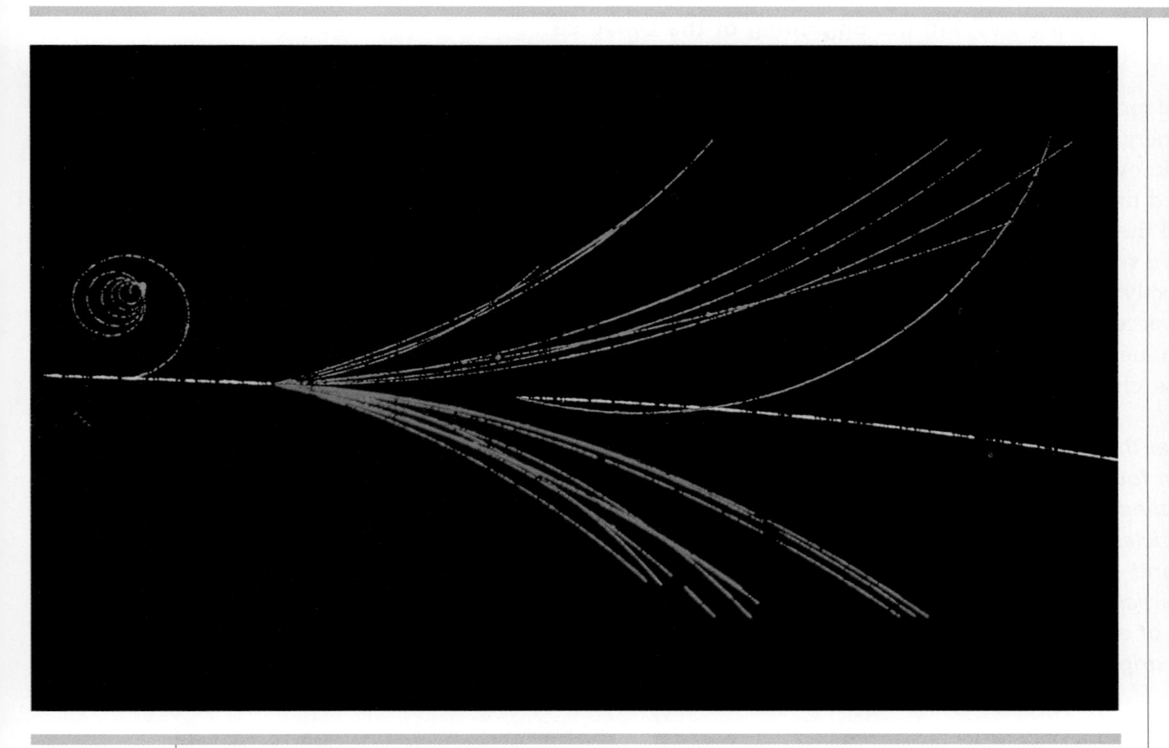

TRACKS IN A BUBBLE CHAMBER PRODUCED BY AN INCOMING HIGH-ENERGY PROTON (YELLOW), INCIDENT FROM THE LEFT, COLLIDING WITH A PROTON AT REST. THE SMALL GREEN SPIRAL IS AN ELECTRON KNOCKED OUT OF AN ATOM. IT CURVES TO THE LEFT BECAUSE OF AN EXTERNAL MAGNETIC FIELD IN THE CHAMBER. THE COLLISION PRODUCES SEVEN NEGATIVE PARTICLES (BLUE), ALL π^-; A NEUTRAL PARTICLE Λ^0 THAT LEAVES NO TRACK; AND NINE POSITIVE PARTICLES (RED) INCLUDING SEVEN π^+, A K^+, AND A PROTON. THE Λ^0 TRAVELS IN THE ORIGINAL DIRECTION OF THE INCOMING PROTON BEFORE DECAYING INTO A PROTON (YELLOW) AND A π^- (PURPLE). *(© Lawrence Livermore Laboratory/Science Photo Library/Photo Researchers.)*

How do you determine the energy of particle interactions?
(See Example 41-1.)

Elementary Particles and the Beginning of the Universe

Items that we encounter in everyday life are made of atoms. In some sense, atoms are the building blocks of nature. However, we know that atoms are not the most fundamental constituents of matter. With the discovery of the electron by J. J. Thomson (1897), the Bohr theory of the nuclear atom (1913), and the discovery of the neutron (1932), it became clear that atoms and even nuclei have considerable structure. Indeed the once simple picture of particle physics in which there were just four "elementary" particles—the proton, neutron, electron, and photon—has become much more complex.

Since the 1950s, enormous sums of money have been spent constructing particle accelerators of greater and greater energies in hopes of finding particles predicted by various theories. At present, we know of several hundred particles that at one time or another have been considered to be elementary, and research teams at the giant accelerator laboratories around the world are searching for and finding new particles. Some of these particles have such short lifetimes (of the order of 10^{-23} s) that they can be detected only indirectly. Many particles are observed only during nuclear reactions using high-energy accelerators. In addition to the usual particle properties of mass, charge, and spin, new properties have been found and given whimsical names such as strangeness, charm, color, topness, and bottomness.

In this chapter, we will first look at the various ways of classifying the multitude of particles that have been found. We will then describe the current theory of elementary particles, called the standard model, *in which all matter in nature—from the exotic particles produced in the giant accelerator laboratories to ordinary grains of sand—is considered to be constructed from just two families of elementary particles, leptons and quarks. In the final section, we will use our knowledge of elementary particles to discuss the big bang theory which describes the origin of the universe.*

41-1 HADRONS AND LEPTONS

All the different forces observed in nature, from ordinary friction to the tremendous forces involved during supernova explosions, can be understood in terms of the four basic interactions: (1) the strong nuclear interaction (also called the hadronic interaction), (2) the electromagnetic interaction, (3) the weak (nuclear) interaction, and (4) the gravitational interaction. The four basic interactions provide a convenient structure for the classification of particles. Some particles participate in all four interactions, whereas other particles participate in only some of the interactions. For example, all particles participate in gravitational interaction, the weakest of the interactions. All particles that have electric charge participate in the electromagnetic interaction.

Particles that interact by the strong interaction are called **hadrons.** There are two kinds of hadrons: **baryons,** which have spin $\frac{1}{2}$ or $\frac{3}{2}$ or $\frac{5}{2}$, etc., and **mesons,** which have spin 0 or 1 or 2, etc. Baryons, which include nucleons, are the most massive of the elementary particles. Mesons have intermediate masses, between the mass of the electron and the mass of the proton. Particles that decay by the strong interaction have very short lifetimes, of the order of 10^{-23} s, which is about the time it takes light to travel a distance equal to the diameter of a nucleus. On the other hand, particles that decay by the weak interaction have much longer lifetimes, of the order of 10^{-10} s. Table 41-1 lists some of the properties of those hadrons that are stable against decay by the strong interaction.

The Super-Kamiokande detector, built in Japan in 1996 as a joint Japanese–American experiment, is essentially a water tank the size of a large cathedral installed in a deep zinc mine 1 mile inside a mountain. When neutrinos pass through the tank, one of the neutrinos occasionally collides with an atom, sending blue light through the water to an array of detectors. This photograph shows the detector wall and top that have approximately 9000 photomultiplier tubes that help detect the neutrinos. Experimental results reported in June 1998 were evidence that the mass of the neutrino cannot be zero. *(ICCR (Institute for Cosmic Ray Research), The University of Tokyo.)*

Hadrons are rather complicated entities and have complex structures. If we use the term *elementary particle* to mean a point particle that has no structure and is not constructed from some more elementary entities, then hadrons are not elementary particles. It is now believed that all hadrons are composed of more fundamental entities called *quarks,* which, as far as we know, are truly elementary particles.

Particles that participate in the weak interaction but not in the strong interaction are called **leptons.** These include electrons, muons, and neutrinos, which are all less massive than the lightest hadron. The word *lepton,* meaning "light particle," was chosen to reflect the relatively small mass of the particles. However, the most recently discovered lepton, the *tau,* found by Martin Lewis Perl in 1975, has a mass of 1784 MeV/c^2, nearly twice the mass of the proton (938 MeV/c^2), so we now have a "heavy lepton." In addition, the word *muon,* short for mu-meson, is something of a misnomer. The muon is not now categorized as a meson, so it is best to refer to it

Table 41-1 Hadrons That Are Stable Against Decay via the Strong Nuclear Interaction

Name	Symbol	Mass MeV/c^2	Spin, $\hbar$	Charge, e	Antiparticle	Mean Lifetime, s	Typical Decay Products*
Baryons							
Nucleon	p (proton)	938.3	$\frac{1}{2}$	+1	p^-	Infinite	
	n (neutron)	939.6	$\frac{1}{2}$	0	$\bar{n}$	930	$p + e^- + \bar{\nu}_e$
Lambda	Λ^0	1116	$\frac{1}{2}$	0	$\overline{\Lambda}^0$	2.5×10^{-10}	$p + \pi^-$
Sigma†	Σ^+	1189	$\frac{1}{2}$	+1	$\overline{\Sigma}^-$	0.8×10^{-10}	$n + \pi^+$
	Σ^0	1193	$\frac{1}{2}$	0	$\overline{\Sigma}^0$	10^{-20}	$\Lambda^0 + \gamma$
	Σ^-	1197	$\frac{1}{2}$	−1	$\overline{\Sigma}^+$	1.7×10^{-10}	$n + \pi^-$
Xi	Ξ^0	1315	$\frac{1}{2}$	0	$\overline{\Xi}^0$	3.0×10^{-10}	$\Lambda^0 + \pi^0$
	Ξ^-	1321	$\frac{1}{2}$	−1	$\overline{\Xi}^+$	1.7×10^{-10}	$\Lambda^0 + \pi^-$
Omega	Ω^-	1672	$\frac{3}{2}$	−1	Ω^+	1.3×10^{-10}	$\Xi^0 + \pi^-$
Mesons							
Pion	π^+	139.6	0	+1	π^-	2.6×10^{-8}	$\mu^+ + \nu_\mu$
	π^0	135	0	0	π^0	0.8×10^{-16}	$\gamma + \gamma$
	π^-	139.6	0	−1	π^+	2.6×10^{-8}	$\mu^- + \bar{\nu}_\mu$
Kaon‡	K^+	493.7	0	+1	K^-	1.24×10^{-8}	$\pi^+ + \pi^0$
	K^0	497.7	0	0	$\overline{K}^0$	0.88×10^{-10}	$\pi^+ + \pi^-$
						and	
						5.2×10^{-8}	$\pi^+ + e^- + \bar{\nu}_e$
Eta	η^0	549	0	0		2×10^{-19}	$\gamma + \gamma$

* Other decay modes also occur for most particles.
† The Σ^0 is included here for completeness even though it does decay via the strong interaction.
‡ The K^0 has two distinct lifetimes, sometimes referred to as K^0_{short} and K^0_{long}. All other particles have a unique lifetime.

as a muon and not as a mu-meson. As far as we know, leptons are point particles that have no structure and can be considered to be truly elementary in the sense that they are not composed of other particles.

There are six leptons. They are the electron and the electron neutrino, the muon and the muon neutrino, and the tau and the tau neutrino. (Each of the leptons has an antiparticle.) The masses of the electron, the muon, and the tau are quite different. The mass of the electron is 0.511 MeV/c^2, the mass of the muon is 106 MeV/c^2, and the mass of the tau is 1784 MeV/c^2. The standard model predicts that neutrinos, like photons, do not have mass. However, there is now strong evidence that their mass, though very small, is greater than zero. During the late 1990s, experiments using a detector in Japan called the Super-Kamiokande (Super-K) found that neutrinos emitted from the Sun arrived on Earth in much smaller numbers than the numbers that are predicted from the fusion processes in the Sun. This result can be explained if the mass of the neutrino is not zero.* In addition, a neutrino mass as small as a few eV/c^2 would have great cosmological significance. The answer to the question of whether the universe will continue to expand indefinitely or will reach a maximum size and begin to contract depends on the total mass in the universe. Thus, the answer could depend on whether the mass of the neutrino is actually zero or is merely small, because the cosmic density of each species of neutrino is ~100 per cm^3.

* The connection between the shortfall of solar-neutrino detections and the mass of the neutrino is elucidated in "On Morphing Neutrinos and Why They Must Have Mass" by Eugene Hecht, *The Physics Teacher* 41 (2003): 164–168.

(a)

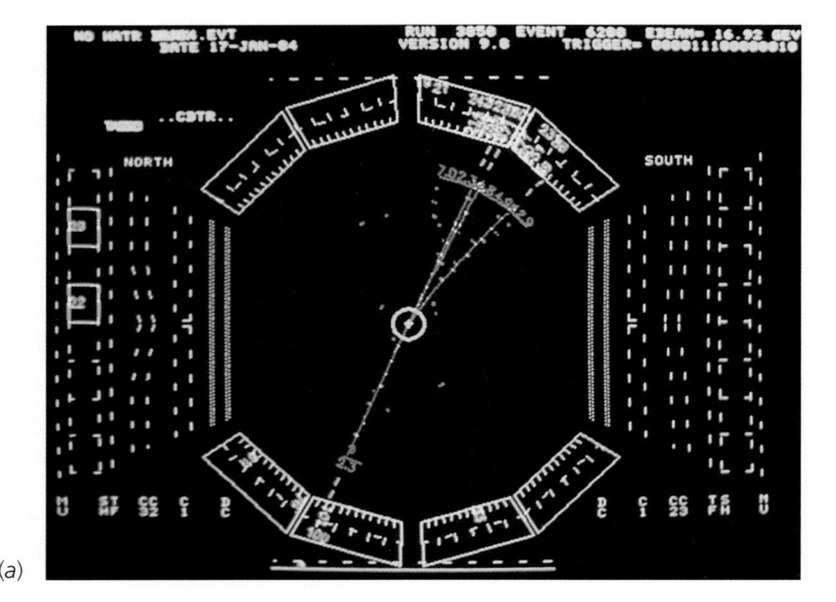

(b)

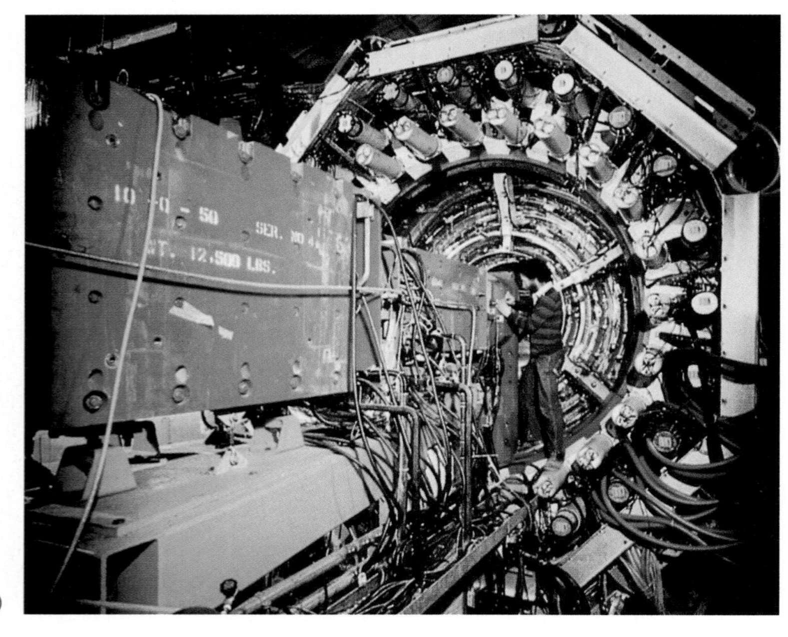

(*a*) A computer display of the production and decay of a τ_1 and τ_2 pair. An electron and a positron annihilate at the center marked by the yellow cross, producing a τ^+ and τ^- pair, which travel in opposite directions, but quickly decay while still inside the beam pipe (yellow circle). The τ^+ decays into two invisible neutrinos and a μ^+, which travels toward the bottom left. Its track in the drift chamber is calculated by a computer and indicated in red. It penetrates the lead–argon counters outlined in purple and is detected at the blue dot near the bottom blue line that marks the end of a muon detector. The τ^- decays into three charged pions (red tracks moving upward) plus invisible neutrinos. (*b*) The Mark I detector, built by a team from the Stanford Linear Accelerator Center (SLAC) and the Lawrence Berkeley Laboratory, became famous for many discoveries, including the J/ψ meson and the τ lepton. Tracks of particles are recorded by wire spark chambers wrapped in concentric cylinders around the beam pipe extending out to the ring where physicist Carl Friedberg has his right foot. Beyond this are two rings of protruding tubes, housing photomultipliers that view various scintillation counters. The rectangular magnets at the left guide the counterrotating beams that collide in the center of the detector. *((a) Science Photo Library/Photo Researchers. (b) © Lawrence Berkeley Laboratory/ Science Photo Library/Photo Researchers.)*

The observation of electron neutrinos from the supernova 1987A puts an upper limit on the masses of the neutrinos. Because the velocity of a particle that has mass depends on its energy, the arrival time of a burst of neutrinos that have mass from a supernova would be spread out in time. The fact that the electron neutrinos from the 1987 supernova all arrived at Earth within 13 s of one another results in an upper limit of about 16 eV/c^2 for their mass. Note that an upper limit does not imply that the mass is not zero. Measurements of the relative number of muon neutrinos and electron neutrinos entering the huge, underground Super-K detector suggest that at least one type of neutrino can oscillate between types (for example, between a mu neutrino and a tau neutrino). Further measurements of antineutrinos from nuclear reactors strongly show that all three types of neutrinos oscillate between types and thus have mass. Measurements made in Japan, using the *Kam*ioka *L*iquid Scintillator *A*nti-*N*eutrino *D*etector (KamLAND), show that oscillations from one species of neutrino to another species of neutrino can be observed over path lengths as short as 180 km (Figure 41-1).

The magnetic field is a combination of the doughnut-shaped magnetic field due to the windings of the toroid and the self-field due to the current of the circulating plasma. The break-even point has almost been achieved using magnetic confinement, but we are still a long way from building a practical fusion reactor.

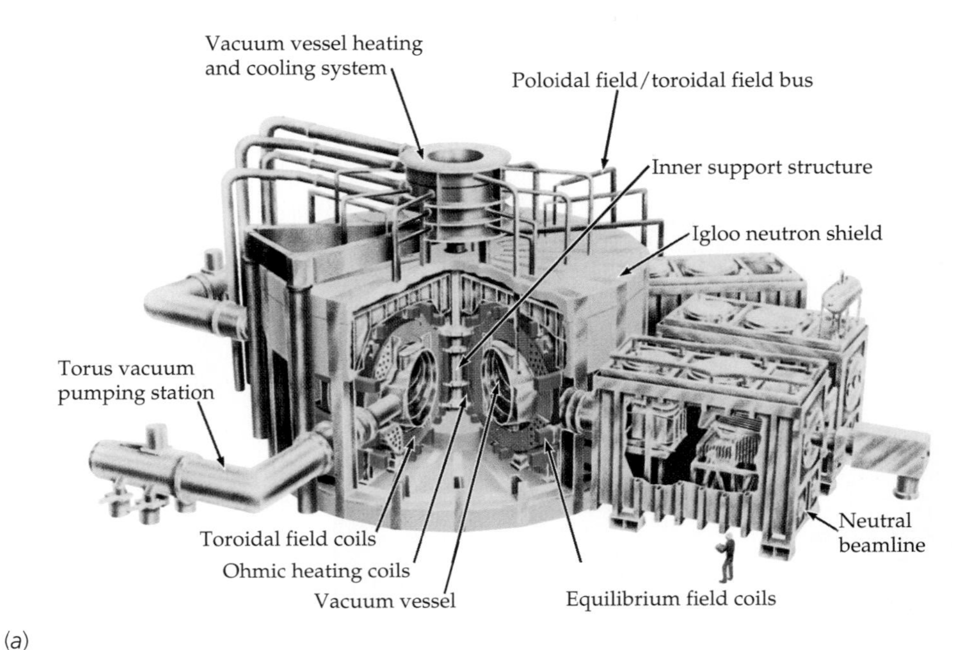

(*a*)

(*b*)

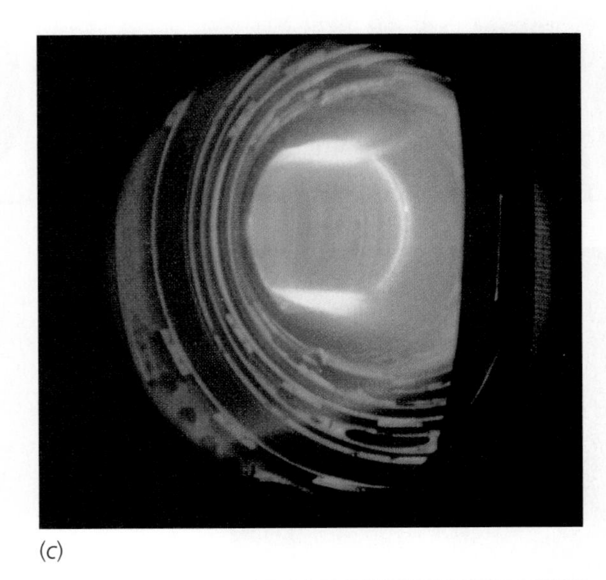

(*c*)

(*a*) Schematic of the Tokamak Fusion Test Reactor (TFTR). The toroidal coils, surrounding the doughnut-shaped vacuum vessel, are designed to conduct current for 3-s pulses, separated by waiting times of 5 min. Pulses peak at 73 000 A, producing a magnetic field of 5.2 T. This magnetic field is the principal means of confining the deuterium–tritium plasma that circulates within the vacuum vessel. Current for the pulses is delivered by converting the rotational energy of two 600-ton flywheels. Sets of poloidal coils, perpendicular to the toroidal coils, carry an oscillating current that generates a current through the confined plasma itself, heating it ohmically. Additional poloidal fields help stabilize the confined plasma. Between four and six neutral-beam injection systems (only one of which is shown in the schematic) are used to inject high-energy deuterium atoms into the deuterium–tritium plasma, heating beyond what could be obtained ohmically, ultimately to the point of fusion. (*b*) The TFTR itself. The diameter of the vacuum vessel is 7.7 m. (*c*) An 800-kA plasma, lasting 1.6 s, as it discharges within the vacuum vessel. *((All) Courtesy of the Princeton Plasma Physics Laboratory.)*

In a second scheme, called **inertial confinement,** a pellet of solid deuterium and tritium is bombarded from all sides by intense pulsed laser beams of energies of the order of 10^4 J lasting about 10^{-8} s. (Intense beams of ions are also used.) Computer simulation studies indicate that the pellet should be compressed to approximately 10^4 times its normal density and heated to a temperature greater than 10^8 K. This should produce approximately 10^6 J of fusion energy in 10^{-11} s, which is so brief that confinement is achieved by inertia alone.

Because the break-even point is just barely being achieved in magnetic-confinement fusion, and because the building of a fusion reactor involves many practical problems that have not yet been solved, the availability of fusion to meet our energy needs is not expected for at least several decades. However, fusion holds great promise as an energy source for the future.

(*a*)

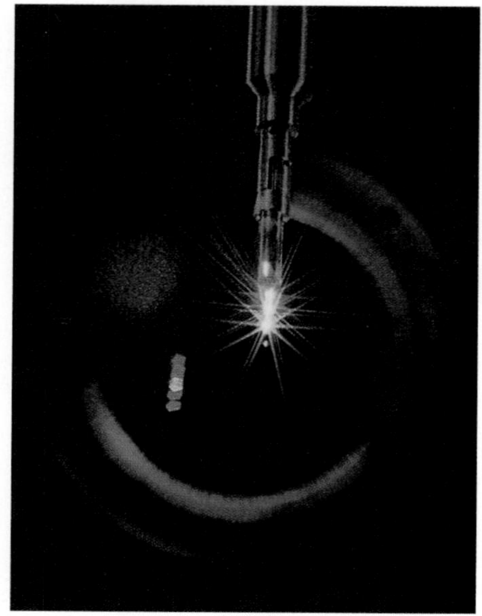

(b)

(*a*) The Nova target chamber, an aluminum sphere approximately 5 m in diameter, inside which 10 beams from the world's most powerful laser converge onto a hydrogen-containing pellet 0.5 mm in diameter. (*b*) The resulting fusion reaction is visible as a tiny star, lasting 10^{-10} s, releasing 10^{13} neutrons. *((All) Courtesy of the Lawrence Livermore National Laboratory/U.S. Department of Energy.)*

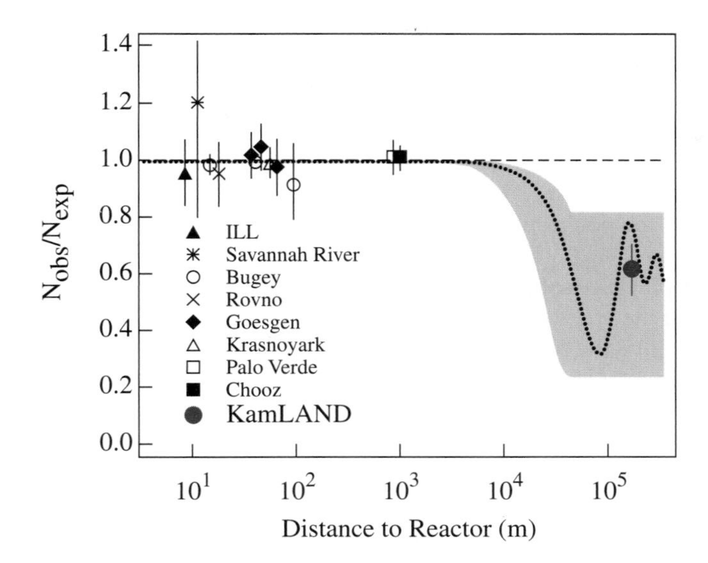

FIGURE 41-1 First evidence for antineutrino disappearance. The ratio of the number of antineutrinos observed N_{obs} to the number that one would expect to observe N_{exp} (assuming no neutrino oscillations) is plotted versus distance to the nearest antineutrino sources. The KamLAND site is 180 km from nearby antineutrino sources (nuclear reactors), while the other eight detector sites are less than 1.0 km from nearby nuclear reactors. For those eight sites, $N_{obs}/N_{exp} = 1.0$, which is what is expected assuming no neutrino oscillations. However, the KamLAND detector found $N_{obs}/N_{exp} = 0.6$. This result is strong evidence that while neutrinos do not oscillate in significant numbers while traveling over path lengths of less than 1.0 km, they do oscillate in significant numbers while traveling over path lengths only a few orders of magnitude longer than 1 km. *(© Lawrence Berkeley Laboratory/ Science Photo Library/Photo Researchers.)*

41-2 SPIN AND ANTIPARTICLES

One important characteristic of a particle is its intrinsic spin angular momentum. We have already discussed the fact that the electron has a quantum number m_s that corresponds to the z component of its intrinsic spin characterized by the quantum number $s = \frac{1}{2}$. Protons, neutrons, neutrinos, and the various other particles that also have an intrinsic spin characterized by the quantum number $s = \frac{1}{2}$ are called **spin-$\frac{1}{2}$ particles.** Particles that have spin $\frac{1}{2}$ (or $\frac{3}{2}$, $\frac{5}{2}$, etc.) are called fermions and obey the exclusion principle. Particles such as pions and other mesons have zero spin or integral spin ($s = 0, 1, 2$, etc.). Those particles are called bosons and do not obey the exclusion principle. That is, any number of those particles can be in the same quantum state.

Spin-$\frac{1}{2}$ particles are described by the Dirac equation, which is an extension of the Schrödinger equation that includes special relativity. One feature of Paul Dirac's theory, proposed in 1927, is the prediction of the existence of antiparticles. In special relativity, the energy of a particle is related to the mass and the momentum of the particle by $E = \pm\sqrt{p^2c^2 + m^2c^4}$ (Equation 39-27). We usually choose the positive solution and dismiss the negative-energy solution with a physical argument. However, the Dirac equation requires the existence of wave functions that correspond to the negative-energy states. Dirac got around this difficulty by postulating that all the negative-energy states were

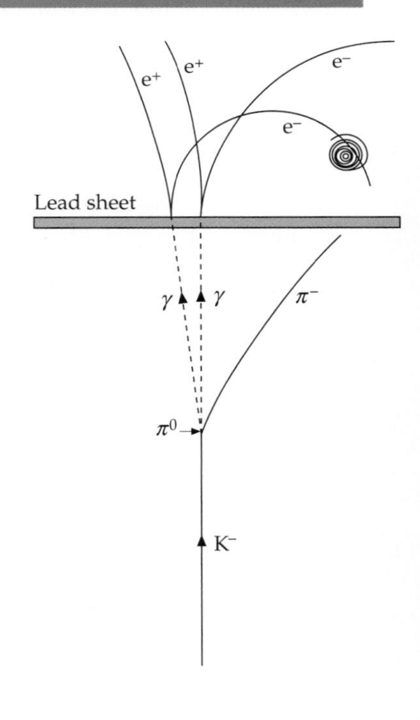

A negative kaon (K^-) enters a bubble chamber from the bottom and decays into a π^-, which moves off to the right, and a π^0, which immediately decays into two photons whose paths are indicated by the dashed lines in the drawing. Each photon interacts in the lead sheet, producing an electron–positron pair. The spiral at the right is another electron that has been knocked out of an atom in the chamber. (Other extraneous tracks have been removed from the photograph.) *(Figure 4 from "First Results from KamLAND: Evidence for Reactor Antineutrino Disappearance" by the KamLAND Collaboration,* Physical Review Letters, *Vol. 90, No. 2, December 17, 2003. Copyright © 2003 The American Physical Society. Reprinted with permission.)*

filled and would therefore not be observable. Only holes in the "infinite sea" of negative-energy states would be observed. For example, a hole in the negative sea of electron energy states would appear as a particle identical to the electron except having positive charge. When such a particle came in the vicinity of an electron the two particles would annihilate, releasing two photons having a minimum total energy of $2m_ec^2$, where m_e is the mass of the electron. This interpretation received little attention until a particle with just those properties, called the positron, was discovered in 1932 by Carl Anderson.

Antiparticles are never created alone but always in particle–antiparticle pairs. In the creation of an electron–positron pair by a photon, the energy of the photon must be at least as great as the rest energy of the electron plus the rest energy of the positron, which is $2m_ec^2 \approx 1.02$ MeV. Although the positron is stable, it has only a short-term existence in our universe because of the large supply of electrons in matter. The fate of a positron is annihilation according to the reaction

$$e^+ + e^- \rightarrow \gamma + \gamma \qquad 41\text{-}1$$

The probability of this reaction is large only if the positron and electron are moving slowly relative to one another. In the center-of-mass reference frame, the momentum of the two particles prior to annihilation is zero, so two photons moving in opposite directions are needed to conserve linear momentum.

The fact that we call electrons *particles* and positrons *antiparticles* does not imply that positrons are less fundamental than electrons. It merely reflects the nature of our universe. If our matter were made up of negative protons and positive electrons, then positive protons and negative electrons would suffer quick annihilation and would be called antiparticles.

An aerial view of the European Laboratory for Particle Physics (CERN) just outside of Geneva, Switzerland. The large circle shows the Large Electron–Positron collider (LEP) tunnel, which is 27 km in circumference. The irregular dashed line is the border between France and Switzerland. *(Richard Ehrlich.)*

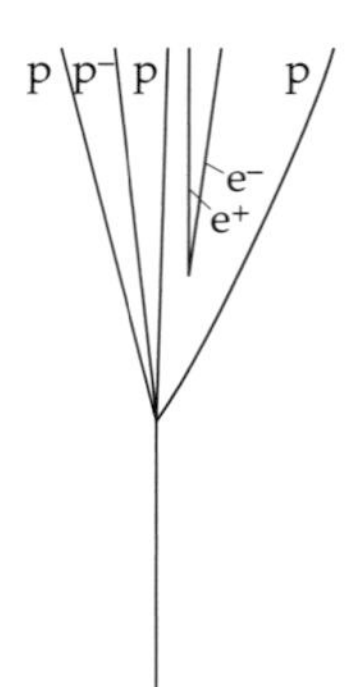

FIGURE 41-2 Bubble-chamber tracks that show the creation of a proton–antiproton pair in the collision of an incident 25-GeV proton with a stationary proton in liquid hydrogen. *(CERN.)*

The antiproton (p^-) was discovered in 1955 by Emilio Segrè and Owen Chamberlain using a beam of protons in the Bevatron at Berkeley to produce the reaction*

$$p^+ + p^+ \rightarrow p^+ + p^+ + p^+ + p^- \qquad 41\text{-}2$$

The creation of a proton–antiproton pair (Figure 41-2) requires kinetic energy of at least $2m_pc^2 = 1877\text{ MeV} = 1.877\text{ GeV}$ in the zero-momentum reference frame in which the two protons approach each other with equal and opposite momenta. In the laboratory frame in which one of the protons is initially at rest, the kinetic energy of the incoming proton must be at least $6m_pc^2 = 5.63\text{ GeV}$ (see Problem 38 of Chapter 39). This energy was not available in laboratories before the development of high-energy accelerators in the 1950s. Antiprotons annihilate with protons to produce two gamma rays in a reaction similar to the reaction in Equation 41-1.

The tunnel of the proton–antiproton collider at CERN. The same bending magnets and focusing magnets can be used for protons or antiprotons moving in opposite directions. The rectangular box in the foreground is a focusing magnet, and the next four boxes are the bending magnets. *(CERN.)*

* The antiproton is sometimes denoted by $\bar{p}$ rather than p^-. For neutral particles, such as the neutron, the bar must be used to denote the antiparticle. Thus, the antineutron is denoted by $\bar{n}$. The electron and proton are often denoted by e and p without the minus sign or plus sign superscripts.

Example 41-1 Proton–Antiproton Annihilation

Context-Rich

You have been reading about nuclear physics and particle interactions. In particular, you have been looking at the reaction $p^+ + p^- \longrightarrow \gamma + \gamma$ (proton–antiproton annihilation). You wonder if the photons produced are visible to the human eye if the two protons are initially at rest. Are the photons visible to the human eye?

PICTURE If the photons are visible, they should have wavelengths in the visible range (400 nm to 800 nm.) Because the proton and the antiproton are at rest, conservation of momentum requires that the two photons created during their annihilation have equal and opposite momenta and therefore equal energies, frequencies, and wavelengths. Conservation of energy implies that the photons have a combined energy equal to the rest energy of the proton plus the rest energy of the antiproton (approximately 938 MeV each).

SOLVE

1. Set the total energy of the two photons, $2E_\gamma$, equal to the rest energy of the proton plus antiproton and solve for E_γ:

$$2E_\gamma = 2m_pc^2$$

so

$$E_\gamma = m_pc^2 = 938\text{ MeV}$$

2. Set the energy of the photon equal to $hf = hc/\lambda$ and solve for the wavelength λ:

$$E_\gamma = hf = \frac{hc}{\lambda}$$

$$\lambda = \frac{hc}{E_\gamma} = \frac{1240\text{ eV}\cdot\text{nm}}{938\text{ MeV}}$$

$$= 1.32 \times 10^{-6}\text{ nm} = 1.32\text{ fm}$$

3. Compare this wavelength with the wavelengths of visible light:

The photons are *not* in the visible spectrum.

CHECK In Chapter 36, we found that the energies of photons in the visible spectrum are equal to only a few electron volts. It is not a surprise to find that photons that have energies on the order of 10^9 electron volts are not in the visible spectrum.

TAKING IT FURTHER The wavelength of the photons produced by proton–antiproton annihilation is more than eight orders of magnitude less than 400 nm—the shortest wavelength in the visible spectrum.

41-3 THE CONSERVATION LAWS

One adage is "anything that can happen does." If a conceivable decay or reaction does not occur, there must be a reason. The reason is usually expressed in terms of a conservation law. The conservation of energy rules out the decay of any particle for which the total mass of the decay products would be greater than the initial mass of the particle before decay. The conservation of linear momentum requires that when an electron and a positron at rest annihilate, two photons must be emitted. Angular momentum must also be conserved during a reaction or a decay. A fourth conservation law that restricts the possible particle decays and reactions is the conservation of electric charge. The net electric charge before a decay or a reaction must equal the net charge after the decay or the reaction.

There are two additional conservation laws that are important in the reactions and the decays of elementary particles: the conservation of baryon number and the conservation of lepton number. Consider the proposed decay

$$p \rightarrow \pi^0 + e^+$$

where π is the symbol for the pion (pi-meson). This decay would conserve charge, energy, angular momentum, and linear momentum, but it does not occur. It does not conserve either lepton number or baryon number. (The proton p is a baryon, the positron e^+ is a lepton, and the π^0 is a meson.) The conservation of lepton number and baryon number implies that whenever a lepton or a baryon is created, an antiparticle of the same type is also created. We assign the **lepton number** $L = +1$ to all leptons, $L = -1$ to all antileptons, and $L = 0$ to all other particles. Similarly, the **baryon number** $B = +1$ is assigned to all baryons, $B = -1$ to all antibaryons, and $B = 0$ to all other particles. The sum of the baryon numbers and the sum of the lepton numbers cannot change during a reaction or a decay. The conservation of baryon number along with the conservation of energy implies that the least massive baryon, the proton, must be stable.

CONCEPT CHECK 41-1

Why is it that the conservation of baryon number along with the conservation of energy implies that the proton, which is the least massive baryon, must be stable?

The conservation of lepton number implies that the neutrino emitted during the β decay of the free neutron is an antineutrino:

$$\mathrm{n} \rightarrow \mathrm{p}^+ + \mathrm{e}^- + \bar{\nu}_\mathrm{e} \qquad 41\text{-}3$$

The fact that neutrinos and antineutrinos are different is illustrated by an experiment in which ^{37}Cl is bombarded with an intense antineutrino beam from the decay of reactor neutrons (reactor neutrons are fission products that are produced in nuclear reactors). If neutrinos and antineutrinos were the same, we would expect the following reaction:

$$^{37}_{17}\mathrm{Cl} + \bar{\nu}_\mathrm{e} \rightarrow {}^{37}_{18}\mathrm{Ar} + \mathrm{e}^- \qquad 41\text{-}4$$

This reaction is not observed. However, if protons are bombarded with antineutrinos, the reaction

$$\mathrm{p} + \bar{\nu}_\mathrm{e} \rightarrow \mathrm{n} + \mathrm{e}^+ \qquad 41\text{-}5$$

is observed. Note that the sum of the lepton numbers is -1 on the left side of the reaction equation in Equation 41-4 and is $+1$ on the right side of the reaction equation. But the sum of the lepton numbers is -1 on both sides of the reaction equation in Equation 41-5.

Not only are neutrinos and antineutrinos distinct particles, but the neutrinos associated with electrons are distinct from the neutrinos associated with muons. Electron-like leptons (e and ν_e), muon-like leptons (μ and ν_μ), and tau-like leptons (τ and ν_τ) are each separately conserved, so we assign separate lepton numbers L_e, L_μ, and L_τ to the particles. The leptons and their lepton numbers are listed in Table 41-2.

Table 41-2 Lepton Numbers

	L_e	L_μ	L_τ
e^-	+1	0	0
ν_e	+1	0	0
e^+	−1	0	0
$\bar{\nu}_e$	−1	0	0
μ^-	0	+1	0
ν_μ	0	+1	0
μ^+	0	−1	0
$\bar{\nu}_\mu$	0	−1	0
τ^-	0	0	+1
ν_τ	0	0	+1
τ^+	0	0	−1
$\bar{\nu}_\tau$	0	0	−1

Example 41-2 What Laws Are Being Violated? *Conceptual*

What conservation laws (if any) are violated by the following proposed decays: (*a*) $n \rightarrow p + \pi^-$, (*b*) $\Lambda^0 \rightarrow p^- + \pi^+$, and (*c*) $\mu^- \rightarrow e^- + \gamma$? ($\Lambda^0$ is the symbol for the lambda-zero particle.)

PICTURE All reactions must separately conserve energy, electric charge, baryon number, electron lepton number, muon lepton number, and tau lepton number.

SOLVE

(*a*) There are no leptons in this decay, so there is no problem with the conservation of lepton number. The net charge is zero before the decay and after the decay, so charge is conserved. Also, the baryon number is +1 both before and after the decay. However, rest energy of the proton (938.3 MeV) plus the rest energy of the pion (139.6 MeV) is greater than the rest energy of the neutron (939.6 MeV). In the rest frame of the neutron, the energy prior to the reaction (the rest energy of the neutron) is less than the total rest energy following the reaction.

This decay does not conserve energy.

(*b*) Again, there are no leptons involved, and the net charge is zero before the decay and after the decay. Also, the rest energy of the lambda-zero (1116 MeV) is greater than the rest energy of the antiproton (938.3 MeV) plus the rest energy of the pion (139.6 MeV), so in the rest frame of the lambda-zero the energy prior to the reaction (the rest energy of the lambda-zero) is greater than the total rest energy following the reaction. Energy could be conserved, with the loss in rest energy equal to the gain in kinetic energy of the decay products. There are no leptons in the reaction, so all three lepton numbers are conserved. The baryon number is +1 for the lambda particle and −1 for the antiproton and zero for the pi-meson.

This decay does not conserve baryon number.

(*c*) The μ^- has a muon lepton number (L_μ) equal to +1 and an electron lepton number (L_e) equal to 0, the e^- has $L_\mu = 0$ and $L_e = +1$, and the γ has $L_\mu = L_e = 0$.

This reaction does not conserve either muon lepton number or electron lepton number.

TAKING IT FURTHER The muon does decay by $\mu^- \rightarrow e^- + \bar{\nu}_e + \nu_\mu$, which does conserve both muon lepton numbers and electron lepton numbers.

There are some conservation laws that are not universal but apply only to certain kinds of interactions. In particular, there are quantities that are conserved during decays and reactions that occur by the strong interaction but not during decays or reactions that occur by the weak interaction. One of these quantities that is particularly important is **strangeness,** introduced by M. Gell-Mann and K. Nishijima in 1952 to explain the strange behavior of some of the heavy baryons and mesons. Consider the reaction

$$p + \pi^- \rightarrow \Lambda^0 + K^0 \qquad 41\text{-}6$$

where K is the symbol for the kaon (K-meson). The proton and the pion interact by the strong interaction. Both the Λ^0 and K^0 decay into hadrons

$$\Lambda^0 \rightarrow p + \pi^- \qquad 41\text{-}7$$

and

$$K^0 \rightarrow \pi^+ + \pi^- \qquad 41\text{-}8$$

However, the decay times for both the Λ^0 and K^0 are of the order of 10^{-10} s, which is characteristic of the weak interaction, rather than 10^{-23} s, which would be expected for the strong interaction. Other particles showing similar behavior were called **strange particles.** These particles are always produced in pairs, even when

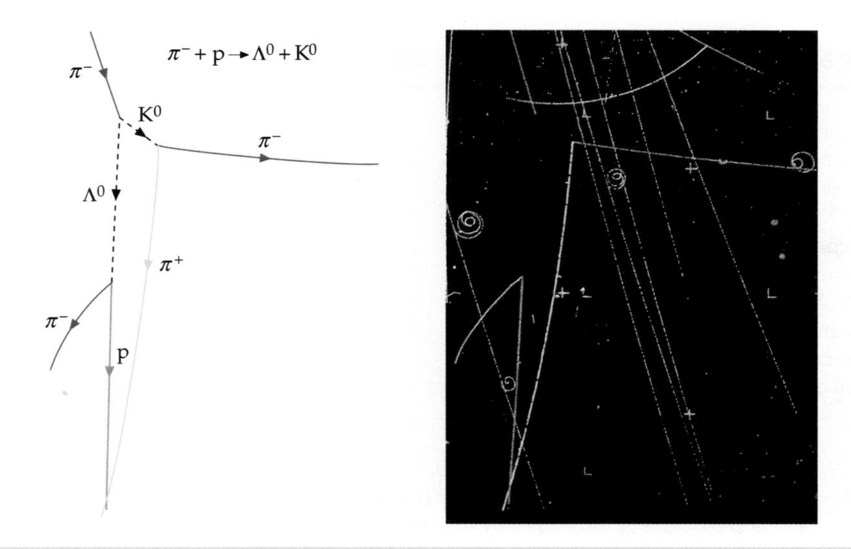

An early photograph of bubble-chamber tracks at the Lawrence Berkeley Laboratory, showing the production and the decay of two particles that have nonzero strangeness, the K^0 and the Λ^0. These neutral particles are identified by the tracks of their electrically charged decay particles. The lambda particle was named because of the similarity of the tracks of its decay particles to the uppercase Greek letter lambda (Λ). (The blue tracks are particles not involved in the reaction of Equation 41-6.) *(© Lawrence Berkeley Laboratory/Science Photo Library/Photo Researchers.)*

all other conservation laws are met. This behavior is described by assigning a new property called strangeness to the particles. During reactions and decays that occur by the strong interaction, strangeness is conserved. During reactions and decays that occur by the weak interaction, the strangeness can only change by ± 1. The strangeness of the ordinary hadrons—the nucleons and pions—was arbitrarily taken to be zero. The strangeness of the K^0 was arbitrarily chosen to be $+1$. The strangeness of the Λ^0 particle must then be -1 so that strangeness is conserved during the reaction described by Equation 41-6. The strangeness of other particles could then be assigned by looking at their various reactions and decays. During reactions and decays that occur by the weak interaction, the strangeness can change by ± 1.

Figure 41-3 shows the masses of the baryons and the mesons that are stable against decay by the strong interaction versus strangeness. We can see from this figure that the particles cluster in multiplets of one, two, or three particles of approximately equal mass, and that the strangeness of a multiplet of particles is related to the *center of charge* of the multiplet.

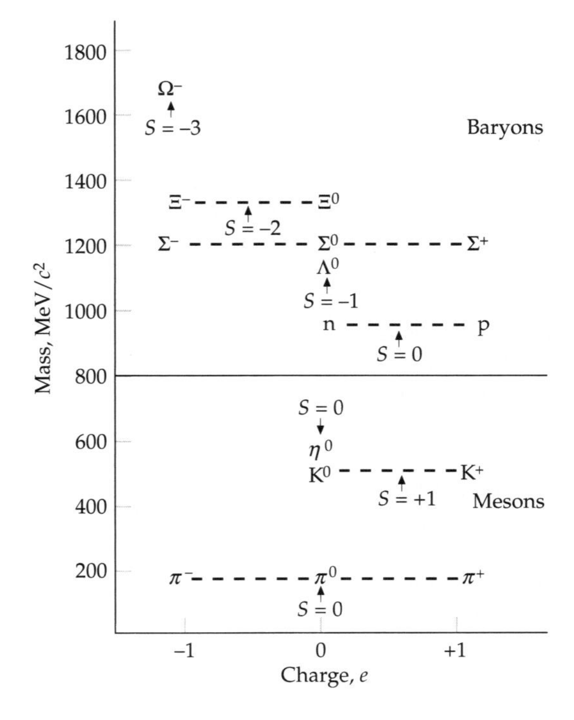

FIGURE 41-3 The strangeness of hadrons shown on a plot of mass versus charge. The strangeness of a baryon–charge multiplet is related to the number of places on the plot that the center of charge of the multiplet is displaced from that of the nucleon doublet. For each "displacement" of e, the strangeness changes by ± 1. For mesons, the strangeness is related to the number of places the center of charge is displaced from that of the pion triplet. Because of the unfortunate original assignment of $+1$ for the strangeness of K-mesons (kaons), all of the baryons that are stable against decay by the strong interaction have negative or zero strangeness.

Example 41-3 Strong Interaction, Weak Interaction, or No Interaction *Conceptual*

State whether the following decays can occur by the strong interaction, by the weak interaction, or not at all: (*a*) $\Sigma^+ \rightarrow \mathrm{p} + \pi^0$, (*b*) $\Sigma^0 \rightarrow \Lambda^0 + \gamma$, and (*c*) $\Xi^0 \rightarrow \mathrm{n} + \pi^0$, where Σ, Λ and Ξ are the symbols for the sigma, lambda and xi particles respectively.

PICTURE We first note that the mass of each decaying particle is greater than the mass of the decay products, so none of the three reactions are in conflict with the principle of conservation of energy. In addition, no leptons are involved in any of the three decays, and charge and baryon number are both conserved during all three reactions. The decay will occur by the strong interaction if strangeness is conserved (if $\Delta S = 0$). If $\Delta S = \pm 1$, the decay will occur via the weak interaction. If $|\Delta S| > 1$, the decay will not occur.

SOLVE

(*a*) From Figure 41-3, we can see that the strangeness of the Σ^+ is -1, whereas the strangeness of both the proton and the pion is zero.

This decay is possible by the weak interaction but not by the strong interaction. It is, in fact, one of the decay modes of the Σ^+ particle which has a lifetime of the order of 10^{-10} s.

(*b*) The strangeness of both the Σ^0 and Λ^0 is -1, whereas the strangeness of the photon is zero.

This decay can proceed by the strong interaction. It is, in fact, the dominant mode of decay of the Σ^0 particle which has a lifetime of approximately 10^{-20} s.

(*c*) The strangeness of the Ξ^0 is -2, whereas the strangeness of both the neutron and the pion is zero.

Because strangeness cannot change by 2 during a decay or a reaction, this decay does not occur.

41-4 QUARKS

Leptons appear to be truly elementary particles in that they do not break down into smaller entities and they seem to have no measurable size or structure. Hadrons, on the other hand, are complex particles that have size and structure, and they decay into other hadrons. Furthermore, at the present time, there are only six known leptons, whereas there are many more hadrons. Except for the Σ^0 particle, Table 41-1 includes only hadrons that are stable against decay by the strong interaction. Hundreds of other hadrons have been discovered; their properties, such as charge, spin, mass, strangeness, and decay schemes, have been measured.

The most important advance in our understanding of elementary particles was the quark model proposed by M. Gell-Mann and G. Zweig in 1963 in which all hadrons consist of combinations of two or three truly elementary particles called **quarks.*** In the original model, quarks came in three types, called **flavors,** labeled *u*, *d*, and *s* (for *up*, *down*, and *strange*). An unusual property of quarks is that they carry fractional electron charges. The charge of the *u* quark is $+\frac{2}{3}e$ and the charge of the *d* and *s* quarks is $\frac{1}{3}e$. Each quark has spin $\frac{1}{2}$ and a baryon number of $\frac{1}{3}$. The strangeness of the *u* and *d* quark is 0, and the strangeness of the *s* quark is -1. Each quark has an antiquark that has the opposite electric charge, baryon number, and strangeness. Baryons consist of three quarks (or three antiquarks for baryons that are antiparticles), whereas mesons consist of a quark and an antiquark, giving mesons baryon numbers $B = 0$, as required. The proton consists of the combination *uud* and the neutron consists of the combination *udd*. Baryons that have a strangeness $S = -1$ have one *s* quark. All the particles listed in Table 41-1 can be

* The name *quark* was chosen by M. Gell-Mann from a quotation from *Finnegan's Wake* by James Joyce.

constructed from these three quarks and three antiquarks.† The great strength of the quark model is that all the allowed combinations of three quarks or quark–antiquark pairs result in known hadrons. Strong evidence for the existence of quarks inside a nucleon is provided by high-energy scattering experiments called *deep inelastic scattering.* During these experiments, a nucleon is bombarded with electrons, muons, or neutrinos of energies from 15 GeV to 200 GeV. Analyses of particles scattered at large angles indicate that inside the nucleon are three spin-$\frac{1}{2}$ particles of sizes much smaller than that of the nucleon. These experiments are analogous to Rutherford's scattering of α particles by atoms in which the presence of a tiny nucleus in the atom was inferred from the large-angle scattering of the α particles.

In 1967, a fourth quark was proposed to explain some discrepancies between experimental determinations of certain decay rates and calculations based on the quark model. The fourth quark is labeled c for a new property called **charm.** Like strangeness, charm is conserved during strong interactions but changes by ± 1 in weak interactions. In 1975, a new heavy meson called the ψ meson was discovered that has the properties expected of a $c\bar{c}$ combination. Since then, other mesons that have combinations such as $c\bar{d}$ and $\bar{c}d$, as well as baryons having the charmed quark, have been discovered. Two more quarks labeled t and b (for *t*op and *b*ottom) were proposed in the 1970s. In 1977, a massive new meson called the Υ (upsilon) meson or **bottomonium,** which is considered to have the quark combination $b\bar{b}$, was discovered. The top quark was observed in 1995. The properties of the six quarks are listed in Table 41-3.

The six quarks and six leptons (and their antiparticles) are thought to be the fundamental elementary particles of which all matter is composed. Table 41-4 lists the masses of the fundamental particles. In this table, the masses given for neutrinos are upper limits. The masses given for quarks are educated guesses. There is strong experimental evidence for the existence of each of these particles.

Table 41-3 Properties of Quarks and Antiquarks

Flavor	Spin	Charge	Baryon Number	Strangeness	Charm	Topness	Bottomness
Quarks							
u (up)	$\frac{1}{2}\hbar$	$+\frac{2}{3}e$	$+\frac{1}{3}$	0	0	0	0
d (down)	$\frac{1}{2}\hbar$	$-\frac{1}{3}e$	$+\frac{1}{3}$	0	0	0	0
s (strange)	$\frac{1}{2}\hbar$	$-\frac{1}{3}e$	$+\frac{1}{3}$	−1	0	0	0
c (charmed)	$\frac{1}{2}\hbar$	$+\frac{2}{3}e$	$+\frac{1}{3}$	0	+1	0	0
t (top)	$\frac{1}{2}\hbar$	$+\frac{2}{3}e$	$+\frac{1}{3}$	0	0	+1	0
b (bottom)	$\frac{1}{2}\hbar$	$-\frac{1}{3}e$	$+\frac{1}{3}$	0	0	0	+1
Antiquarks							
$\bar{u}$	$\frac{1}{2}\hbar$	$-\frac{2}{3}e$	$-\frac{1}{3}$	0	0	0	0
$\bar{d}$	$\frac{1}{2}\hbar$	$+\frac{1}{3}e$	$-\frac{1}{3}$	0	0	0	0
$\bar{s}$	$\frac{1}{2}\hbar$	$+\frac{1}{3}e$	$-\frac{1}{3}$	+1	0	0	0
$\bar{c}$	$\frac{1}{2}\hbar$	$-\frac{2}{3}e$	$-\frac{1}{3}$	0	−1	0	0
$\bar{t}$	$\frac{1}{2}\hbar$	$-\frac{2}{3}e$	$-\frac{1}{3}$	0	0	−1	0
$\bar{b}$	$\frac{1}{2}\hbar$	$+\frac{1}{3}e$	$-\frac{1}{3}$	0	0	0	−1

† The correct quark combinations of hadrons are not always obvious, because of the symmetry requirements on the total wave function. For example, the π^0 meson is represented by a linear combination of $u\bar{u}$ and $d\bar{d}$.

Table 41-4 Masses of Fundamental Particles

Particle	Mass
Quarks	
u (up)	336 MeV/c^2
d (down)	338 MeV/c^2
s (strange)	540 MeV/c^2
c (charmed)	1500 MeV/c^2
t (top)	174 000 MeV/c^2
b (bottom)	4500 MeV/c^2
Leptons	
e^- (electron)	0.511 MeV/c^2
ν_e (electron neutrino)	< 2.2 eV/c^2
μ^- (muon)	105.659 MeV/c^2
ν_μ (muon neutrino)	< 0.17 MeV/c^2
τ^- (tau)	1784 MeV/c^2
ν_τ (tau neutrino)	< 28 MeV/c^2

Example 41-4 Given the Constituent Quark Species, Identify the Particle *Conceptual*

What are the properties of the particles made up of the following quarks: (*a*) $u\bar{u}$, (*b*) $\bar{u}d$, (*c*) dds, and (*d*) uss?

PICTURE Baryons are made up of three quarks, whereas mesons consist of a quark and an antiquark. We add the electric charges of the quarks to find the total charge of the hadron. We also find the strangeness of the hadron by adding the strangeness of the quarks.

SOLVE

(*a*) Because $u\bar{d}$ is a quark-antiquark combination, it has baryon number 0 and is therefore a meson. There is no strange quark here (that is, $S = 0$), so the strangeness of the meson is zero. The charge of the up quark is $+\frac{2}{3}e$ and the charge of the antidown quark is $+\frac{1}{3}e$, so the charge of the meson is $+1e$.

The quark combination $u\bar{d}$ is the π^+ meson.

(*b*) The particle $\bar{u}d$ is also a meson that has zero strangeness. Its electric charge is $-\frac{2}{3}e + \left(-\frac{1}{3}e\right) = -1e$.

The quark combination $\bar{u}d$ is the π^- meson.

(*c*) The particle dds is a baryon that has strangeness -1 because it has one strange quark. Its electric charge is $-\frac{1}{3}e - \frac{1}{3}e - \frac{1}{3}e = -1e$.

The quark combination dds is the Σ^- particle.

(*d*) The particle uss is a baryon that has strangeness -2. Its electric charge is $+\frac{2}{3}e - \frac{1}{3}e - \frac{1}{3}e = 0$.

The quark combination uss is the Ξ^0 particle.

QUARK CONFINEMENT

Despite considerable experimental effort, no isolated quark has ever been observed. It is now believed that it is impossible to obtain an isolated quark. Although the force between quarks is not known, it is believed that the potential energy of two quarks increases with increasing separation distance so that an infinite amount of energy would be needed to separate the quarks completely.

This would be true, for example, if the force of attraction between two quarks remains constant or increases with separation distance, rather than decreasing with increasing separation distance as is the case for other fundamental forces, such as the electric force between two charges, the gravitational force between two masses, and the strong nuclear force between two hadrons.

When a large amount of energy is added to a quark system, such as a nucleon, a quark–antiquark pair is created and the original quarks remain confined within the original system. Because quarks cannot be isolated, but are always bound together to form a baryon or a meson, the mass of a quark cannot be accurately known, which is why the masses listed in Table 41-4 are merely educated guesses.

41-5 FIELD PARTICLES

In addition to the six fundamental leptons and six fundamental quarks, there are other particles, called *field particles,* or *field quanta,* that are associated with the forces exerted by one elementary particle on another. In **quantum electrodynamics,** the electromagnetic field of a single charged particle is described by **virtual photons** that are continuously being emitted and reabsorbed by the particle. If we put energy into the system by accelerating the charge, some of these virtual photons are shaken off and become real, observable photons. The photon is said to mediate the electromagnetic interaction. Each of the four basic interactions can be described via mediating field particles.

The field quantum associated with the gravitational interaction, called the **graviton,** has not yet been observed. The gravitational *charge* analogous to electric charge is mass.

The weak interaction is thought to be mediated by three field quanta called **vector bosons:** W^+, W^-, and Z^0. These particles were predicted by Sheldon Glashow, Abdus Salam, and Steven Weinberg in a theory called the *electroweak theory,* which we discuss in the next section. The W and Z particles were first observed in 1983 by a group of over a hundred scientists led by Carlo Rubbia using the high-energy accelerator at CERN in Geneva, Switzerland. The masses of the $W^\pm$ particles (about 80 GeV/c^2) and the Z particle (about 91 GeV/c^2) measured during this experiment were in excellent agreement with those predicted by the electroweak theory. (The W^- particle is the antiparticle of the W^+ particle, so they must have identical masses.)

The field quanta associated with the strong force between quarks are called **gluons.** Isolated gluons have not been observed experimentally. The *charge* responsible for the strong interactions comes in three varieties, labeled *red, green,* and *blue* (analogous with the three primary colors), and the strong charge is called the **color charge.** The field theory for strong interactions, analogous to quantum electrodynamics for electromagnetic interactions, is called **quantum chromodynamics (QCD).**

Table 41-5 lists the bosons responsible for mediating the basic interactions.

Table 41-5 Bosons That Mediate the Basic Interactions

Interaction	Boson	Spin	Mass	Electric Charge
Strong	g (gluon)	1	0	0
Weak	$W^\pm$	1	80.22 GeV/c^2	$\pm 1e$
	Z^0	1	91.19 GeV/c^2	0
Electromagnetic	γ (photon)	1	0	0
Gravitational	Graviton†	2	0	0

† Not yet observed.

41-6 THE ELECTROWEAK THEORY

In the **electroweak theory,** the electromagnetic and weak interactions are considered to be two different manifestations of a more fundamental electroweak interaction. At very high energies ($\gg 100$ GeV), the electroweak interaction would be mediated by four bosons. From symmetry considerations, these would be a triplet consisting of W^+, W^0, and W^-, all of equal mass, and a singlet B^0 of some other mass. Neither the W^0 nor the B^0 would be observed directly, but one linear combination of the W^0 and the B^0 would be the Z^0 and another would be the photon. At ordinary energies, the symmetry is broken. This leads to the separation of the electromagnetic interaction mediated by the massless photon and the weak interaction mediated by the W^+, W^-, and Z^0 particles. The fact that the photon is massless and that the W and Z particles have masses of the order of 100 GeV/c^2 shows that the symmetry assumed in the electroweak theory does not exist at lower energies.

The symmetry-breaking mechanism is called a **Higgs field,** which requires a new boson, the **Higgs boson,** whose rest energy is expected to be of the order of 1 TeV (1 TeV $= 10^{12}$ eV). The Higgs boson has not yet been observed. Calculations show that Higgs bosons (if they exist) should be produced in head-on collisions between protons of energies of the order of a few TeV. Such energies are not presently available, but the Large Hadron Collider, a particle accelerator under construction near Geneva, Switzerland, is scheduled to come on line late in 2007. The LHC is designed to produce head-on proton–proton collisions for which each proton has an energy of 7 TeV.

41-7 THE STANDARD MODEL

The combination of the quark model, electroweak theory, and quantum chromodynamics is called the **standard model.** In this model, the fundamental particles are the leptons and quarks, each of which comes in six flavors, as shown in Table 41-4; the force carriers are the photon, the $W^\pm$ and Z particles, and the gluons (of which there are eight types). The leptons and quarks are all spin-$\frac{1}{2}$ fermions, which obey the exclusion principle, and the force carriers are integral-spin bosons, which do not obey the exclusion principle. Every interaction in nature is due to one of the four basic interactions: strong, electromagnetic, weak, and gravitational. A particle experiences one of the basic interactions if it carries a charge associated with that interaction. Electric charge is the familiar charge that we have studied previously. Weak charge, also called flavor charge, is carried by leptons and quarks. The charge associated with the strong interaction is called color charge and is carried by quarks and gluons but not by leptons. The charge associated with the gravitational force is mass. It is important to note that the photon, which mediates the electromagnetic interaction, does not carry electric charge. Similarly, the $W^\pm$ and Z particles, which mediate the weak interaction, do not carry weak charge. However, the gluons, which mediate the strong interaction, do carry color charge. This fact is related to the confinement of quarks as discussed in Section 41-4.

All matter is made up of leptons or quarks. There are no known composite particles consisting of leptons bound together by the weak force. Leptons exist only as isolated particles. Hadrons (baryons and mesons) are composite particles consisting of quarks bound together by the color charge. A result of QCD theory is that only color-neutral combinations of quarks are allowed. Three quarks of

Table 41-6 Properties of the Basic Interactions

	Gravitational	Weak	Electromagnetic	Strong: Fundamental	Strong: Residual
Acts on	Mass	Flavor	Electric charge	Color charge	
Particles experiencing	All	Quarks, leptons	Electrically charged	Quarks, gluons	Hadrons
Particles mediating	Graviton	$W^{\pm}, Z$	γ	Gluons	Mesons
Strength for two quarks at 10^{-18} m†	10^{-41}	0.8	1	25	(not applicable)
Strength for two protons in nucleus†	10^{-36}	10^{-7}	1	(not applicable)	20

† Strengths are relative to electromagnetic strength.

different colors can combine to form color-neutral baryons, such as the neutron and the proton. Mesons each have a quark and an antiquark and are also color-neutral. Excited states of hadrons are considered to be different particles. For example, the Δ^+ particle is an excited state of the proton. Both are made up of the *uud* quarks, but the proton is in the ground state and has spin $\frac{1}{2}$ and a rest energy of 938 MeV, whereas the Δ^+ particle is in the first excited state and has spin $\frac{3}{2}$ and a rest energy of 1232 MeV. The two *u* quarks can be in the same spin state in the Δ^+ without violating the exclusion principle, because they have different color. All baryons eventually decay to the lightest baryon, the proton. That the proton does not decay is consistent with the conservation of energy and conservation of baryon number.

The strong interaction has two parts, the fundamental interaction or color interaction and what is called the *residual strong interaction.* The fundamental interaction is responsible for the force exerted by one quark on another quark and is mediated by gluons. The residual strong interaction is responsible for the force between color-neutral nucleons, such as the neutron and the proton. This force is due to the residual strong interactions between the color-charged quarks that make up the nucleons and can be viewed as being mediated by the exchange of mesons. The residual strong interaction between color-neutral nucleons can be thought of as analogous to the residual electromagnetic interaction between neutral atoms that bind them together to form molecules. Table 41-6 lists some of the properties of the basic interactions.

For each particle, there is an antiparticle. A particle and its antiparticle have identical mass and spin but opposite electric charge. For leptons, the lepton numbers L_e, L_μ, and L_τ of the antiparticles are the negatives of the corresponding lepton numbers for the particles. For example, the lepton number for the electron is $L_e = +1$, and the lepton number for the positron is $L_e = -1$. For hadrons, the baryon number, strangeness, charm, topness, and bottomness are the sums of those quantities for the quarks that make up the hadron. The number of each antiparticle is the negative of the number for the corresponding particle. For example, the lambda particle Λ^0, which is made up of the *uds* quarks, has $B = 1$ and $S = -1$, whereas its antiparticle $\overline{\Lambda}^0$, which is made up of the $\overline{u}\overline{d}\overline{s}$ quarks, has $B = -1$ and $S = +1$. A particle such as the photon γ or the Z^0 particle that has zero electric charge; $B = 0$, $L = 0$, $S = 0$; and zero charm, topness, and bottomness is its own antiparticle. Note that the K^0 meson ($d\bar{s}$) has a zero value for all of these quantities except strangeness, which is +1. Its antiparticle, the $\bar{K}^0$ meson ($\bar{d}s$), has strangeness −1, which makes it distinct from the K^0. The π^+ ($u\bar{d}$) and π^- ($\bar{u}d$) are somewhat special in that they have electric charge but zero values for L, B, and S. They are antiparticles of each other, but because there is no conservation law for mesons, it is impossible to say which is the particle and which is the antiparticle. Similarly, the W^+ and W^- are antiparticles of each other.

GRAND UNIFICATION THEORIES

With the success of the electroweak theory, attempts have been made to combine the strong, electromagnetic, and weak interactions in various **grand unification theories (GUTs).** In one of these theories, leptons and quarks are considered to be two aspects of a single class of particles. Under certain conditions, a quark could change into a lepton and vice versa, even though this would appear to violate the conservation of lepton number and baryon number. One of the exciting predictions of this theory is that the proton is not stable but merely has a very long lifetime of the order of 10^{32} y. Such a long lifetime makes proton decay difficult to observe. However, projects are ongoing in which detectors monitor very large numbers of protons in search of an event indicating the decay of a proton.

41-8 THE EVOLUTION OF THE UNIVERSE

In the presently accepted model, the universe began with a singular cataclysmic event called the **big bang** and is expanding. The first evidence that the universe is expanding was the astronomer Edwin Powell Hubble's discovery of the relation between the redshifts in the spectra of galaxies and their distances from us. This relation is illustrated in Figure 41-4 for a group of spiral galaxies used by astronomers for calibrating distances. Provided that the redshift is due to the Doppler effect, the recession velocity v of a galaxy is related to its distance r from us by Hubble's law,

$$v = Hr \qquad 41\text{-}9$$

where H is the **Hubble constant.** In principle, the value of H is easy to obtain because it relies on the direct calculation of v from redshift measurements. However, astronomical distances are very challenging to measure, and they have been determined for only a fraction of the 10^{10} or so galaxies in the observable universe.

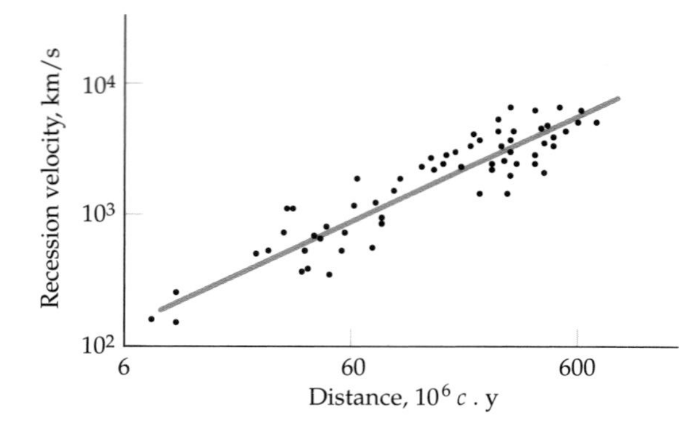

FIGURE 41-4 A plot of the recession velocities of individual galaxies versus distance.

Thus, the value of H changes as distance calibration data are refined. The currently accepted value of the Hubble constant is about

$$H = \frac{23\ \text{km/s}}{10^6\,c \cdot \text{y}} \qquad 41\text{-}10$$

Hubble's law tells us that the galaxies are all rushing away from us, and those galaxies that are the farthest away are moving the fastest. However, there is no reason why our location should be special. An observer in any galaxy would make the same observations and compute the same Hubble constant. Thus, Hubble's law suggests that all of the galaxies are receding from each other at an average speed of 23 km/s per $10^6 c \cdot \text{y}$ of separation. In other words, the universe is expanding. Notice that the basic dimension of H is reciprocal time. The quantity $1/H$ is called the **Hubble age** and equals about 1.3×10^{10} y. This would correspond to the age of the universe if the gravitational pull on the receding galaxies were ignored.

Example 41-5 Using Hubble's Law

Redshift measurements of a galaxy in the constellation Virgo yield a recession velocity of 1200 km/s. How far is it to that galaxy?

PICTURE We calculate the distance from Hubble's law.

SOLVE

Use Hubble's law to find r:

$$r = \frac{v}{H} = (1200\ \text{km/s})\frac{10^6\,c \cdot \text{y}}{23\ \text{km/s}} = \boxed{52 \times 10^6\,c \cdot \text{y}}$$

PRACTICE PROBLEM 41-1 Show that $1/H = 1.3 \times 10^{10}$ y.

THE 2.7-*K* BACKGROUND RADIATION

In investigating ways of accounting for the cosmic abundance of atoms that are heavier than hydrogen atoms, cosmologists recognized that nucleosynthesis in stars could explain the abundance of atoms heavier than helium atoms but could not by itself explain the abundance of helium atoms. Helium must therefore have been formed during the big bang. To synthesize an amount of helium sufficient to account for its present abundance, the big bang would have to have occurred at an extremely high initial temperature to provide the necessary reaction rate before fusion was shut down by the decreasing density of the very rapid initial expansion. The high temperature implies a corresponding thermal (blackbody) radiation field that would cool as the expansion progressed. Theoretical analysis predicted that from the estimated time of the big bang to the present, the remnants of the radiation field should have cooled to a temperature of about 3 K, corresponding to a blackbody spectrum with peak wavelength λ_{max} in the microwave region. In 1965, the predicted cosmic background radiation was discovered by Arno Penzias and Robert Wilson at the Bell Labs. Since this landmark discovery, careful analysis has established that the temperature of the background field is 2.7281 K and has shown that it has an isotropic distribution in space.

THE BIG BANG

The singular event that initiated the expansion of the universe is thought to have been a huge explosion. The four interactions of nature (strong, electromagnetic, weak, and gravitational) initially were unified into a single interaction. Physicists have been successful in developing theoretical descriptions that unify the first three interactions, but a theory of quantum gravity, needed for the extreme densities of the single-interaction period, does not yet exist. Consequently, until the cooling universe "froze" or "condensed out" the gravitational interaction at approximately 10^{-43} s after the big bang, when the temperature was still 10^{32} K, we have no means of describing what was occurring. At this point, the average energy of the particles created would have been about 10^{19} GeV. As the universe continued to cool below 10^{32} K, the three interactions other than gravity remained unified and are described by the grand unification theories (GUTs). Quarks and leptons were indistinguishable and particle quantum numbers were not conserved. It was during this period that a slight excess of quarks over antiquarks occurred, roughly 1 part in 10^9, that ultimately resulted in the predominance of matter over antimatter that we now observe in the universe.

At 10^{-35} s, the universe had expanded sufficiently to cool to approximately 10^{27} K, at which point another phase transition occurred as the strong interaction condensed out of the GUTs group, leaving only the electromagnetic and weak interactions still unified as the **electroweak interaction.** During this period, the previously free quarks in the dense mixture of roughly equal numbers of quarks, leptons, their antiparticles, and photons began to combine into hadrons and their antiparticles, including the nucleons. By the time the universe had cooled to approximately 10^{13} K, at about $t = 10^{-6}$ s, the hadrons had mostly disappeared. This is because 10^{13} K corresponds to $kT \sim 1$ GeV, which is the minimum energy needed to create nucleons and antinucleons from the photons present by the reactions

$$\gamma \rightarrow p^+ + p^- \qquad 41\text{-}11a$$

and

$$\gamma \rightarrow n^+ + \bar{n} \qquad 41\text{-}11b$$

The particle–antiparticle pairs annihilated and there was no new production to replace them. Only the slight earlier excess of quarks over antiquarks led to a slight excess of protons and neutrons over their antiparticles. The annihilations resulted in photons and leptons, and after about $t = 10^{-4}$ s, those particles in roughly equal numbers dominated the universe. This was the **lepton era.** At about $t = 10$ s, the temperature had fallen to 10^{10} K ($kT \sim 1$ MeV). Further expansion and cooling dropped the average photon energy below the energy needed to form an electron–positron pair. Annihilation then removed all of the positrons as it had the antiprotons and antineutrons earlier, leaving only the small excess of electrons arising from charge conservation, and the **radiation era** began. The particles present were primarily photons and neutrinos.

Within a few more minutes, the temperature dropped sufficiently to enable fusing protons and neutrons to form nuclei that were not immediately photodisintegrated. The nuclei of deuterium, helium, and lithium were produced during this **nucleosynthesis period,** but the rapid expansion soon dropped the temperature too low for the fusion to continue and the formation of heavier elements had to await the birth of stars.

A long time later, when the temperature had dropped to about 3000 K as the universe grew to about 1/1000 of its present size, kT dropped below typical atomic ionization energies and atoms were formed. By then, the expansion had redshifted the radiation field so that the total radiation energy was about equal to the energy represented by the remaining mass. As expansion and cooling continued, the energy of the steadily redshifting radiation declined at a steady rate until, at $t = 10^{10}$ y (now), matter came to dominate the universe, with its energy density exceeding that of the 2.7-K radiation remaining from the big bang by a factor of about 1000.

Summary

TOPIC	RELEVANT EQUATIONS AND REMARKS
1. Basic Interactions	There are four basic interactions: strong, electromagnetic, weak, and gravitational.
Strong	The *charge* associated with the strong interaction is called color. Quarks and gluons have color and experience the strong interaction. Hadrons (baryons and mesons) experience a residual strong interaction resulting from the fundamental strong interaction between the quarks that make up the hadrons. Decay times by the strong interaction are typically 10^{-23} s.
Electromagnetic	All particles that have electric charge experience the force due to the electromagnetic interaction.
Weak	The *charge* associated with the weak interaction is called flavor. Quarks and leptons have flavor and experience the weak interaction. Decay times by the weak interaction are typically 10^{-10} s.
Gravitational	The *charge* associated with the gravitational interaction is called mass.
2. Fundamental Particles	There are two families of fundamental particles, leptons and quarks, each having six members. It is thought that these particles have no size and no internal structure.
Leptons	Leptons are spin-$\frac{1}{2}$ fermions: the electron e and its neutrino ν_e, the muon μ and its neutrino ν_μ, and the tau τ and its neutrino ν_τ. The electron, muon, and tau have mass, electric charge, and flavor, but not color; so they participate in the gravitational, electromagnetic, and weak interactions, but not the strong interaction. The neutrinos have flavor but no electric charge and no color. They have a very small mass.
Quarks	There are six quarks, called up u, down d, strange s, charmed c, top t, and bottom b. Each is a spin-$\frac{1}{2}$ fermion. The quarks participate in all of the basic interactions. Because they are always confined in mesons or baryons, their masses can only be estimated.
3. Hadrons	Hadrons are composite particles that are made up of quarks. There are two types of hadrons, baryons and mesons. Baryons, which include the neutron and proton, are fermions of half-integral spin consisting of three quarks. Mesons, which include pions and kaons, have zero or integral spin. Hadrons interact with each other by the residual strong interaction.
4. Field Particles	In addition to the six fundamental leptons and six fundamental quarks, there are field particles that are associated with the basic interactions. *Interaction* — *Field Particle* Gravitational — Graviton (not yet observed) Electromagnetic — Photon Weak — W^+, W^-, Z^0 Strong — Gluons
5. The Conservation Laws	Some quantities, such as energy, linear momentum, electric charge, angular momentum, baryon number, and each of the three lepton numbers, are strictly conserved during all reactions and decays. Others, such as strangeness and charm, are conserved during reactions and decays that proceed by the strong interaction but not in those that proceed by the weak interaction.
6. Particles and Antiparticles	Particles and their antiparticles have identical masses but opposite values for their other properties, such as charge, lepton number, baryon number, and strangeness. Particle–antiparticle pairs can be produced during various nuclear reactions if the energy available is greater than $2mc^2$, where m is the mass of the particle.
7. Hubble's Law	Hubble's law relates the recession velocity of a galaxy, determined from the redshift of its spectrum, to the distance of the galaxy from us: $v = Hr$ 41-9 where the Hubble constant H = 23 km/s per million light-years. From Hubble's law, we conclude that the universe is expanding and that the expansion began approximately $1/H$ years ago.

TOPIC	RELEVANT EQUATIONS AND REMARKS
8. **The Big Bang**	According to the model currently used to describe the evolution of the universe, the universe began with a big bang approximately 10^{10} years ago. The big bang model is supported by substantial experimental observations, including the isotropic, 2.7-K background blackbody radiation spectrum.

Answer to Concept Check

41-1 A proton is a baryon that has a baryon number (B) equal to 1, and all particles that are not baryons have $B = 0$. If a proton decays, conservation of baryon number implies that the decay products must contain a minimum of one baryon. In addition, conservation of energy implies that the rest mass of the decay products cannot be greater than the rest mass of the proton. Because there are no baryons that have a rest mass less than the rest mass of the proton, the proton cannot decay without either violating conservation of baryon number, conservation of energy, or both.

Problems

In a few problems, you are given more data than you actually need; in a few other problems, you are required to supply data from your general knowledge, outside sources, or informed estimate.

Interpret as significant all digits in numerical values that have trailing zeros and no decimal points.

- • Single-concept, single-step, relatively easy
- •• Intermediate-level, may require synthesis of concepts
- ••• Challenging
- SSM Solution is in the *Student Solutions Manual*
- Consecutive problems that are shaded are paired problems.

CONCEPTUAL PROBLEMS

1 • How are baryons and mesons similar? How are they different?

2 • The muon and the pion have nearly the same masses. How do the particles differ?

3 • How can you tell whether a decay proceeds by the strong interaction or the weak interaction? SSM

4 • True or false:

(*a*) All baryons are hadrons.
(*b*) All hadrons are baryons.

5 • True or false: All mesons are spin-$\frac{1}{2}$ particles.

6 • A particle that is made of exactly two quarks is (*a*) a meson, (*b*) a baryon, (*c*) a lepton, (*d*) either a meson or a baryon, but definitely not a lepton.

7 • Have any quark–antiquark combinations whose electric charge is not an integer multiplied by the fundamental charge e been observed?

8 • True or false:

(*a*) A lepton is a combination of three quarks.
(*b*) The typical times for decays by the weak interaction are orders of magnitude longer than the typical times for decays by the strong interaction.
(*c*) The muon and the pion are both mesons.

9 • True or false:

(*a*) Electrons interact with protons by the strong interaction.
(*b*) Strangeness is not conserved in reactions involving the weak interactions.
(*c*) Neutrons have zero charm.

ESTIMATION AND APPROXIMATION

10 •• Grand unification theories predict that the proton has a long but finite lifetime. Current experiments based on detecting the decay of protons in water infer that this lifetime is at least 10^{32} years. Assume 10^{32} years is, in fact, the mean lifetime of the proton. Estimate the expected time between proton decays that occur in the water of a filled Olympic-size swimming pool. An Olympic-size swimming pool is 100 m × 25 m × 2.0 m. Give your answer in days.

11 •• Table 41-6 lists some properties of the four fundamental interactions. To better understand the significance of this table, confirm the ratio of the numerical entries in the second and fourth column of the last row of the table by estimating the ratio of the electromagnetic force to the gravitational force between two protons of a nucleus.

SPIN AND ANTIPARTICLES

12 • Two pions at rest annihilate according to the reaction $\pi^+ + \pi^- \rightarrow \gamma + \gamma$. (*a*) Why must the energies of the two γ rays be equal? (*b*) Find the energy of each γ ray. (*c*) Find the wavelength of each γ ray.

13 • Find the minimum energy of the photon needed for the following pair-production reactions: (*a*) $\gamma \rightarrow \pi^+ + \pi^-$, (*b*) $\gamma \rightarrow p + p^-$, and (*c*) $\gamma \rightarrow \mu^- + \mu^+$.

THE CONSERVATION LAWS

14 • State which of the following decays or reactions violate one or more of the conservation laws, and give the law or laws violated in each case: (*a*) $p^+ \rightarrow n + e^+ + \bar{\nu}_e$, (*b*) $n \rightarrow p^+ + \pi^-$, (*c*) $e^+ + e^- \rightarrow \gamma$, (*d*) $p^+ + p^- \rightarrow \gamma + \gamma$, and (*e*) $\bar{\nu}_e + p^+ \rightarrow n + e^+$.

15 • Determine the change in strangeness in each reaction that follows, and state whether the each decay can proceed by the strong interaction, by the weak interaction, or not at all: (*a*) $\Omega^- \rightarrow \Xi^0 + \pi^-$, (*b*) $\Xi^0 \rightarrow p + \pi^- + \pi^0$, and (*c*) $\Lambda^0 \rightarrow p + \pi^-$.

16 • Determine the change in strangeness for each decay, and state whether each decay can proceed by the strong interaction, by the weak interaction, or not at all: (*a*) $\Omega^- \rightarrow \Lambda^0 + K^-$ and (*b*) $\Xi^0 \rightarrow p + \pi^-$.

17 • Determine the change in strangeness for each decay, and state whether each decay can proceed by the strong interaction, by the weak interaction, or not at all: (*a*) $\Omega^- \rightarrow \Lambda^0 + \bar{\nu}_e + e^-$ and (*b*) $\Sigma^+ \rightarrow p + \pi^0$.

18 • (*a*) Which of the following decays of the τ particle is possible?

$$\tau \rightarrow \mu^- + \bar{\nu}_\mu + \nu_\tau$$
$$\tau \rightarrow \mu^- + \nu_\mu + \bar{\nu}_\tau$$

(*b*) Explain why the other decay is not possible. (*c*) Calculate the kinetic energy of the decay products for the decay that is possible.

19 •• Using Table 41-2 and the laws of conservation of charge number, baryon number, strangeness, and spin, identify the unknown particle, symbolized by (?), in each of the following reactions: (*a*) $p + \pi^- \rightarrow \Sigma^0 + (?)$, (*b*) $p + p \rightarrow \pi^+ + n + K^+ + (?)$, and (*c*) $p + K^- \rightarrow \Xi^- + (?)$ SSM

20 •• Test the following decays for violation of the conservation of energy, electric charge, baryon number, and lepton number: (*a*) $n \rightarrow \pi^+ + \pi^- + \mu^+ + \mu^-$ and (*b*) $\pi^0 \rightarrow e^+ + e^- + \gamma$. Assume that linear momentum and angular momentum are conserved. State which conservation laws (if any) are violated in each decay.

QUARKS

21 • Find the baryon number, charge, and strangeness for the following quark combinations and identify the hadron: (*a*) *uud*, (*b*) *udd*, (*c*) *uus*, (*d*) *dds*, (*e*) *uss*, and (*f*) *dss*.

22 • Find the baryon number, charge, and strangeness for the following quark combinations: (*a*) $u\bar{d}$, (*b*) $\bar{u}d$, (*c*) $u\bar{s}$, and (*d*) $\bar{u}s$.

23 • The Δ^{++} particle is a baryon that decays by the strong interaction. Its strangeness, charm, topness, and bottomness are all zero. What combination of quarks gives a particle that has those properties?

24 • Find a possible combination of quarks that gives the correct values for electric charge, baryon number, and strangeness for (*a*) K^+ and (*b*) K^0.

25 • The D^+ meson has zero strangeness, but it has charm of +1. (*a*) What is a possible quark combination that will give the correct properties for the particle? (*b*) Repeat Part (*a*) for the D^- meson, which is the antiparticle of the D^+ meson.

26 • Find a possible combination of quarks that gives the correct values for electric charge, baryon number, and strangeness for (*a*) K^- (the K^- is the antiparticle of the K^+) and (*b*) $\bar{K}^0$.

27 •• Find a possible quark combination for the following particles: (*a*) Λ^0, (*b*) p^-, and (*c*) Σ^-. SSM

28 •• Find a possible quark combination for the following particles: (*a*) $\bar{n}$, (*b*) Ξ^0, and (*c*) Σ^+.

29 •• Find a possible quark combination for the following particles: (*a*) Ω^- and (*b*) Ξ^-.

30 •• State the properties of the particles made up of the following quarks: (*a*) *ddd*, (*b*) $u\bar{c}$, (*c*) $u\bar{b}$, and (*d*) $\overline{sss}$.

THE EVOLUTION OF THE UNIVERSE

31 • A galaxy is receding from Earth at 2.5 percent the speed of light. Estimate the distance from Earth to the galaxy. SSM

32 • Estimate the speed of a galaxy that is $12 \times 10^9\, c \cdot y$ away from us.

33 •• The Doppler frequency shift for a light from a source that is receding from a stationary receiver is given by $f' = f_0\sqrt{(1-\beta)/(1+\beta)}$, where $\beta = v/c$ (Equation 39-16*b*). Show that the Doppler wavelength shift for light is $\lambda' = \lambda_0\sqrt{(1+\beta)/(1-\beta)}$.

34 •• The red line in the spectrum of atomic hydrogen is frequently referred to as the Hα line, and it has a wavelength of 656.3 nm. Using Hubble's law and the Doppler equation for light from Problem 33, determine the wavelength of the Hα line in the spectrum emitted from galaxies at distances of (*a*) $5.00 \times 10^6\, c \cdot y$, (*b*) $5.00 \times 10^8\, c \cdot y$, and (*c*) $5.00 \times 10^9\, c \cdot y$ from Earth.

GENERAL PROBLEMS

35 • (*a*) What conditions are necessary for a particle and its antiparticle to be identical? (*b*) Find the quark combination of both the particle and the antiparticle of both the π^0 and the Ξ^0 particles. (*c*) Of the π^0 and the Ξ^0 particles, which, if any, is its own antiparticle?

36 •• The red line in the spectrum of atomic hydrogen is frequently referred to as the *H*α line, and it has a wavelength of 656.3 nm. Light from a distant galaxy shows a redshift of the *H*α line of hydrogen to a wavelength of 1458 nm. (*a*) What is the recessional velocity of the galaxy? (*b*) Estimate the distance to the galaxy.

37 •• (*a*) In terms of the quark model, show that the reaction $\pi^0 \rightarrow \gamma + \gamma$ does not violate any conservation laws. (*b*) Which conservation law is violated by the reaction $\pi^0 \rightarrow \gamma$? SSM

38 •• Test the following decays for violation of the conservation of energy, electric charge, baryon number, and lepton number: (*a*) $\Lambda^0 \rightarrow p + \pi^-$, (*b*) $\Sigma^- \rightarrow n + p^-$, and (*c*) $\mu^- \rightarrow e^- + \bar{\nu}_e + \nu_\mu$. Assume that linear momentum and angular momentum are conserved. State which conservation laws (if any) are violated in each decay.

39 •• Consider the following high-energy particle reaction: $p + p \rightarrow \Lambda^0 + K^0 + p + (?)$, where (?) represents an unknown particle. During this reaction, stationary protons are bombarded with a beam of high-energy protons. (*a*) Use the laws of conservation of charge number, baryon number, strangeness (Table 41-2), and spin to determine the unknown particle. (*b*) Calculate the *Q* value for the reaction. (*c*) The threshold kinetic energy K_{th} for this reaction is given by $K_{th} = -\frac{1}{2}Q(m_p + m_p + M_1 + M_2 + M_3 + M_4)/m_p$, where M_1, M_2, M_3, and M_4 are the masses of the reaction products. Find K_{th}.

40 ••• In this problem, you will calculate the difference in the time of arrival of two neutrinos of different energy from a supernova that is 170 000 light-years away. Let the energies of the neutrinos be $E_1 = 20$ MeV and $E_2 = 5$ MeV, and assume that the mass of a neutrino is 2.0 eV/c^2. Because the total energies of the neutrinos is so much greater than their rest energies, the neutrinos have speeds that are very nearly equal to c and energies that are approximately $E \approx pc$. (*a*) If t_1 and t_2 are the times that the neutrinos with speeds u_1 and u_2 take to travel a distance x, show that $\Delta t = t_2 - t_1 = x(u_1 - u_2)/u_1u_2 \approx (x\,\Delta u)/c^2$. (*b*) The speed of a neutrino of mass m and total energy E can be found from $E = mc^2/[1 - (u^2/c^2)]^{1/2}$ (Equation 39-24). Show that when $E \gg mc^2$, the speed ratio u/c is given approximately by $u/c \approx 1 - \frac{1}{2}(mc^2/E)^2$. (*c*) Use the results from Part (*a*) and Part (*b*) to calculate $u_1 - u_2$ for the energies and mass given, and calculate Δt from the result from Part (*a*) for $x = 170\,000\ c \cdot \text{y}$. (*d*) Repeat the calculation in Part (*c*) using 20 eV/c^2 for the neutrino mass.

41 ••• A Λ^0 at rest decays by the reaction $\Lambda^0 \rightarrow \text{p} + \pi^-$. (*a*) Calculate the total kinetic energy of the decay products. (*b*) Find the ratio of the kinetic energy of the pion to the kinetic energy of the proton. (*c*) Find the kinetic energies of the proton and the pion for the decay.

42 ••• A Σ^0 particle at rest decays by the reaction $\Sigma^0 \rightarrow \Lambda^0 + \gamma$. (*a*) What is the total energy (total energy includes rest energy) of the decay products? (*b*) Assuming that the kinetic energy of the Λ^0 is negligible compared with the energy of the photon, calculate the approximate momentum of the photon. (*c*) Use your result from Part (*b*) to calculate the kinetic energy of the Λ^0. (*d*) Use your result from Part (*c*) to obtain a better estimate of the momentum and the energy of the photon.

Appendix A

SI Units and Conversion Factors

Base Units*

Length	The *meter* (m) is the distance traveled by light in a vacuum in 1/299,792,458 s.
Time	The *second* (s) is the duration of 9,192,631,770 periods of the radiation corresponding to the transition between the two hyperfine levels of the ground state of the ^{133}Cs atom.
Mass	The *kilogram* (kg) is the mass of the international standard body preserved at Sèvres, France.
Mole	The *mole* (mol) is the amount of substance of a system which contains as many elementary entities as there are atoms in 0.012 kilogram of carbon 12.
Current	The *ampere* (A) is that constant current which, if maintained in two straight parallel conductors of infinite length, of negligible circular cross section, and placed 1 m apart in vacuum would produce between these conductors a force equal to 2×10^{-7} N/m of length.
Temperature	The *kelvin* (K) is 1/273.16 of the thermodynamic temperature of the triple point of water.
Luminous intensity	The *candela* (cd) is the luminous intensity in a given direction, of a source that emits monochromatic radiation of frequency 540×10^{12} hertz and that has a radiant intensity in that direction of 1/683 watt/steradian.

* These definitions are found on the Internet at http://physics.nist.gov/cuu/Units/current.html

Derived Units

Force	newton (N)	$1 \text{ N} = 1 \text{ kg} \cdot \text{m/s}^2$
Work, energy	joule (J)	$1 \text{ J} = 1 \text{ N} \cdot \text{m}$
Power	watt (W)	$1 \text{ W} = 1 \text{ J/s}$
Frequency	hertz (Hz)	$1 \text{ Hz} = \text{cy/s}$
Charge	coulomb (C)	$1 \text{ C} = 1 \text{ A} \cdot \text{s}$
Potential	volt (V)	$1 \text{ V} = 1 \text{ J/C}$
Resistance	ohm (Ω)	$1 \, \Omega = 1 \text{ V/A}$
Capacitance	farad (F)	$1 \text{ F} = 1 \text{ C/V}$
Magnetic field	tesla (T)	$1 \text{ T} = 1 \text{ N/(A} \cdot \text{m)}$
Magnetic flux	weber (Wb)	$1 \text{ Wb} = 1 \text{ T} \cdot \text{m}^2$
Inductance	henry (H)	$1 \text{ H} = 1 \text{ J/A}^2$

Conversion Factors

Conversion factors are written as equations for simplicity; relations marked with an asterisk are exact.

Length

1 km = 0.6215 mi

1 mi = 1.609 km

1 m = 1.0936 yd = 3.281 ft = 39.37 in

*1 in = 2.54 cm

*1 ft = 12 in = 30.48 cm

*1 yd = 3 ft = 91.44 cm

1 lightyear = 1 $c \cdot$y = 9.461×10^{15} m

*1 Å = 0.1 nm

Area

*1 m^2 = 10^4 cm^2

1 km^2 = 0.3861 mi^2 = 247.1 acres

*1 in^2 = 6.4516 cm^2

1 ft^2 = 9.29×10^{-2} m^2

1 m^2 = 10.76 ft^2

*1 acre = 43 560 ft^2

1 mi^2 = 640 acres = 2.590 km^2

Volume

*1 m^3 = 10^6 cm^3

*1 L = 1000 cm^3 = 10^{-3} m^3

1 gal = 3.785 L

1 gal = 4 qt = 8 pt = 128 oz = 231 in^3

1 in^3 = 16.39 cm^3

1 ft^3 = 1728 in.^{3} = 28.32 L
= 2.832×10^4 cm^3

Time

*1 h = 60 min = 3.6 ks

*1 d = 24 h = 1440 min = 86.4 ks

1 y = 365.24 d = 3.156×10^7 s

Speed

*1 m/s = 3.6 km/h

1 km/h = 0.2778 m/s = 0.6215 mi/h

1 mi/h = 0.4470 m/s = 1.609 km/h

1 mi/h = 1.467 ft/s

Angle and Angular Speed

*π rad = 180°

1 rad = 57.30°

1° = 1.745×10^{-2} rad

1 rev/min = 0.1047 rad/s

1 rad/s = 9.549 rev/min

Mass

*1 kg = 1000 g

*1 tonne = 1000 kg = 1 Mg

1 u = 1.6605×10^{-27} kg
= 931.49 MeV/c^2

1 kg = 6.022×10^{26} u

1 slug = 14.59 kg

1 kg = 6.852×10^{-2} slug

Density

*1 g/cm^3 = 1000 kg/m^3 = 1 kg/L

(1 g/cm^3)g = 62.4 lb/ft^3

Force

1 N = 0.2248 lb = 10^5 dyn

*1 lb = 4.448222 N

(1 kg)g = 2.2046 lb

Pressure

*1 Pa = 1 N/m^2

*1 atm = 101.325 kPa = 1.01325 bars

1 atm = 14.7 lb/in^2 = 760 mmHg
= 29.9 inHg = 33.9 ftH$_2$O

1 lb/in^2 = 6.895 kPa

1 torr = 1 mmHg = 133.32 Pa

1 bar = 100 kPa

Energy

*1 kW$\cdot$h = 3.6 MJ

*1 cal = 4.1840 J

1 ft$\cdot$lb = 1.356 J = 1.286×10^{-3} Btu

*1 L$\cdot$atm = 101.325 J

1 L$\cdot$atm = 24.217 cal

1 Btu = 778 ft$\cdot$lb = 252 cal = 1054.35 J

1 eV = 1.602×10^{-19} J

1 u$\cdot c^2$ = 931.49 MeV

*1 erg = 10^{-7} J

Power

1 horsepower = 550 ft$\cdot$lb/s = 745.7 W

1 Btu/h = 2.931×10^{-4} kW

1 W = 1.341×10^{-3} horsepower
= 0.7376 ft$\cdot$lb/s

Magnetic Field

*1 T = 10^4 G

Thermal Conductivity

1 W/(m$\cdot$K) = 6.938 Btu$\cdot$in/(h$\cdot$ft$^2\cdot$F°)

1 Btu$\cdot$in/(h$\cdot$ft$^2\cdot$F°) = 0.1441 W/(m$\cdot$K)

Appendix B

Numerical Data

Terrestrial Data

Free-fall acceleration g	
Standard value (at sea level at 45° latitude)*	9.806 65 m/s^2; 32.1740 ft/s^2
At equator*	9.7804 m/s^2
At poles*	9.8322 m/s^2
Mass of Earth M_E	5.97 × 10^{24} kg
Radius of Earth R_E, mean	6.37 × 10^6 m; 3960 mi
Escape speed $\sqrt{2R_E g}$	1.12 × 10^4 m/s; 6.95 mi/s
Solar constant†	1.37 kW/m^2
Standard temperature and pressure (STP):	
Temperature	273.15 K (0.00°C)
Pressure	101.325 kPa (1.00 atm)
Molar mass of air	28.97 g/mol
Density of air (STP), ρ_{air}	1.217 kg/m^3
Speed of sound (STP)	331 m/s
Heat of fusion of H_2O (0°C, 1 atm)	333.5 kJ/kg
Heat of vaporization of H_2O (100°C, 1 atm)	2.257 MJ/kg

* Measured relative to Earth's surface.

† Average power incident normally on 1 m^2 outside Earth's atmosphere at the mean distance from Earth to the Sun.

Astronomical Data*

Earth	
Distance to moon, mean†	3.844 × 10^8 m; 2.389 × 10^5 mi
Distance to the Sun, mean†	1.496 × 10^{11} m; 9.30 × 10^7 mi; 1.00 AU
Orbital speed, mean	2.98 × 10^4 m/s
Moon	
Mass	7.35 × 10^{22} kg
Radius	1.737 × 10^6 m
Period	27.32 d
Acceleration of gravity at surface	1.62 m/s^2
Sun	
Mass	1.99 × 10^{30} kg
Radius	6.96 × 10^8 m

* Additional solar-system data is available from NASA at <http://nssdc.gsfc.nasa.gov/planetary/planetfact.html>.

† Center to center.

Physical Constants*

Gravitational constant	G	$6.6742(10) \times 10^{-11}\ \mathrm{N \cdot m^2/kg^2}$
Speed of light	c	$2.997\,924\,58 \times 10^{8}\ \mathrm{m/s}$
Fundamental charge	e	$1.602\,176\,453(14) \times 10^{-19}\ \mathrm{C}$
Avogadro's number	N_A	$6.022\,141\,5(10) \times 10^{23}$ particles/mol
Gas constant	R	$8.314\,472(15)\ \mathrm{J/(mol \cdot K)}$
		$1.987\,2065(36)\ \mathrm{cal/(mol \cdot K)}$
		$8.205\,746(15) \times 10^{-2}\ \mathrm{L \cdot atm/(mol \cdot K)}$
Boltzmann constant	$k = R/N_A$	$1.380\,650\,5(24) \times 10^{-23}\ \mathrm{J/K}$
		$8.617\,343(15) \times 10^{-5}\ \mathrm{eV/K}$
Stefan-Boltzmann constant	$\sigma = (\pi^2/60)k^4/(\hbar^3 c^2)$	$5.670\,400(40) \times 10^{-8}\ \mathrm{W/(m^2 k^4)}$
Atomic mass constant	$m_u = \frac{1}{12}m(^{12}\mathrm{C})$	$1.660\,538\,86(28) \times 10^{-27}\ \mathrm{kg} = 1\mathrm{u}$
Magnetic constant (permeability of free space)	μ_0	$4\pi \times 10^{-7}\ \mathrm{N/A^2}$
		$1.256\,637 \times 10^{-6}\ \mathrm{N/A^2}$
Electric constant (permittivity of free space)	$\epsilon_0 = 1/(\mu_0 C^2)$	$8.854\,187\,817 \ldots \times 10^{-12}\ \mathrm{C^2/(N \cdot m^2)}$
Coulomb constant	$k = 1/(4\pi\epsilon_0)$	$8.987\,551\,788 \ldots \times 10^{9}\ \mathrm{N \cdot m^2/C^2}$
Planck's constant	h	$6.626\,0693(11) \times 10^{-34}\ \mathrm{J \cdot s}$
		$4.135\,667\,43(35) \times 10^{-15}\ \mathrm{eV \cdot s}$
	$\hbar = h/2\pi$	$1.054\,571\,68(18) \times 10^{-34}\ \mathrm{J \cdot s}$
		$6.582\,119\,15(56) \times 10^{-16}\ \mathrm{eV \cdot s}$
Mass of electron	m_e	$9.109\,382\,6(16) \times 10^{-31}\ \mathrm{kg}$
		$0.510\,998\,918(44)\ \mathrm{MeV}/c^2$
Mass of proton	m_p	$1.672\,621\,71(29) \times 10^{-27}\ \mathrm{kg}$
		$938.272\,029(80) \times \mathrm{MeV}/c^2$
Mass of neutron	m_n	$1.674\,927\,28(29) \times 10^{-27}\ \mathrm{kg}$
		$939.565\,360(81)\ \mathrm{MeV}/c^2$
Bohr magneton	$m_B = eh/2m_e$	$9.274\,009\,49(80) \times 10^{-24}\ \mathrm{J/T}$
		$5.788\,381\,804(39) \times 10^{-5}\ \mathrm{eV/T}$
Nuclear magneton	$m_n = eh/2m_p$	$5.050\,783\,43(43) \times 10^{-27}\ \mathrm{J/T}$
		$3.152\,451\,259(21) \times 10^{-8}\ \mathrm{eV/T}$
Magnetic flux quantum	$\phi_0 = h/2e$	$2.067\,833\,72(18) \times 10^{-15}\ \mathrm{T \cdot m^2}$
Quantized Hall resistance	$R_K = h/e^2$	$2.581\,280\,7449(86) \times 10^{4}\ \Omega$
Rydberg constant	R_H	$1.097\,373\,156\,8525(73) \times 10^{7}\ \mathrm{m^{-1}}$
Josephson frequency-voltage quotient	$K_J = 2e/h$	$4.835\,978\,79(41) \times 10^{14}\ \mathrm{Hz/V}$
Compton wavelength	$\lambda_C = h/m_e c$	$2.426\,310\,238(16) \times 10^{-12}\ \mathrm{m}$

* The values for these and other constants may be found on the Internet at http://physics.nist.gov/cuu/Constants/index.html. The numbers in parentheses represent the uncertainties in the last two digits. (For example, 2.044 43(13) stands for 2.044 43 ± 0.000 13.) Values without uncertainties are exact, including those values with ellipses (such as the value of pi is exactly 3.1415. . .).

For additional data, see the following tables in the text.

Geometry and Trigonometry

$C = \pi d = 2\pi r$	definition of π
$A = \pi r^2$	area of circle
$V = \frac{4}{3}\pi r^3$	spherical volume
$A = \partial V/\partial r = 4\pi r^2$	spherical surface area
$V = A_{\text{base}} L = \pi r^2 L$	cylindrical volume
$A = \partial V/\partial r = 2\pi r L$	cylindrical surface area

$o = h \sin\theta$
$a = h \cos\theta$

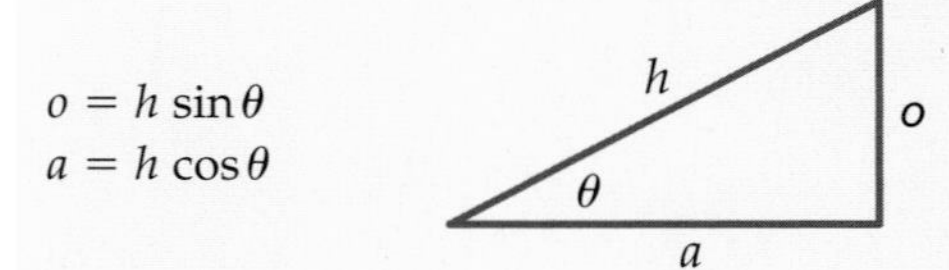

$\sin^2\theta + \cos^2\theta = 1$

$\sin(A \pm B) = \sin A \cos B \pm \cos A \sin B$

$\cos(A \pm B) = \cos A \cos B \mp \sin A \sin B$

$\sin A \pm \sin B = 2 \sin[\frac{1}{2}(A \pm B)] \cos[\frac{1}{2}(A \mp B)]$

$\sin\theta \equiv y$
$\cos\theta \equiv x$
$\tan\theta \equiv \frac{y}{x}$

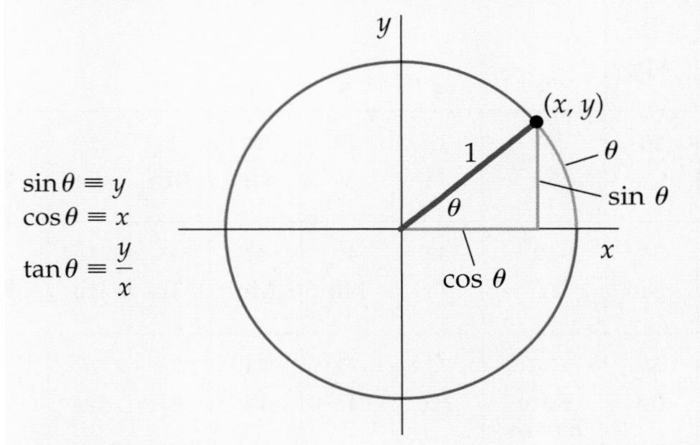

If $|\theta| << 1$, then
$\cos\theta \approx 1$ and $\tan\theta \approx \sin\theta \approx \theta$ (θ in radians)

Quadratic Formula

If $ax^2 + bx + c = 0$, then $x = \dfrac{-b \pm \sqrt{b^2 - 4ac}}{2a}$

Binomial Expansion

If $|x| < 1$, then $(1 + x)^n =$

$$1 + nx + \frac{n(n-1)}{2!}x^2 + \frac{n(n-1)(n-2)}{3!}x^3 + \ldots$$

If $|x| << 1$, then $(1 + x)^n \approx 1 + nx$

Differential Approximation

If $\Delta F = F(x + \Delta x) - F(x)$ and if $|\Delta x|$ is small,

then $\Delta F \approx \dfrac{dF}{dx}\Delta x$.

Appendix C

Periodic Table of Elements*

1	2	3	4	5	6	7	8	9	10	11	12	13	14	15	16	17	18
1 **H**																	2 **He**
3 **Li**	4 **Be**											5 **B**	6 **C**	7 **N**	8 **O**	9 **F**	10 **Ne**
11 **Na**	12 **Mg**											13 **Al**	14 **Si**	15 **P**	16 **S**	17 **Cl**	18 **Ar**
19 **K**	20 **Ca**	21 **Sc**	22 **Ti**	23 **V**	24 **Cr**	25 **Mn**	26 **Fe**	27 **Co**	28 **Ni**	29 **Cu**	30 **Zn**	31 **Ga**	32 **Ge**	33 **As**	34 **Se**	35 **Br**	36 **Kr**
37 **Rb**	38 **Sr**	39 **Y**	40 **Zr**	41 **Nb**	42 **Mo**	43 **Tc**	44 **Ru**	45 **Rh**	46 **Pd**	47 **Ag**	48 **Cd**	49 **In**	50 **Sn**	51 **Sb**	52 **Te**	53 **I**	54 **Xe**
55 **Cs**	56 **Ba**	57–71 **Rare Earths**	72 **Hf**	73 **Ta**	74 **W**	75 **Re**	76 **Os**	77 **Ir**	78 **Pt**	79 **Au**	80 **Hg**	81 **Tl**	82 **Pb**	83 **Bi**	84 **Po**	85 **At**	86 **Rn**
87 **Fr**	88 **Ra**	89–103 Actinides	104 **Rf**	105 **Db**	106 **Sg**	107 **Bh**	108 **Hs**	109 **Mt**	110 **Ds**	111 **Rg**							

Rare Earths (Lanthanides)	57 **La**	58 **Ce**	59 **Pr**	60 **Nd**	61 **Pm**	62 **Sm**	63 **Eu**	64 **Gd**	65 **Tb**	66 **Dy**	67 **Ho**	68 **Er**	69 **Tm**	70 **Yb**	71 **Lu**
Actinides	89 **Ac**	90 **Th**	91 **Pa**	92 **U**	93 **Np**	94 **Pu**	95 **Am**	96 **Cm**	97 **Bk**	98 **Cf**	99 **Es**	100 **Fm**	101 **Md**	102 **No**	103 **Lr**

*The 1–18 group designation has been recommended by the International Union of Pure and Applied Chemistry (IUPAC). Elements with atomic numbers 112, 114, and 116 have been reported but not fully authenticated as of September 2003. From http://www.iupac.org/reports/periodic_table/IUPAC_Periodic_Table-3Oct05.pdf

Atomic Numbers and Atomic Masses*

Atomic Number	Name	Symbol	Mass	Atomic Number	Name	Symbol	Mass
1	Hydrogen	H	1.00794(7)	57	Lanthanum	La	138.90547(7)
2	Helium	He	4.002602(2)	58	Cerium	Ce	140.116(1)
3	Lithium	Li	6.941(2)	59	Praseodymium	Pr	140.90765(2)
4	Beryllium	Be	9.012182(3)	60	Neodymium	Nd	144.242(3)
5	Boron	B	10.811(7)	61	Promethium	Pm	[145]
6	Carbon	C	12.0107(8)	62	Samarium	Sm	150.36(2)
7	Nitrogen	N	14.0067(2)	63	Europium	Eu	151.964(1)
8	Oxygen	O	15.9994(3)	64	Gadolinium	Gd	157.25(3)
9	Fluorine	F	18.9984032(5)	65	Terbium	Tb	158.92535(2)
10	Neon	Ne	20.1797(6)	66	Dysprosium	Dy	162.500(1)
11	Sodium	Na	22.98976928(2)	67	Holmium	Ho	164.93032(2)
12	Magnesium	Mg	24.3050(6)	68	Erbium	Er	167.259(3)
13	Aluminum	Al	26.9815386(8)	69	Thulium	Tm	168.93421(2)
14	Silicon	Si	28.0855(3)	70	Ytterbium	Yb	173.04(3)
15	Phosphorus	P	30.973762(2)	71	Lutetium	Lu	174.967(1)
16	Sulfur	S	32.065(5)	72	Hafnium	Hf	178.49(2)
17	Chlorine	Cl	35.453(2)	73	Tantalum	Ta	180.94788(2)
18	Argon	Ar	39.948(1)	74	Tungsten	W	183.84(1)
19	Potassium	K	39.0983(1)	75	Rhenium	Re	186.207(1)
20	Calcium	Ca	40.078(4)	76	Osmium	Os	190.23(3)
21	Scandium	Sc	44.955912(6)	77	Iridium	Ir	192.217(3)
22	Titanium	Ti	47.867(1)	78	Platinum	Pt	195.084(9)
23	Vanadium	V	50.9415(1)	79	Gold	Au	196.966569(4)
24	Chromium	Cr	51.9961(6)	80	Mercury	Hg	200.59(2)
25	Manganese	Mn	54.938045(5)	81	Thallium	Tl	204.3833(2)
26	Iron	Fe	55.845(2)	82	Lead	Pb	207.2(1)
27	Cobalt	Co	58.933195(5)	83	Bismuth	Bi	208.98040(1)
28	Nickel	Ni	58.6934(2)	84	Polonium	Po	[209]
29	Copper	Cu	63.546(3)	85	Astatine	At	[210]
30	Zinc	Zn	65.409(4)	86	Radon	Rn	[222]
31	Gallium	Ga	69.723(1)	87	Francium	Fr	[223]
32	Germanium	Ge	72.64(1)	88	Radium	Ra	[226]
33	Arsenic	As	74.92160(2)	89	Actinium	Ac	[227]
34	Selenium	Se	78.96(3)	90	Thorium	Th	232.03806(2)
35	Bromine	Br	79.904(1)	91	Protactinium	Pa	231.03588(2)
36	Krypton	Kr	83.798(2)	92	Uranium	U	238.02891(3)
37	Rubidium	Rb	85.4678(3)	93	Neptunium	Np	[237]
38	Strontium	Sr	87.62(1)	94	Plutonium	Pu	[244]
39	Yttrium	Y	88.90585(2)	95	Americium	Am	[243]
40	Zirconium	Zr	91.224(2)	96	Curium	Cm	[247]
41	Niobium	Nb	92.90638(2)	97	Berkelium	Bk	[247]
42	Molybdenum	Mo	95.94(2)	98	Californium	Cf	[251]
43	Technetium	Tc	[98]	99	Einsteinium	Es	[252]
44	Ruthenium	Ru	101.07(2)	100	Fermiun	Fm	[257]
45	Rhodium	Rh	102.90550(2)	101	Mendelevium	Md	[258]
46	Palladium	Pd	106.42(1)	102	Nobelium	No	[259]
47	Silver	Ag	107.8682(2)	103	Lawrencium	Lr	[262]
48	Cadmium	Cd	112.411(8)	104	Rutherfordium	Rf	[261]
49	Indium	In	114.818(3)	105	Dubnium	Db	[262]
50	Tin	Sn	118.710(7)	106	Seaborgium	Sg	[266]
51	Antimony	Sb	121.760(1)	107	Bohrium	Bh	[264]
52	Tellurium	Te	127.60(3)	108	Hassium	Hs	[277]
53	Iodine	I	126.90447(3)	109	Meitnerium	Mt	[268]
54	Xenon	Xe	131.293(6)	110	Darmstadtium	Ds	[271]
55	Cesium	Cs	132.9054519(2)	111	Roentgenium	Rg	[272]
56	Barium	Ba	137.327(7)				

* IUPAC 2005 standard atomic weights (mean relative atomic masses) as approved at the 43rd IUPAC General Assembly in Beijing, China, in August 2005 are listed with uncertainties in the last figure in parentheses. From http://www.iupac.org/reports/periodic_table/IUPAC_Periodic_Table-3Oct05.pdf

Math Tutorial

In this tutorial, we review some of the basic results of algebra, geometry, trigonometry, and calculus. In many cases, we merely state results without proof. Table M-1 lists some mathematical symbols.

Table M-1 Mathematical Symbols

Symbol	Meaning
$=$	is equal to
$\neq$	is not equal to
$\approx$	is approximately equal to
$\sim$	is of the order of
$\propto$	is proportional to
$>$	is greater than
$\geq$	is greater than or equal to
$\gg$	is much greater than
$<$	is less than
$\leq$	is less than or equal to
$\ll$	is much less than
Δx	change in x
$\lvert x \rvert$	absolute value of x
n!	$n(n-1)(n-2)\ldots 1$
$\sum$	sum
lim	limit
$\Delta t \rightarrow 0$	Δt approaches zero
$\frac{dx}{dt}$	derivative of x with respect to t
$\frac{\partial x}{\partial t}$	partial derivative of x with respect to t
$\int$	integral

M-1 SIGNIFICANT DIGITS

Many numbers we work with in science are the result of measurement and are therefore known only within a degree of uncertainty. This uncertainty should be reflected in the number of digits used. For example, if you have a 1-meter-long rule with scale spacing of 1 cm, you know that you can measure the height of a box to within a fifth of a centimeter or so. Using this rule, you might find that the box height is 27.0 cm. If there is a scale with a spacing of 1 mm on your rule, you might perhaps measure the box height to be 27.03 cm. However, if there is a scale with a spacing of 1 mm on your rule, you might not be able to measure the height more accurately than 27.03 cm because the height might vary by 0.01 cm or so, depending on which part of the box you measure the height at. When you write down that the height of the box is 27.03 cm, you are stating that your best estimate of the height is 27.03 cm, but you are not claiming that it is exactly 27.030000 . . . cm high. The four digits in 27.03 cm are called **significant digits.** Your measured length, 2.703 m, has four significant digits. Significant digits are also called significant figures.

The number of significant digits in an answer to a calculation will depend on the number of significant digits in the given data. When you work with numbers that have uncertainties, you should be careful not to include more digits than the certainty of measurement warrants. *Approximate* calculations (order-of-magnitude estimates) always result in answers that have only one significant digit or none. When you multiply, divide, add, or subtract numbers, you must consider the accuracy of the results. Listed below are some rules that will help you determine the number of significant digits of your results.

1. When multiplying or dividing quantities, the number of significant digits in the final answer is no greater than that in the quantity with the fewest significant digits.
2. When adding or subtracting quantities, the number of decimal places in the answer should match that of the term with the smallest number of decimal places.
3. Exact values have an unlimited number of significant digits. For example, a value determined by counting, such as 2 tables, has no uncertainty and is an exact value. In addition, the conversion factor 0.0254000 . . . m/in is an exact value because 1.000 . . . inches is exactly equal to 0.0254000 . . . meters. (The yard is, by definition, equal to exactly 0.9144 meters, and 0.9144 divided by 36 is exactly equal to 0.0254.)
4. Sometimes zeros are significant and sometimes they are not. If a zero is before a leading nonzero digit, then the zero is not significant. For example, the number 0.00890 has three significant digits. The first three zeroes are not significant digits but are merely markers to locate the decimal point. Note that the zero after the nine is significant.
5. Zeros that are between nonzero digits are significant. For example, 5603 has four significant digits.
6. The number of significant digits in numbers with trailing zeros and no decimal point is ambiguous. For example 31000 could have as many as five significant digits or as few as two significant digits. To prevent ambiguity, you should report numbers by using scientific notation or by using a decimal point.

Example M-1 Finding the Average of Three Numbers

Find the average of 19.90, −7.524, and −11.8179.

PICTURE You will be adding 3 numbers and then dividing the result by 3. The first number has three significant digits, the second number has four, and the third number has five.

SOLVE

1. Sum the three numbers.

$$19.90 + (-7.524) + (-11.8179) = 0.55\textit{81}$$

2. If the problem only asked for the sum of the three numbers, we would round the answer to the least number of decimal places among all the numbers being added. However, we must divide this intermediate result by 3, so we use the intermediate answer with the two extra digits (italicized and red).

$$\frac{0.55\textit{81}}{3} = 0.18\textit{60333}\ldots$$

3. Only two of the digits in the intermediate answer, 0.18*60333*. . . , are significant digits, so we must round this number to get our final answer. The number 3 in the denominator is a whole number and has an unlimited number of significant digits. Thus, the final answer has the same number of significant digits as the numerator, which is 2.

The final answer is $\boxed{0.19.}$

CHECK The sum in step 1 has two significant digits following the decimal point, the same as the number being summed with the least number of significant digits after the decimal point.

PRACTICE PROBLEMS

1. $\dfrac{5.3\text{ mol}}{22.4\text{ mol/L}}$

2. 57.8 m/s − 26.24 m/s

M-2 EQUATIONS

An **equation** is a statement written using numbers and symbols to indicate that two quantities, written on either side of an equals sign (=), are equal. The quantity on either side of the equal sign may consist of a single term, or of a sum or difference of two or more **terms.** For example, the equation $x = 1 - (ay + b)/(cx - d)$ contains three terms, x, 1 and $(ay + b)/(cx - d)$.

You can perform the following operations on equations:

1. The same quantity can be added to or subtracted from each side of an equation.
2. Each side of an equation can be multiplied or divided by the same quantity.
3. Each side of an equation can be raised to the same power.

These operations are meant to be applied to each *side* of the equation rather than each term in the equation. (Because multiplication is distributive over addition, operation 2—and only operation 2—of the preceding operations also applies term by term.)

Caution: *Division by zero is forbidden at* any *stage in solving an equation; results (if any) would be invalid.*

Adding or Subtracting Equal Amounts

To find x when $x - 3 = 7$, add 3 to both sides of the equation: $(x - 3) + 3 = 7 + 3$; thus, $x = 10$.

Multiplying or Dividing by Equal Amounts

If $3x = 17$, solve for x by dividing both sides of the equation by 3; thus, $x = \frac{17}{3}$, or 5.7.

Example M-2 Simplifying Reciprocals in an Equation

Solve the following equation for x:

$$\frac{1}{x} + \frac{1}{4} = \frac{1}{3}$$

Equations containing reciprocals of unknowns occur in geometric optics and in electric circuit analysis—for example, in finding the net resistance of parallel resistors.

PICTURE In this equation, the term containing x is on the same side of the equation as a term not containing x. Furthermore, x is found in the denominator of a fraction.

SOLVE

1. Subtract $\frac{1}{4}$ from each side:

$$\frac{1}{x} = \frac{1}{3} - \frac{1}{4}$$

2. Simplify the right side of the equation by using the lowest common denominator:

$$\frac{1}{x} = \frac{1}{3} - \frac{1}{4} = \frac{4}{12} - \frac{3}{12} = \frac{4-3}{12} = \frac{1}{12} \quad \text{so} \quad \frac{1}{x} = \frac{1}{12}$$

3. Multiply both sides of the equation by $12x$ to determine the value of x:

$$12\cancel{x}\frac{1}{\cancel{x}} = \cancel{12}x\frac{1}{\cancel{12}}$$

$$\boxed{12} = x$$

CHECK Substitute 12 for x in the left side of original equation.

$$\frac{1}{x} + \frac{1}{4} = \frac{1}{12} + \frac{3}{12} = \frac{4}{12} = \frac{1}{3}$$

PRACTICE PROBLEMS Solve each of the following for x:

3. $(7.0\ \text{cm}^3)x = 18\ \text{kg} + (4.0\ \text{cm}^3)x$
4. $\frac{4}{x} + \frac{1}{3} = \frac{3}{x}$

M-3 DIRECT AND INVERSE PROPORTIONS

When we say variable quantities x and y are **directly proportional,** we mean that as x and y change, the ratio x/y is constant. To say that two quantities are proportional is to say that they are directly proportional. When we say variable quantities x and y are **inversely proportional,** we mean that as x and y change, the ratio xy is constant.

Relationships of direct and inverse proportion are common in physics. Objects moving at the same velocity have momenta directly proportional to their masses. The ideal-gas law ($PV = nRT$) states that pressure P is directly proportional to (absolute) temperature T, when volume V remains constant and is inversely proportional to volume, when temperature remains constant. Ohm's law ($V = IR$) states that the voltage V across a resistor is directly proportional to the electric current in the resistor when the resistance R remains constant.

CONSTANT OF PROPORTIONALITY

When two quantities are directly proportional, the two quantities are related by a *constant of proportionality*. If you are paid for working at a regular rate R in dollars per day, for example, the money m you earn is directly proportional to the time t you work; the rate R is the constant of proportionality that relates the money earned in dollars to the time worked t in days:

$$\frac{m}{t} = R \qquad \text{or} \qquad m = Rt$$

If you earn \$400 in 5 days, the value of R is \$400/(5 days) = \$80/day. To find the amount you earn in 8 days, you could perform the calculation

$$m = (\$80/\text{day})(8\text{ days}) = \$640$$

Sometimes the constant of proportionality can be ignored in proportion problems. Because the amount you earn in 8 days is $\frac{8}{5}$ times what you earn in 5 days, this amount is

$$m = \frac{8}{5}(\$400) = \$640$$

Example M-3 Painting Cubes

You need 15.4 mL of paint to cover one side of a cube. The area of one side of the cube is 426 cm². What is the relation between the volume of paint needed and the area to be covered? How much paint do you need to paint one side of a cube in which the one side has an area of 503 cm²?

PICTURE To determine the amount of paint for the side whose area is 503 cm², you will need to set up a proportion.

SOLVE

1. The volume V of paint needed increases in proportion to the area A to be covered.

   $\boxed{V \text{ and } A \text{ are directly proportional.}}$

   That is, $\frac{V}{A} = k$ or $V = kA$

   where k is the proportionality constant

2. Determine the value of the proportionality constant using the given values $V_1 = 15.4$ mL and $A_1 = 426$ cm²:

   $$k = \frac{V_1}{A_1} = \frac{15.4\text{ mL}}{426\text{ cm}^2} = 0.0361\text{ mL/cm}^2$$

3. Determine the volume of paint needed to paint a side of a cube whose area is 503 cm² using the proportionality constant in step 1:

   $$V_2 = kA_2 = (0.0361\text{ mL/cm}^2)(503\text{ cm}^2) = \boxed{18.2\text{ mL}}$$

CHECK Our value for V_2 is greater than the value for V_1, as expected. The amount of paint needed to cover an area equal to 503 cm^2 should be greater than the amount of paint needed to cover an area of 426 cm^2 because 503 cm^2 is larger than 426 cm^2.

PRACTICE PROBLEMS

5. A cylindrical container holds 0.384 L of water when full. How much water would the container hold if its radius were doubled and its height remained unchanged?
Hint: The volume of a right circular cylinder is given by $V = \pi r^2 h$, *where* r *is its radius and* h *is its height. Thus,* V *is directly proportional to* r^2 *when* h *remains constant.*
6. For the container in Practice Problem 5, how much water would the container hold if both its height and its radius were doubled? How much water would the container hold if its radius were doubled and its height remained unchanged?
Hint: The volume V *of a right circular cylinder is given by* $V = \pi r^2 h$, *where* r *is its radius and* h *is its height.*

M-4 LINEAR EQUATIONS

A **linear equation** is an equation of the form $x + 2y - 4z = 3$. That is, an equation is linear if each term either is constant or is the product of a constant and a variable raised to the first power. Such equations are said to be linear because the plots of these equations form straight lines or planes. The equations of direct proportion between two variables are linear equations.

GRAPH OF A STRAIGHT LINE

A linear equation relating y and x can always be put into the standard form

$$y = mx + b \qquad \text{M-1}$$

where m and b are constants that may be either positive or negative. Figure M-1 shows a graph of the values of x and y that satisfy Equation M-1. The constant b, called the **y intercept,** is the value of y at $x = 0$. The constant m is the **slope** of the line, which equals the ratio of the change in y to the corresponding change in x. In the figure, we have indicated two points on the line, (x_1, y_1) and (x_2, y_2), and the changes $\Delta x = x_2 - x_1$ and $\Delta y = y_2 - y_1$. The slope m is then

$$m = \frac{y_2 - y_1}{x_2 - x_1} = \frac{\Delta y}{\Delta x}$$

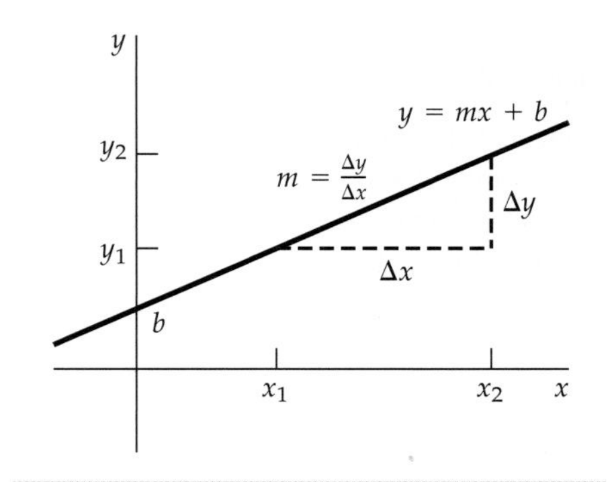

FIGURE M-1 Graph of the linear equation $y = mx + b$, where b is the y intercept and $m = \Delta y/\Delta x$ is the slope.

If x and y are both unknown in the equation $y = mx + b$, there are no unique values of x and y that are solutions to the equation. Any pair of values (x_1, y_1) on the line in Figure M-1 will satisfy the equation. If we have two equations, each with the same two unknowns x and y, the equations can be solved simultaneously for the unknowns. Example M-4 shows how simultaneous linear equations can be solved.

Example M-4 Using Two Equations to Solve for Two Unknowns

Find any and all values of x and y that simultaneously satisfy

$$3x - 2y = 8 \qquad \text{M-2}$$

and

$$y - x = 2 \qquad \text{M-3}$$

PICTURE Figure M-2 shows a graph of the two equations. At the point where the lines intersect, the values of x and y satisfy both equations. We can solve two simultaneous equations by first solving either equation for one variable in terms of the other variable and then substituting the result into the other equation.

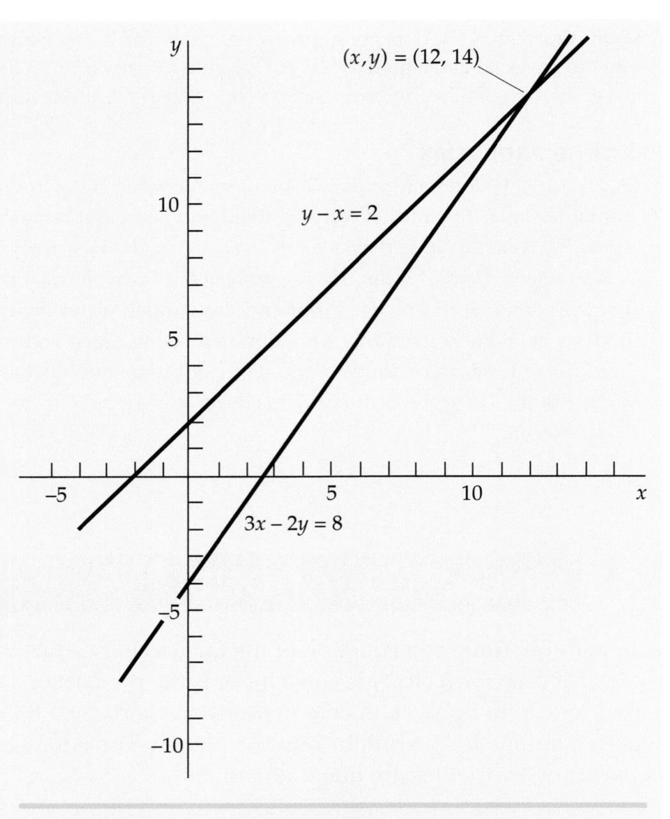

FIGURE M-2 Graph of Equations M-2 and M-3. At the point where the lines intersect, the values of x and y satisfy both equations.

SOLVE

1. Solve Equation M-3 for y: $y = x + 2$
2. Substitute this value for y into Equation M-2: $3x - 2(x + 2) = 8$
3. Simplify the equation and solve for x: $3x - 2x - 4 = 8$

   $x - 4 = 8$

   $x = \boxed{12}$
4. Use your solution for x and one of the given equations to find the value of y: $y - x = 2$, where $x = 12$

   $y - 12 = 2$

   $y = 2 + 12 = \boxed{14}$

CHECK An alternative method is to multiply one equation by a constant such that one of the unknown terms is eliminated when the equations are added or subtracted. We can multiply through Equation M-3 by 2

$$2(y - x) = 2(2)$$
$$2y - 2x = 4$$

and add the result to Equation M-2 and solve for x:

$$\cancel{2y} - 2x = 4$$
$$3x - \cancel{2y} = 8$$
$$3x - 2x = 12 \Rightarrow x = 12$$

Substitute into Equation M-3 and solve for y:

$$y - 12 = 2 \Rightarrow y = 14$$

PRACTICE PROBLEMS

7. True or false: $xy = 4$ is a linear equation.
8. At time $t = 0.0$ s, the position of a particle moving along the x axis at a constant velocity is $x = 3.0$ m. At $t = 2.0$ s, the position is $x = 12.0$ m. Write a linear equation showing the relation of x to t.
9. Solve the following pair of simultaneous equations for x and y:

$$\frac{5}{4}x + \frac{1}{3}y = 30$$
$$y - 5x = 20$$

M-5 QUADRATIC EQUATIONS AND FACTORING

A **quadratic equation** is an equation of the form $ax^2 + bxy + cy^2 + ex + fy + g = 0$, where x and y are variables and a, b, c, e, f, and g are constants. In each term of the equation the powers of the variables are integers that sum to 2, 1, or 0. The designation *quadratic equation* usually applies to an equation of one variable that can be written in the standard form

$$ax^2 + bx + c = 0 \qquad \text{M-4}$$

where a, b, and c are constants. The quadratic equation has two solutions or **roots**—values of x for which the equation is true.

FACTORING

We can solve some quadratic equations by **factoring.** Very often terms of an equation can be grouped or organized into other terms. When we factor terms, we look for multipliers and multiplicands—which we now call **factors**—that will yield two or more new terms as a product. For example, we can find the roots of the quadratic equation $x^2 - 3x + 2 = 0$ by factoring the left side, to get $(x - 2)(x - 1) = 0$. The roots are $x = 2$ and $x = 1$.

Factoring is useful for simplifying equations and for understanding the relationships between quantities. You should be familiar with the multiplication of the factors $(ax + by)(cx + dy) = acx^2 + (ad + bc)xy + bdy^2$.

You should readily recognize some typical factorable combinations:

1. Common factor: $2ax + 3ay = a(2x + 3y)$
2. Perfect square: $x^2 - 2xy + y^2 = (x - y)^2$ (If the expression on the left side of a quadratic equation in standard form is a perfect square, the two roots will be equal.)
3. Difference of squares: $x^2 - y^2 = (x + y)(x - y)$

Also, look for factors that are prime numbers (2, 5, 7, etc.) because these factors can help you factor and simplify terms quickly. For example, the equation $98x^2 - 140 = 0$ can be simplified because 98 and 140 share the common factor 2. That is, $98x^2 - 140 = 0$ becomes $2(49x^2 - 70) = 0$, so we have $49x^2 - 70 = 0$.

This result can be further simplified because 49 and 70 share the common factor 7. Thus, $49x^2 - 70 = 0$ becomes $7(7x^2 - 10) = 0$, so we have $7x^2 - 10 = 0$.

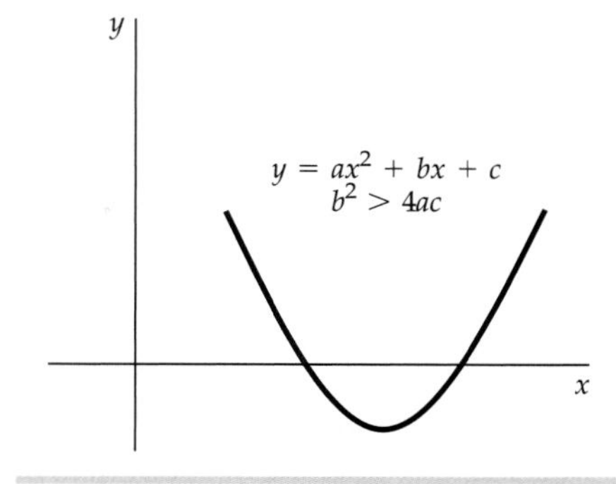

FIGURE M-3 Graph of y versus x when $y = ax^2 + bx + c$ for the case $b^2 > 4ac$. The two values of x for which $y = 0$ satisfy the quadratic equation (Equation M-4).

THE QUADRATIC FORMULA

Not all quadratic equations can be solved by factoring. However, *any* quadratic equation in the standard form $ax^2 + bx + c = 0$ can be solved by the **quadratic formula,**

$$x = \frac{-b \pm \sqrt{b^2 - 4ac}}{2a} = -\frac{b}{2a} \pm \frac{1}{2a}\sqrt{b^2 - 4ac} \qquad \text{M-5}$$

When b^2 is greater than $4ac$, there are two solutions corresponding to the + and − signs. Figure M-3 shows a graph of y versus x where $y = ax^2 + bx + c$. The curve, a **parabola,** crosses the x axis twice. (The simplest representation of a parabola in (x, y) coordinates is an equation of the form $y = ax^2 + bx + c$.) The two roots of this equation are the values for which $y = 0$; that is, they are the *x intercepts.*

When b^2 is less than $4ac$, the graph of y versus x does not intersect the x axis, as is shown in Figure M-4; there are still two roots, but they are not real numbers (see the discussion of complex numbers beginning on page M-19). When $b^2 = 4ac$, the graph of y versus x is tangent to the x axis at the point $x = -b/2a$; the two roots are each equal to $-b/2a$.

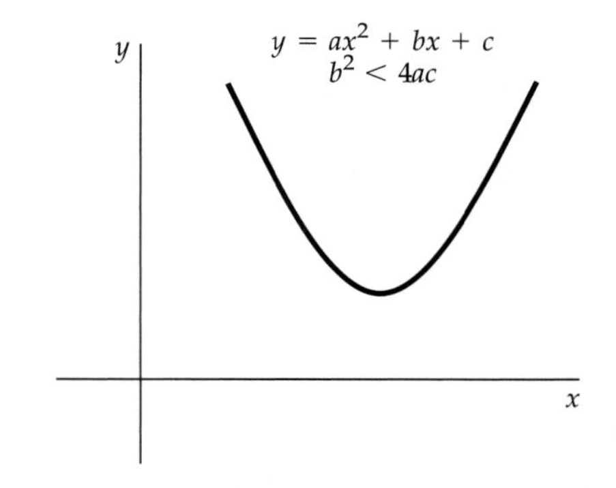

FIGURE M-4 Graph of y versus x when $y = ax^2 + bx + c$ for the case $b^2 < 4ac$. In this case, there are no real values for x for which $y = 0$.

Example M-5 Factoring a Second-Degree Polynomial

Factor the expression $6x^2 + 19xy + 10y^2$.

PICTURE We examine the coefficients of the terms to see whether the expression can be factored without resorting to more advanced methods. Remember that the multiplication $(ax + by)(cx + dy) = acx^2 + (ad + bc)xy + bdy^2$.

SOLVE

1. The coefficient of x^2 is 6 which can be factored two ways:

$$ac = 6$$
$$3 \cdot 2 = 6 \quad \text{or} \quad 6 \cdot 1 = 6$$

2. The coefficient of y^2 is 10 which can also be factored two ways:

$$bd = 10$$
$$5 \cdot 2 = 10 \quad \text{or} \quad 10 \cdot 1 = 10$$

3. List the possibilities for a, b, c, and d in a table. Include a column for $ad + bc$.
   If $a = 3$, then $c = 2$, and vice versa. In addition, if $a = 6$, then $c = 1$, and vice versa. For each value of a there are four values for b.

a	b	c	d	$ad + bc$
3	5	2	2	16
3	2	2	5	19
3	10	2	1	23
3	1	2	10	32
2	5	3	2	19
2	2	3	5	16
2	10	3	1	32
2	1	3	10	23
6	5	1	2	17
6	2	1	5	32
6	10	1	1	16
6	1	1	10	61
1	5	6	2	32
1	2	6	5	17
1	10	6	1	61
1	1	6	10	16

4. Find a combination such that $ad + bc = 19$. As you can see from the table there are two such combinations, and each gives the same results:

$$ad + bc = 19$$
$$3 \cdot 5 + 2 \cdot 2 = 19$$

5. Use the combination in the second row of the table to factor the expression in question:

$$6x^2 + 19xy + 10y^2 = (3x + 2y)(2x + 5y)$$

CHECK As a check, expand $(3x + 2y)(2x + 5y)$.

$$(3x + 2y)(2x + 5y) = 6x^2 + 15xy + 4xy + 10y^2 = 6x^2 + 19xy + 10y^2$$

The combination in the fifth row of the table also gives the step-4 result.

PRACTICE PROBLEMS

10. Show that the combination in the fifth row of the table also gives the step-4 result.
11. Factor $2x^2 - 4xy + 2y^2$.
12. Factor $2x^4 + 10x^3 + 12x^2$.

M-6 EXPONENTS AND LOGARITHMS

EXPONENTS

The notation x^n stands for the quantity obtained by multiplying x times itself n times. For example, $x^2 = x \cdot x$ and $x^3 = x \cdot x \cdot x$. The quantity n is called the **power,** or the **exponent,** of x (the **base**). Listed below are some rules that will help you simplify terms that have exponents.

1. When two powers of x are multiplied, the exponents are added:

$$(x^m)(x^n) = x^{m+n} \qquad \text{M-6}$$

   Example: $x^2x^3 = x^{2+3} = (x \cdot x)(x \cdot x \cdot x) = x^5$.

2. Any number (except 0) raised to the 0 power is defined to be 1:

$$x^0 = 1 \qquad \text{M-7}$$

3. Based on rule 2,

$$x^n x^{-n} = x^0 = 1$$

$$x^{-n} = \frac{1}{x^n} \qquad \text{M-8}$$

4. When two powers are divided, the exponents are subtracted:

$$\frac{x^n}{x^m} = x^n x^{-m} = x^{n-m} \qquad \text{M-9}$$

5. When a power is raised to another power, the exponents are multiplied:

$$(x^n)^m = x^{nm} \qquad \text{M-10}$$

6. When exponents are written as fractions, they represent the roots of the base. For example,

$$x^{1/2} \cdot x^{1/2} = x$$

so

$$x^{1/2} = \sqrt{x} \qquad (x > 0)$$

Example M-6 Simplifying a Quantity That Has Exponents

Simplify $\frac{x^4 x^7}{x^8}$.

PICTURE According to rule 1, when two powers of x are multiplied, the exponents are added. Rule 4 states that when two powers are divided, the exponents are subtracted.

SOLVE

1. Simplify the numerator $x^4 x^7$ using rule 1. $\qquad x^4 x^7 = x^{4+7} = x^{11}$

2. Simplify $\frac{x^{11}}{x^8}$ using rule 4: $\qquad \frac{x^{11}}{x^8} = x^{11} x^{-8} = x^{11-8} = x^3$

CHECK Use the value $x = 2$ to determine if your answer is correct.

$$\frac{2^4 2^7}{2^8} = 2^3 = 8$$

$$\frac{2^4 2^7}{2^8} = \frac{(16)(128)}{256} = \frac{2048}{256} = 8$$

PRACTICE PROBLEMS

13. $(x^{1/18})^9 =$
14. $x^6 x^0 =$

LOGARITHMS

Any positive number can be expressed as some power of any other positive number except one. If y is related to x by $y = a^x$, then the number x is said to be the **logarithm** of y to the **base** a, and the relation is written

$$x = \log_a y$$

Thus, logarithms are *exponents*, and the rules for working with logarithms correspond to similar laws for exponents. Listed below are some rules that will help you simplify terms that have logarithms.

1. If $y_1 = a^n$ and $y_2 = a^m$, then

$$y_1 y_2 = a^n a^m = a^{n+m}$$

Correspondingly,

$$\log_a y_1 y_2 = \log_a a^{n+m} = n + m = \log_a a^n + \log_a a^m = \log_a y_1 + \log_a y_2 \qquad \text{M-11}$$

It then follows that

$$\log_a y^n = n \log_a y \qquad \text{M-12}$$

2. Because $a^1 = a$ and $a^0 = 1$,

$$\log_a a = 1 \qquad \text{M-13}$$

and

$$\log_a 1 = 0 \qquad \text{M-14}$$

There are two bases in common use: logarithms to base 10 are called **common logarithms,** and logarithms to base e (where $e = 2.718\ldots$) are called **natural logarithms.**

In this text, the symbol ln is used for natural logarithms and the symbol log, without a subscript, is used for common logarithms. Thus,

$$\log_e x = \ln x \quad \text{and} \quad \log_{10} x = \log x \qquad \text{M-15}$$

and $y = \ln x$ implies

$$x = e^y \qquad \text{M-16}$$

Logarithms can be changed from one base to another. Suppose that

$$z = \log x \qquad \text{M-17}$$

Then

$$10^z = 10^{\log x} = x \qquad \text{M-18}$$

Taking the natural logarithm of both sides of Equation M-18, we obtain

$$z \ln 10 = \ln x$$

Substituting $\log x$ for z (see Equation M-17) gives

$$\ln x = (\ln 10)\log x \qquad \text{M-19}$$

Example M-7 Converting Between Common Logarithms and Natural Logarithms

The steps leading to Equation M-19 show that, in general, $\log_b x = (\log_b a)\log_a x$, and thus that conversion of logarithms from one base to another requires only multiplication by a constant. Describe the mathematical relation between the constant for converting common logarithms to natural logarithms and the constant for converting natural logarithms to common logarithms.

PICTURE We have a general mathematical formula for converting logarithms from one base to another. We look for the mathematical relation by exchanging a for b and vice versa in the formula.

SOLVE

1. You have a formula for converting logarithms from base a to base b:
$$\log_b x = (\log_b a)\log_a x$$
2. To convert from base b to base a, exchange all a for b and vice versa:
$$\log_a x = (\log_a b)\log_b x$$
3. Divide both sides of the equation in step 1 by $\log_a x$:
$$\frac{\log_b x}{\log_a x} = \log_b a$$
4. Divide both sides of the equation in step 2 by $(\log_a b)\log_a x$:
$$\frac{1}{\log_a b} = \frac{\log_b x}{\log_a x}$$
5. The results show that the conversion factors $\log_b a$ and $\log_a b$ are reciprocals of one other:
$$\frac{1}{\log_a b} = \log_b a$$

CHECK For the value of $\log_{10} e$, your calculator will give 0.43429. For ln 10, your calculator will give 2.3026. Multiply 0.43429 by 2.3026; you will get 1.0000.

PRACTICE PROBLEMS

15. Evaluate $\log_{10}1000$.
16. Evaluate $\log_2 5$

M-7 GEOMETRY

The properties of the most common **geometric figures**—bounded shapes in two or three dimensions whose lengths, areas, or volumes are governed by specific ratios—are a basic analytical tool in physics. For example, the characteristic ratios within triangles give us the laws of *trigonometry* (see the next section of this tutorial), which in turn give us the theory of vectors, essential in analyzing motion in two or more dimensions. Circles and spheres are essential for understanding, among other concepts, angular momentum and the probability densities of quantum mechanics.

BASIC FORMULAS IN GEOMETRY

Circle The ratio of the circumference of a circle to its diameter is a number π, which has the approximate value

$$\pi = 3.141\,592$$

The circumference C of a circle is thus related to its diameter d and its radius r by

$$C = \pi d = 2\pi r \qquad \text{circumference of circle} \qquad \text{M-20}$$

The area of a circle is (Figure M-5)

$$A = \pi r^2 \qquad \text{area of circle} \qquad \text{M-21}$$

Parallelogram The area of a parallelogram is the base b times the height h (Figure M-6):

$$A = bh$$

The area of a triangle is one-half the base times the height (Figure M-7)

$$A = \frac{1}{2}bh$$

Sphere A sphere of radius r (Figure M-8) has a surface area given by

$$A = 4\pi r^2 \qquad \text{surface area of sphere} \qquad \text{M-22}$$

and a volume given by

$$V = \frac{4}{3}\pi r^3 \qquad \text{volume of sphere} \qquad \text{M-23}$$

Cylinder A cylinder of radius r and length L (Figure M-9) has surface area (not including the end faces) of

$$A = 2\pi rL \qquad \text{surface of cylinder} \qquad \text{M-24}$$

and volume of

$$V = \pi r^2 L \qquad \text{volume of cylinder} \qquad \text{M-25}$$

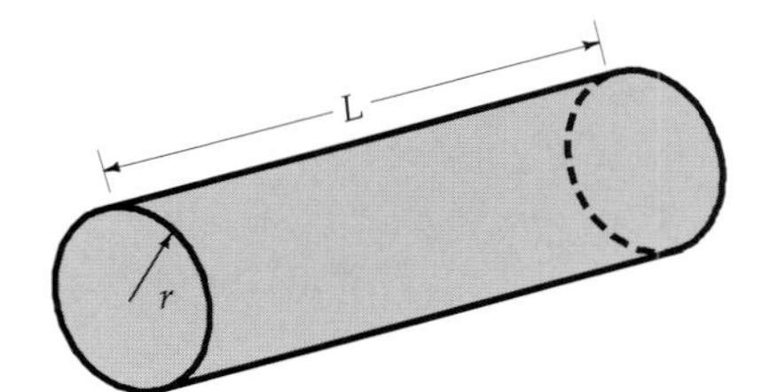

FIGURE M-9 Surface area (not including the end faces) and the volume of a cylinder.

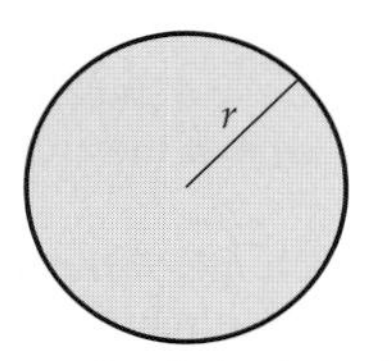

FIGURE M-5 Area of a circle.

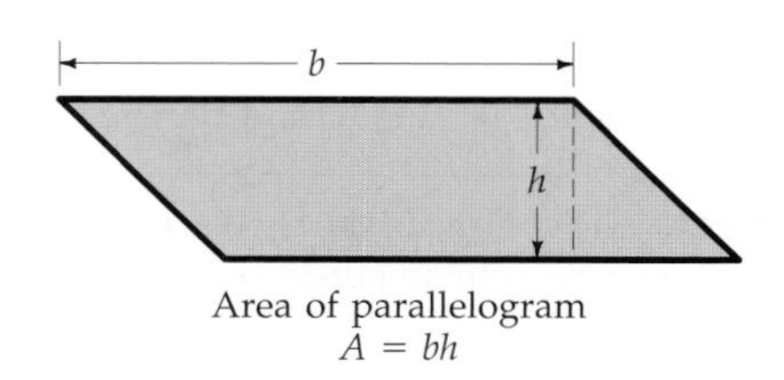

FIGURE M-6 Area of a parallelogram.

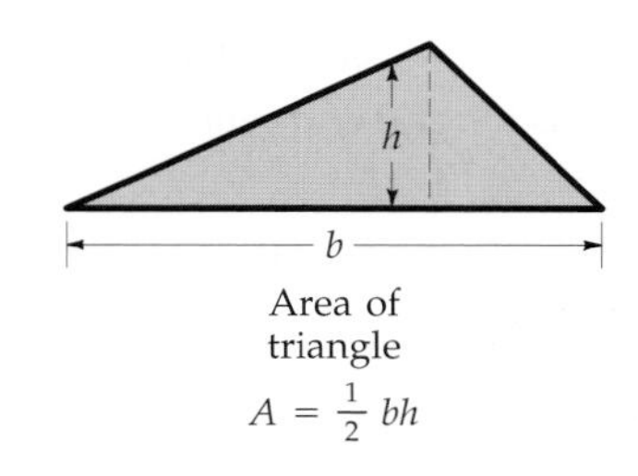

FIGURE M-7 Area of a triangle.

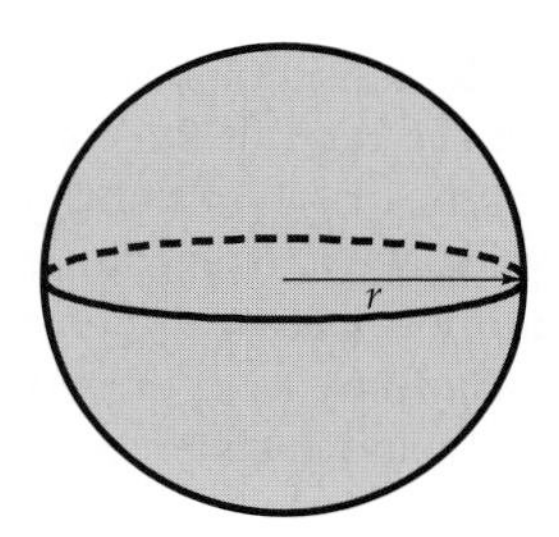

FIGURE M-8 Surface area and volume of a sphere.

Example M-8 Calculating the Mass of a Spherical Shell

An aluminum spherical shell has an outer diameter of 40.0 cm and an inner diameter of 39.0 cm. Find the volume of the aluminum in this shell.

PICTURE The volume of the aluminum in the spherical shell is the volume that remains when we subtract the volume of the inner sphere having $d_i = 2r_i = 39.0$ cm from the volume of the outer sphere having $d_o = 2r_o = 40.0$ cm.

SOLVE

1. Subtract the volume of the sphere of radius r_i from the volume of the sphere of radius r_o:
   $$V = \frac{4}{3}\pi r_o^3 - \frac{4}{3}\pi r_i^3 = \frac{4}{3}\pi(r_o^3 - r_i^3)$$
2. Substitute 20.0 cm for r_o and 19.5 cm for r_i:
   $$V = \frac{4}{3}\pi[(20.0\text{ cm})^3 - (19.5\text{ cm})^3] = \boxed{2.45 \times 10^3\text{ cm}^3}$$

CHECK The volume of the shell is expected to be the same order of magnitude as the volume of a hollow cube with an outside edge length of 40.0 cm and an inside edge length of 39.0 cm. The volume of such a hollow cube is $(40.0\text{ cm})^3 - (39.0\text{ cm})^3 = 4.68 \times 10^3\text{ cm}^3$. The step-2 result meets the expectation that the volume of the shell is the same order of magnitude as the volume of the hollow cube.

PRACTICE PROBLEMS

17. Find the ratio between the volume V and the surface A of a sphere of radius r.
18. What is the area of a cylinder that has a radius that is 1/3 its length?

M-8 TRIGONOMETRY

Trigonometry, which gets its name from Greek roots meaning "triangle" and "measure," is the study of some important mathematical functions, called **trigonometric functions.** These functions are most simply defined as ratios of the sides of right triangles. However, these right-triangle definitions are of limited use because they are valid only for angles between zero and 90°. However, the validity of the right-triangle definitions can be extended by defining the trigonometric functions in terms of the ratio of the coordinates of points on a circle of unit radius drawn centered at the origin of the xy plane.

In physics, we first encounter trigonometric functions when we use vectors to analyze motion in two dimensions. Trigonometric functions are also essential in the analysis of any kind of periodic behavior, such as circular motion, oscillatory motion, and wave mechanics.

ANGLES AND THEIR MEASURE: DEGREES AND RADIANS

The size of an angle formed by two intersecting straight lines is known as its **measure.** The standard way of finding the measure of an angle is to place the angle so that its **vertex,** or point of intersection of the two lines that form the angle, is at the center of a circle located at the origin of a graph that has Cartesian coordinates and one of the lines extends rightward on the positive x axis. The distance traveled *counterclockwise* on the circumference from the positive x axis to reach the intersection of the circumference with the other line defines the measure of the angle. (Traveling clockwise to the second line would simply give us a negative measure; to illustrate basic concepts, we position the angle so that the smaller rotation will be in the counterclockwise direction.)

The most familiar unit for expressing the measure of an angle is the **degree,** which equals 1/360 of the full distance around the circumference of the circle. For greater precision, or for smaller angles, we either show degrees plus minutes (′)

and seconds (″), with 1′ = 1°/60 and 1″ = 1′/60 = 1°/3600; or show degrees as an ordinary decimal number.

For scientific work, a more useful measure of an angle is the **radian** (rad). Again, place the angle with its vertex at the center of a circle and measure counterclockwise rotation around the circumference. The measure of the angle in radians is then defined as the length of the circular arc from one line to the other divided by the radius of the circle (Figure M-10). If s is the arc length and r is the radius of the circle, the angle θ measured in radians is

$$\theta = \frac{s}{r} \qquad \text{M-26}$$

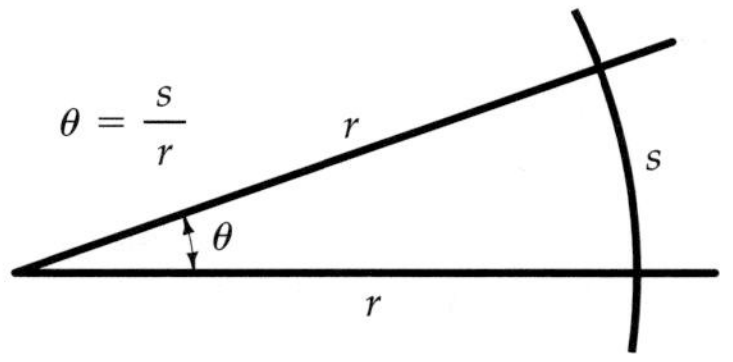

FIGURE M-10 The angle θ in radians is defined to be the ratio s/r, where s is the arc length intercepted on a circle of radius r.

Because the angle measured in radians is the ratio of two lengths, it is dimensionless. The relation between radians and degrees is

$$360° = 2\pi \text{ rad}$$

or

$$1 \text{ rad} = \frac{360°}{2\pi} = 57.3°$$

Figure M-11 shows some useful relations for angles.

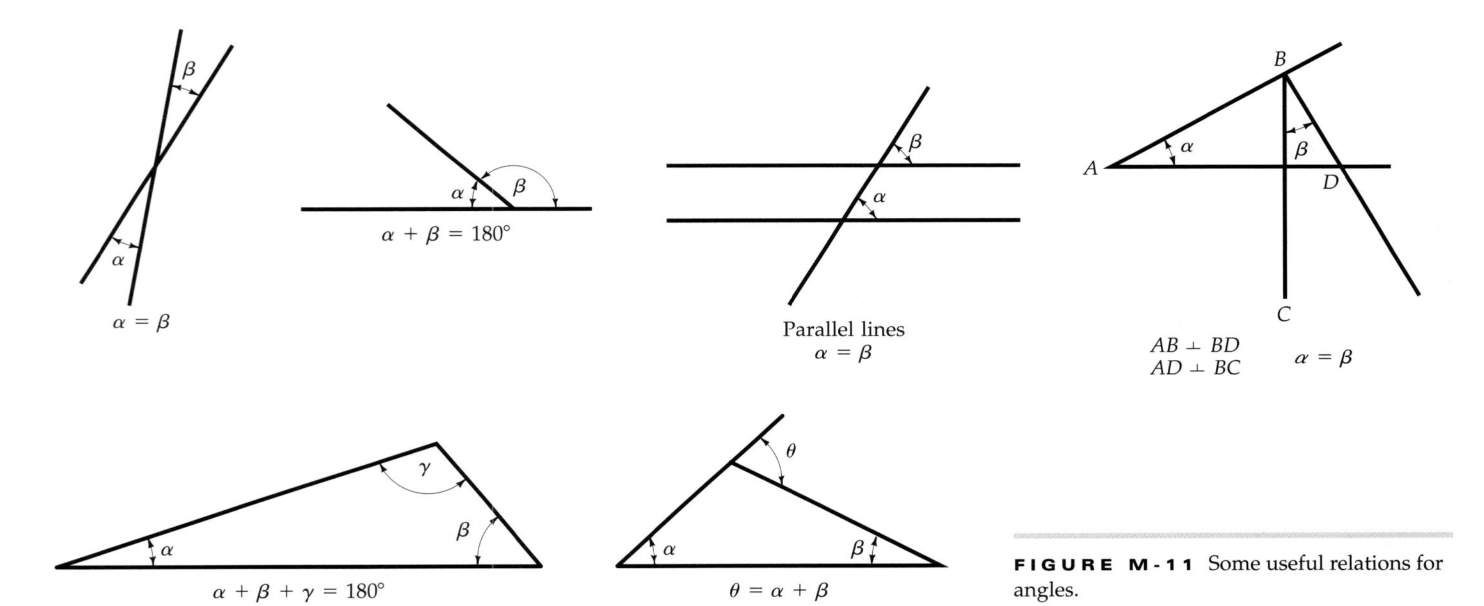

FIGURE M-11 Some useful relations for angles.

THE TRIGONOMETRIC FUNCTIONS

Figure M-12 shows a right triangle formed by drawing the line BC perpendicular to AC. The lengths of the sides are labeled a, b, and c. The right-triangle definitions of the trigonometric functions $\sin\theta$ (the **sine**), $\cos\theta$ (the **cosine**), and $\tan\theta$ (the **tangent**) for an acute angle θ are

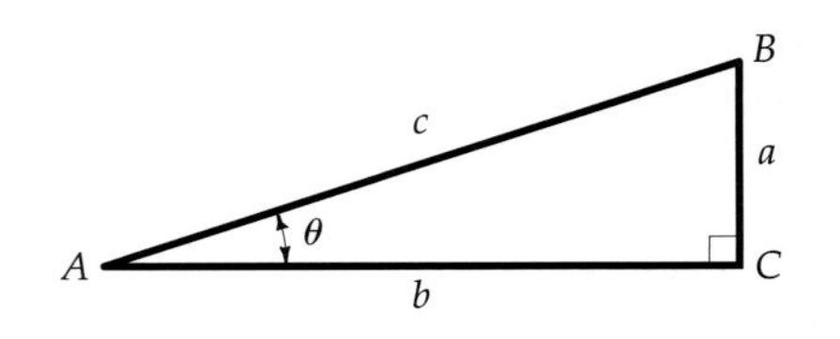

FIGURE M-12 A right triangle with sides of length a and b and a hypotenuse of length c.

$$\sin\theta = \frac{a}{c} = \frac{\text{Opposite side}}{\text{Hypotenuse}} \qquad \text{M-27}$$

$$\cos\theta = \frac{b}{c} = \frac{\text{Adjacent side}}{\text{Hypotenuse}} \qquad \text{M-28}$$

$$\tan\theta = \frac{a}{b} = \frac{\text{Opposite side}}{\text{Adjacent side}} = \frac{\sin\theta}{\cos\theta} \qquad \text{M-29}$$

(**Acute angles** are angles whose positive rotation around the circumference of a circle measures less than 90°, or $\pi/2$.) Three other trigonometric functions—the

secant (sec), the **cosecant** (csc), and the **cotangent** (cot), defined as the reciprocals of these functions—are

$$\sec\theta = \frac{c}{b} = \frac{1}{\cos\theta} \qquad \text{M-30}$$

$$\csc\theta = \frac{c}{a} = \frac{1}{\sin\theta} \qquad \text{M-31}$$

$$\cot\theta = \frac{b}{a} = \frac{1}{\tan\theta} = \frac{\cos\theta}{\sin\theta} \qquad \text{M-32}$$

The angle θ, whose sine is x, is called the arcsine of x, and is written $\sin^{-1} x$. That is, if

$$\sin\theta = x$$

then

$$\theta = \arcsin x = \sin^{-1} x \qquad \text{M-33}$$

The arcsine is the inverse of the sine. The inverse of the cosine and tangent are defined similarly. The angle whose cosine is y is the arccosine of y. That is, if

$$\cos\theta = y$$

then

$$\theta = \arccos y = \cos^{-1} y \qquad \text{M-34}$$

The angle whose tangent is z is the arctangent of z. That is, if

$$\tan\theta = z$$

then

$$\theta = \arctan z = \tan^{-1} z \qquad \text{M-35}$$

TRIGONOMETRIC IDENTITIES

We can derive several useful formulas, called **trigonometric identities,** by examining relationships between the trigonometric functions. Equations M-30 through M-32 list three of the most obvious identities, formulas expressing some trigonometric functions as reciprocals of others. Almost as easy to discern are identities derived from the **Pythagorean theorem,**

$$a^2 + b^2 = c^2 \qquad \text{M-36}$$

(Figure M-13 illustrates a graphic proof of the theorem.) Simple algebraic manipulation of Equation M-36 gives us three more identities. First, if we divide each term in Equation M-36 by c^2, we obtain

$$\frac{a^2}{c^2} + \frac{b^2}{c^2} = 1$$

or, from the definitions of $\sin\theta$ (which is a/c) and $\cos\theta$ (which is b/c)

$$\sin^2\theta + \cos^2\theta = 1 \qquad \text{M-37}$$

Similarly, we can divide each term in Equation M-36 by a^2 or b^2 and obtain

$$1 + \cot^2\theta = \csc^2\theta \qquad \text{M-38}$$

and

$$1 + \tan^2\theta = \sec^2\theta \qquad \text{M-39}$$

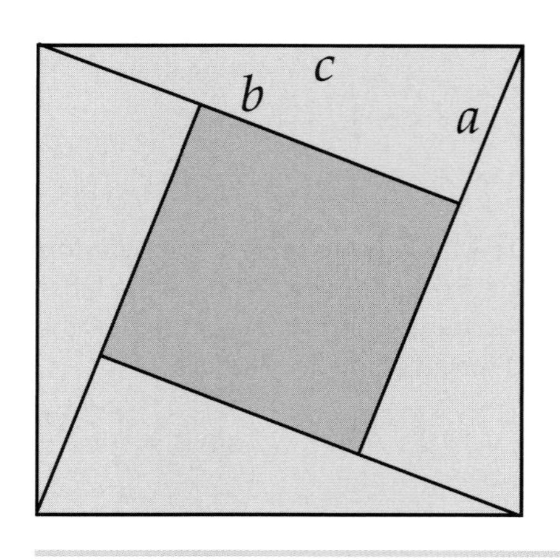

FIGURE M-13 When this figure was first published, the letters were absent and it was accompanied by the single word "Behold!" Using the drawing, establish the Pythagorean theorem ($a^2 + b^2 = c^2$).

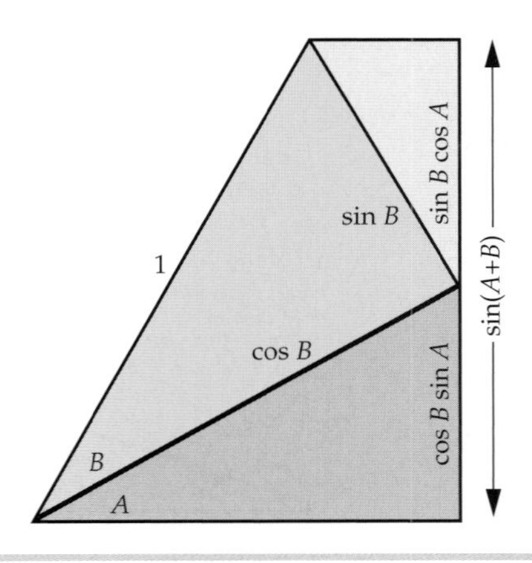

FIGURE M-14 Using this drawing, establish the identity $\sin(A + B) = \sin A \cos B + \cos A \sin B$. You can also use it to establish the identity $\cos(A + B) = \cos A \cos B - \sin A \sin B$. Try it.

Table M-2 Trigonometric Identities

$$\sin(A \pm B) = \sin A \cos B \pm \cos A \sin B$$

$$\cos(A \pm B) = \cos A \cos B \mp \sin A \sin B$$

$$\tan(A \pm B) = \frac{\tan A \pm \tan B}{1 \mp \tan A \tan B}$$

$$\sin A \pm \sin B = 2 \sin\left[\frac{1}{2}(A \pm B)\right]\cos\left[\frac{1}{2}(A \mp B)\right]$$

$$\cos A + \cos B = 2 \cos\left[\frac{1}{2}(A + B)\right]\cos\left[\frac{1}{2}(A - B)\right]$$

$$\cos A - \cos B = 2 \sin\left[\frac{1}{2}(A + B)\right]\sin\left[\frac{1}{2}(B - A)\right]$$

$$\tan A \pm \tan B = \frac{\sin(A \pm B)}{\cos A \cos B}$$

$$\sin^2 \theta + \cos^2 \theta = 1;\ \sec^2 \theta - \tan^2 \theta = 1;\ \csc^2 \theta - \cot^2 \theta = 1$$

$$\sin 2\theta = 2 \sin \theta \cos \theta$$

$$\cos 2\theta = \cos^2 \theta - \sin^2 \theta = 2 \cos^2 \theta - 1 = 1 - 2 \sin^2 \theta$$

$$\tan 2\theta = \frac{2 \tan \theta}{1 - \tan^2 \theta}$$

$$\sin \frac{1}{2}\theta = \pm\sqrt{\frac{1 - \cos \theta}{2}};\ \cos \frac{1}{2}\theta = \pm\sqrt{\frac{1 + \cos \theta}{2}};\ \tan \frac{1}{2}\theta = \pm\sqrt{\frac{1 - \cos \theta}{1 + \cos \theta}}$$

Table M-2 lists these last three and many more trigonometric identities. Notice that they fall into four categories: functions of sums or differences of angles, sums or differences of squared functions, functions of double angles (2θ), and functions of half angles $\left(\frac{1}{2}\theta\right)$. Notice that some of the formulas contain paired alternatives, expressed with the signs $\pm$ and $\mp$; in such formulas, remember to always apply the formula with either all the "upper" or all the "lower" alternatives. Figure M-14 shows a graphic proof of the first two sum-of-angle identities.

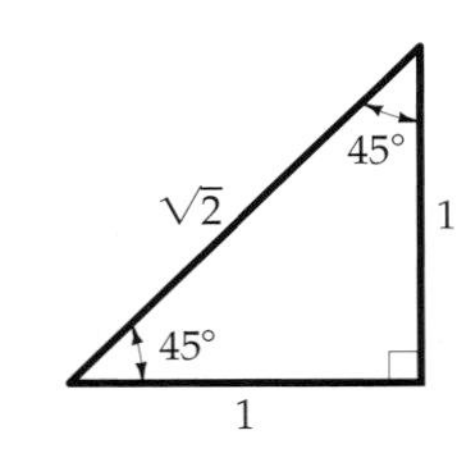

FIGURE M-15 An isosceles right triangle.

SOME IMPORTANT VALUES OF THE FUNCTIONS

Figure M-15 is a diagram of an *isosceles* right triangle (an isosceles triangle is a triangle with two equal sides), from which we can find the sine, cosine, and tangent of 45°. The two acute angles of this triangle are equal. Because the sum of the three angles in a triangle must equal 180° and the right angle is 90°, each acute angle must be 45°. For convenience, let us assume that the equal sides each have a length of 1 unit. The Pythagorean theorem gives us a value for the hypotenuse of

$$c = \sqrt{a^2 + b^2} = \sqrt{1^2 + 1^2} = \sqrt{2}\text{ units}$$

We calculate the values of the functions as follows:

$$\sin 45° = \frac{a}{c} = \frac{1}{\sqrt{2}} = 0.707 \quad \cos 45° = \frac{b}{c} = \frac{1}{\sqrt{2}} = 0.707 \quad \tan 45° = \frac{a}{b} = \frac{1}{1} = 1$$

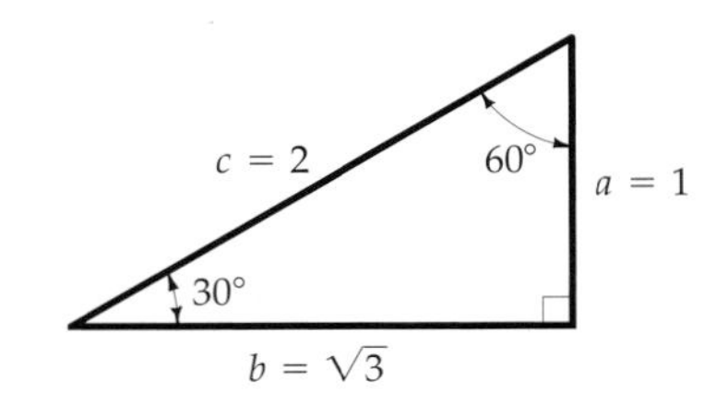

FIGURE M-16 A 30°–60° right triangle.

Another common triangle, a 30°–60° right triangle, is shown in Figure M-16. Because this particular right triangle is in effect half of an *equilateral triangle* (a 60°–60°–60° triangle, or a triangle having three equal sides and three equal angles), we can see that the sine of 30° must be exactly 0.5 (Figure M-17). The equilateral triangle must have all sides equal to c, the hypotenuse of the 30°–60° right triangle. Thus, side a is one-half the length of the hypotenuse, and so

$$\sin 30° = \frac{1}{2}$$

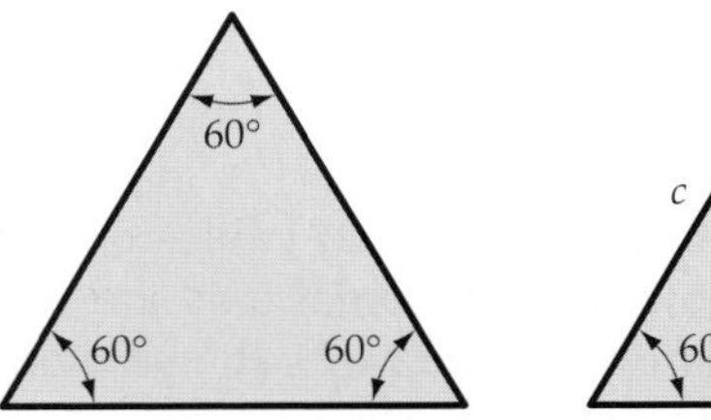

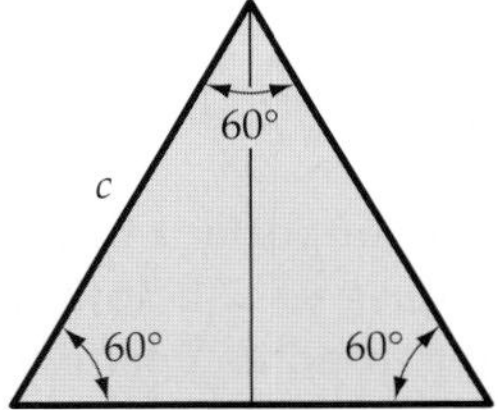

FIGURE M-17 (a) An equilateral triangle. (b) An equilateral triangle that has been bisected to form two 30°–60° right triangles.

To find the other ratios within the 30–60° right triangle, let us assign a value of 1 to the side opposite the 30° angle. Then

$$c = \frac{1}{0.5} = 2 \qquad b = \sqrt{c^2 - a^2} = \sqrt{2^2 - 1^2} = \sqrt{3}$$

$$\cos 30° = \frac{b}{c} = \frac{\sqrt{3}}{2} = 0.866 \qquad \tan 30° = \frac{a}{b} = \frac{1}{\sqrt{3}} = 0.577$$

$$\sin 60° = \frac{b}{c} = \cos 30° = 0.866 \qquad \cos 60° = \frac{a}{c} = \sin 30° = \frac{1}{2}$$

$$\tan 60° = \frac{b}{a} = \frac{\sqrt{3}}{1} = 1.732$$

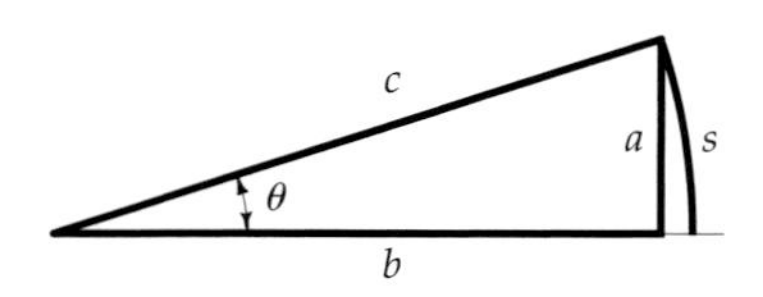

FIGURE M-18 For small angles, $\sin\theta = a/c$, $\tan\theta = a/b$, and the angle $\theta = s/c$ are all approximately equal.

SMALL-ANGLE APPROXIMATION

For small angles, the length a is nearly equal to the arc length s, as can be seen in Figure M-18. The angle $\theta = s/c$ is therefore nearly equal to $\sin\theta = a/c$:

$$\sin\theta \approx \theta \qquad \text{for small values of } \theta \qquad \text{M-40}$$

Similarly, the lengths c and b are nearly equal, so $\tan\theta = a/b$ is nearly equal to both θ and $\sin\theta$ for small values of θ:

$$\tan\theta \approx \sin\theta \approx \theta \qquad \text{for small values of } \theta \qquad \text{M-41}$$

Equations M-40 and M-41 hold only if θ is measured in radians. Because $\cos\theta = b/c$, and because these lengths are nearly equal for small values of θ, we have

$$\cos\theta \approx 1 \qquad \text{for small values of } \theta \qquad \text{M-42}$$

Figure M-19 shows graphs of θ, $\sin\theta$, and $\tan\theta$ versus θ for small values of θ. If accuracy of a few percent is needed, small-angle approximations can be used only for angles of about a quarter of a radian (or about 15°) or less. Below this value, as the angle becomes smaller, the approximation $\theta \approx \sin\theta \approx \tan\theta$ is even more accurate.

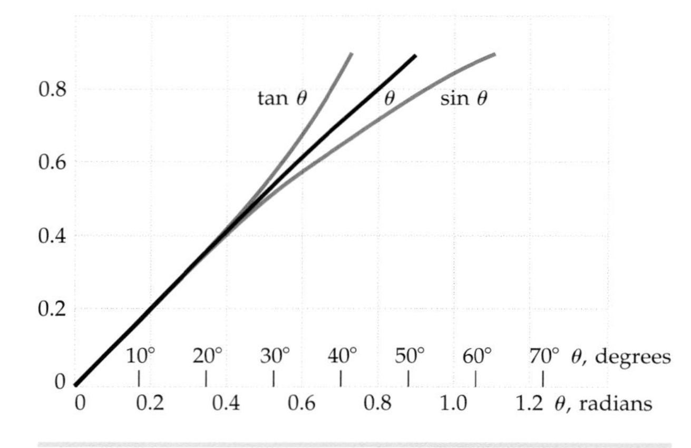

FIGURE M-19 Graphs of $\tan\theta$, θ, and $\sin\theta$ versus θ for small values of θ.

TRIGONOMETRIC FUNCTIONS AS FUNCTIONS OF REAL NUMBERS

So far we have illustrated the trigonometric functions as properties of angles. Figure M-20 shows an *obtuse* angle with its vertex at the origin and one side along the x axis. The trigonometric functions for a "general" angle such as this are defined by

$$\sin\theta = \frac{y}{c} \qquad \text{M-43}$$

$$\cos\theta = \frac{x}{c} \qquad \text{M-44}$$

$$\tan\theta = \frac{y}{x} \qquad \text{M-45}$$

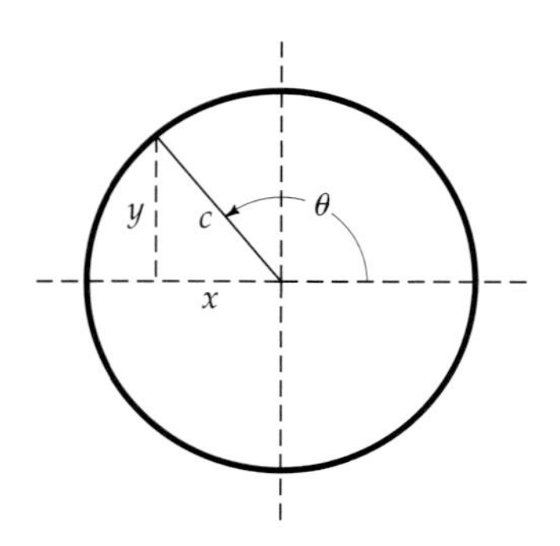

FIGURE M-20 Diagram for defining the trigonometric functions for an obtuse angle.

It is important to remember that values of x to the left of the vertical axis and values of y below the horizontal axis are negative; c in the figure is always regarded as positive. Figure M-21 shows plots of the general sine, cosine, and tangent functions versus θ. The sine function has a period of 2π rad. Thus, for any value of θ, $\sin(\theta + 2\pi) = \sin\theta$, and so forth. That is, when an angle changes by 2π rad, the function returns to its original value. The tangent function has a

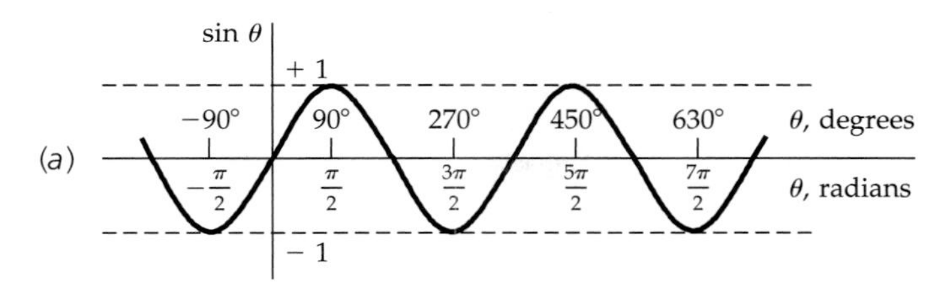

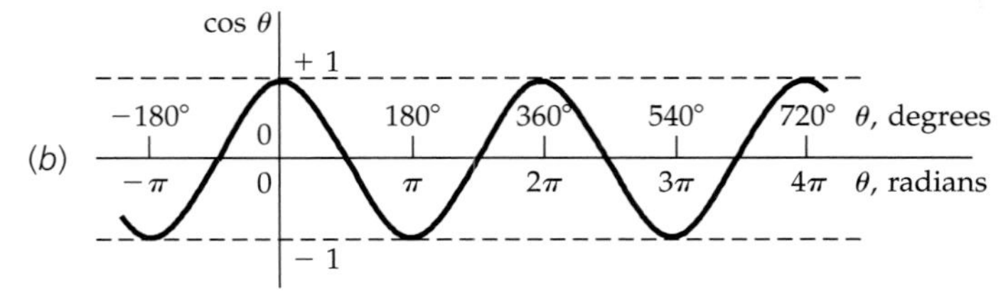

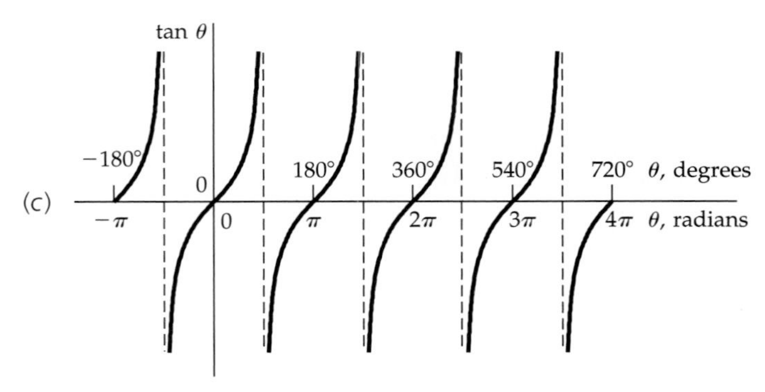

FIGURE M-21 The trigonometric functions sin θ, cos θ, and tan θ versus θ.

period of π rad. Thus, $\tan(\theta + \pi) = \tan \theta$, and so forth. Some other useful relations are

$$\sin(\pi - \theta) = \sin \theta \qquad \text{M-46}$$

$$\cos(\pi - \theta) = -\cos \theta \qquad \text{M-47}$$

$$\sin(\tfrac{1}{2}\pi - \theta) = \cos \theta \qquad \text{M-48}$$

$$\cos(\tfrac{1}{2}\pi - \theta) = \sin \theta \qquad \text{M-49}$$

Because the radian is dimensionless, it is not hard to see from the plots in Figure M-21 that the trigonometric functions are functions of all real numbers. The functions can also be expressed as power series in θ. The series for sin θ and cos θ are

$$\sin \theta = \theta - \frac{\theta^3}{3!} + \frac{\theta^5}{5!} - \frac{\theta^7}{7!} + \cdots \qquad \text{M-50}$$

$$\cos \theta = 1 - \frac{\theta^2}{2!} + \frac{\theta^4}{4!} - \frac{\theta^6}{6!} + \cdots \qquad \text{M-51}$$

When θ is small, good approximations are obtained using only the first few terms in the series.

Example M-9 Cosine of a Sum

Using the suitable trigonometric identity from Table M-2, find $\cos(135° + 22°)$. Give your answer in four significant figures.

PICTURE As long as all angles are given in degrees, there is no need to convert to radians, because all operations are numerical values of the functions. Be sure, however, that your calculator is in degree mode. The suitable identity is $\cos(A \pm B) = \cos A \cos B \mp \sin A \sin B$, where the upper signs are appropriate.

SOLVE

1. Write the trigonometric identity for the cosine of a sum, with $A = 135°$ and $B = 22°$:

   $\cos(135° + 22°) = (\cos 135°)(\cos 22°) - (\sin 135°)(\sin 22°)$

2. Using a calculator, find cos 135°, sin 135°, cos 22°, and sin 22°:

   $\cos 135° = -0.7071$ $\quad$ $\sin 135° = 0.7071$

   $\cos 22° = 0.9272$ $\quad$ $\sin 22° = 0.3746$

3. Enter the values in the formula and calculate the answer:

   $$\cos(135° + 22°) = (-0.7071)(0.9272) - (0.7071)(0.3746) = -0.9205$$

CHECK The calculator shows that the $\cos(135° + 22°) = \cos(157°) = -0.9205$.

PRACTICE PROBLEMS

19. Find $\sin\theta$ and $\cos\theta$ for the right triangle shown in Figure M-12 in which $a = 4$ cm and $b = 7$ cm. What is the value for θ?
20. Find $\sin\theta$ where $\theta = 8.2°$. Is your answer consistent with the small-angle approximation?

M-9 THE BINOMIAL EXPANSION

A **binomial** is an expression consisting of two terms joined by a plus sign or a minus sign. The **binomial theorem** states that a binomial raised to a power can be written, or *expanded*, as a series of terms. If we raise the binomial $(1 + x)$ to a power n, the binomial theorem takes the form

$$(1 + x)^n = 1 + nx + \frac{n(n-1)}{2!}x^2 + \frac{n(n-1)(n-2)}{3!}x^3 + \cdots \qquad \text{M-52}$$

The series is valid for any value of n if $|x|$ is less than 1. The binomial expansion is very useful for approximating algebraic expressions, because when $|x| < 1$, the higher-order terms in the sum are small. (The order of a term is the power of x in the term. Thus, the terms explicitly shown in Equation M-52 are of order 0, 1, 2, and 3.) The series is particularly useful in situations where $|x|$ is small compared with 1; then each term is *much* smaller than the previous term and we can drop all but the first two or three terms in the expansion. If $|x|$ is much less than 1, we have

$$(1 + x)^n \approx 1 + nx, \qquad |x| \ll 1 \qquad \text{M-53}$$

The binomial expansion is used in deriving many formulas of calculus that are important in physics. A well-known use in physics of the approximation in Equation M-53 is the proof that relativistic kinetic energy reduces to the classic formula when the velocity of a particle is very small compared with the velocity of light c.

Example M-10 Using the Binomial Expansion to Find a Power of a Number

Use Equation M-53 to find an approximate value for the square root of 101.

PICTURE The number 101 readily suggests a binomial, namely, $(100 + 1)$. To approximate the answer using the binomial expansion, we must manipulate the expression to get a binomial consisting of 1 and a term less than 1.

SOLVE

1. Write $(101)^{1/2}$ to give an expression $(1 + x)^n$ in which x is much less than 1:

   $$(101)^{1/2} = (100 + 1)^{1/2} = (100)^{1/2}(1 + 0.01)^{1/2} = 10(1 + 0.01)^{1/2}$$

2. Use Equation M-53 with $n = \frac{1}{2}$ and $x = 0.01$ to expand $(1 + 0.01)^{1/2}$:

   $$(1 + 0.01)^{1/2} = 1 + \tfrac{1}{2}(0.01) + \frac{\frac{1}{2}\left(-\frac{1}{2}\right)}{2}(0.01)^2 + \cdots$$

3. Because $|x| \ll 1$, we expect the magnitude of terms of order 2 and higher to be significantly smaller than the magnitude of the first-order term. Approximate the binomial (1) by keeping only the zeroth and first-order terms, and (2) by keeping only the first 3 terms:

Keeping only the zeroth and first-order terms gives

$$(1 + 0.01)^{1/2} \approx 1 + \tfrac{1}{2}(0.01) = 1 + 0.005\,000\,0 = 1.005\,000\,0$$

Keeping only the zeroth, first-, and second-order terms gives

$$(1 + 0.01)^{1/2} \approx 1 + \tfrac{1}{2}(0.01) + \frac{\tfrac{1}{2}(-\tfrac{1}{2})}{2}(0.01)^2 \approx 1 + 0.005\,000\,0 - 0.000\,012\,5 = 1.004\,987\,5$$

4. Substitute these results into the equation in step 1:

Keeping only the zeroth and first-order terms gives

$$(101)^{1/2} = 10(1 + 0.01)^{1/2} \approx \boxed{10.050\,000}$$

Keeping only the zeroth, first-, and second-order terms gives

$$(101)^{1/2} = 10(1 + 0.01)^{1/2} \approx \boxed{10.049\,875}$$

CHECK We therefore expect our answer to be correct to within about 0.001%. The value of $(101)^{1/2}$, to eight figures, is 10.049 876. This differs from 10.050 000 by 0.000 124, or about one part in 10^5, and differs from 10.049 875 by about one part in 10^7.

PRACTICE PROBLEMS For the following, calculate the answer keeping the zeroth and first-order terms in the binomial series (Equation M-53), find the answer using your calculator, and show the percentage discrepancy between the two values:

21. $(1 + 0.001)^{-4}$
22. $(1 - 0.001)^{40}$

M-10 COMPLEX NUMBERS

Real numbers are all numbers, from $-\infty$ to $+\infty$, that can be *ordered*. We know that, given two real numbers, one is always equal to, greater than, or less than the other. For example, $3 > 2$, $1.4 < \sqrt{2} < 1.5$, and $3.14 < \pi < 3.15$. A number that *cannot* be ordered is $\sqrt{-1}$; we cannot measure the size of this number, and so it makes no sense to say, for example, that $3 \times \sqrt{-1}$ is greater than or less than $2 \times \sqrt{-1}$. The earliest mathematicians who dealt with numbers containing $\sqrt{-1}$ referred to these numbers as *imaginary* numbers because they could not be used to measure or count something. In mathematics the symbol i is used to represent $\sqrt{-1}$.

Equation M-5, the quadratic formula, applies to equations of the form

$$ax^2 + bx + c = 0$$

The formula shows that there are no real roots when $b^2 < 4ac$. There are, however, still two roots. Each root is a number containing two terms: a real number, and a multiple of $i = \sqrt{-1}$. The multiple of i is called an **imaginary number,** and i is called the **unit imaginary.**

A general **complex number** z can be written

$$z = a + bi \qquad \text{M-54}$$

where a and b are real numbers. The quantity a is called the real part of z, or Re(z), and the quantity b is called the imaginary part of z, or Im(z). We can represent a complex number z as a point in a plane, called the complex plane, as shown in Figure M-22, where the x axis is the **real axis** and the y axis is the **imaginary axis.** We can also use the relations $a = r\cos\theta$ and $b = r\sin\theta$ from Figure M-22 to write the complex number z in **polar coordinates** (a system in which a point is designated by the counterclockwise angle of rotation θ and the distance r in the direction of θ):

$$z = r\cos\theta + ir\sin\theta \qquad \text{M-55}$$

where $r = \sqrt{a^2 + b^2}$ is called the **magnitude** of z.

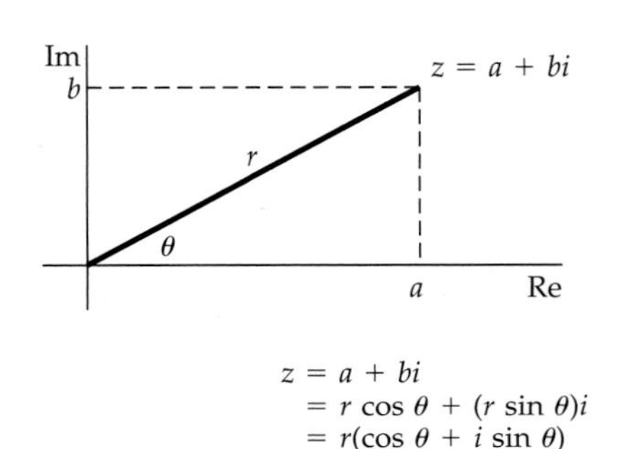

FIGURE M-22 Representation of a complex number in a plane. The real part of the complex number is plotted along the horizontal axis, and the imaginary part is plotted along the vertical axis.

When complex numbers are added or subtracted, the real and imaginary parts are added or subtracted separately:

$$z_1 + z_2 = (a_1 + ib_1) + (a_2 + ib_2) = (a_1 + a_2) + i(b_1 + b_2) \qquad \text{M-56}$$

However, when two complex numbers are multiplied, each part of one number is multiplied by each part of the other number:

$$z_1 z_2 = (a_1 + ib_1)(a_2 + ib_2) = a_1a_2 + i^2b_1b_2 + i(a_1b_2 + a_2b_1)$$
$$= a_1a_2 - b_1b_2 + i(a_1b_2 + a_2b_1) \qquad \text{M-57}$$

where we have used $i^2 = -1$.

The **complex conjugate** z^* of the complex number z is that number obtained by replacing i with $-i$ when writing z. If $z = a + ib$, then

$$z^* = (a + ib)^* = a - ib \qquad \text{M-58}$$

(When a quadratic equation has complex roots, the roots are **conjugate complex numbers,** in the form $a \pm bi$.) The product of a complex number and its complex conjugate equals the square of the magnitude of the number:

$$zz^* = (a + ib)(a - ib) = a^2 + b^2 = r^2 \qquad \text{M-59}$$

A particularly useful function of a complex number is the exponential $e^{i\theta}$. Using an expansion for e^x, we have

$$e^{i\theta} = 1 + i\theta + \frac{(i\theta)^2}{2!} + \frac{(i\theta)^3}{3!} + \frac{(i\theta)^4}{4!} + \cdots$$

Using $i^2 = -1$, $i^3 = -i$, $i^4 = +1$, and so forth, and separating the real parts from the imaginary parts, this expansion can be written

$$e^{i\theta} = \left(1 - \frac{\theta^2}{2!} + \frac{\theta^4}{4!} - \cdots\right) + i\left(\theta - \frac{\theta^3}{3!} + \cdots\right)$$

Comparing this result with Equations M-50 and M-51, we can see that

$$e^{i\theta} = \cos\theta + i\sin\theta \qquad \text{M-60}$$

Using this result, we can express a general complex number as an exponential:

$$z = a + ib = r\cos\theta + ir\sin\theta = re^{i\theta} \qquad \text{M-61}$$

If $z = x + iy$, where x and y are real variables, then z is called a **complex variable.**

COMPLEX VARIABLES IN PHYSICS

Complex variables are often used in formulas describing AC circuits: the impedance of a capacitor or an inductor includes a real part (the resistance) and an imaginary part (the reactance). (There are alternative ways, however, of analyzing AC circuits—such as rotating vectors called *phasors*—that do not require assigning imaginary values.) Complex variables are also important in the study of harmonic waves through Fourier analysis and synthesis. The time-dependent Schrödinger equation contains a complex-valued function of position and time.

Example M-11 Finding a Power of a Complex Number

Calculate $(1 + 3i)^4$ by using the binomial expansion.

PICTURE The expression is of the form $(1 + x)^n$. Because n is a positive integer, the expansion is valid for any value of x, and all terms, other than those of order n or lower must equal zero.

SOLVE

1. Write out the expansion of $(1 + 3i)^4$ to show the terms up through the fouth-order term:

$$1 + 4\cdot 3i + \frac{4(3)}{2!}(3i)^2 + \frac{4(3)(2)}{3!}(3i)^3 + \frac{4(3)(2)(1)}{4!}(3i)^4$$

2. Evaluate each term, remembering that $i^2 = -1$, $i^3 = -i$, and $i^4 = +1$: $1 + 12i - 54 - 108i + 81$

3. Show the result in the form $a + bi$: $(1 + 3i)^4 = \boxed{28 - 96i}$

CHECK We can solve the problem algebraically to show that the answer is correct. We first square $(1 + 3i)$ and then square the result, to get $(1 + 3i)^4$:

$$(1 + 3i)^2 = 1 \cdot 1 + 2 \cdot 1 \cdot 3i + (3i)^2 = 1 + 6i - 9 = -8 + 6i$$

$$(-8 + 6i)^2 = (-8)(-8) + 2(-8)(6i) + (6i)^2 = 64 - 96i - 36 = 28 - 96i$$

PRACTICE PROBLEMS Express in the form $a + bi$:

23. $e^{i\pi}$

24. $e^{i\pi/2}$

M-11 DIFFERENTIAL CALCULUS

Calculus is the branch of mathematics that allows us to deal with instantaneous rates of change of functions and variables. From the equation of a function—say, x as a function of t—we can always find x for a particular t, but with the methods of calculus you can go much further. You can know where x will have certain properties, such as a maximum or a minimum value, without having to try endless values of t. With calculus, if given the proper data, you can find, for example, the location of maximum stress on a beam, or the velocity or position of a falling object at a time t, or the energy a falling object has acquired at the time of impact. The principles of calculus are derived from examining functions at the infinitesimal level—analyzing how, say, x will change when the change in t becomes vanishingly small. We start with **differential calculus,** in which we determine the *limit* of the rate of change of x with respect to t as the change in t becomes closer and closer to zero.

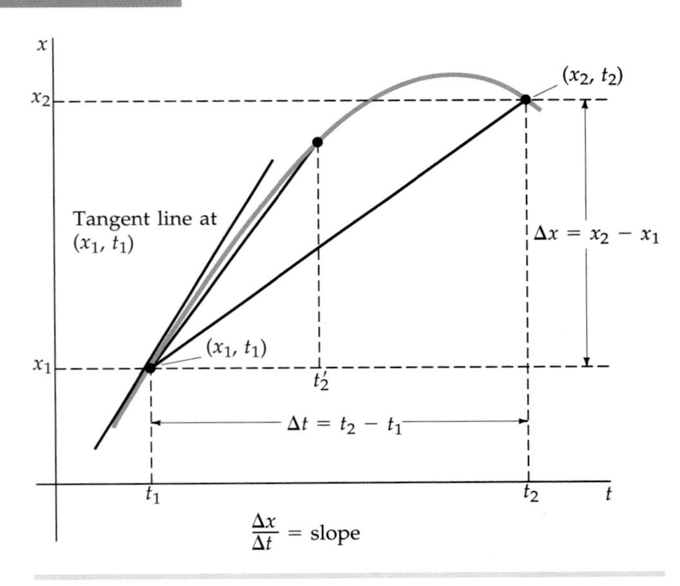

FIGURE M-23 Graph of a typical function $x(t)$. The points (x_1, t_1) and (x_2, t_2) are connected by a straight line. The slope of this line is $\Delta x/\Delta t$. As the time interval beginning at t_1 is decreased, the slope for that interval approaches the slope of the line tangent to the curve at time t_1, which is the derivative of x with respect to t.

Figure M-23 is a graph of x versus t for a typical function $x(t)$. At a particular value $t = t_1$, x has the value of x_1, as indicated. At another value t_2, x has the value x_2. The change in t, $t_2 - t_1$, is written $\Delta t = t_2 - t_1$; and the corresponding change in x is written $\Delta x = x_2 - x_1$. The ratio $\Delta x/\Delta t$ is the slope of the straight line connecting (x_1, t_1) and (x_2, t_2). If we take the limit as t_2 approaches t_1 (as Δt approaches zero) the slope of the line connecting (x_1, t_1) and (x_2, t_2) approaches the slope of the line that is tangent to the curve at the point (x_1, t_1). The slope of this tangent line is equal to the **derivative** of x with respect to t and is written dx/dt:

$$\frac{dx}{dt} = \lim_{\Delta t \to 0} \frac{\Delta x}{\Delta t} \qquad \text{M-62}$$

(When we find the derivative of a function, we say that we are **differentiating** the function; and the very small "dx" and "dt" elements are called **differentials** of x and t, respectively.) The derivative of a function of t is another function of t. If x is a constant and does not change, the graph of x versus t is a horizontal line with zero slope. The derivative of a constant is thus zero. In Figure M-24, x is not constant but is proportional to t:

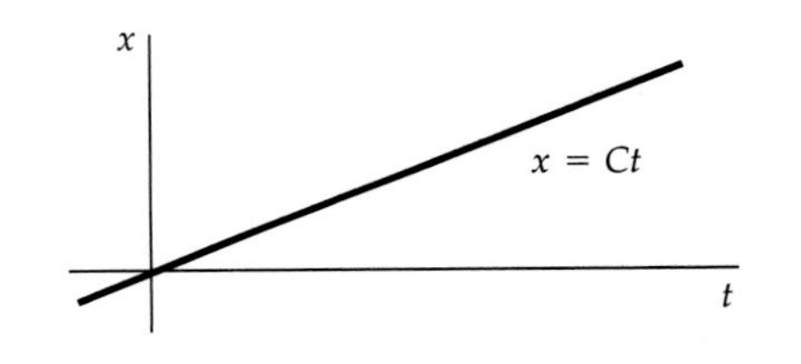

FIGURE M-24 Graph of the linear function $x = Ct$. This function has a constant slope C.

$$x = Ct$$

This function has a constant slope equal to C. Thus the derivative of Ct is C. Table M-3 lists some properties of derivatives and the derivatives of some particular functions that occur often in physics. It is followed by comments aimed at making these properties and rules clearer. More detailed discussion can be found in most calculus textbooks.

Table M-3 Properties of Derivatives and Derivatives of Particular Functions

Linearity

1. The derivative of a constant C times a function $f(t)$ equals the constant times the derivative of the function:

$$\frac{d}{dt}[Cf(t)] = C\frac{df(t)}{dt}$$

2. The derivative of a sum of functions equals the sum of the derivatives of the functions:

$$\frac{d}{dt}[f(t) + g(t)] = \frac{df(t)}{dt} + \frac{dg(t)}{dt}$$

Chain rule

3. If f is a function of x and x is in turn a function of t, the derivative of f with respect to t equals the product of the derivative of f with respect to x and the derivative of x with respect to t:

$$\frac{d}{dt}f(x(t)) = \frac{df}{dx}\frac{dx}{dt}$$

Derivative of a product

4. The derivative of a product of functions $f(t)g(t)$ equals the first function times the derivative of the second plus the second function times the derivative of the first:

$$\frac{d}{dt}[f(t)g(t)] = f(t)\frac{dg(t)}{dt} + g(t)\frac{df(t)}{dt}$$

Reciprocal derivative

5. The derivative of t with respect to x is the reciprocal of the derivative of x with respect to t, assuming that neither derivative is zero:

$$\frac{dt}{dx} = \left(\frac{dx}{dt}\right)^{-1} \quad \text{if} \quad \frac{dt}{dx} \neq 0 \quad \text{and} \quad \frac{dx}{dt} \neq 0$$

Derivatives of particular functions

6. If C is a constant, then $dC/dt = 0$.

7. $\dfrac{d(t^n)}{dt} = nt^{n-1}$ If n is constant.

8. $\dfrac{d}{dt}\sin \omega t = \omega \cos \omega t$ If ω is constant.

9. $\dfrac{d}{dt}\cos \omega t = -\omega \sin \omega t$ If ω is constant.

10. $\dfrac{d}{dt}\tan \omega t = \omega \sin^2 \omega t$ If ω is constant.

11. $\dfrac{d}{dt}e^{bt} = be^{bt}$ If b is constant.

12. $\dfrac{d}{dt}\ln bt = \dfrac{1}{t}$ If b is constant.

COMMENTS ON RULES 1 THROUGH 5

Rules 1 and 2 follow from the fact that the limiting process is linear. We can understand rule 3, the chain rule, by multiplying $\Delta f/\Delta t$ by $\Delta x/\Delta x$ and noting that as Δt approaches zero, Δx also approaches zero. That is,

$$\lim_{\Delta t\to 0}\frac{\Delta f}{\Delta t} = \lim_{\Delta t\to 0}\left(\frac{\Delta f}{\Delta t}\frac{\Delta x}{\Delta x}\right) = \lim_{\Delta t\to 0}\left(\frac{\Delta f}{\Delta x}\frac{\Delta x}{\Delta t}\right) = \left(\lim_{\Delta x\to 0}\frac{\Delta f}{\Delta x}\right)\left(\lim_{\Delta t\to 0}\frac{\Delta x}{\Delta t}\right) = \frac{df}{dx}\frac{dx}{dt}$$

where we have used that the limit of the product is equal to product of the limits.

Rule 4 is not immediately apparent. The derivative of a product of functions is the limit of the ratio

$$\frac{f(t+\Delta t)g(t+\Delta t) - f(t)g(t)}{\Delta t}$$

If we add and subtract the quantity $f(t+\Delta t)g(t)$ in the numerator, we can write this ratio as

$$\frac{f(t+\Delta t)g(t+\Delta t) - f(t+\Delta t)g(t) + f(t+\Delta t)g(t) - f(t)g(t)}{\Delta t}$$
$$= f(t+\Delta t)\left[\frac{g(t+\Delta t) - g(t)}{\Delta t}\right] + g(t)\left[\frac{f(t+\Delta t) - f(t)}{\Delta t}\right]$$

As Δt approaches zero, the terms in square brackets become $dg(t)/dt$ and $df(t)/dt$, respectively, and the limit of the expression is

$$f(t)\frac{dg(t)}{dt} + g(t)\frac{df(t)}{dt}$$

Rule 5 follows directly from the definition:

$$\frac{dx}{dt} = \lim_{\Delta t\to 0}\frac{\Delta x}{\Delta t} = \lim_{\Delta x\to 0}\left(\frac{\Delta t}{\Delta x}\right)^{-1} = \left(\frac{dt}{dx}\right)^{-1}$$

COMMENTS ON RULE 7

We can obtain this important result using the binomial expansion. We have

$$f(t) = t^n$$

$$f(t+\Delta t) = (t+\Delta t)^n = t^n\left(1+\frac{\Delta t}{t}\right)^n$$
$$= t^n\left[1 + n\frac{\Delta t}{t} + \frac{n(n-1)}{2!}\left(\frac{\Delta t}{t}\right)^2 + \frac{n(n-1)(n-2)}{3!}\left(\frac{\Delta t}{t}\right)^3 + \cdots\right]$$

Then

$$f(t+\Delta t) - f(t) = t^n\left[n\frac{\Delta t}{t} + \frac{n(n-1)}{2!}\left(\frac{\Delta t}{t}\right)^2 + \cdots\right]$$

and

$$\frac{f(t+\Delta t) - f(t)}{\Delta t} = nt^{n-1} + \frac{n(n-1)}{2!}t^{n-2}\Delta t + \cdots$$

The next term omitted from the last sum is proportional to $(\Delta t)^2$, the following to $(\Delta t)^3$, and so on. Each term except the first approaches zero as Δt approaches zero. Thus

$$\frac{df}{dt} = \lim_{\Delta x\to 0}\frac{f(t+\Delta t) - f(t)}{\Delta t} = nt^{n-1}$$

COMMENTS ON RULES 8 TO 10

We first write $\sin \omega t = \sin\theta$ with $\theta = \omega t$ and use the chain rule,

$$\frac{d\sin\theta}{dt} = \frac{d\sin\theta}{d\theta}\frac{d\theta}{dt} = \omega\frac{d\sin\theta}{d\theta}$$

We then use the trigonometric formula for the sine of the sum of two angles θ and $\Delta\theta$:

$$\sin(\theta + \Delta\theta) = \sin \Delta\theta \cos \theta + \cos \Delta\theta \sin \theta$$

Because $\Delta\theta$ is to approach zero, we can use the small-angle approximations

$$\sin \Delta\theta \approx \Delta\theta \qquad \text{and} \qquad \cos \Delta\theta \approx 1$$

Then

$$\sin(\theta + \Delta\theta) \approx \Delta\theta \cos \theta + \sin \theta$$

and

$$\frac{\sin(\theta + \Delta\theta) - \sin \theta}{\Delta\theta} \approx \cos \theta$$

Similar reasoning can be applied to the cosine function to obtain rule 9.

Rule 10 is obtained by writing $\tan \theta = \sin \theta / \cos \theta$ and applying rule 4 along with rules 8 and 9:

$$\begin{aligned}\frac{d}{dt}(\tan \theta) &= \frac{d}{dt}(\sin \theta)(\cos \theta)^{-1} = \sin \theta \frac{d}{dt}(\cos \theta)^{-1} + \frac{d(\sin \theta)}{dt}(\cos \theta)^{-1} \\ &= \sin \theta(-1)(\cos \theta)^{-2}(-\sin \theta) + (\cos \theta)(\cos \theta)^{-1} \\ &= \frac{\sin^2 \theta}{\cos^2 \theta} + 1 = \tan^2 \theta + 1 = \sec^2 \theta\end{aligned}$$

To obtain rule 10, let $\theta = \omega t$ and use the chain rule.

COMMENTS ON RULE 11

Again we use the chain rule

$$\frac{de^\theta}{dt} = \frac{b\, de^\theta}{b\, dt} = b\frac{de^\theta}{d(bt)} = b\frac{de^\theta}{d\theta} \qquad \text{with} \qquad \theta = bt$$

and the series expansion for the exponential function:

$$e^{\theta + \Delta\theta} = e^\theta e^{\Delta\theta} = e^\theta\left[1 + \Delta\theta + \frac{(\Delta\theta)^2}{2!} + \frac{(\Delta\theta)^3}{3!} + \cdots\right]$$

Then

$$\frac{e^{\theta + \Delta\theta} - e^\theta}{\Delta\theta} = e^\theta + e^\theta\frac{\Delta\theta}{2!} + e^\theta\frac{(\Delta\theta)^2}{3!} + \cdots$$

As $\Delta\theta$ approaches zero, the right side of this equation approaches e^θ.

COMMENTS ON RULE 12

Let

$$y = \ln bt$$

Then

$$e^y = bt \Rightarrow t = \frac{1}{b}e^y$$

Then, using rule 11, we obtain

$$\frac{dt}{dy} = \frac{1}{b}e^y \therefore \frac{dt}{dy} = t$$

Then, using rule 5, we obtain

$$\frac{dy}{dt} = \left(\frac{dt}{dy}\right)^{-1} = \frac{1}{t}$$

SECOND- AND HIGHER-ORDER DERIVATIVES; DIMENSIONAL ANALYSIS

Once we have differentiated a function, we can differentiate the resulting derivative as long as terms remain to differentiate. A function such as $x = e^{bt}$ can be differentiated indefinitely: $dx/dt = be^{bt}$ (this function differentiates to give b^2e^{bt}, and so on).

Consider velocity and acceleration. We can define velocity as the rate of change of position of a particle, or dx/dt, and acceleration as the rate of change of velocity, or the *second* derivative of x with respect to t, written dx^2/dt^2. If a particle moves at a constant velocity, then dx/dt will equal a constant. The acceleration, however, will be zero: having constant velocity is the same as having no acceleration, and the derivative of a constant is zero. Now consider a falling object, subject to the constant acceleration of gravity: the velocity itself will be time-dependent, so the *second* derivative, dx^2/dt^2, will be a constant.

The *physical dimensions* of a derivative with respect to a variable are those that would result if the original function of the variable were divided by a value of the variable. For example, the dimension of an equation in which one term is x (for position) is that of length (L); the dimensions of the derivative of x with respect to time t are those of velocity (L/T), and the dimensions of dx^2/dt^2 are those of acceleration (L/T^2).

Example M-12 Position, Velocity, and Acceleration

Find the first and the second derivative of $x = \frac{1}{2}at^2 + bt + c$, where a, b, and c are constants. The function gives the position (in m) of a particle in one dimension, where t is the time (in s), a is acceleration (in m/s^2), b is velocity (in m/s) at a time $t = 0$, and c is the position (in m) of the particle at $t = 0$.

PICTURE Both the first and the second derivatives are sums of terms; for each differentiation we take the derivative of each term separately and add the results.

SOLVE

1. To find the first derivative, first compute the derivative of the first term:
$$\frac{d(\frac{1}{2}at^2)}{dt} = \left(\frac{1}{2}a\right)2t^1 = at$$

2. Compute the first derivative of the second and third terms:
$$\frac{d(bt)}{dt} = b, \quad \frac{d(c)}{dt} = 0$$

3. Add these results:
$$\frac{dx}{dt} = at + b$$

4. To compute the second derivative, repeat the process for the result in step 3:
$$\frac{d^2x}{dt^2} = a + 0 = a$$

CHECK The physical dimensions show that the answer is plausible. The original function is an equation for position; all terms are in meters—the units of t^2 and t cancel the units of s^2 and s in the constants a and b, respectively. In the function for dx/dt, all terms are similarly in m/s: the constant c has differentiated to zero, and the unit for t cancels one of the units for s in the constant a. In the function for dx^2/dt^2, only the acceleration constant remains; as expected, its dimensions are L/T^2.

PRACTICE PROBLEMS

25. Find dy/dx for $y = \frac{5}{8}x^3 - 24x - \frac{5}{8}$.

26. Find dy/dt for $y = ate^{bt}$, where a and b are constants.

SOLVING DIFFERENTIAL EQUATIONS USING COMPLEX NUMBERS

A **differential equation** is an equation in which the derivatives of a function appear as variables. It is an equation in which the variables are related to each other through their derivatives. Consider an equation of the form

$$a\frac{d^2x}{dt^2} + b\frac{dx}{dt} + cx = A\cos\omega t \qquad \text{M-63}$$

that represents a physical process, such as a damped harmonic oscillator driven by a sinusoidal force, or a series *RLC* combination being driven by a sinusoidal potential drop. Although each of the parameters in Equation M-63 is a real number, the time-dependent cosine term suggests that we might find the steady-state solution to this equation by introducing complex numbers. We first construct the "parallel" equation

$$a\frac{d^2y}{dt^2} + b\frac{dy}{dt} + cy = A\sin\omega t \qquad \text{M-64}$$

Equation M-64 has no physical meaning of its own, and we have no interest in solving it. However, it is of use in solving Equation M-63. After multiplying through Equation M-64 by the unit imaginary i, we add Equation M-64 and Equation M-63 to obtain

$$\left(a\frac{d^2x}{dt^2} + ai\frac{d^2y}{dt^2}\right) + \left(b\frac{dx}{dt} + bi\frac{dy}{dt}\right) + (cx + ciy) = A\cos\omega t + Ai\sin\omega t$$

We next combine terms to get

$$a\frac{d^2(x + iy)}{dt^2} + b\frac{d(x + iy)}{dt} + c(x + iy) = A(\cos\omega t + i\sin\omega t) \qquad \text{M-65}$$

which is valid because the derivative of a sum is equal to the sum of the derivatives. We simplify our result by defining $z = x + iy$ and by using the identity $e^{i\omega t} = \cos\omega t + i\sin\omega t$. Substituting these into Equation M-65, we obtain

$$a\frac{d^2z}{dt^2} + b\frac{dz}{dt} + cz = Ae^{i\omega t} \qquad \text{M-66}$$

which we now solve for z. Once z is obtained, we can solve for x using $x = \mathrm{Re}(z)$.

Because we are looking only for the steady-state solution for Equation M-65, we can assume its solution is of the form $x = x_0\cos(\omega t - \phi)$, where ϕ is a constant. This is equivalent to assuming that the solution to Equation M-66 is of the form $z = \eta e^{i\omega t}$, where η, pronounced eta (like beta without the b), is a constant complex number. Then $dz/dt = i\omega z$, $d^2z/dt^2 = -\omega^2 z$, and $e^{i\omega t} = z/\eta$. Substituting these into Equation M-65 gives

$$-a\omega^2 z + i\omega bz + cz = A\frac{z}{\eta}$$

Dividing both sides of this equation by z and solving for η gives

$$\eta = \frac{A}{-a\omega^2 + i\omega b + c}$$

Expressing the denominator in polar form gives

$$(-a\omega^2 + c) + i\omega b = \sqrt{(-a\omega^2 + c)^2 + \omega^2b^2}\, e^{i\phi}$$

where $\tan\phi = \omega^2b^2/(-a\omega^2 + c)$. Thus,

$$\eta = \frac{A}{\sqrt{(-a\omega^2 + c)^2 + \omega^2b^2}}e^{-i\phi}$$

so

$$z = \eta e^{i\omega t} = \frac{A}{\sqrt{(-a\omega^2 + c)^2 + \omega^2 b^2}} e^{i(\omega t - \phi)}$$

$$= \frac{A}{\sqrt{(-a\omega^2 + c)^2 + \omega^2 b^2}}[\cos(\omega t - \phi) + i\sin(\omega t - \phi)] \qquad \text{M-67}$$

It follows that

$$x = \text{Re}(z) = \frac{A}{\sqrt{(-a\omega^2 + c)^2 + \omega^2 b^2}}\cos(\omega t - \phi) \qquad \text{M-68}$$

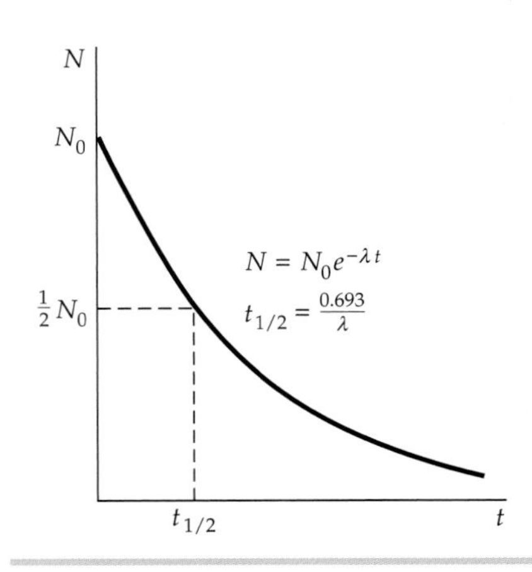

FIGURE M-25 Graph of N versus t when N decreases exponentially. The time $t_{1/2}$ is the time it takes for N to decrease by one-half.

THE EXPONENTIAL FUNCTION

An **exponential function** is a function of the form a^{bx}, where $a > 0$ and b are constants. The function is usually written as e^{cx}, where c is constant.

When the rate of change of a quantity is proportional to the quantity itself, the quantity increases or decreases exponentially, depending on the sign of the proportionality constant. An example of an *exponentially* decreasing function is nuclear decay. If N is the number of radioactive nuclei at some time, then the change dN in some very small time interval dt will be proportional to N and to dt:

$$dN = -\lambda N\,dt$$

where λ is the *decay constant* (not to be confused with the decay rate dN/dt, which decreases exponentially). The function N satisfying this equation is

$$N = N_0 e^{-\lambda t} \qquad \text{M-69}$$

where N_0 is the value of N at time $t = 0$. Figure M-25 shows N versus t. A characteristic of exponential decay is that N decreases by a constant factor in a given time interval. The time interval for N to decrease to half its original value is its *half-life* $t_{1/2}$. The half-life is obtained from Equation M-69 by setting $N = \frac{1}{2}N_0$ and solving for the time. This gives

$$t_{1/2} = \frac{\ln 2}{\lambda} = \frac{0.693}{\lambda} \qquad \text{M-70}$$

An example of *exponential increase* is population growth. If the number of organisms is N, the change in N after a very small time interval dt is given by

$$dN = +\lambda N\,dt$$

where λ is now the *growth constant*. The function N satisfying this equation is

$$N = N_0 e^{\lambda t} \qquad \text{M-71}$$

(Note the change of sign in the exponent.) A graph of this function is shown in Figure M-26. An exponential increase can be characterized by a doubling time T_2, which is related to λ by

$$T_2 = \frac{\ln 2}{\lambda} = \frac{0.693}{\lambda} \qquad \text{M-72}$$

Very often, we know population growth as an annual percentage increase and wish to calculate the doubling time. In this case, we find T_2 (in years) from the equation

$$T_2 = \frac{69.3}{r} \qquad \text{M-73}$$

where r is the percent per year. For example, if the population increases by 2 percent per year, the population will double every $69.3/2 \approx 35$ years. Table M-4 lists some useful relations for exponential and logarithmic functions.

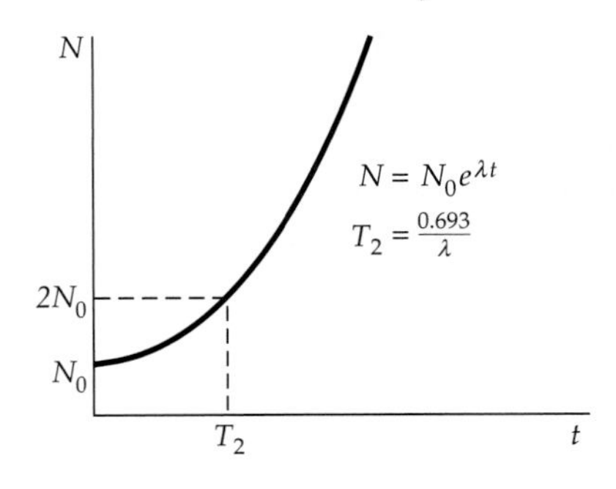

FIGURE M-26 Graph of N versus t when N increases exponentially. The time T_2 is the time it takes for N to double.

Table M-4 Exponential and Logarithmic Functions

$e = 2.718\,28$

$e^0 = 1$

If $y = e^x$, then $x = \ln y$.

$e^{\ln x} = x$

$e^x e^y = e^{(x+y)}$

$(e^x)^y = e^{xy} = (e^y)^x$

$\ln e = 1$; $\ln 1 = 0$

$\ln xy = \ln x + \ln y$

$\ln\frac{x}{y} = \ln x - \ln y$

$\ln e^x = x$; $\ln a^x = x\ln a$

$\ln x = (\ln 10)\log x$

$= 2.30\,26\log x$

$\log x = (\log e)\ln x = 0.434\,29\ln x$

$e^x = 1 + x + \frac{x^2}{2!} + \frac{x^3}{3!} = \ldots$

$\ln(1 + x) = x - \frac{x^2}{2} + \frac{x^3}{3} - \frac{x^4}{4}$

Example M-13 Radioactive Decay of Cobalt-60

The half-life of cobalt-60 (^{60}Co) is 5.27 y. At $t = 0$ you have a sample of ^{60}Co that has a mass equal to 1.20 mg. At what time t (in years) will 0.400 mg of the sample of ^{60}Co have decayed?

PICTURE When we derived the half-life in exponential decay, we set $N/N_0 = 1/2$. In this example, we are to find the time at which two-thirds of a sample remains, and so the ratio N/N_0 will be 0.667.

SOLVE

1. Express the ratio N/N_0 as an exponential function: $$\frac{N}{N_0} = 0.667 = e^{-\lambda t}$$
2. Take the reciprocal of both sides: $$\frac{N_0}{N} = 1.50 = e^{\lambda t}$$
3. Solve for t: $$t = \frac{\ln 1.50}{\lambda} = \frac{0.405}{\lambda}$$
4. The decay constant is related to the half-life by $\lambda = (\ln 2)/t_{1/2}$ (Equation M-70). Substitute $(\ln 2)/t_{1/2}$ for λ and evaluate the time: $$t = \frac{\ln 1.5}{\ln 2}t_{1/2} = \frac{\ln 1.5}{\ln 2} \times 5.27\text{ y} = 3.08\text{ y}$$

CHECK It takes 5.27 y for the mass of a sample of ^{60}Co to decrease to 50 percent of its initial mass. Thus, we expect it to take less than 5.27 y for the sample to lose 33.3 percent of its mass. Our step-4 result of 3.08 y is less than 5.27 y, as expected.

PRACTICE PROBLEMS

27. The discharge time constant τ of a capacitor in an RC circuit is the time in which the capacitor discharges to e^{-1} (or 0.368) times its charge at $t = 0$. If $\tau = 1$ s for a capacitor, at what time t (in seconds) will it have discharged to 50.0% of its initial charge?
28. If the coyote population in your state is increasing at a rate of 8.0% a decade and continues increasing at the same rate indefinitely, in how many years will it reach 1.5 times its current level?

M-12 INTEGRAL CALCULUS

Integration can be considered the inverse of differentiation. If a function $f(t)$ is *integrated*, a function $F(t)$ is found for which $f(t)$ is the derivative of $F(t)$ with respect to t.

THE INTEGRAL AS AN AREA UNDER A CURVE; DIMENSIONAL ANALYSIS

The process of finding the area under a curve on the graph illustrates integration. Figure M-27 shows a function $f(t)$. The area of the shaded element is approximately $f_i\,\Delta t_i$, where f_i is evaluated anywhere in the interval Δt_i. This approximation is highly accurate if Δt_i is very small. The total area under some stretch of the curve is found by summing all the area elements it covers and taking the limit as each Δt_i approaches zero. This limit is called the **integral** of f over t and is written

$$\int f\,dt = \text{area}_i = \lim_{\Delta t_i \to 0} \sum_i f_i \Delta t_i \qquad \text{M-74}$$

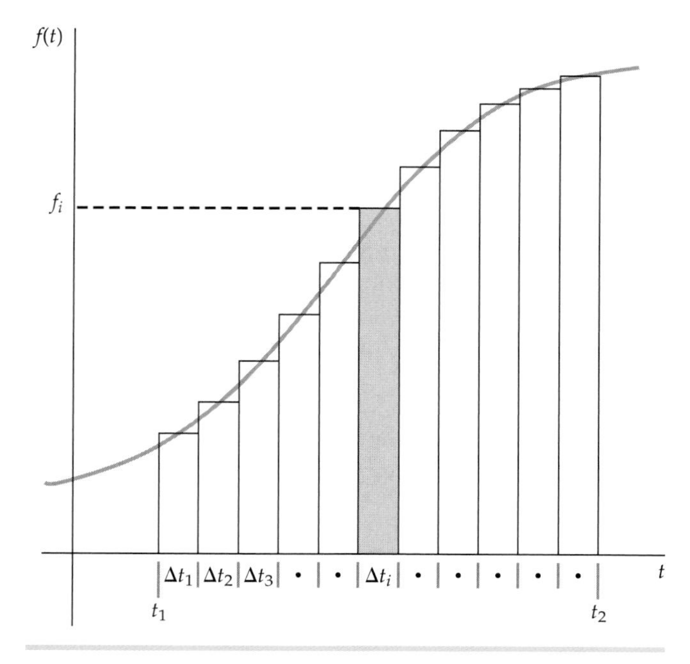

FIGURE M-27 A general function $f(t)$. The area of the shaded element is approximately $f_i\Delta t_i$, where f_i is evaluated anywhere in the interval.

The *physical dimensions* of an integral of a function $f(t)$ are found by multiplying the dimensions of the *integrand* (the function being integrated) and the dimensions of the integration variable t. For example, if the integrand is a velocity function

$v(t)$ (dimensions L/T) and the integration variable is time t), the dimension of the integral is $L = (L/T) \times T$. That is, the dimensions of the integral are those of velocity times time.

Let

$$y = \int_{t_1}^{t} f\, dt \qquad \text{M-75}$$

The function y is the area under the f-versus-t curve from t_1 to a general value t. For a small interval Δt, the change in the area Δy is approximately $f\,\Delta t$:

$$\Delta y \approx f\,\Delta t$$

$$f \approx \frac{\Delta y}{\Delta t}$$

If we take the limit as Δt approaches 0, we can see that f is the derivative of y:

$$f = \frac{dy}{dt} \qquad \text{M-76}$$

INDEFINITE INTEGRALS AND DEFINITE INTEGRALS

When we write

$$y = \int f\, dt \qquad \text{M-77}$$

we are showing y as an **indefinite integral** of f over t. To evaluate an indefinite integral, we find the function y whose derivative is f. Because that function could contain a constant term that differentiated to zero, we include as our final term a **constant of integration** C. If we are integrating the function over a known segment—such as t_1 to t_2 in Figure M-27—we can find a **definite integral,** eliminating the unknown constant C:

$$\int_{t_1}^{t_2} f\, dt = y(t_2) - y(t_1) \qquad \text{M-78}$$

Table M-5 lists some important integration formulas. More extensive lists of integration formulas can be found in any calculus textbook or by searching for "table of integrals" on the Internet.

Table M-5 Integration Formulas†

1. $\int A\, dt = At$
2. $\int At\, dt = \frac{1}{2}At^2$
3. $\int At^n\, dt = A\frac{t^{n+1}}{n+1},\quad n \neq -1$
4. $\int At^{-1}\, dt = A\ln|t|$
5. $\int e^{bt}\, dt = \frac{1}{b}e^{bt}$
6. $\int \cos \omega t\, dt = \frac{1}{\omega}\sin \omega t$
7. $\int \sin \omega t\, dt = -\frac{1}{\omega}\cos \omega t$
8. $\int_0^{\infty} e^{-ax}\, dx = \frac{1}{a}$
9. $\int_0^{\infty} e^{-ax^2}\, dx = \frac{1}{2}\sqrt{\frac{\pi}{a}}$
10. $\int_0^{\infty} xe^{-ax^2}\, dx = \frac{2}{a}$
11. $\int_0^{\infty} x^2e^{-ax^2}\, dx = \frac{1}{4}\sqrt{\frac{\pi}{a^3}}$
12. $\int_0^{\infty} x^3e^{-ax^2}\, dx = \frac{4}{a^2}$
13. $\int_0^{\infty} x^4e^{-ax^2}\, dx = \frac{3}{8}\sqrt{\frac{\pi}{a^5}}$

† In these formulas, A, b, and ω are constants. In formulas 1 through 7, an arbitrary constant C can be added to the right side of each equation. The constant a is greater than zero.

Example M-14 Integrating Equations of Motion

A particle is moving at a constant acceleration a. Write a formula for position x at time t given that the position and velocity are x_0 and v_0 at time $t = 0$.

PICTURE Velocity v is the derivative of x with respect to time t, and acceleration is the derivative of v with respect to t. We should be able to write a function $x(t)$ by performing two integrations.

SOLVE

1. Integrate a with respect to t to find the v as a function of t. The a can be factored from the integrand because a is constant:

$$v = \int a\, dt = a\int dt$$

$$v = at + C_1$$

where C_1 represents a times the constant of integration.

2. The velocity $v = v_0$ when $t = 0$

$$v_0 = 0 + C_1 \Rightarrow C_1 = v_0$$

so $v = v_0 + at$

3. Integrate v with respect to t to find x as a function of t:

$$x = \int v\,dt = \int (v_0 + at)\,dt = \int v_0\,dt + \int at\,dt$$

$$x = v_0 \int dt + a \int t\,dt = v_0 t + \tfrac{1}{2}at^2 + C_2$$

where C_2 represents the combined constants of integration.

4. The position $x = x_0$ when $t = 0$

$$x_0 = 0 + 0 + C_2$$

so $x = x_0 + v_0 t + \tfrac{1}{2}at^2$

CHECK Differentiate the step-4 result twice to get the acceleration

$$v = \frac{dx}{dt} = \frac{d}{dt}(x_0 + v_0 t + \tfrac{1}{2}at^2) = 0 + v_0 + at$$

$$a = \frac{dv}{dt} = \frac{d}{dt}(v_0 + at) = a$$

PRACTICE PROBLEMS

29. $\int_3^6 3\,dx =$

30. $V = \int_5^8 \pi r^2\,dL =$

Answers to Practice Problems

1. 0.24 L
2. 31.6 m/s
3. 6.0 kg/cm^3
4. -3
5. 1.54 L
6. 3.07 L
7. False
8. $x = (4.5\text{ m/s})t + 3.0\text{ m}$
9. $x = 8,\ y = 60$
11. $2(x - y)^2$
12. $x^2(2x + 4)(x + 3)$
13. $x^{1/2}$
14. x^6
15. 3
16. ~ 2.322
17. $V/A = \frac{1}{3}r$
18. $A = \frac{2}{3}\pi L^2$
19. $\sin\theta = 0.496,\ \cos\theta = 0.868,\ \theta = 29.7°$
20. $\sin 8.2° = 0.1426,\ 8.2° = 0.1431$ rad
21. 0.996, 0.996 00, close to 0%
22. 0.96, 0.960 77, $\ll 1\%$
23. $-1 + 0i = -1$
24. $0 + i = i$
25. $dy/dx = \frac{5}{24}x^2 - 24$
26. $dy/dt = ae^{bt}(bt + 1)$
27. 0.693 s
28. 51 y
29. 9
30. $3\pi r^2$

Answers to Odd-Numbered End-of-Chapter Problems

Differences in the last figure can result from differences in rounding the input data.

Chapter 34

1 (*c*)

3 (*a*)

5 (*a*) True, (*b*) True, (*c*) True

7 (*c*)

9 According to quantum theory, the average value of many measurements of the same quantity will yield the expectation value of that quantity. However, any single measurement may differ from the expectation value.

11 2.48 pm, 2%

13 (*a*)

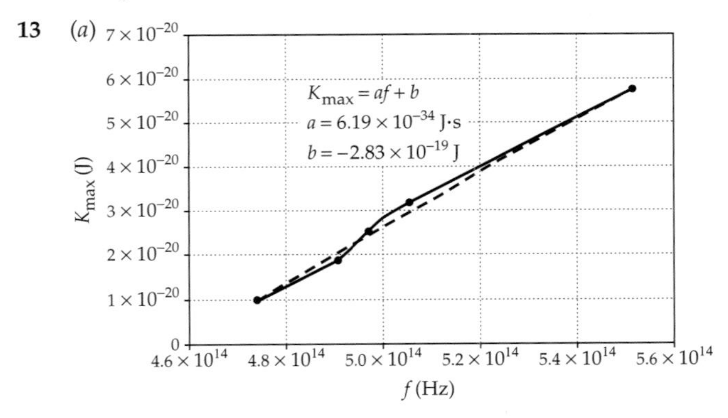

(*b*) 1.77 eV (*c*) cesium

15 (*a*) 4.14×10^{-7} eV, (*b*) 3.72×10^{-9} eV

17 (*a*) 12.4 keV, (*b*) 1.24 GeV

19 $1.95 \times 10^{16}\ \mathrm{s^{-1}}$

21 (*a*) 4.13 eV, (*b*) 2.10 eV, (*c*) 0.78 eV, (*d*) 590 nm

23 (*a*) 653 nm, 4.58×10^{14} Hz, (*b*) 3.06 eV, (*c*) 1.64 eV

25 1.2 pm

27 0.18 nm

29 9.32×10^{-24} kg · m/s, 1.80×10^{-23} kg · m/s

31 2.9 nm

33 (*a*) $p_e = 2.09 \times 10^{-22}$ N · s, $p_p = 8.97 \times 10^{-21}$ N · s, $p_\alpha = 8.97 \times 10^{-21}$ N · s
(*b*) $\lambda_e = 3.17$ pm, $\lambda_p = 73.9$ fm, $\lambda_\alpha = 37.0$ fm

35 20.2 fm

37 0.17 nm

39 4.6 pm

41 (*a*) $E_1 = 205$ MeV, $E_2 = 818$ MeV, $E_3 = 1.84$ GeV

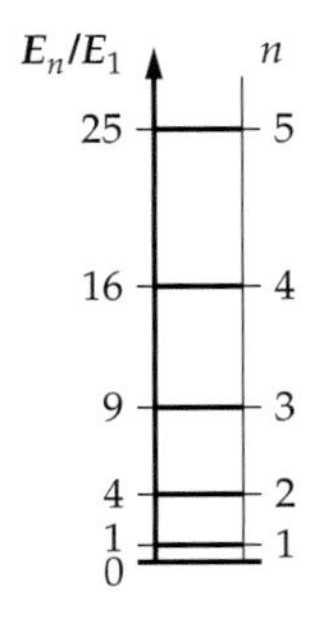

(*b*) $\lambda_{2\to1} = 2.02$ fm, (*c*) $\lambda_{3\to2} = 1.21$ fm, (*d*) $\lambda_{3\to1} = 0.758$ fm

43 (*a*) 0, (*b*) 1, (*c*) 0.002

45 (*a*) $L/2$, (*b*) $0.321L^2$

47 (*a*) $1/\sqrt{2}$, (*b*) 0.865

49 (*a*) 0.500, (*b*) 0.402, (*c*) 0.750

51 (*b*) For large values of n, the result agrees with the classical value of $L^2/3$ given in Problem 50.

53 $\langle x \rangle = 0$, $\langle x^2 \rangle = L^2\left[\frac{1}{12} - \frac{1}{2\pi^2}\right]$

55 (*a*) 3.10 eV, (*b*) 6.24×10^{16} eV, (*c*) 2.08×10^{16}

57 (*a*) 1 μm, 10^{-16} kg · m/s, (*b*) 2×10^{11}

59 0.2 keV

61 7×10^3 km

63 (*a*) 92 mW/m², (*b*) 3×10^4

67 1.3 MeV. The energy of the most energetic electron is approximately 2.5 times the rest energy of an electron.

69 1.04 eV, 554 nm

71 (*b*) 0.2% (*c*) Classically, the energy is continuous. For very large values of n, the energy difference between adjacent levels is infinitesimal.

73 (*a*) 6.2×10^{-4} eV/s, (*b*) 53 min

Chapter 35

1 (*a*)

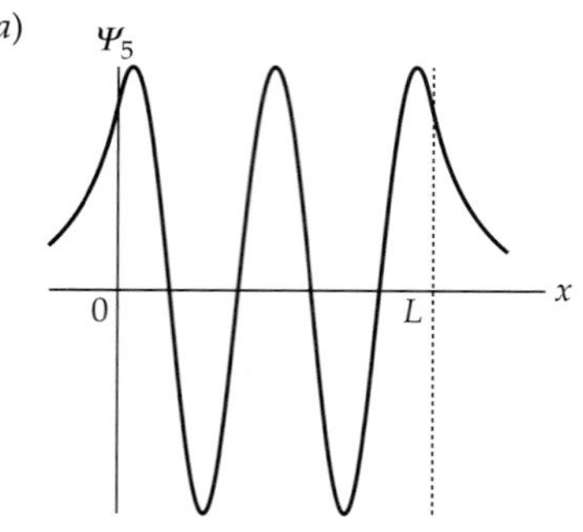

(*b*)

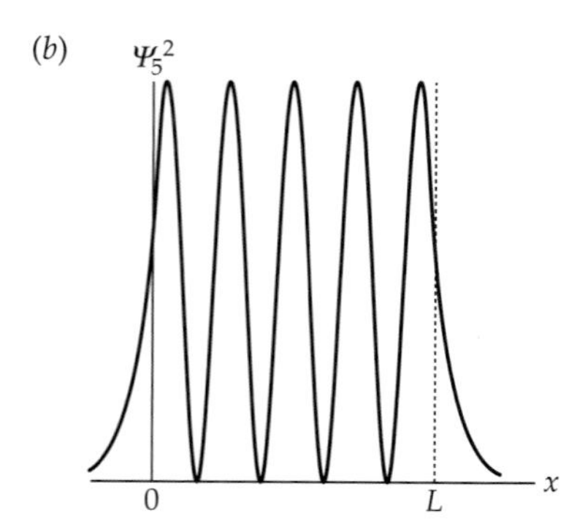

9 (*a*) 9.5 nm, (*b*) 4.1 meV

11 $\Delta x \, \Delta p = \dfrac{\hbar}{2}$

13 (*b*)

Cell	Content/Formula	Algebraic Form
A2	1.0	α
B2	(1−SQRT((A2−1)/A2))/(1+SQRT((A2−1)/A2))^2	$\left(\dfrac{1-\sqrt{\dfrac{\alpha-1}{\alpha}}}{1+\sqrt{\dfrac{\alpha-1}{\alpha}}}\right)^2$
C2	1 − B2	$1-\left(\dfrac{1-\sqrt{\dfrac{\alpha-1}{\alpha}}}{1+\sqrt{\dfrac{\alpha-1}{\alpha}}}\right)^2$

	A	B	C
1	α	*R*	*T*
2	1.0	1.000	0.000
3	1.2	0.298	0.702
4	1.4	0.298	0.802
5	1.6	0.149	0.851
18	4.2	0.036	0.964
19	4.4	0.034	0.966
20	4.6	0.032	0.968
21	4.8	0.031	0.969
22	5.0	0.029	0.971

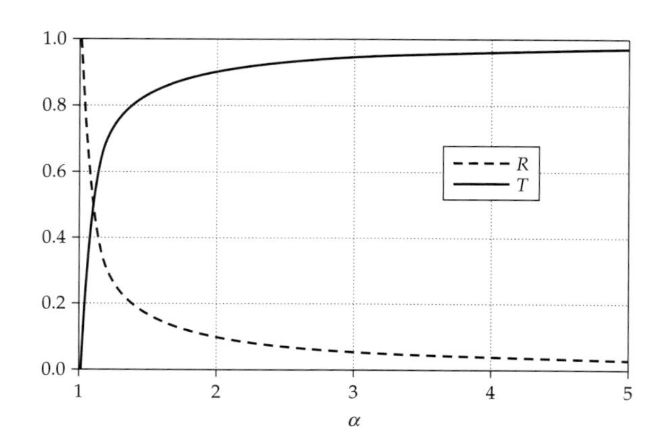

15 (*a*) 10^{-17}, (*b*) 10^{-2}

17 (*a*) $r_{1\,4.0\,\text{MeV}} = 66$ fm, $r_{1\,7.0\,\text{MeV}} = 38$ fm

(*b*) $T_{4.0\,\text{MeV}} \approx 10^{-51}$, $T_{7.0\,\text{MeV}} \approx 10^{-38}$

19 (*a*)

n_1	1	1	1	1	1	1	1	1	1	1
n_2	1	1	1	2	1	2	2	1	2	3
n_3	1	2	3	1	4	2	3	5	4	1
E	21	24	29	33	36	36	41	45	48	53

(*b*) (1, 1, 4) and (1, 2, 2)

(*c*) $\psi(1, 1, 4) = A \sin\left(\dfrac{\pi}{L_1}x\right) \sin\left(\dfrac{\pi}{2L_1}y\right) \sin\left(\dfrac{\pi}{L_1}z\right)$

21 (*a*) $\psi(x,y) = A \sin\dfrac{n\pi}{L}x \sin\dfrac{m\pi}{L}y$

(*b*) $E_{nm} = \dfrac{h^2}{8mL^2}\left(n^2 + m^2\right)$

(*c*) (1, 2) and (2, 1)

(*d*) (1, 7), (7, 1), and (5, 5)

23 $E_{110\text{ bosons}} = \dfrac{5h^2}{4mL^2}$

29 $E_0 = \dfrac{5h^2}{mL^2}$, $E_1 = E_2 = \dfrac{21h^2}{4mL^2}$

31 (*b*) $\langle x^2 \rangle = \dfrac{2}{L}\left(\dfrac{L^3}{24} - \dfrac{L^3}{4n^2\pi^2}\cos n\pi\right)$

35 $A_2 = \sqrt[4]{\dfrac{8m\omega_0}{h}}$

Chapter 36

1 Examination of Figure 36-4 indicates that as n increases, the spacing of adjacent energy levels decreases.

3 (*a*)

5 (*d*)

7 (*a*)

9 The energy of a bound isolated system that consists of two oppositely charged particles, such as an electron and a proton, depends only upon the principle quantum number n. For sodium, which consists of 12 charged particles, the energy of an $n = 3$ electron depends upon the degree to which the wave function of the electron penetrates the $n = 1$ and $n = 2$ electron shells. An electron in a 3s ($n = 3, \ell = 0$) state penetrates these shells to a greater degree than does an electron in a 3p ($n = 3, \ell = 1$) state, so a 3s electron has less energy (is more tightly bound) than is a 3p electron. In hydrogen, however, the wave function of an electron in the $n = 3$ shell cannot penetrate any other electron shells because no other electron shells exist. Thus, an electron in the 3s state in hydrogen has the same energy as an electron in the 3p state in hydrogen.

11 In conformity with the exclusion principle, the total number of electrons that can be accommodated in states of quantum number n is n^2 (see Problem 48). The fact that closed shells correspond to $2n^2$ electrons indicates that there is another quantum number that can have two possible values.

13 (*a*) phosphorus, (*b*) chromium

15 (*d*)

17 The optical spectrum of any atom is due to the configuration of its outer-shell electrons. Ionizing the next atom in the periodic table gives you an ion with the same number of outer-shell electrons, and almost the same nuclear charge. Hence, the spectra should be very similar.

21 (*a*) 10^5, (*b*) 10^3, (*c*) 5.08×10^4

23 (*a*) 103 nm, (*b*) 97.3 nm

25 (*a*) 1.51 eV, 821 nm
(*b*) 0.661 eV, 1880 nm, 0.967 eV, 1280 nm, 1.13 eV, 1100 nm

```
   6→3          5→3                                   4→3
---|--------------|--------------------------------------|-----
1100 nm     1280 nm                                1880 nm
```

27 (*b*) $1.096850 \times 10^7\ \text{m}^{-1}$, $1.097448 \times 10^7\ \text{m}^{-1}$, R_H and $R_{H\,\text{approx}}$ agree to three significant figures.
(*c*) 0.0545%

29 (*a*) $1.49 \times 10^{-34}\ \text{J} \cdot \text{s}$, (*b*) $-1, 0, +1$

(*c*)

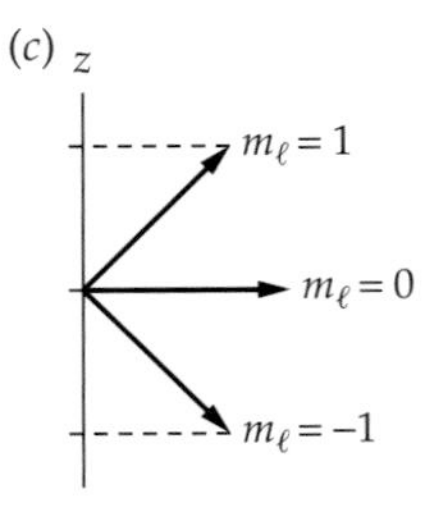

31 (*a*) 0, 1, 2, (*b*) For $\ell = 0$, $m_\ell = 0$. For $\ell = 1$, $m_\ell = -1, 0, +1$. For $\ell = 2$, $m_\ell = -2, -1, 0, +1, +2$. (*c*) 18

33 (*a*) 45.0°, (*b*) 26.6°, (*c*) 8.05°

35 (*a*) $6\hbar^2$, (*b*) $4\hbar^2$, (*c*) $2\hbar^2$

37 (*a*) 4

(*b*)

n	ℓ	m_ℓ	(n, ℓ, m_ℓ)
2	0	0	(2, 0, 0)
2	1	−1	(2, 1, −1)
2	1	0	(2, 1, 0)
2	1	1	(2, 1, 1)

39 (*a*) $\psi_{200}(a_0) = \dfrac{0.0605}{a_0^{3/2}}$, (*b*) $\left[\psi_{200}(a_0)\right]^2 = \dfrac{0.00366}{a_0^3}$

(*c*) $P(a_0) = \dfrac{0.0460}{a_0}$

41 (*a*) 9.20×10^{-4}, (*b*) 0

47 0.323

49 $\ell = 0$ or 1

51

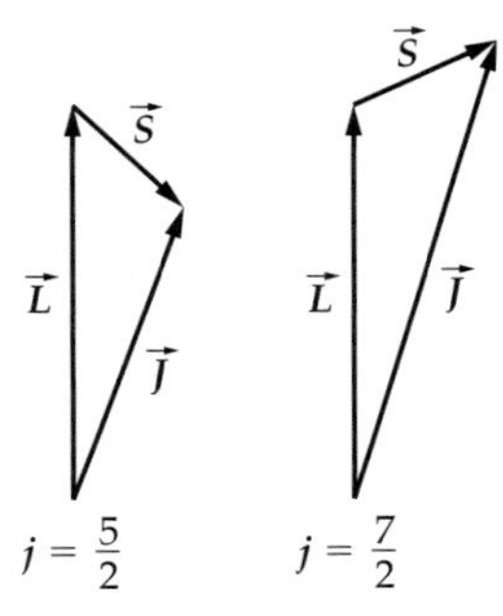

53 (*c*)

55 (*a*) $L_z = -2\hbar, -\hbar, 0, \hbar, 2\hbar$, (*b*) $L_z = -3\hbar, -2\hbar, -\hbar, 0, \hbar, 2\hbar, 3\hbar$

57 (*a*) 2s or 2p, (*b*) $1s^22s^22p^63p$, (*c*) 1s2s

59 (*a*) 0.0610 nm, 0.0578 nm, (*b*) 0.0542 nm

61 (*a*) 1.00 nm, (*b*) 0.155 nm

63 $n_i = 4$ to $n_f = 1$

65

λ, nm	n_i	n_f
164	3	2
230.6	9	3
541	7	4

67 (*a*) 1.6179 eV, 1.6106 eV, (*b*) 0.00730 eV, (*c*) 63.0 T
(*b*) No

71 (*a*) 1.06 GHz, (*b*) 28.4 cm, microwave

73 (*a*) $1.097075 \times 10^7\ \text{m}^{-1}$, (*b*) 0.179 nm

75 (*a*) $1.097074 \times 10^7\ \text{m}^{-1}$, (*b*) 0.0600 nm, (*c*) 0.238 nm

Chapter 37

1 Because the center of charge of the positive Na ion does not coincide with the center of charge for the negative Cl ion, the NaCl molecule has a permanent dipole moment. Hence, it is a polar molecule.

3 Neon occurs naturally as Ne, not Ne_2. Neon is a noble gas. Atoms of noble gases have a closed shell electron configuration.

5 The diagram would consist of a nonbonding ground state with no vibrational or rotational states for ArF (similar to the upper curve in Figure 37-4) but for ArF* there should be a bonding excited state with a definite minimum with respect to internuclear separation and several vibrational states as in the excited state curve of Figure 37-13.

7 The effective force constant from Example 37-4 is 1.85×10^3 N/m. This value is about 25% larger than the given value of the force constant of the suspension springs on a typical automobile.

9 For H_2, the concentration of negative charge between the two protons holds the protons together. In the H_2^+ ion, there is only one electron that is shared by the two positive charges such that most of the electronic charge is again between the two protons. However, in the H_2^+ ion the negative charge between the protons is not as effective as the larger negative charge between them in the H_2 molecule, and the protons should be farther apart. The experimental values support this argument. For H_2, $r_0 = 0.074$ nm, while for H_2^+, $r_0 = 0.106$ nm.

11 For more than two atoms in the molecule, there will be more than just one frequency of vibration because more relative motions are possible. In advanced mechanics, these are known as normal modes of vibration.

13 $\ell \approx 2 \times 10^{30}$, $E_{0r} \approx 5 \times 10^{-65}$ J

15 0.947 nm

17 0.44 eV

19 You should agree. The potential energy curve is shown in the following diagram. The turning points for vibrations of energy E_1 and E_2 are at the values of r where the energies equal $U(r)$. The average value of r for the vibrational levels E_1 and E_2 are labeled $r_{1\,av}$ and $r_{2\,av}$. Note that the estimate of $r_{1\,av}$ is force midway between $r_{1\,min}$ and $r_{1\,max}$. The potential is like a special spring that has a greater force constant for compressions than it has for extensions. The period of a spring-and-mass oscillator is inversely proportional to the square root of the spring constant, so our "special spring" spends more time in extension than in compression. As a result, $r_{1\,av}$ will be greater than the equilibrium radius. This argument can be extended to explain why $r_{2\,av}$ is greater than $r_{1\,av}$. It is because the "force constant" for extension, which can be estimated by taking the average slope of the potential energy curve in the region to the right of the equilibrium position, is greater for $E = E_2$ than for $E = E_1$. It is also because the "force constant" for compression is greater for $E = E_2$ than for $E = E_1$. It follows that $r_{2\,av}$ is greater than $r_{1\,av}$. Because $r_{2\,av}$ is greater than $r_{1\,av}$, it follows that as the vibrational energy of a diatomic molecule increases, the average separation of the atoms of the molecule increases and, hence, the solid expands with heating.

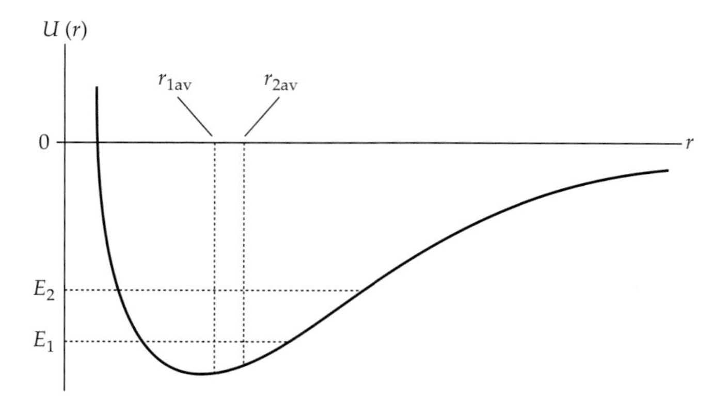

21 (a) $U_e = -6.64$ eV, (b) $E_{d\,calc} = 5.70$ eV, (c) $U_{rep} = 0.63$ eV

23 0.121 nm

25 41

27 5.6 meV

29 (a) 0.179 eV, (b) 3×10^{-47} kg·m², (c) 0.1 nm

31 (a) 1.45×10^{-46} kg·m², 0.239 meV

(b)

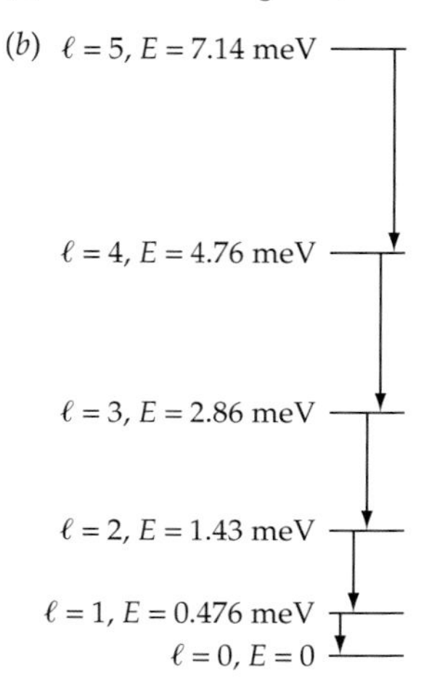

$\Delta E_{54} = 2.38$ meV, $\Delta E_{43} = 1.90$ meV, $\Delta E_{32} = 1.43$ meV, $\Delta E_{21} = 1.25$ meV, $\Delta E_{10} = 0.476$ meV

(c) $\lambda_{10} = 2600\ \mu$m, $\lambda_{21} = 1300\ \mu$m, $\lambda_{32} = 867\ \mu$m, $\lambda_{43} = 650\ \mu$m, $\lambda_{54} = 520\ \mu$m, microwave

33 $\mu_{H^{35}Cl} = 0.972$ u, $\mu_{H^{37}Cl} = 0.974$ u,

$\frac{\Delta\mu}{\mu} = 0.00150$, $\Delta f/f = 0.0012$, in fair agreement (about 20% difference) with the calculated result. Note that Δf is difficult to determine precisely from Figure 37-17.

35 0.955 meV

37 1.55 kN/m

39 $r_0 = a$, $U_{min} = -U_0$, $r_0 = 0.074$ nm, $U_0 = 4.52$ eV

41 $F_x = -\frac{dU}{dx} \propto \frac{1}{x^4}$

43 (a) $\frac{1\text{ eV}}{\text{molecule}} = 23.0$ kcal/mol

(b) 98.2 kcal/mol

Chapter 38

1 The energy lost by the electrons in collision with the ions of the crystal lattice appears as thermal energy throughout the crystal.

3 (*a*) potassium and nickel (*b*) 3.1 V

5 The resistivity of brass at 4 K is almost entirely due to the residual resistance (the resistance due to impurities and other imperfections of the crystal lattice). In brass, the zinc ions act as impurities in copper. In pure copper, the resistivity at 4 K is due to its residual resistance. The residual resistance is very low if the copper is very pure.

7 The resistivity of copper increases with increasing temperature; the resistivity of (pure) silicon decreases with increasing temperature because the number density of charge carriers increases.

9 (*b*)

11 The excited electron is the motion of the electron in the conduction band and contributes to the current. A hole is left in the valence band allowing the positive hole to move through the band, also the motion of the hole contributes to the current.

13 (*c*)

15

V (V)	1/slope (Ω)
−20	∞
+0.2	40
+0.4	20
+0.6	10
+0.8	5

17 2.07 g/cm^3

19 (*a*) −10.6 eV, (*b*) 2.83%

21 (*a*) 0.123 $\mu\Omega \cdot$m, (*b*) 70.7 n$\Omega \cdot$m

23 (*a*) $n_{Ag} = 5.86 \times 10^{22}$ electrons/cm^3
(*b*) $n_{Ag} = 5.90 \times 10^{22}$ electrons/cm^3. Both these results agree with the values in Table 38-1.

25 4.0

27 (*a*) 1.07×10^6 m/s (*b*) 1.39×10^6 m/s
(*c*) 1.89×10^6 m/s

29 (*a*) 4.22 eV (*b*) 2.85 eV

31 (*a*) 5.90×10^{28} e/m^3, (*b*) 5.50 eV, (*c*) 212, (*d*) The ratio E_F/kT is equal to 212 at $T = 300$ K. The Fermi energy is the energy of the most energetic conduction electron when the crystal is at absolute zero. Because no two conduction electrons can occupy the same state, the Fermi energy is quite high compared with kT. The kT energy is the energy the average conduction electron would have when the crystal is at temperature T if the electrons did not obey the exclusion principle.

33 3.82×10^{10} N/m^2 = 3.77×10^5 atm

35 0.192 J/(mol · K)

37 (*a*) 66 nm, (*b*) 1.8×10^{-4} nm^2

39 1.09 μm

41 180 nm

43 116 K

45 $a_{B\,Si} = 3$ nm, $a_{B\,Ge} = 8$ nm

47 37.1 nm, 38.7 nm. The mean free paths agree to within about 4%.

49

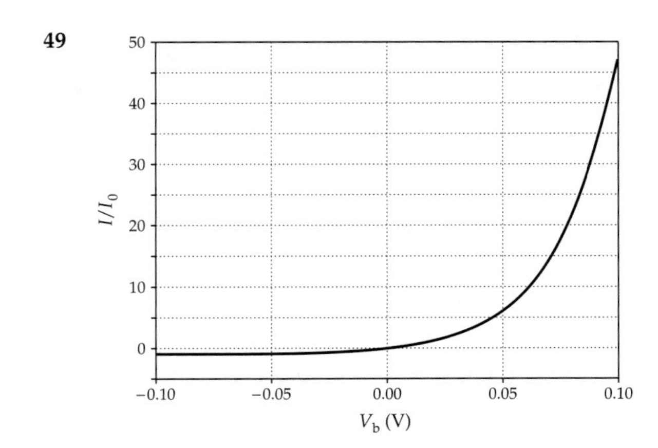

51 250

53 (*a*)

Conduction band
Fermi level
Valence band
p-side
n-side

(*b*)

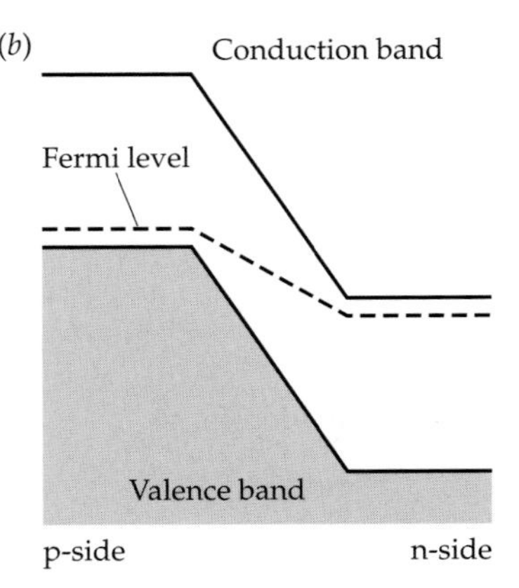

55 The charge carriers are holes and the semiconductor is *p*-type. 1.0×10^{23} m^{-3}

57 (*a*) 2.17 meV, $E_g \approx 0.8E_{g\,measured}$, (*b*) 0.454 mm

59 2.0×10^{18}

61 (*a*) 5.51 eV, (*b*) 3.31 eV, (*c*) 1.08×10^6 m/s

63 1

67 0.60

71 1.07

73 (*a*) 5.51×10^{-3}, (*b*) 1.84×10^{-2}

75 4.35×10^{14} Hz

Chapter 39

1 (*a*)

3 (*a*) True, (*b*) True, (*c*) False, (*d*) True, (*e*) False, (*f*) False, (*g*) True

5 Although $\Delta y = \Delta y'$, $\Delta t \neq \Delta t'$. Consequently, $\Delta y/\Delta t \neq \Delta y'/\Delta t'$.

7 (*a*) 0.946, (*b*) $1.23 \times 10^{10}\ c \cdot y$

9 (*a*) 0.98 km. The width of the beam is unchanged. (*b*) 9.6×10^{7} m, (*c*) 0.10 μm

11 (*a*) 0.91*c*, (*b*) $22\ c \cdot y$, (*c*) 101 y

13 (*a*) 0.385 μs, (*b*) 0.193 μs, (*c*) 0.998*c*

15 1.85×10^{4} y

17 (*a*) 1.76 μs, (*b*) 6.32 μs, (*c*) 3.1 μs, (*d*) 1.70 km

19 4.4 μs

21 (*a*) 2.10 μs, (*b*) 2.59 μs, (*c*) 0.49 μs, (*d*) 2.59 μs, (*e*) 4.36 h, (*f*) 18.8 h

23 2.22×10^{7} m/s

27 11 ms

29 (*a*) $u_x = v$ and $u_y = \dfrac{c}{\gamma}$

31 (*a*) 0.976, (*b*) 0.997*c*

33 66.7%

37 (*a*) 290 MeV, (*b*) 629 MeV

39 (*a*) 0.943*c*, (*b*) 3.0 MeV, (*c*) 2.8 MeV/*c*, (*d*) 4.1 MeV/c^2, (*e*) 0.9 MeV

43 0.999*c*

45 (*a*) $-0.50c$, S' moves in the $-x$ direction, (*b*) 1.7 y

47 281 MeV

49 (*a*) $v = -\dfrac{E}{Mc}$, (*b*) $d = -\dfrac{LE}{Mc^2}$

51 $K_{th} = 6m_pc^2$ in agreement with Problem 40.

Chapter 40

1 (*a*) ^{15}N, ^{16}N, (*b*) ^{54}Fe, ^{55}Fe, (*c*) ^{117}Sn, ^{119}Sn

3 Generally, decay by α emission leaves the daughter nucleus neutron rich, i.e. above the line of stability. The daughter nucleus therefore tends to decay via β^- emission which converts a nuclear neutron to a proton.

5 It would make the dating unreliable because the current concentration of ^{14}C is not equal to that at some earlier time.

7 The probability for neutron capture by the fissionable nucleus is large only for slow (thermal) neutrons. The neutrons emitted during the fission process are fast (high energy) neutrons and must be slowed to thermal neutrons before they are likely to be captured by another fissionable nucleus.

9 (*a*) β^+, (*b*) β^-

11 (*a*) True (given an unlimited supply of ^{238}U), (b) False, (c) True, (d) False

13

Material Property	Ratio (order of magnitude)
Mass density	10^{15}
Half-life	10^{15}
Nuclear masses	2

15 (*a*) $E_b = 92.2$ MeV, $E_b/A = 7.68$ MeV
(*b*) $E_b = 492$ MeV, $E_b/A = 8.79$ MeV
(*c*) $E_b = 1802$ MeV, $E_b/A = 7.57$ MeV

17 (*a*) 3.0 fm, (*b*) 4.6 fm, (*c*) 7.0 fm

19 (*a*) $E_{thermal} = 4.11 \times 10^{-21}$ J $= 25.7$ meV, (*b*) 2.22 km/s, (*c*) 10.1 min

23 (*a*) 5 min, (*b*) 250 Bq

25 (*a*) 200 s, (*b*) $3.5 \times 10^{-3}\ s^{-1}$, (*c*) 125 Bq

27 (*a*) 500 Bq, 250 Bq, (*b*) $N_0 = 1.0 \times 10^{6}$, $N_{2.4\ min} = 5.2 \times 10^{5}$, (*c*) 12 min

29 (*a*) $4.5 \times 10^{3}\ \alpha/s$, (*b*) 5.3×10^{4} y

31 $^{239}_{94}Pu \rightarrow {}^{235}_{92}U + {}^{4}_{2}\alpha + Q$, $Q = 5.24$ MeV, $K_\alpha = 5.15$ MeV, $K_{^{235}U} = 89.2$ keV

33 (*a*) $\lambda = 0.133\ h^{-1}$, $t_{1/2} = 5.20$ h, (*b*) $N_0 = 3.11 \times 10^{6}$

35 (*a*) 1.13 mCi, (*b*) 0.898 mCi

37 About 15 g

39 7.0×10^{8} y

41 (*a*) -0.764 MeV, (*b*) 3.27 MeV

43 (*a*) 0.156 MeV, (*b*) The masses given are for atoms, not nuclei, so the atomic masses are too large by the atomic number multiplied by the mass of an electron. For the given nuclear reaction, the mass of the carbon atom is too large by $6m_e$ and the mass of the nitrogen atom is too large by $7m_e$. Subtracting $6m_e$ from both sides of the reaction equation leaves an extra electron mass on the right. Not including the mass of the beta particle (electron) is mathematically equivalent to explicitly subtracting $1m_e$ from the right side of the equation.

45 $1.56 \times 10^{19}\ s^{-1}$

47 208 MeV

49 3.2×10^{10} J

51 (*c*) $3.7 \times 10^{38}\ s^{-1}$, 5.0×10^{10} y

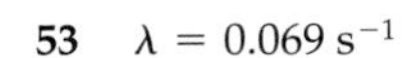

53 $\lambda = 0.069\ \text{s}^{-1}$

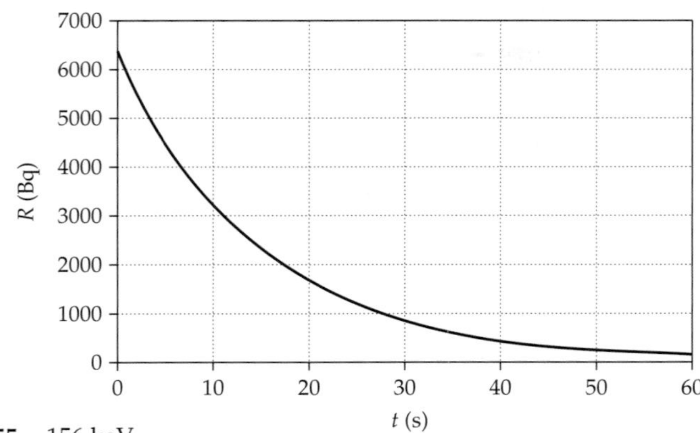

55 156 keV

59 6.7×10^3 Bq

61 6.3 L

63 (*a*) 23 MeV, (*b*) 4.2 GeV, (*c*) 1.3 GeV

65 (*a*)

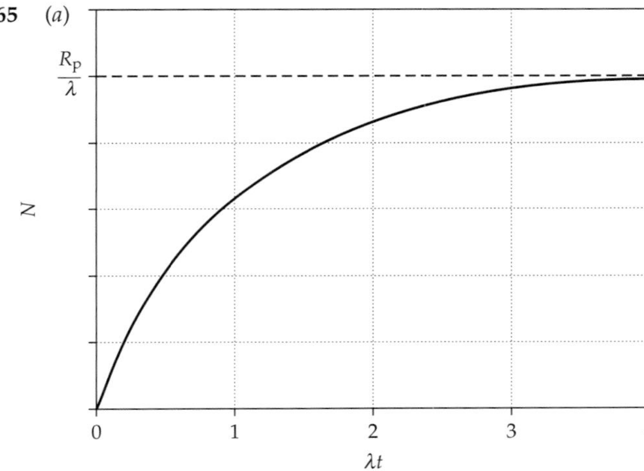

(*b*) 8.7×10^4

67 (*a*) 4.00 fm, (*b*) 310 MeV/c, (*d*) 310 MeV

69 (*a*) 1.188 MeV/c, (*b*) 752 eV, (*c*) 0.0962%

71 (*b*) 55

73 (*d*)

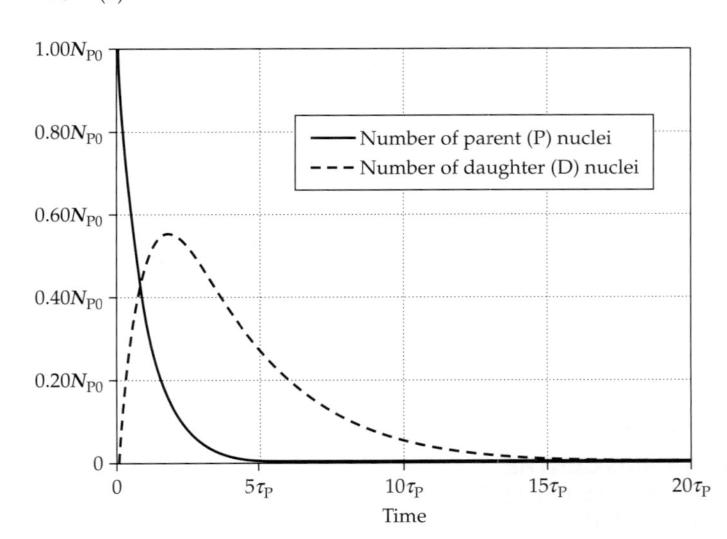

Chapter 41

1

Similarities	Differences
Baryons and mesons are hadrons, i.e., they participate in the strong interaction. Both are composed of quarks.	Baryons consist of three quarks and are fermions. Mesons consist of two quarks and are bosons. Baryons have baryon number +1 or −1. Mesons have baryon number 0.

3 A decay process involving the strong interaction has a very short lifetime (~10^{-23} s), whereas decay processes that proceed via the weak interaction have lifetimes of order 10^{-10} s.

5 False

7 No; from Table 41-3 it is evident that any quark–antiquark combination always results in an integral or zero charge.

9 (*a*) False, (*b*) True, (*c*) True

11 $\dfrac{F_{\text{em}}}{F_{\text{grav}}} = 1.24 \times 10^{36}$

13 (*a*) 279.2 MeV, (*b*) 1877 MeV, (*c*) 211.3 MeV

15 (*a*) Because $\Delta S = +1$, the reaction can proceed via the weak interaction, (*b*) Because $\Delta S = +2$, the reaction is not allowed, (*c*) Because $\Delta S = +1$, the reaction can proceed via the weak interaction.

17 (*a*) Because $\Delta S = +2$, the reaction is not allowed, (*b*) Because $\Delta S = +1$, the reaction can proceed via the weak interaction.

19 (*a*) K^0, (*b*) Σ^0 or Λ^0, (*c*) K^+

21

	Combination	B	Q	S	Hadron
(*a*)	*uud*	1	+1	0	p^+
(*b*)	*udd*	1	0	0	n
(*c*)	*uus*	1	+1	−1	Σ^+
(*d*)	*dds*	1	−1	−1	Σ^-
(*e*)	*uss*	1	0	−2	Ξ^0
(*f*)	*dss*	1	−1	−2	Ξ^-

23 From Table 41-3 we see that to satisfy the properties of charge number equal to +2 and strangeness, charm, topness, and bottomness all equal to zero, the quark combination must be *uuu*.

25 (*a*) $c\bar{d}$, (*b*) $\bar{c}d$

27 (*a*) *uds*, (*b*) $\bar{u}\bar{u}\bar{d}$, (*c*) *dds*

29 (*a*) *sss*, (*b*) *ssd*

31 $3.3 \times 10^8\, c \cdot \text{y}$

35 (*a*) Baryon number and lepton numbers are conserved quantities. A particle and its antiparticle must have baryon numbers that add to zero and lepton numbers that add to zero. Thus, for a particle and its antiparticle to be identical, its baryon number and all three of its lepton numbers must equal zero. This means it cannot be a lepton or a baryon, so it must be a meson. A particle and its antiparticle have the

complementary quark content. That is, if each quark in a particle is replaced by its antiquark, then the resulting entity is the antiparticle of the particle.
(*b*) The quark combination for the π^0 is a linear combination of $u\overline{u}$ and $d\overline{d}$ and the quark combination for the $\overline{\pi}^0$ is a linear combination of $\overline{u}u$ and $\overline{d}d$. The quark combination for the Ξ^0 is uss and that of the $\overline{\Xi}^0$ is $\overline{u}\,\overline{s}\,\overline{s}$.
(*c*) The π^0 is a meson with quark content of a linear combination of $u\overline{u}$ and $d\overline{d}$, so the π^0 is its own antiparticle.
The Ξ^0 is a baryon. As is explained in the answer to Part (*a*), a baryon cannot be its own antiparticle.

37 (*a*) The u and $\overline{u}$ annihilate, resulting in the photons.
(*b*) Two or more photons are required to conserve linear momentum.

39 (*a*) π^+, (*b*) -815 MeV, (*c*) 1.98 GeV

41 (*a*) 38 MeV, (*b*) 6.72, (*c*) 5 MeV, 33 MeV

Index

If a page number is followed by an *n*, the entry is in a footnote.

Physical Constants*

Atomic mass constant	$m_u = \frac{1}{12}m(^{12}C)$	$1\ u = 1.660\ 538\ 86(28) \times 10^{-27}$ kg
Avogadro's number	N_A	$6.022\ 1415(10) \times 10^{23}$ particles/mol
Boltzmann constant	$k = R/N_A$	$1.380\ 6505(24) \times 10^{-23}$ J/K $8.617\ 343(15) \times 10^{-5}$ eV/K
Bohr magneton	$m_B = e\hbar/(2m_e)$	$9.274\ 009\ 49(80) \times 10^{-24}$ J/T = $5.788\ 381\ 804(39) \times 10^{-5}$ eV/T
Coulomb constant	$k = 1/(4\pi\epsilon_0)$	$8.987\ 551\ 788\ldots \times 10^{9}\ N \cdot m^2/C^2$
Compton wavelength	$\lambda_C = h/(m_e c)$	$2.426\ 310\ 238(16) \times 10^{-12}$ m
Fundamental charge	e	$1.602\ 176\ 53(14) \times 10^{-19}$ C
Gas constant	R	$8.314\ 472(15)\ J/(mol \cdot K)$ = $1.987\ 2065(36)\ cal/(mol \cdot K)$ = $8.205\ 746(15) \times 10^{-2}\ L \cdot atm/(mol \cdot K)$
Gravitational constant	G	$6.6742(10) \times 10^{-11}\ N \cdot m^2/kg^2$
Mass of electron	m_e	$9.109\ 3826(16) \times 10^{-31}$ kg = $0.510\ 998\ 918(44)\ MeV/c^2$
Mass of proton	m_p	$1.672\ 621\ 71(29) \times 10^{-27}$ kg = $938.272\ 029(80)\ MeV/c^2$
Mass of neutron	m_n	$1.674\ 927\ 28(29) \times 10^{-27}$ kg = $939.565\ 360(81)\ MeV/c^2$
Magnetic constant (permeability of free space)	μ_0	$4\pi \times 10^{-7}\ N/A^2$
Electric constant (permittivity of free space)	ϵ_0	$= 1/(\mu_0 c^2) = 8.854\ 187\ 817\ldots \times 10^{-12}\ C^2/(N \cdot m^2)$
Planck's constant	h	$6.626\ 0693(11) \times 10^{-34}\ J \cdot s$ = $4.135\ 667\ 43(35) \times 10^{-15}\ eV \cdot s$
	$\hbar = h/(2\pi)$	$1.054\ 571\ 68(18) \times 10^{-34}\ J \cdot s$ = $6.582\ 119\ 15(56) \times 10^{-16}\ eV \cdot s$
Speed of light	c	$2.997\ 924\ 58 \times 10^{8}$ m/s
Stefan-Boltzmann constant	σ	$5.670\ 400(40) \times 10^{-8}\ W/(m^2 \cdot K^4)$

*The values for these and other constants can be found in Appendix B as well as on the Internet at http://physics.nist.gov/cuu/Constants/index.html. The numbers in parentheses represent the uncertainties in the last two digits. (For example, 2.044 43(13) stands for 2.044 43 ± 0.000 13.) Values without uncertainties are exact. Values with ellipses are exact (like the number π = 3.1415. . .), but are not completely specified.

Derivatives and Definite Integrals

$$\frac{d}{dx}\sin ax = a\cos ax \qquad \int_0^\infty e^{-ax}\,dx = \frac{1}{a} \qquad \int_0^\infty x^2 e^{-ax^2}\,dx = \frac{1}{4}\sqrt{\frac{\pi}{a^3}}$$

$$\frac{d}{dx}\cos ax = -a\sin ax \qquad \int_0^\infty e^{-ax^2}\,dx = \frac{1}{2}\sqrt{\frac{\pi}{a}} \qquad \int_0^\infty x^3 e^{-ax^2}\,dx = \frac{4}{a^2}$$

$$\frac{d}{dx}e^{ax} = ae^{ax} \qquad \int_0^\infty xe^{-ax^2}\,dx = \frac{2}{a} \qquad \int_0^\infty x^4 e^{-ax^2}\,dx = \frac{3}{8}\sqrt{\frac{\pi}{a^5}}$$

The a in the six integrals is a positive constant.

Vector Products

$\vec{A} \cdot \vec{B} = AB\cos\theta$ $\qquad$ $\vec{A} \times \vec{B} = AB\sin\theta\,\hat{n}$ ($\hat{n}$ obtained using right-hand rule)

For additional data, see the following tables in the text.

Geometry and Trigonometry

$C = \pi d = 2\pi r$ definition of π
$A = \pi r^2$ area of circle
$V = \frac{4}{3}\pi r^3$ spherical volume
$A = \partial V/\partial r = 4\pi r^2$ spherical surface area
$V = A_{\text{base}}L = \pi r^2 L$ cylindrical volume
$A = \partial V/\partial r = 2\pi r L$ cylindrical surface area

$o = h \sin\theta$
$a = h \cos\theta$

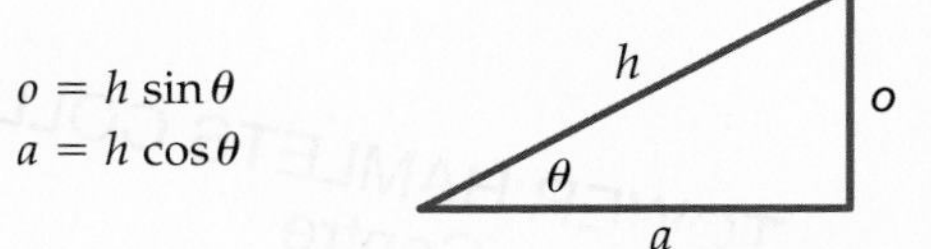

$$\sin^2\theta + \cos^2\theta = 1$$
$$\sin(A \pm B) = \sin A \cos B \pm \cos A \sin B$$
$$\cos(A \pm B) = \cos A \cos B \mp \sin A \sin B$$
$$\sin A \pm \sin B = 2 \sin[\tfrac{1}{2}(A \pm B)] \cos[\tfrac{1}{2}(A \mp B)]$$

$\sin\theta \equiv y$
$\cos\theta \equiv x$
$\tan\theta \equiv \dfrac{y}{x}$

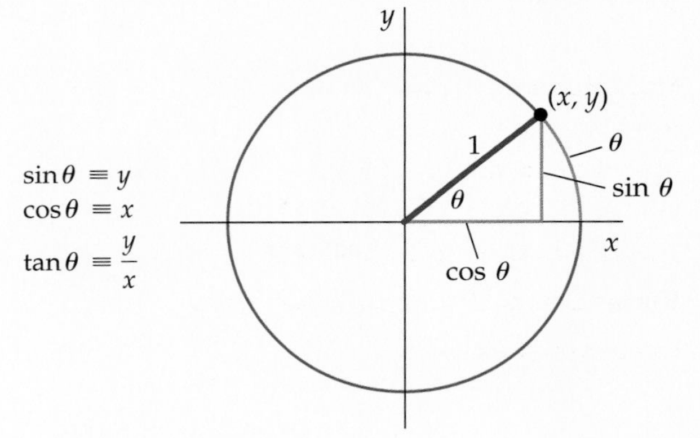

If $|\theta| << 1$, then
$\cos\theta \approx 1$ and $\tan\theta \approx \sin\theta \approx \theta$ (θ in radians)

Quadratic Formula

If $ax^2 + bx + c = 0$, then $x = \dfrac{-b \pm \sqrt{b^2 - 4ac}}{2a}$

Binomial Expansion

If $|x| < 1$, then $(1 + x)^n =$

$$1 + nx + \frac{n(n-1)}{2!}x^2 + \frac{n(n-1)(n-2)}{3!}x^3 + \ldots$$

If $|x| << 1$, then $(1 + x)^n \approx 1 + nx$

Differential Approximation

If $\Delta F = F(x + \Delta x) - F(x)$ and if $|\Delta x|$ is small,

then $\Delta F \approx \dfrac{dF}{dx}\Delta x.$